Biodiversity and Ecosystem Services on Post-Industrial Land

Biodiversity and Ecosystem Services on Post-Industrial Land

Edited by

Vimal Chandra Pandey
CSIR-National Botanical Research Institute
Lucknow, India

WILEY

This edition first published 2025

Registered Offices
John Wiley & Sons, Inc., 111 River Street, Hoboken, NJ 07030, USA
John Wiley & Sons Ltd, The Atrium, Southern Gate, Chichester, West Sussex, PO19 8SQ, UK

For details of our global editorial offices, customer services, and more information about Wiley products visit us at www.wiley.com.

Wiley also publishes its books in a variety of electronic formats and by print-on-demand. Some content that appears in standard print versions of this book may not be available in other formats.

Library of Congress Cataloging-in-Publication Data
Names: Pandey, Vimal Chandra, editor.
Title: Biodiversity and ecosystem services on post-industrial land / edited by Vimal Chandra Pandey.
Description: Hoboken, NJ : Wiley, 2025. | Includes bibliographical references and index.
Identifiers: LCCN 2024023401 (print) | LCCN 2024023402 (ebook) | ISBN 9781394187386 (hardback) | ISBN 9781394187393 (adobe pdf) | ISBN 9781394187409 (epub)
Subjects: LCSH: Restoration ecology. | Abandoned mined lands reclamation. | Biodiversity. | Soil remediation.
Classification: LCC QH541.15.R45 B558 2025 (print) | LCC QH541.15.R45 (ebook) | DDC 333.73/153–dc23/eng/20240809
LC record available at https://lccn.loc.gov/2024023401
LC ebook record available at https://lccn.loc.gov/2024023402

Cover Design: Wiley
Cover Image: © Anna Kucherova/Shutterstock

Set in 9.5/12.5pt STIXTwoText by Straive, Pondicherry, India

Printed and bound by CPI Group (UK) Ltd, Croydon, CR0 4YY

C9781394187386_151024

Contents

List of Contributors

Austin E. Abah
Department of Animal and
Environmental Biology
University of Port Harcourt
Nigeria

Shubham Abhishek
CSIR-Central Institute of Mining and
Fuel Research, Barwa Road
Dhanbad, Jharkhand
India

Academy of Scientific and Innovative
Research (AcSIR)
Ghaziabad
India

Fernando A. Alvarez
Administration Department
Universidad Católica del Norte
Antofagasta
Chile

Valeria Ancona
Water Research Institute of the Italian
National Research Council
Bari
Italy

Arnab Banerjee
Department of Environmental Science
Sant Gahira Guru Vishwavidyalaya
Ambikapur, Chhattisgarh
India

Kiran Bargali
Department of Botany, DSB Campus
Kumaun University
Nainital, Uttarakhand
India

Surendra Singh Bargali
Department of Botany, DSB Campus
Kumaun University
Nainital, Uttarakhand
India

Nicolas Bélanger
Laboratoire sur la science des données
Université du Québec (TÉLUQ)
Montréal, Québec
Canada

Simon Bilodeau-Gauthier
Direction de la recherche forestière
Ministère des Ressources naturelles
et des Forêts
Québec, Québec
Canada

CK Dotaniya
Department of Soil Science and
Agricultural Chemistry
SKRAU
Bikaner
India

ML Dotaniya
ICAR-Directorate of Rapeseed-Mustard Research
Bharatpur
India

RK Doutaniya
Department of Agronomy
SKN College of Agriculture
Jobner
India

Sanat Kumar Dwibedi
Odisha University of Agriculture and Technology
Bhubaneswar
India

Amlan Kumar Ghosh
Department of Soil Science and Agricultural Chemistry
Institute of Agricultural Sciences, BHU
Varanasi
India

Laurence Grimond
Laboratoire sur la science des données
Université du Québec (TÉLUQ)
Montréal, Québec
Canada

Anna Grobelak
Faculty of Infrastructure and Environment
Czestochowa University of Technology
Czestochowa
Poland

Ksenija Jakovljević
Department of Ecology
Institute for Biological Research 'Siniša Stanković' - National Institute of the Republic of Serbia
University of Belgrade
Serbia

Manoj Kumar Jhariya
Department of Farm Forestry
Sant Gahira Guru Vishwavidyalaya
Ambikapur, Chhattisgarh
India

Rim Khlifa
Laboratoire sur la science des données
Université du Québec (TÉLUQ)
Montréal, Québec
Canada

Aneta Kowalska
Faculty of Science and Technology
Jan Dlugosz University
Czestochowa
Poland

Amit Kumar
School of Hydrology and Water Resources
Nanjing University of Information Science and Technology
Nanjing
China

Kuldeep Kumar
ICAR Indian Institute of Soil and Water Conservation
Dehradun
RS Kota
India

Rupesh Kumar
Jindal Global Business School
O.P. Jindal Global University
Sonipat, Haryana
India

Santosh Kumar
ICAR-Indian Agricultural Research Institute
Jharkhand
India

Elizabeth J. Lam
Chemical Engineering Department
Universidad Católica del Norte
Antofagasta
Chile

Manju Lata
University of Rajasthan
Jaipur
India

Jonnada Likhita
Department of Fruit Science
Rani Lakshmi Bai Central
Agricultural University
Jhansi, Uttar Pradesh
India

Justyna Likus-Cieślik
Department of Ecological Engineering
and Forest Hydrology, Faculty of Forestry
University of Agriculture in Krakow
Krakow
Poland

Eberechukwu M. Maduike
Department of Animal and
Environmental Biology
University of Port Harcourt
Nigeria

Ebhin Masto
Environment, Emission and CRM Section
CSIR - Central Institute of
Mining and Fuel Research
(Digwadih Campus), PO; FRI
Dhanbad, Jharkhand
India

HM Meena
ICAR-Central Arid Zone Research Institute
Jodhpur
India

Tomica Mišljenović
Institute of Botany and Botanical Garden
Faculty of Biology
University of Belgrade
Serbia

Italo L. Montofré
Mining Business School, ENM
Universidad Católica del Norte
Antofagasta
Chile

Mining and Metallurgical Engineering
Department
Universidad Católica del Norte
Antofagasta
Chile

Aroloye O. Numbere
Department of Animal and
Environmental Biology
University of Port Harcourt
Nigeria

Victoria C. Obinna
Department of Animal and
Environmental Biology
University of Port Harcourt
Nigeria

Marek Pająk
Department of Ecological Engineering
and Forest Hydrology
Faculty of Forestry
University of Agriculture in Krakow
Krakow
Poland

Bhanu Pandey
CSIR-Central Institute of Mining and
Fuel Research, Barwa Road
Dhanbad, Jharkhand
India

Vimal Chandra Pandey
CSIR-National Botanical Research Institute
Lucknow, Uttar Pradesh
India

Marcin Pietrzykowski
Department of Ecological Engineering and
Forest Hydrology, Faculty of Forestry
University of Agriculture in Krakow
Krakow
Poland

Chandini Pradhan
Department of Soil Science and
Agricultural Chemistry
Institute of Agricultural Sciences, BHU
Varanasi
India

Abhishek Raj
Department of Forestry,
Pt. Deendayal Upadhyay College of
Horticulture and Forestry
Dr. Rajendra Prasad Central
Agriculture University
Samastipur
Bihar
India

Dragana Ranđelović
Institute for Technology of Nuclear and
Other Mineral Raw Materials
Serbia

David Rivest
Département des sciences naturelles
Université du Québec en Outaouais
Gatineau, Québec
Canada

Katarzyna Sadowy
SGH Warsaw School of Economics
Warsaw
Poland

Gaurav Sharma
Department of Fruit Science
Rani Lakshmi Bai Central
Agricultural University
Jhansi, Uttar Pradesh
India

Shivali Sharma
Department of Fruit Science
Rani Lakshmi Bai Central
Agricultural University
Jhansi, Uttar Pradesh
India

Sunny Sharma
Department of Horticulture
School of Agriculture
Lovely Professional University
Phagwara, Punjab
India

Sameer Shekhar
CSIR-Central Institute of Mining and
Fuel Research, Barwa Road
Dhanbad, Jharkhand
India

AO Shirale
ICAR-Indian Institute of Soil Science
Bhopal
India

Saurabh Shukla
Faculty of Civil Engineering
Institute of Technology, Shri Ramswaroop
Memorial University
Lucknow
India

Edyta Sierka
Institute of Biology, Biotechnology and
Environmental Protection
University of Silesia in Katowice
Katowice
Poland

Preeti Singh
ICAR-Indian Agricultural Research Institute
Jharkhand
India

Siddharth Singh
CSIR-Central Institute of Mining and Fuel Research, Barwa Road
Dhanbad, Jharkhand
India

Academy of Scientific and Innovative Research (AcSIR)
Ghaziabad
India

Barbara Stalmachová
Department of Environmental Engineering
Technical University of Ostrava
Ostrava-Poruba
Czech Republic

D.D. Tewari
Gayatri Vidyapeeth P.G. College
Rishi Bhoomi, Risia, Bahraich
Uttar Pradesh
India

Bartłomiej Woś
Department of Ecological Engineering and Forest Hydrology, Faculty of Forestry
University of Agriculture in Krakow
Krakow
Poland

DK Yadav
ICAR-Indian Institute of Soil Science
Bhopal
India

RK Yadav
Agriculture University Kota
India

Shailesh Kumar Yadav
Department of Environmental Science
Sant Gahira Guru Vishwavidyalaya
Ambikapur, Chhattisgarh
India

About the Editor

Dr. Vimal Chandra Pandey is, an applied research scientist, internationally recognized for his research in the area of phytomanagement of polluted sites. Dr. Pandey is listed as the World's Top 2% Scientists, announced by Stanford University, California, United States, and published by Elsevier BV, 2020, 2021, 2022, and 2023. His research focuses mainly on the remediation and management of polluted lands, using ecologically and socio-economically valuable plants, to regain ecosystem services and support a bio-based economy as phytoproducts. Dr. Pandey's research interests also lie in fostering phytoremediation for utilizing polluted lands and thereby attaining the United Nations Sustainable Development Goals. His phytoremediation work has led to the extension of phytoremediation beyond its traditional application. He is now engaged in exploring commercial phytoremediation with the least risk, minimum input, and low maintenance. Dr. Pandey worked at CSIR-National Botanical Research Institute, Babasaheb Bhimrao Ambedkar University, and the Council of Science and Technology, Uttar Pradesh (CSTUP), Lucknow, India. He is the recipient of several awards/honors/fellowships. Dr. Pandey is a member of the IUCN Commission on Ecosystem Management and the National Academy of Sciences, India. Dr. Pandey has published over 114 scientific articles/book chapters in peer-reviewed journals/books. He is also the author and editor of several books published by Elsevier, Springer, Wiley, and CRC Press, with several more forthcoming. Dr. Pandey is associate editor/editor/academic editor/board member of prestigious journals such as *Land Degradation and Development*; *Restoration Ecology*; *Ecological Processes*; *Environment, Development, and Sustainability*; *Ambio*; *Environmental Management*; *Discover Sustainability*; *Bulletin of Environmental Contamination and Toxicology*; and *PLOS ONE* by Wiley/Springer/PLOS. He also works/worked as a guest editor for many reputed international journals.

Foreword

It is of immense pleasure that I accept the gracious invitation from Dr. Vimal Chandra Pandey to contribute introductory statements to the notable work entitled *Biodiversity and Ecosystem Services on Post-Industrial Land*. Continuous increase in post-industrial land over the world is a serious threat in terms of environmental degradation; the proposed book is an innovative and timely solution to address this global issue. Such industrial activities have several negative impacts on the surrounding environment, including air pollution (i.e., smog and soot), water pollution (i.e., gases, chemicals, heavy metals, radioactive materials, etc.), soil pollution (industrial wastes dumps destroy the soil fertility), greenhouse gas emissions (it releases into the environment by the burning of coal fossil fuels that trap heat and contribute to global climate change), acid rain (coal burning is a major cause of acid rain). Thus, industrial activities and post-industrial land pose a threat to human health and earth's natural ecosystems.

Therefore, the eco-restoration of post-industrial land worldwide needs wide-ranging participants, including academia, policymakers, private companies, entrepreneurs, practitioners, and financial institutions alike to discuss and explore the possibilities of integration of biodiversity and ecosystem services into their eco-restoration approach. There are natural processes that lead to ecosystem recovery; these processes may be slow, namely in initial stages of ecosystem development but affect ecosystem development in long run and lead to formation of sustainable and viable ecosystems. Eco restoration approaches try to use and enhance these natural processes. Revegetated/rehabilitated post-industrial land helps to achieve the goal of eco-restoration. It improves substrate characteristics of the post-industrial land, which are the basic needs for the re-establishment of desired plant species on such post-industrial landscapes. Therefore, ecologically and socio-economically valuable plant species must be screened among diverse geographical limits and should be included in the restoration of post-industrial land for ecological balance so that a self-sustaining ecosystem can be established on such land. In this direction, the use of native plant species for plantation is the most viable tool for restoring post-industrial land at a quicker pace.

This edited book is well-timed with up-to-date information that offers a cutting-edge synthesis of scientific, experiential, and established knowledge as a single source on different aspects of ecological restoration on post-industrial land.

I congratulate the editor, Dr. Vimal Chandra Pandey, for bringing out this valuable book published by a renowned publisher Wiley. The book consists of fourteen chapters covering various aspects of biodiversity and ecosystem services on post-industrial land. I believe this book will be a remarkable asset for researchers, environmentalists, plant scientists, eco-engineers, practitioners, industry professionals, eco-planners, policymakers, entrepreneurs, and other stakeholders alike, as well as would provide future directives in R&D to the field of restoration ecology.

08 August 2024

Dr. Jan Frouz
Professor and Director
Environmental Centre, Charles University, Prague, Czech Republic

Preface

Globally, the mining and industrial activities are well documented to have vivid environmental impacts. In addition, these activities destroy natural ecosystems and historical sites with high value, creating large post-industrial areas (fly ash deposits, post-mining land, red mud dumpsites, oil drilling sites, brownfield land, etc.). Moreover, they can directly impact human health, e.g., through the air and water pollution. Therefore, the post-mining and post-industrial lands have gained intensive attention among restoration scientists and applied ecologists aiming to transform such sites into risk-less biodiversity strongholds. By declaring the UN Decade on Ecosystem Restoration, such efforts have been widened to implement providing ecosystem services to meet the needs of the 21st century and to attain partial goals of UN-SDGs. Therefore, the restoration of post-mining and post-industrial lands on the global scale needs wide-ranging participants from different countries and from different disciplines, including academia, private companies, policymakers, financial institutions, consultancy firms, local farmers, NGOs, entrepreneurs, practitioners, and other stakeholders alike to discuss the integration of biodiversity and ecosystem services into their ecological restoration. By engaging in ecological restoration, everyone should be involved toward resetting our relationship with nature.

Ecological restoration aims at enhancing the biodiversity and ecosystem services, mostly by implementing natural processes into rehabilitation of disturbed sites. The first goal at post-mining and post-industrial lands are often associated with the improvement of substrate characteristics (i.e., physicochemical and biological), which are basic needs for the establishment of everlasting vegetation. The second goal should be focused on the re-establishment of rich communities of target plant species with biodiversity and their ecosystem services, i.e., pedogenesis, nutrient cycling, habitat for other organisms, food plants for herbivores, carbon sequestration, and aesthetic value. Here I want to explore biodiversity and ecosystem services of rehabilitated post-industrial land. The species diversity is a useful and often easily measured criterion in restoration projects, although it should be often accompanied by measures of potential for threatened species. Using the proper biodiversity indicators in mining and restoration projects, together with setting the realistic and beneficial conservation aims, allows focusing on the selected aspects of diverse field of ecological restoration and makes the entire process more efficient. During the ecological restoration of lands disturbed by mining and industrial activities, restoration projects must include ecologically and socio-economically benefits. Therefore, managing biodiversity during ecological restoration of such lands is a major challenge because this may influence the restoration target, which is partially dependent on the values of the stakeholders.

This book will offer how the biodiversity and ecosystem services concepts have been integrated into the restoration and reclamation of post-industrial lands. Finally, I believe this book will open wider conceptual insights and well-designed case studies on any general aspects on ecological restoration and biodiversity of such disturbed sites.

Vimal Chandra Pandey

Acknowledgments

I sincerely pay gratitude to *Frank Weinreich* (Publisher) and *Mary Angelin Rose* (Production Editor) from Wiley for their excellent support, guidance, and coordination during the production of this fascinating project. I thank all the reviewers for their careful and insightful review of the book chapters. The editor is highly thankful to Professor *Jan Frouz*, Director, Environmental Centre, Charles University, Prague, Czech Republic for writing the foreword at short notice.

1

Mining Sustainability

A Reality in Arid Zones

Elizabeth J. Lam[1], Italo L. Montofré[2,3], and Fernando A. Alvarez[4]

[1] *Chemical Engineering Department, Universidad Católica del Norte, Antofagasta, Chile*
[2] *Mining Business School, ENM, Universidad Católica del Norte, Antofagasta, Chile*
[3] *Mining and Metallurgical Engineering Department, Universidad Católica del Norte, Antofagasta, Chile*
[4] *Administration Department, Universidad Católica del Norte, Antofagasta, Chile*

1.1 Introduction

Mining, one of man's oldest and most important activities (Dubiński 2013; Candeias et al. 2018), has depended on extracted products since the beginning of civilization. Due to the crucial need for materials for industrial and human development, its participation in the economic structure of mining countries and the global economy has greatly increased in the last few decades, currently being one of the world's main economic activities (Lodhia 2018; Lam et al. 2021).

The mining process is based on the extraction and processing of materials of interest (Whitworth et al. 2022). It consists of consecutive operational stages: prospection; exploration; project assessment, development, and construction; resource production and exploitation; and operational closing. These stages may cause environmental and social impacts (Haddaway et al. 2019).

Ore extraction processes have changed with time, forcing the generation of more efficient processes environmentally and community-friendly. In addition, current trends in the extraction of ore with an increasingly lower grade pose new challenges for waste management, considering the increased number of wastes from mining units. This decreased ore grade is accompanied by a tendency to make new explorations and start new mining operations to balance decreased production, by further extracting and processing ore. This results in increased ore drilling, loading, and transport, thus increasing concentrator capacity and leaching processes. Also, low-grade ore makes it necessary to transform it into smaller particles, therefore increasing grinding and its associated energy requirements (Lagos et al. 2018).

Biodiversity and Ecosystem Services on Post-Industrial Land, First Edition.
Edited by Vimal Chandra Pandey.

Analogically, the production and exploitation stages cause environmental impacts such as industrial emission of sulfur dioxide and nitrogen oxides, high water consumption for industrial processes, and massive mining wastes (MMWs) such as tailings, residues, slags, sterile material, and solid wastes, among others (Lam et al. 2017, 2020a; Zhou et al. 2021). In addition, mine closing has become one of the most important stages on a global basis since a plan is needed from mining companies to mitigate the negative effects of the extractive mining industry, ensuring the physical and chemical stability of installations and mining wastes, according to the legislation on a country basis. In Chile, the closing plan must safeguard people's lives, health, safety, and the environment, according to law 20.551, regulating the closing and post-closing phases of mining sites (Lam et al. 2018).

Impacts from mining processes can be classified into six main groups: (i) soil quality, (ii) flora and fauna, (iii) air quality, (iv) water resources, (v) socioeconomic conditions, and (vi) climate change. These impacts make it necessary to face a series of challenges, even more so when the demand for most minerals may increase in the next few years. In addition, considering sustainable energy transition, the mining industry must create new technological developments to exploit increasingly complex ore bodies, accept the requirement to decrease water and energy resources, optimize waste management in the context of a circular economy (CE), and respond to climate change effects due to increased greenhouse gas emissions (Valenta et al. 2019).

Owing to the spatial distribution of resources, mining activity can be developed in different geographic regions, each of them with particular challenges depending on environmental conditions and the nature of mining activity, considering the type of ore and the processing method. These challenges are particularly relevant in arid zones, where impacts are notorious, considering ecosystem fragility and water scarcity (Liu et al. 2019). For these reasons, the mining industry in arid zones faces different challenges to make production sustainable, particularly the supply of critical agents such as water and energy, along with ecosystem preservation. This challenge is faced by using and controlling sustainability indicators for arid zone mining. At present, many factors could be considered sustainability indicators, such as water consumption, electric and fossil energy consumption, greenhouse gas emissions, land distribution, waste management, and finally, environmental and social impact.

In a different ambit, mining activity evolution has caused an impact on the economic development of many countries such as Chile, where, apart from contributing to the economic development of mining regions, it also influences the development of human beings, who are constantly requiring various mining raw materials and input for creating products and technology for man's progress (Qi 2020). Among mining countries, Chile is a world leader in the production of copper, iodine, rhenium, lithium, molybdenum, boron, silver, and gold (Ministerio de Minería 2022). This brings about great economic benefits, but, on the other hand, it produces a negative environmental impact associated with risks affecting the environment and the population (Castro and Sánchez 2003).

In Chile, mining production became massive in 2001–2015. At the same time, mines aged owing to a decreased ore grade. So, mining activity increased by starting new operations mainly in the country's northern zone, characterized by arid environmental conditions. These zones show scarce rainfall and vegetation, a dry climate, and high differences in temperature between day and night, making it particularly difficult to obtain resources such as water and energy, thus putting ecosystem stability at risk. Hence, sustainable mining must be developed, involving rational natural resource consumption so that annual production cannot jeopardize resources for future generations (Lagos et al. 2018).

1.2 Mining in Chile

1.2.1 Economic Benefits of Chilean Mining

Most Chilean mining industries are located in the northern zone, characterized by extreme climatic conditions. Exploitation, apart from economic benefits, has important impacts on the environment and ecosystems. So, it is essential to address future mining by focusing on sustainability, considering three basic pillars: economic, environmental, and social. Therefore, efforts must be directed to actions for maximizing profits, according to available resources, and preserving natural resources for proper ecosystem functioning and equilibrium. Both aspects must be in agreement with the demands and needs of local and surrounding populations, fostering citizen dialog, commitment, and participation in fair and equitable decision-making (Aznar-Sánchez et al. 2019).

1.2.2 Negative Impacts of Chilean Mining

Chilean mining produces great volumes of waste, by extracting, using, and processing ore, which is discarded or accumulated, occupying big spaces. It is estimated that more than one million tons of waste from ore concentration and more than two million tons of sterile material are generated daily. The decreased grade of ore deposits currently exploited and those projected requires treating increasingly higher amounts of ore tonnages to keep or increase production levels. In this context, the amount of waste to be disposed of either as sterile material or tailings will increase. Tailings are estimated to increase twice by 2035. According to Art. 23 D.S 148, 2003 (Sanitary Rules for Dangerous Waste Management), from the Ministry of Health, sterile material, low-grade ore, leaching wastes, tailings, and debris make up the so-called MMWs, which result from mining operations (Pérez et al. 2021). Tailings are deposited near mining sites. They can be defined as finely ground material piles consisting of a mixture of gangue, rock fragments, sediments, etc. Tailings are not originally considered "toxic"; however, they can acquire this characteristic when combined with water. Examples of this are arsenic, copper, zinc, lead, etc. (Adiansyah et al. 2015; Peña-Ortega et al. 2019).

1.2.3 New Mining Process Redefinition

On a country's basis, four technological challenges are defined: facing increasing water and surface scarcity, minimizing infiltration impact and ensuring ore deposit stability; promoting the change of wastes into assets; and fostering community inclusion and agreement. By facing these challenges, new mining is expected to not only be concerned about economic benefits from mineral production and exploitation but also committed to incorporating these sustainability elements in their processes and become an industry more friendly with the environment and the different habitats sharing the territory intervened.

In this new view of mining processes, another great concern is the responsible management of materials with low economic value, which are not of great interest for commercialization and are discarded in the so-called dumps, originating from overload removal to access materials that are extracted because of their greater economic attraction. Owing to their heterogeneity, particularly in terms of granulometry, they show basic differences with other MMWs such as tailings, which have been further studied (Lam et al. 2020b). In the mining process, overloaded material is removed via drilling, blasting, transport, and downloading (Hartman and Mutmansky 2002). The characteristics of these operations make deposited material quite heterogeneous. No data are available on their initial characterization since they are of little economic interest. Nonetheless, many dumps have been studied in the last few years, considering problems such as acidity generation, slope stability, and erosion.

The accumulation of great amounts of waste highly variable in components and concentration has a negative impact on the environment after the operations close (post-closing phase) since, in the medium and long term, wastes represent a potential risk and danger due to the adverse change of the natural environment, produced by particulate material, metals, metalloids, contaminant soil, and surface and underground water migration.Both mining projects in exploitation before the passing of Law 20.551 on mining operation closing and new projects must propose a closing plan including activities, measures, and actions mainly directed at maintaining the physical and chemical stabilization of installations. This, added to the increasing community concern about the impact that mining activity could have on the environment and the exposed population's health, forces mining companies to take measures, according to the legislation, and meet the requirements of communities and authorities.

Some MMW potentially risky substances are Cu, Zn, As, Fe, Hg, Pb, Cd, organic and inorganic cyanides, hydrocarbons, blasting wastes, containers and vessels contaminated with these elements, etc. Due to their potential risk, they must be considered in the new definition of more sustainable mining, preventing the eventual threat they represent from becoming a more appealing option for mining companies. These substances should remain as accumulations because, although not economically useful, they make up a potential risk to the surrounding environment. As to heavy metals, they cannot be degraded and, therefore, have a cumulative effect on the environment (Lam et al. 2018; Ghosh and Singh 2005; Kahlon et al. 2018) since they are potentially high contaminants in the food chain and have a dangerous effect on both biological properties and functional and taxonomic soil diversity (Vacca et al. 2012). In this context, soils contaminated with heavy metals put at risk the environment and people's health (Roy and McDonald 2015; Lam et al. 2018) due to biomagnification (Zenker et al. 2014). Thus, it is important to remediate these sites; however, owing to the complexity of these projects, they usually involve high mining costs. Therefore, these processes must be improved by adding new technology with attainable economic results for companies.

MMWs are characterized by high contents of minerals, chemicals, metalloids, and wastes accumulated along the life cycle of a mining project. They represent a potential risk for the environment mainly due to: (i) Heavy metal leaching in underground water, particularly in acid conditions (Puga et al. 2016); (ii) discharge and infiltration of drainage or material that could contaminate surface water resources; (iii) MMW may lose cohesion and collapse over towns and cities; and (iv) climatic conditions such as the wind could cause dragging, releasing contaminants into the atmosphere or settling and contaminating the soil again.

Among MMWs are tailings which, on certain conditions, are exposed to the potential generation of mining acid drainage by exposing sulfur minerals to air, water, microbial processes, and oxidation. This drainage represents one of the main environmental impacts mining must currently face (Broughton and Robertson 1992).

On a national and world basis, data on the remediation and restoration of sites affected by mining is scarce and relatively new. Although there are technologies, they are costly and limited for use since their efficiency and efficacy depend on factors such as particular site conditions (climate, ecosystems, phreatic level, earthquakes), MMW physical and biological characterization, soil availability and use, remediation time (rehabilitation or restoration), potential risks associated or imminent danger, and other related negative and positive impacts. So, in MMW remediation, each case is particular and must, therefore, be studied in itself.

To prevent and control MMW's negative impacts on people and the environment, they must be managed appropriately. The bioavailability of contaminants (metals, metalloids, etc.) plays a key role in the remediation of a contaminated site. The use of products, technically known as amendments, may decrease or increase contaminant bioavailability, mobility, and pH. Greater contaminant mobility results in a greater transfer from the substrate containing it, thus improving removal and facilitating absorption, either by vegetal species or soil washing. The opposite process is also possible, i.e., using amendments to reduce contaminant mobility to avoid transport into the trophic chain through plant absorption and underground water leaching (Bolan et al. 2014). Amendments are used for producing minerals containing less bioavailable contaminants. This is possible due to the chemical reaction between contaminants and the aggregated product, forming highly insoluble compounds. In this case, the mechanisms involved may be mineral surface adsorption, formation of stable complexes with organic ligands, precipitation (e.g., as salts), and cation exchange capacity (CEC) of soils (Venegas et al. 2016). This technology is applicable in situ and ex situ. Application in situ is a cost-efficient and environmentally friendly measure, as compared with contaminated substrate removal and disposal on another site (Puga et al. 2015).

Apart from being a positive soil remediation strategy, amendments are a potential alternative for reusing wastes (Venegas et al. 2016). Among amendments to reduce soil metal availability are phosphate fertilizers, organic material or biosolids, iron or manganese hydroxides, natural or artificial clay minerals, and mixtures of them.

In addition, MMW volume and component diversity and concentration make it necessary to assess a second processing line to change wastes into assets both technically and economically.

1.3 Chile: Arid Zone Mining

1.3.1 Mining Impacts on Arid Zones

In arid zones, representing more than 40% of the earth's surface, precipitation does not balance surface evapotranspiration. These zones show an extremely fragile ecological environment quite sensitive to human activities and global climate changes (Shao et al. 2022).

Chilean mining mainly occurs in the country's northern zone. This region, characterized by extreme aridity, faces continuous impacts proper of geographic and ecological units on extreme arid conditions and reduced vegetal coverage. The surface is the result of numerous processes dominated by aridity, scarce precipitation, and wind. The region is also characterized by a 10 mm or less mean annual precipitation rate and up to 10 mm day^{-1} mean evaporation rates (Wels et al. 2004).

The increasing demand for Cu and related products has increased low-grade ore processing, with large amounts of associated wastes. Currently, thousands of million tons can be found as Cu tailings in copper-producing countries. The national mining industry produces 5.8 million tons of fine copper, generating 530 million tons of tailings annually (Villa Gomez et al. 2022; Rubinos et al. 2021). In Chile, metallic mining-originating tailings are mainly related to Cu and Au. According to the last survey from the National Service of Geology and Mining (SERNAGEOMIN, for its acronym in Spanish) in August 2020, there are 757 tailings in Chile, located in 10 of the 15 regions: 173 are abandoned, 112 are active, 5 in construction, and 467 inactive.

Mining companies are responsible for designing a closing plan, according to Law 20.551. Many abandoned tailings were in such a state when the law was passed in November 2012. So, the state is responsible for them. Of the 112 active tailings, 32 are extremely big in size, i.e., bigger than the authorized 10 million m^3, while 72 are medium and small in size. Large-scale mining companies are estimated to deposit about 15.6 thousand million of m^3 tailings at the end of their lifetime, corresponding to 99.5% of the country's total (SERNAGEOMIN 2020). According to SERNAGEOMIN, large-scale mining operations had deposited 46.7% of the total accepted by the end of their lifetime in 2020. Meanwhile, medium- and small-scale mining companies had deposited 68.7% of the volume authorized by that time. These data allow predicting that the latter are currently in the middle of their lifetime and, thus, in full operation. The problem of tailings can not only be described by the number of tailings but also by their volume reaching 16 840 million of m^3 as the total volume accepted, with deposits being distributed in 67 of the 346 countries' communes. Tailings are unstable in themselves and usually concentrate big amounts of toxic substances (Kossoff et al. 2014). So far, technical and economic resources have been unfortunately insufficient to dispose of tailings safely on a national basis. This involves risks such as: (i) Physical collapse and destabilization caused by earthquakes, poor compaction, and excess fines in sands, among others; (ii) chemical destabilization due to the reaction of solids in water, toxic element solubilization, and mining acid drainage; and (iii) Environmental dispersion and pollution owing to wind currents and erosion. Many tailings are found in arid and semiarid regions, where erosion is greater since, the lower the humidity, the greater the number of particles with a high content of metals and soluble phases emitted by the action of the wind.

On the other hand, the effect of a mining operation on water resources is probably the most significant. To assess this effect, it is necessary to determine if water sources close to the mining area will remain appropriate for human consumption and if water quality will be enough for the regional flora and fauna. Another risk associated with tailings is the potential generation of mining acid drainage, resulting from the reaction of sulfide minerals in the tailing with air and water, which can even be catalyzed by the presence of certain bacteria (Lam et al. 2020a).

With time, this acid leaches metals and other contaminants present in tailings and wastes, resulting in an acid solution rich in sulfates and metals (e.g., cadmium, copper, lead, arsenic, etc.). Even in the absence of an acid environment, the leaching of some minerals such as arsenic, selenium, and others is still a possibility (Peña-Ortega et al. 2019; Khobragade 2020).

In this context, soil erosion, which is also greater in arid regions, may have several effects on the areas close to mining operations. Erosion refers to surface material loss due to water and wind. In the case of water, erosion may occur because of rain or runoff, where soil particle strength can be overcome by contact with tensile forces. As a whole, owing to particle dragging by the action of water, nearby water bodies can be contaminated, although infrequently, particularly during rainfall, storms, or snow events. These events have historically caused changes in water habitats, along with fines and sediment accumulation in water bodies. Rainfall amount and speed, infiltration rate through the soil, the amount of vegetation in the area, slope length, the separation between the flow origin point and the deposition point, and in general, operational structures for controlling the phenomenon are factors influencing the water erosion process.

The greatest erosion sources in mining operations include "open pits," heap leach, sterile rock piles, access roads, and tailings dams. Also, impacts vary qualitatively or quantitatively, depending on several factors such as geology, vegetation, topography, climate, and water properties and proximity. Considering the amounts and characteristics of the minerals exposed in operations, it is impossible to generalize sediment loads. Again, there is a variety of potential impacts that could emerge, having negative effects in the short and long terms.

Increased levels of particulate material in surface waters may cause acute and/or permanent toxic effects on fish and other species. Minerals associated with deposited sediments may decrease runoff pH, moving and infiltrating heavy metals into the underground or nearby water bodies. This may produce persistent environmental contamination and also soil pH decrease. In turn, this could cause vegetation and habitat loss. In addition, changes in erosion may affect the physical environment. For example, runoff rate and volume may increase, resulting in inundations, canal undermining, structural damage, etc. In extreme cases, where air erosion deposits acid particles, native vegetation may be destroyed, increasing erosion produced by runoff and, thus, soil removal (Khobragade 2020; Li et al. 2020).

The effects of air quality, mainly produced by atmospheric emissions, can also be discussed. Although these emissions may occur during any stage of the mining process, they generally originate during construction and operation and in waste piles due to dispersion caused by the wind. Most air contaminants from these operations are particulate material carried by the wind due to sources such as material transport, wind erosion, waste piles, dumps, and roads. In addition, exhaust emissions from moving vehicles such as cars, trucks, and others, together with blasts and ore processing, may increase the levels of particulate material. This is particularly important because after entering the atmosphere, it undergoes several physical and chemical changes before getting in contact with a receptor, which may affect human health and the environment. As a whole, mining activity may cause huge environmental damage, particularly by air contamination during operations since each stage of the mining process may emit contaminants as particulate material, including heavy metals, carbon monoxide, sulfur dioxide, and NOx gases (Khobragade 2020; Li et al. 2020).

Water cannot be considered a self-renewable and low-cost natural resource in most arid regions. So, new supply sources are needed. In this sense, sea water desalination is a technology greatly accepted for arid zones owing to the exhaustion of conventional water resources. Desalination processes in regions under water stress produce negative effects on the sea, heavy energy consumption, and solid wastes during the operational phase (polyamide membranes), among others.

Therefore, it is clear that, although mining is undoubtedly necessary for man's development, it is also destructive because, on the one hand, it can produce health, employment, roads, and infrastructure, but on the other hand, it may cause several negative effects on the environment and people's health on a general and specific basis. It is also important to consider people's perceptions because they frequently feel that their rights are infringed, resulting in social discomfort and eventually conflicts between parties. This usually occurs because studies on environmental impact underestimate or totally ignore its effects. Hence, informed citizen participation must be taken into account for making decisions about these studies.

1.3.2 Sustainable Mining Challenges

As described above, mining exploitation is highly important (He et al. 2017; Ma et al. 2019). So, using different practices is a must for reducing environmental impact and maximizing social and economic benefits for local communities and the environment (Mahapatra and Ghosh 2021). These practices must be directed to minimize water consumption, reduce contamination, and promote ecosystem restoration after exploitation (Gonzalez et al. 2019).

Thus, the big challenges faced by the mining sector are water and energy scarcity since they are insufficient to meet present demands (Leiva González and Onederra 2022).

Considering the main proposals made in different studies, the most important challenges can be summarized as follows:

a) **Water scarcity:** This is one of the most critical challenges due to scarce rainfall in arid zones and the impact of climate change. However, scarcity resulting from the chemical quality of water must also be considered since water contamination due to the lack of environmental regulation before 1994 restricts its use not only for mining activity but also for agriculture.

   The use of water has originated in different social and environmental conflicts in the last two decades since it is extracted mainly from millenary aquifers hydraulically connected with important Andes wetlands (Carvalho 2017). In arid zones, using it for mining affects water availability for local communities and agriculture. Thus, different practices and technologies have been implemented for sustainable mining in these zones, e.g., water recycling and use in the mining process and the exploration of alternative water sources (Jiang et al. 2019; Oyarzún and Oyarzún 2011).

According to predictions, total water consumption for Chilean copper mining will increase by 2.1% annual average in 2031 (COCHILCO 2020), while energy consumption will increase by 3.05% annual average in the same year (Abarca and Lucero, 2024). Therefore, the mining industry is currently focusing on the use of sea water and renewable energies to expand its operations. However, an efficient energy plan is needed to avoid increasing energy consumption (Leiva González and Onederra 2022).

b) **Extreme climate:** Extreme temperatures and the absence of rainfall make mining more difficult and costly, increasing workers' health and safety risks (Pearce et al. 2011).
c) **Soil degradation:** Ore extraction and infrastructure construction can damage the soil and reduce its capacity for sustaining vegetal and animal life. Some studies propose restoring soils, including planting and artificial wetlands (Moreno et al. 2007; Freedman et al. 2014).
d) **Impact on nearby communities:** Communities must get involved in mining planning and operation to make sure that their needs and concerns are taken into account. Collaboration and dialog with local communities, government authorities, and other interested parties are essential for promoting sustainable mining and ensuring community and environmental development (Lam Esquenazi et al. 2019).
e) **Environmental contamination:** Effects and potential risks associated with atmospheric contamination by particles and toxic emissions produced by smelting processes (Lam et al. 2016) are challenges to be faced, along with soil stabilization and mining acid drainage associated with physical and chemical substance penetration in the soil, which create a huge environmental footprint caused by tailings storage (Adiansyah et al. 2015). In addition, technologies to decrease carbon footprint and greenhouse gas emissions must be implemented, using renewable energies and optimizing processes (Mahapatra and Ghosh 2021). Also, environmental restoration programs must be established to recover ecosystems and foster biodiversity in areas affected by mining (Wassenaar et al. 2013). Likewise, it is important to reduce environmental impact in mining life cycles, preventing and mitigating negative operational impacts (Pactwa et al. 2021).
f) **Social challenge:** Mining workers and local communities are the most affected by environmental mining impacts such as atmospheric contamination, scarce surface and underground water resources, and a large number of solid wastes, e.g., mine tailings and rock debris damaging the soil (Aznar-Sánchez et al. 2019).

   Likewise, it is important to keep in mind the future automatization era for mining processes. This will bring about a decrease in the number of jobs and a deterioration of local and nearby communities' life quality expectations (Carvalho 2017).
g) **Economic challenge:** The ultimate objective of the mining industry is to make profits for the parties involved, particularly the community, by reducing costs and improving productivity and efficiency. Wealth can be generated by creating jobs, paying taxes, and contributing to GNP. In addition, foreign capital is attracted using investment, thus contributing to the country's wealth. Hence, an equative wealth distribution among the regions involved must be guaranteed (Aznar-Sánchez et al. 2019).

The mining industry should share a common view of sustainability; coordinate and integrate tools and methods to develop sustainable products for making decisions and implementing widely used initiatives; and, finally, highlight the importance of efficient communication among parties (Hallstedt et al. 2013).

According to the above, Gorman and Dzombak (2018) indicated, *In the future, we are expected to meet the needs of the present without compromising the ability of future generations to meet their own needs.*

1.3.3 Arid Zone Mining Sustainability Indicators

Considering future mining production, the current scenario of the mining industry must take into account environmental resource conservation and negative anthropic factors that may put at risk resources for future generations.

In the last two decades, mining production in Chile has represented 33% of the world's production. Due to this massive demand, it is affected by aging, a phenomenon occurring when ore extraction results in production loss. This is attributed to open-pit and underground mine expansion and low-grade ore. Therefore, there may be an increase in water and energy consumption, greenhouse gas emissions, and waste (Ghorbani and Kuan 2017).

To ensure resource preservation for future mining operations, indicators are classified into the three most influential: economic, environmental, and social.

Mining expansion involves a greater distance for ore and personnel transport. So, greater fossil energy consumption is needed. As to electrical energy, the greater the mining expansion, the greater the electricity consumption because more electrical energy is needed to feed networks owing to depth, as in an underground mine.

Likewise, the greater the consumption of fossil and electrical energy, the greater the greenhouse gas emissions, which influence climate change causing a negative effect on ecosystems (Covert et al. 2016). Greenhouse gas emission indicators are used for determining emission sources (Thomas et al. 2000).

Water consumption is also an indicator because, owing to its scarcity in arid zones, it must be managed efficiently to prevent exhaustion. Another indicator is soil distribution, i.e., its organization and porosity, the latter determining permeability and capacity to retain water and nutrients.

Environmental impact is a relevant indicator of its negative effects on water, air, and biodiversity. Finally, social indicators refer to the life quality of communities close to the mining production area. It deals with the actions taken by companies to fulfill their social, environmental, and economic responsibilities, taking charge of the whole operational process, from start-up to post-closing, by properly managing wastes (Xiao et al. 2020).

Sustainability indicators are used for identifying critical points where natural resources are affected and, later, designing strategies to ensure environmental resource conservation.

In general, it is important to use several indicators for environmental sustainability and also work for the continuous improvement of mining practices to minimize impact on the environment and local communities. Also, arid zone mining shows special conditions because resources are scarce and difficult to access. So, using them inefficiently leads to exhaustion and higher costs.

Finally, indicators are related to each other since, e.g., if water is misused, it affects the soil, the environment, and people's quality of life.

1.3.4 Technological Developments

For efficient water use and reuse, the sector is always looking for new alternatives to control this resource, among them, recovery from tailings and mining wastes mainly generated by copper processing.

Many important mining operations in Chile, which process a large amount of sulfide minerals, are located in extremely dry areas. In some of them, more than 100 000 metric tons are produced daily generating a great number of tailings (Cacciuttolo and Valenzuela 2022), which are transported hydraulically to nearby zones. These tailings result from the disposal of sulfide ore concentrations previously subjected to a phase separation process to recover water. Most water used in tailings transport is recovered to be reused in the metallurgical process.

Several technologies are used for dehydrating tailings, avoiding water loss by infiltration, evaporation, and interstitial retention (Bascetin et al. 2016; Lyashenko et al. 2019). Typical tailings disposed are slightly thick, at a 48–52% solid concentration. In this case, the design is adapted to the topography of the disposal area. Tailings are made thicker before disposal, reaching a homogeneous thickness with a non-segregated laminar flow and 80% water recovery (Robinsky 1999). On the other hand, water recovery from tailings by using filtration technology produces a non-saturated tort storing great amounts of material with a supporting structure self-sufficient under compaction, recovering 90% water, reducing areas affected, decreasing physical stability risk, and improving community perception (Cacciuttolo and Scognamillo 2014). In the last few years, pressure and vacuum filtration have improved operations and the development of big filters, reaching up to 50 000 metric tons of filtered tailings per day (Cacciuttolo and Valenzuela 2022; Cacciuttolo et al. 2023).

Current and future trends in water use for mining indicate that it will be captured from seawater (direct or desalinated) and also tap water. Technologies considering mineral characteristics are implemented to classify it as coarse, sand, or fine "overflow." They are transported to independent sites, considering sand pile installation for the coarse fraction and another for storing silts (fines fraction) (Wels and Caldwell 2013).

Water recirculation in the copper concentration process exceeds 75%. The analysis of these technologies allows identifying gaps concerning more reliable equipment for large-scale thickening and filtering processes to recover and reuse wastewater during ore processing. Other related advances are artificial aquifer recharge, flood storage, basin water transfer, and treated sewage water.

On the other hand, considering the location of mining operations, the use of seawater involves high costs and energy expenses. Despite this, an increasing number of mining operations use seawater for their productive processes. Every activity done with seawater must consider measures to avoid damage caused by the corrosion of equipment, pipelines, and others. Aqueduct design uses a type of pipeline that must be treated with chemicals for a 20-year lifetime. However, there are other solutions such as coated pipelines or those particularly designed for this technology that do not require maintenance but could mean high investment costs. Plant location, desalination if needed, pumping stations, and transport pipelines must be taken into account for capturing water from the coast (Tabra and Lange 2014).

Desalination plants will reach $11.3\,m^3\,s^{-1}$ in 20 years, that is, a 289.9% increase. Although mining is responsible for only 3% of the country's water consumption (78% of copper mining companies), most operations are located in the country's northern zone, where only 0.13% of national runoff is present (Montes 2016).

One of the most important limitations of desalination is the high electricity consumption for pumping water toward mining operations. For 3000 m.a.s.l., costs amount to almost US$ 5 per m^3, representing up to 8% of mining costs. Current technological developments are intended to optimize two factors: energy consumption and conversion, the latter for returning the least amount of brine per m^3 of extracted water into the sea.

The total amount of desalinated water used in mining processes is $4.471\,L\,s^{-1}$, while direct sea water totals $3.373\,L\,s^{-1}$. At least, 22 new operations using $9.046\,L\,s^{-1}$ of new desalinated water and $5.715\,L\,s^{-1}$ of direct sea water are projected. As to pretreatment, ultrafiltration shows a maturity degree that makes it competitive, compared to conventional technologies.

In addition, there is a water-energy relationship to face current and future mining challenges by implementing comprehensive projects. In this regard, the study done by COCHILCO "Projected electrical energy consumption in copper mining, 2021–2032" (2022) estimates an electrical energy increase from 25.8 TWh in 2021 to 33.8 TWh in 2032. This increase is attributed to ore concentration, which will consume 22 TWh by itself, representing 65% of the electric consumption of the sector. Water desalination and pumping will also increase significantly, from 1.5 TWh in 2021 to 4.5 TWh in 2032, representing the second most intense electric consumption process. A significant decrease is projected for leaching (mineralogical changes), from 5 TWh in 2021 to 2 TWh in 2032. Meanwhile, a slight increase is projected for smelting, from 1.7 TWh in 2021 to 1.9 TWh in 2032. Underground mining, open-pit mining, and service processes will not exceed 3% consumption in this period.

The climatic characteristics of the Chilean northern zone make it appropriate for implementing non-conventional renewable (solar and wind) energy projects (ERNC, for its acronym in Spanish). In the last few years, great advances in ERNC have been made in copper mining and other areas. An aspect to highlight is the tendency to decrease costs associated with applied technologies and technological developments for energy storage. This has facilitated the current and future integration of this clean energy into the country's energy matrix. A significant increase is projected for the use of renewable energies in the mining sector by:

a) Using them in processes via projects developed by mining companies.
b) Signing Power Purchase Agreements (PPAs) in which the mining company made investments.
c) Signing PPA in which the mining company (customer) asks its supplier to use renewable energies. This is the most frequent option.

ERNC projects have been gradually implemented, considering their intermittence and the fact that mining processes operate continuously. So, energy storage systems must be improved.

As to ERNC distribution, the data provided below are only estimations, given the complex national energy matrix structure that does not allow a direct relationship between a matrix generation unit and the final consumer (Abarca and Lucero 2024.). For copper mining, an increase in ERNC (Renewable Non-Conventional Energy) consumption is estimated from 11.4 TWh in 2021 to 20.6 TWh in 2025; for an expected process consumption that will demand 25.8 TWh in 2021, changing to 33.1 TWh in 2025, which translates to a projected increase in ERNC participation from 44% in 2021 to 66.3% in 2025.

The substitution of all types of fossils by renewable energies contributes to the transit towards carbon neutrality. In this regard, energy accumulation systems play a key role, considering network stabilization, in making possible self-consumption and electric propulsion. The Chilean northern region has climatic characteristics and lithium resources in Atacama Salt Lake to incorporate renewable energies and contribute to the development of lithium-ion batteries by providing lithium salts. Current technological developments in the process industry and governmental policies are being directed to respond to the new energy pattern of a big present and future market.

At present, lithium is considered a critical element for industrial development, mainly due to its impact on green technology implementation (Ambrose and Kendall 2020). Australia is the world's main producer with 48%, followed by Chile with 29%, and Argentina and China with 9% each. In 2019, 71% of production was lithium carbonate, followed by lithium hydroxide with 22%, with hydroxide production being expected to be greater than lithium carbonate (COCHILCO 2020). Considering that other countries will enter the lithium market, Chilean participation in the global market is expected to decrease in 2030. However, based on its consolidated history as an experienced producer, Chile is prepared to expand production, having a relevant participation in the lithium production industry.

In the lithium battery market, lithium carbonate equivalent (LCE) is one of the main components for cathodes, a market competing in lithium carbonate purity, which includes sodium concentration and magnetic impurity content, among others. Producers make important efforts to advance in supplying battery-grade lithium with low sodium and magnetic contents, decreasing the water footprint, and aiming for sustainable mining in arid zones.

To reach carbon neutrality in 2050, Chile fostered the development of several projects in the northern and southernmost parts of the country to produce green hydrogen, taking into account the natural advantages of the territory, such as the world's greatest radiation and remarkable wind resources. These initiatives involve developing green hydrogen production plants, operations starting in December 2025, at the most. At least, 10 MW is required and production is expected to amount to 45 000 tons of green hydrogen annually, thus reducing more than 600 000 CO_2 tons per year and reaching carbon neutrality in 2050. Three of these initiatives were planned and environmentally assessed for Antofagasta Region

Technological innovation projects in northern Chile are driven by a medium-term strategic view involving metallic and industrial mineral resources in an extremely arid zone characterized by increasing energy demand, limited water resources, and higher environmental safety and protection standards.

1.4 Circular Mining and Arid Zone Sustainability

Mining processes produce millions of tons of tailings every year, billions of tons being already stored worldwide. This trend is expected to increase as demand for mineral products increases, with gradually lower-grade mineral resources, involving the extraction of more tons of materials to meet production expectations. Submicrometric ore grinding and classification are inevitable for production. This results in increased concentrations of fines such as clay and silt, accumulating great volumes of mine tailings, increasing destabilization risks, making solid–liquid separation and, therefore, processing water recovery more difficult (Wang et al. 2014).

To produce one ton of copper, it is estimated that 128–196 tons of tailings are generated (Gordon 2002; Bridge 2000). As a reference, 89% of the material from Caren tailings at El Teniente copper mine are classified as fines, including clays and silts, according to ISO 14688- 1 (Smuda et al. 2008). Tonnage, fines, and the chemical and mineralogical characteristics of national copper tailings (SERNAGEOMIN 2020) require the use of technologies to improve the geotechnical properties of deposits and control chemical destabilization. Alternatively, prevention techniques and mine waste recycling are gaining much attention (Mylona et al. 2000; Jiang et al. 2000; Tabelin et al. 2018). Mining wastes can be used in building materials, reducing their volume and producing economic and environmental benefits (Onuaguluchi and Eren 2013; Shi and Kan 2009; Swami et al. 2007).

Hence, the high volume of tailings on a national basis is an opportunity to develop a circular economy model, based on copper tailings reuse as an aggregate for making concrete and mortar. The construction industry, one the greatest on a world basis, is characterized by the use of great volumes of raw materials. The use of mining wastes with high silica content, 60% as an average for Chilean tailings, may contribute to the protection of natural resources and the decrease of environmental impacts (SERNAGEOMIN 2020). Building materials include coarse and fine aggregate. In this regard, tailings substitute only fines partially or totally because the particle diameter is smaller than 1 mm (Gou et al. 2019). However, progress in grinding technology produces increasingly more clay and silt in tailings, which are characterized by a quite rough and irregular surface. Therefore, water density and absorption show great variability, significantly influenced by particle size (Zhao et al. 2014; Argane et al. 2016; Filho et al. 2017). Concerning tailings management, each country follows its own regulations. In Chile, the most relevant are: Rules for Projects on Designing, Building, Operating, and Closing Tailings (D.S. 248, 2007), Regulations for Closing Mining Operations (Law 20.551, 2011), and Rules for Environmental Impact Assessment (D.S. 40, 2013).

As to tailings assessment in Chile, CORFO financed the project "Chile Polimetálico," in which four areas were identified to propose initiatives for recovering non-traditional elements of value (EV): Mining project development of small- and medium-scale natural ore deposits, EV recovery via environmental remediation projects on abandoned mining sites; EV recovery from mining wastes (tailings, rubble, clearings, debris, smelting dust, etc.), and mineral exploration and extraction from polymetallic ore deposits at the sea bottom (Public Well-being Project N°18BPE93816). On the other hand, BHP and Fundación Chile, through its open innovation program, "Expande," is developing the international initiative "BHP Tailings Challenge" to identify enterprises, start-ups, consortiums, research centers, and innovative universities that can help reuse fresh tailings and create innovative business models. The projects developed during the concept testing stage include: natural fertilizer production, tailings mineral content extraction (copper and molybdenum), water recovery, energy generation, geofoam production for the construction industry, sulfide content reduction, silicon waste treatment, special cement making, geopolymer production, use of solar energy for partial fusion agglomeration of dry material at extremely high temperature and agglomerate synthesis for producing pellets, carbon capture and use to produce building materials, soil recovery, green hydrogen production, metal stabilization, insoluble fraction incorporation to produce mineral paper, and, as a whole, changing wastes into assets (https://expandemineria.cl/convocatoria-custom/bhp-tailings-challenge/).

Three alternatives for managing tailings are currently being developed in Chile: the first one deals with assessing the recovery of EV, mainly Cu and Mo, and water; the second one refers to safe final disposal with minimal potential risk; and the third one deals with changing wastes into useful resources, e.g., aggregates for building materials. On the other hand, secondary mining is an opportunity to develop a circular economy model by reusing mining wastes. In this sense, an opportunity may be the great tailings volume on a national basis, which can become a business opportunity if they are used for making mortar and concrete (Lam et al. 2020b).

CE deals with making economic changes and establishing new competitive advantages (Sillanpaa and Ncibi 2019). When market and technology need change, components not previously prioritized become interesting and are considered potential valuable resources (Kinnunen and Kaksonen 2019).

Tayebi-Khorami et al. (2019) point out that CE, when introduced into the mining sector, can improve its development and economic sustainability. Kinnunen and Kaksonen (2019) indicate that mining waste management is based on a linear economy, where wastes are an opportunity to solve the limited supply and increasing demand for metals.

The scientific community perceives that CE's role in the mining industry has not been as relevant as other areas of the industrial chain, that is, from raw material production to consumer products and services. Ranjbari et al. (2021) studied 20 years of academic research associated with CE and waste management to identify the most frequent contributions. Their results reveal that, among the five most interesting, those related to the mining sector do not have a great impact or are not much studied by the scientific community.

A similar situation is observed in industries, as found by Upadhyay et al. (2021), who analyzed sustainability reports from three of the world's five largest mining companies. Their results show that the CE concept is extremely limited, only making implicit references to processes such as recycling.

However, there is an increasing trend in the analysis, experimentation, and understanding of how CE can be implemented in mining, particularly in managing the mining footprint, either as mining wastes, energy overuse, or as an impact on the landscape, the latter historically considered irreparable.

CE is intended to create wealth via sustainable practices. In mining, regardless of the strategy used, the key is reducing process costs and also decreasing and transforming mining wastes.

Apart from the above, the question made by Tayebi-Khorami et al. (2019) is relevant: How can the mining industry create new economic value, minimize its social and environmental impacts, and reduce liability from mining wastes?

Metal extraction uses technologies producing big waste flows such as dumps and stockpiles resulting from the surface removal of sterile material or one with subeconomic mineral concentration, respectively, along with tailings potentially causing contaminated water infiltration and/or collapse of the supporting dam (Sillanpaa and Ncibi 2019).

Tayebi-Khorami et al. (2019) identify five areas which can provide a framework for rethinking mining wastes: social dimensions, geological and environmental aspects, geological and metallurgical specifications, economic drives, and legal implications.

These diverse issues show how complex it is to understand and put into practice synergy among them. Probably, this led researchers to use a methodology based on multidisciplinary data exchange between industry experts and academics, via interviews, workshops, and hierarchical and content analysis, among others (Kinnunen and Kaksonen 2019; Singh et al. 2020; Gedam et al. 2021). Using different approaches, these authors deal with aspects such as opportunities, drivers, needs, and challenges to contribute new knowledge to CE implementation and adherence to the mining industry and waste management.

The analysis here focuses on tailings management. Currently, tailings are the cornerstones for introducing circular mentality into the mining industry because of their great volume and economic value.

As an opportunity, Kinnunen and Kaksonen (2019) point out that gradual grade decrease and the metallurgical process of complex ore deposits make tailings a technically and economically competitive option. However, they warn that challenges mainly point to technological development, knowledge gaps, and legislation. For example, they refer to possible institutional measures using taxes as an incentive or competitive advantage for raw materials from CE-associated processes. Singh et al. (2020) found that the most important barriers are governmental policies and regulations. Likewise, the lack of accurate knowledge of mineralogy and tailings concentrations to allow economic assessment, analogous to a model for making blocks from mining resources, is an uncertainty in avoiding potential extraction and reprocessing.

Although the latter is open to research, there are approaches with scaling potential. For example, Lam et al. (2020b) assessed a semiquantitative tool based on magnetic susceptibility to characterize tailings, potentially applicable to other large mining wastes.

The lack of early mining planning for tailings management during and after the project lifetime fosters research on waste remediation, reutilization, and/or rehabilitation. Phytoremediation, as a low-cost technology, emerges as one of the most promising methods for contaminated multielement tailings remediation, by introducing plants tolerating these contaminants, allowing the effective fixation of these elements in the presence of external physical, chemical, and mechanical agents (Lam et al. 2023). Montofré et al. (2021) assessed alternatives for pretreatment and substrate generation facilitating phytoremediation.

The literature also examines the use of tailings as a raw material for making building materials. However, owing to the composition of rare earth and contaminating metals, leaching compromises its use in civil engineering works. Despite this, the manufacture of bricks made from cement containing tailings as an aggregate is under assessment for their use in dams generated from abandoned tailings (Lam et al. 2020b). Although advances are promising, the scaling capacity of tailings on an industrial basis is still limited because of the high investments needed, which, even in university research centers, are still insufficient (Kinnunen and Kaksonen 2019).

Similarly, de Palacios and Rodríguez (2022) indicate that mining operations are not the best example of EC culture since they respond to reactive-type initiatives, emphasizing that EC application in much more effective during the early stages of mining project design, so as to minimize waste generation and simultaneously maximize their recovery.

Pinchuk et al. (2019) suggest that the underlying reason for not implementing CE is the absence of data, the missing link being the digital economy, Industry 4.0 coordinating data, and material flow.

A contradiction between expected economic profits and reduced dependence on raw materials is far from becoming a reality. Sillanpaa and Ncibi (2019) point out that the reasons for current recycling cannot completely replace primary extraction.

The increasing demand for metals, whose aim is transiting toward clean energies, makes western countries demanding metal supply remain economically and geopolitically vulnerable. This scenario fosters strategies based on recycling, reuse, and extension of metal life cycle, thus reducing primary extraction needs.

The literature indicates that professional multidisciplinary collaboration is essential for materializing CE in the mining industry, while R&D academic units must play a key role in establishing links with it. The use of resources and priorities in the mining industry are usually related to urgent activities or direct operational benefits, disregarding CE innovation/research initiatives.

It is also important to develop not only sustainability symbiosis among industries to create closed circles, where wastes and secondary products from one sector become inputs in another one, but also cross-cutting benchmarking, not only with the mining industry. In this regard, histories of success and failure become relevant for accelerating the change from a linear into a circular economy.

As to tailings, these must be at the base of industrial business strategies. One of the keys to attaining this is their valuation and later transformation into mining assets. The inherent complexity of each ore deposit and its operation indicates that future solutions will be hardly standardly used in the industry; on the contrary, they will have to be particularly adjusted to each ore deposit.

The economic benefits and added wealth behind CE are clear. However, materializing it requires overlapping different investment initiatives, making it difficult to convince decision-makers if economic valuation and return predictions are low. It must be kept in mind that behind every economic venture, the ultimate goal is creating wealth. This gap may be managed by the government via legislation and norms, not only as mandates but also through participation, giving impulse to innovation using incentives such as research and collaboration funds.

Effectively remediating the environmental and social legacy of decades of classical linear mining schemes is a must. The transition toward sustainable mining must involve changes in business models and management culture. Only companies are able to ensure long-term sustainability. By focusing on R&D, they will enjoy a globally competitive advantage in the mining industry in the next years.

1.5 Conclusions

Mining in arid zones poses a series of unique challenges; however, sustainable solutions to help minimize their environmental and social impact are being explored. Water management, soil restoration, and local community participation are key areas where sustainable practices can be implemented. It is important to find innovative and sustainable solutions for mining in arid zones. This will help not only to reduce the negative impact of this industry on mining regions but also help improve relationships with local communities and increase the social acceptance of mining. In brief, this is a complex and challenging issue requiring a sustainable and responsible approach. Mining in arid zones can be an important economic activity if its impact on the environment and local communities is minimized by using a careful approach and paying continuous attention to sustainability.

References

Abarca, R. B., & Lucero, V. G. (2024). Informe de actualización del consumo energético de la minería del cobre al año 2022.

Adiansyah, J.S., Rosano, M., Vink, S., and Keir, G. (2015). A framework for a sustainable approach to mine tailings management: disposal strategies. *Journal of Cleaner Production* 108: 1050–1062.

Ambrose, H. and Kendall, A. (2020). Understanding the future of lithium: part 2, temporally and spatially resolved life-cycle assessment modeling. *Journal of Industrial Ecology* 24 (1): 90–100.

Argane, R., El Adnani, M., Benzaazoua, M. et al. (2016). Geochemical behavior and environmental risks related to the use of abandoned base-metal tailings as construction material in the upper-Moulouya district, Morocco. *Environmental Science and Pollution Research* 23: 598–611.

Aznar-Sánchez, J.A., Velasco-Muñoz, J.F., Belmonte-Ureña, L.J., and Manzano-Agugliaro, F. (2019). Innovation and technology for sustainable mining activity: a worldwide research assessment. *Journal of Cleaner Production* 221: 38–54.

Bascetin, A., Tuylu, S., Adıguzel, D., and Ozdemir, O. (2016). New technologies on mine process tailing disposal. *Journal of Geological Resource and Engineering* 4 (2): 63–72.

Bolan, N., Kunhikrishnan, A., Thangarajan, R. et al. (2014). Remediation of heavy metal (loid) s contaminated soils–to mobilize or to immobilize? *Journal of Hazardous Materials* 266: 141–166.

Bridge, G. (2000). The social regulation of resource access and environmental impact: production, nature and contradiction in the US copper industry. *Geoforum* 31 (2): 237–256.

Broughton, L.M. and Robertson, A.M. (1992). Acid rock drainage from mines – where we are now. *IMM Minerals, Metals and Environment Conference*, February 4–6, Manchester, UK.

Cacciuttolo, C., Pastor, A., Valderrama, P., and Atencio, E. (2023). Process water management and seepage control in tailings storage facilities: engineered environmental solutions applied in Chile and Peru. *Water* 15 (1): 196.

Cacciuttolo, C. and Scognamillo, C. (2014). Improving water recovery with different tailings management technologies. *Proceedings of the 4th International Congress on Water Management in Mining WIM 2014*, Viña del Mar, Chile (10–13 May 2014).

Cacciuttolo, C. and Valenzuela, F. (2022). Efficient use of water in tailings management: new technologies and environmental strategies for the future of mining. *Water* 14 (11): 1741.

Candeias, C., Ávila, P., Coelho, P. and Teixeira, J.P. (2018). Mining activities: health impacts. In: Nriagu, J., Ed., *Encyclopedia of Environmental Health*, Elsevier, Amsterdam, 415–435. https://doi.org/10.1016/B978-0-12-409548-9.11056-5

Carvalho, F.P. (2017). Mining industry and sustainable development: time for change. *Food and Energy security* 6 (2): 61–77.

Castro, S.H. and Sánchez, M. (2003). Environmental viewpoint on small-scale copper, gold and silver mining in Chile. *Journal of Cleaner Production* 11 (2): 207–213.

COCHILCO. (2020). Proyeccion Agua Mineria del Cobre 2019–2030.

Covert, T., Greenstone, M., and Knittel, C.R. (2016). Will we ever stop using fossil fuels? *Journal of Economic Perspectives* 30 (1): 117–138.

Dubiński, J. (2013). Sustainable development of mining mineral resources. *Journal of Sustainable Mining* 12 (1): 1–6.

Filho, J.N.S.A., Da Silva, S.N., Silva, G.C. et al. (2017). Technical and environmental feasibility of interlocking concrete pavers with iron ore tailings from tailings dams. *Journal of Materials in Civil Engineering* 29 (9): 04017104.

Freedman, A., Gross, A., Shelef, O. et al. (2014). Salt uptake and evapotranspiration under arid conditions in horizontal subsurface flow constructed wetland planted with halophytes. *Ecological Engineering* 70: 282–286.

Gedam, V.V., Raut, R.D., de Sousa Jabbour, A.B.L., and Agrawal, N. (2021). Moving the circular economy forward in the mining industry: challenges to closed-loop in an emerging economy. *Resources Policy* 74: 102279.

Ghorbani, Y. and Kuan, S.H. (2017). A review of sustainable development in the Chilean mining sector: past, present and future. *International Journal of Mining, Reclamation and Environment* 31 (2): 137–165.

Ghosh, M. and Singh, S.P. (2005). A review on phytoremediation of heavy metals and utilization of it's by products. *Asian Journal on Energy & Environment* 6 (4): 18.

Gonzalez, A., Aitken, D., Heitzer, C. et al. (2019). Reducing mine water use in arid areas through the use of a byproduct road dust suppressant. *Journal of Cleaner Production* 230: 46–54.

Gordon, R.B. (2002). Production residues in copper technological cycles. *Resources, Conservation and Recycling* 36 (2): 87–106.

Gorman, M.R. and Dzombak, D.A. (2018). A review of sustainable mining and resource management: transitioning from the life cycle of the mine to the life cycle of the mineral. *Resources, Conservation and Recycling* 137: 281–291.

Gou, M., Zhou, L., and Then, N.W.Y. (2019). Utilization of tailings in cement and concrete: a review. *Science and Engineering of Composite Materials* 26 (1): 449–464.

Haddaway, N.R., Cooke, S.J., Lesser, P. et al. (2019). Evidence of the impacts of metal mining and the effectiveness of mining mitigation measures on social–ecological systems in Arctic and Boreal regions: a systematic map protocol. *Environmental Evidence* 8 (1): 1–11.

Hallstedt, S.I., Thompson, A.W., and Lindahl, P. (2013). Key elements for implementing a strategic sustainability perspective in the product innovation process. *Journal of Cleaner Production* 51: 277–288.

Hartman, H.L. and Mutmansky, J.M. (2002). *Introductory Mining Engineering*. John Wiley & Sons.

He, Y., He, X., Liu, Z. et al. (2017). Coal mine subsidence has limited impact on plant assemblages in an arid and semi-arid region of northwestern China. *Ecoscience* 24 (3–4): 91–103.

Jiang, M., Gao, X., Guan, Q. et al. (2019). The structural roles of sectors and their contributions to global carbon emissions: a complex network perspective. *Journal of Cleaner Production* 208: 426–435.

Kahlon, S.K., Sharma, G., Julka, J.M. et al. (2018). Impact of heavy metals and nanoparticles on aquatic biota. *Environmental Chemistry Letters* 16: 919–946.

Khobragade, K. (2020). Impact of mining activity on environment: an overview. *International Journal of Scientific and Research Publications* 10 (5): 784–791.

Kinnunen, P.H.M. and Kaksonen, A.H. (2019). Towards circular economy in mining: opportunities and bottlenecks for tailings valorization. *Journal of Cleaner Production* 228:153–160.

Kossoff, D., Dubbin, W.E., Alfredsson, M. et al. (2014). Mine tailings dams: characteristics, failure, environmental impacts, and remediation. *Applied Geochemistry* 51: 229–245.

Lagos, G., Peters, D., Videla, A., and Jara, J.J. (2018). The effect of mine aging on the evolution of environmental footprint indicators in the Chilean copper mining industry 2001–2015. *Journal of Cleaner Production* 174: 389–400.

Lam, E.J., Gálvez, M.E., Cánovas, M. et al. (2016). Evaluation of metal mobility from copper mine tailings in northern Chile. *Environmental Science and Pollution Research* 23: 11901–11915.

Lam, E.J., Cánovas, M., Gálvez, M.E. et al. (2017). Evaluation of the phytoremediation potential of native plants growing on a copper mine tailing in northern Chile. *Journal of Geochemical Exploration* 182: 210–217.

Lam Esquenazi, E.J., Keith Norambuena, B.F., Montofre Bacigalupo, Í.L., and Gálvez Estay, M.E. (2019). Necessity of intervention policies for tailings identified in the Antofagasta Region, Chile. *Revista internacional de contaminación ambiental* 35 (3): 515–539.

Lam, E.J., Gálvez, M.E., Cánovas, M. et al. (2018). Assessment of the adaptive capacity of plant species in copper mine tailings in arid and semiarid environments. *Journal of Soils and Sediments* 18: 2203–2216.

Lam, E.J., Montofré, I.L., Álvarez, F.A. et al. (2020a). Methodology to prioritize chilean tailings selection, according to their potential risks. *International Journal of Environmental Research and Public Health* 17 (11): 3948.

Lam, E.J., Montofré, Í.L., and Ramírez, Y. (2021). Mine tailings phytoremediation in arid and semiarid environments. In: *Phytorestoration of Abandoned Mining and Oil Drilling Sites*, 115–166. Elsevier.

Lam, E.J., Urrutia, J., Bech, J. et al. (2023). Heavy metal pollution index calculation in geochemistry assessment: a case study on Playa Las Petroleras. *Environmental Geochemistry and Health* 45 (2): 409–426.

Lam, E.J., Zetola, V., Ramírez, Y. et al. (2020b). Making paving stones from copper mine tailings as aggregates. *International Journal of Environmental Research and Public Health* 17 (7): 2448.

Leiva González, J. and Onederra, I. (2022). Environmental management strategies in the copper mining industry in Chile to address water and energy challenges. *Mining* 2 (2): 197–232.

Li, S., Xiao, W., Zhao, Y., and Lv, X. (2020). Incorporating ecological risk index in the multi-process MCRE model to optimize the ecological security pattern in a semi-arid area with intensive coal mining: a case study in northern China. *Journal of Cleaner Production* 247: 119143.

Liu, S., Li, W., Qiao, W. et al. (2019). Effect of natural conditions and mining activities on vegetation variations in arid and semiarid mining regions. *Ecological Indicators* 103: 331–345.

Lodhia, S. (2018). *Mining and Sustainable Development: Current Issues*. Routledge.

Lyashenko, V., Topolnij, F., and Dyatchin, V. (2019). Development of technologies and technical means for storage of waste processing of ore raw materials in the tailings dams. *Technology Audit and Production Reserves* 5 (3 (49)): 33–40.

Ma, K., Zhang, Y., Ruan, M. et al. (2019). Land subsidence in a coal mining area reduced soil fertility and led to soil degradation in arid and semi-arid regions. *International Journal of Environmental Research and Public Health* 16 (20): 3929.

Mahapatra, M. and Ghosh, S.K. (2021). Sustainable mining in arid and semiarid regions: a review. *Arabian Journal of Geosciences* 14 (3): 1–15.

Ministerio de minería (2022). *Informe: Mineria 2050, Política Nacional Minera*. Santiago, Chile: Gobierno de Chile.

Montes, C. (2016). Proyección de consumo de agua en la minería del cobre 2016-2027. Dirección de Estudios y Políticas Públicas Proyección de consumo de agua en la minería del cobre 2016-2027

Montofré, Í.L., Lam, E.J., Ramírez, Y., and Gálvez, M.E. (2021). Evaluation of copper tailing amendments through poultry waste and ammonium nitrate. *Environmental Geochemistry and Health* 43: 2213–2230.

Moreno, D., Pedrocchi, C., Comín, F.A. et al. (2007). Creating wetlands for the improvement of water quality and landscape restoration in semi-arid zones degraded by intensive agricultural use. *Ecological Engineering* 30 (2): 103–111.

Mylona, E., Xenidis, A., and Paspaliaris, I. (2000). Inhibition of acid generation from sulphidic wastes by the addition of small amounts of limestone. *Minerals Engineering* 13 (10–11): 1161–1175.

Onuaguluchi, O. and Eren, O. (2013). Rheology, strength and durability properties of mortars containing copper tailings as a cement replacement material. *European Journal of Environmental and Civil Engineering* 17 (1): 19–31.

Oyarzún, J. and Oyarzún, R. (2011). Sustainable development threats, inter-sector conflicts and environmental policy requirements in the arid, mining rich, northern Chile territory. *Sustainable Development* 19 (4): 263–274.

Pactwa, K., Woźniak, J., and Strempski, A. (2021). Sustainable mining–challenge of Polish mines. *Resources Policy* 74: 101269.

de Palacios, L.D.L.T. and Rodríguez, J.A.E. (2022). In mining, not everything is a circular economy: case studies from recent mining projects in Iberia. *Resources Policy* 78: 102798.

Pearce, T.D., Ford, J.D., Prno, J. et al. (2011). Climate change and mining in Canada. *Mitigation and Adaptation Strategies for Global Change* 16: 347–368.

Peña-Ortega, M., Del Rio-Salas, R., Valencia-Sauceda, J. et al. (2019). Environmental assessment and historic erosion calculation of abandoned mine tailings from a semi-arid zone of northwestern Mexico: insights from geochemistry and unmanned aerial vehicles. *Environmental Science and Pollution Research* 26: 26203–26215.

Pérez, M., Estay, R., and Encina, V. (2021). Sustainable logistic plan for the dynamic distribution of massive mining waste. *Boletín Geológico y Minero* 132 (4): 509.

Pinchuk, A., Tkalenko, N., and Marhasova, V. (2019). Implementation of circular economy elements in the mining regions. In: *E3S Web of Conferences*, vol. 105, 04048. EDP Sciences.

Puga, A.P., Abreu, C.A., Melo, L.C.A. et al. (2015). Cadmium, lead, and zinc mobility and plant uptake in a mine soil amended with sugarcane straw biochar. *Environmental Science and Pollution Research* 22: 17606–17614.

Puga, A.P., Melo, L.C.A., de Abreu, C.A. et al. (2016). Leaching and fractionation of heavy metals in mining soils amended with biochar. *Soil and Tillage Research* 164: 25–33.

Qi, C.C. (2020). Big data management in the mining industry. *International Journal of Minerals, Metallurgy and Materials* 27: 131–139.

Ranjbari, M., Saidani, M., Esfandabadi, Z.S. et al. (2021). Two decades of research on waste management in the circular economy: insights from bibliometric, text mining, and content analyses. *Journal of Cleaner Production* 314: 128009.

Robinsky, E.I. (1999). *Thickened Tailings Disposal in the Mining Industry*. EI Robinsky Associates.

Roy, M. and McDonald, L.M. (2015). Metal uptake in plants and health risk assessments in metal-contaminated smelter soils. *Land Degradation & Development* 26 (8): 785–792.

Rubinos, D.A., Jerez, Ó., Forghani, G. et al. (2021). Geochemical stability of potentially toxic elements in porphyry copper-mine tailings from Chile as linked to ecological and human health risks assessment. *Environmental Science and Pollution Research* 28 (41): 57499–57529.

Shao, W., Wang, Q., Guan, Q. et al. (2022). Distribution of soil available nutrients and their response to environmental factors based on path analysis model in arid and semi-arid area of northwest China. *Science of The Total Environment* 827: 154254.

Shi, H.S. and Kan, L.L. (2009). Leaching behavior of heavy metals from municipal solid wastes incineration (MSWI) fly ash used in concrete. *Journal of Hazardous Materials* 164 (2–3): 750–754.

Sillanpaa, M. and Ncibi, C. (2019). *The Circular Economy: Case Studies About the Transition from the Linear Economy*. Academic Press.

Singh, R.K., Kumar, A., Garza-Reyes, J.A., and de Sá, M.M. (2020). Managing operations for circular economy in the mining sector: an analysis of barriers intensity. *Resources Policy* 69: 101752.

Smuda, J., Dold, B., Spangenberg, J.E., and Pfeifer, H.-R. (2008). Geochemistry and stable isotope composition of fresh alkaline porphyry copper tailings: implications on sources and mobility of elements during transport and early stages of deposition. *Chemical Geology* 256: 62–76.

Swami, R.K., Pundhir, N.K.S., and Mathur, S. (2007). Kimberlite tailings: a road construction material. *Transportation Research Record* 1989 (1): 131–134.

Tabelin, C.B., Igarashi, T., Villacorte-Tabelin, M. et al. (2018). Arsenic, selenium, boron, lead, cadmium, copper, and zinc in naturally contaminated rocks: a review of their sources, modes of enrichment, mechanisms of release, and mitigation strategies. *Science of the Total Environment* 645: 1522–1553.

Tabra, K. and Lange, S. (2014). Active treatment of tailings seepage with focus on sulphate and manganese removal. *Proceedings of the 2nd International Seminar on Tailings Management TAILINGS 2014*, Antofagasta, Chile (6–8 August 2014).

Tayebi-Khorami, M., Edraki, M., Corder, G., and Golev, A. (2019). Re-thinking mining waste through an integrative approach led by circular economy aspirations. *Minerals* 9 (5): 286.

Thomas, C., Rolls, J., and Tennant, T. (2000). *The GHG Indicator: UNEP Guidelines for Calculating Greenhouse Gas Emissions for Businesses and Non-commercial Organisations*, 61. Paris: UNEP.

Upadhyay, A., Laing, T., Kumar, V., and Dora, M. (2021). Exploring barriers and drivers to the implementation of circular economy practices in the mining industry. *Resources Policy* 72: 102037.

Vacca, A., Bianco, M.R., Murolo, M., and Violante, P. (2012). Heavy metals in contaminated soils of the Rio Sitzerri floodplain (Sardinia, Italy): characterization and impact on pedodiversity. *Land Degradation & Development* 23 (4): 350–364.

Valenta, R.K., Kemp, D., Owen, J.R. et al. (2019). Re-thinking complex orebodies: consequences for the future world supply of copper. *Journal of Cleaner Production* 220: 816–826.

Venegas, A., Rigol, A., and Vidal, M. (2016). Changes in heavy metal extractability from contaminated soils remediated with organic waste or biochar. *Geoderma* 279: 132–140.

Villa Gomez, D., Sáez Salgado, E., Mejías, O. et al. (2022). Data integration of critical elements from mine waste in Mexico, Chile and Australia. *Minerals* 12 (2): 122.

Wang, C., Harbottle, D., Liu, Q., and Xu, Z. (2014). Current state of fine mineral tailings treatment: a critical review on theory and practice. *Minerals Engineering* 58: 113–131.

Wassenaar, T.D., Henschel, J.R., Pfaffenthaler, M.M. et al. (2013). Ensuring the future of the Namib's biodiversity: ecological restoration as a key management response to a mining boom. *Journal of Arid Environments* 93: 126–135.

Wels, C. and Caldwell, J. (2013). Challenges in tailings water balance analysis—make-up water, seepage and consolidation. *Proceedings of the 1st Mine Water Solutions in Extreme Environments Conference*, Lima, Peru (15–17 April 2013).

Wels, C., Robertson, A.M., and Madariaga, P.M. (2004). Water recovery study for Pampa Pabellon tailings impoundment, Collahuasi, Chile. *Proceedings of the Eleventh Annual Conference on Tailings and Mine Waste*, Vail, Colorado, USA, 10–13 October 2004.

Whitworth, A.J., Forbes, E., Verster, I. et al. (2022). Review on advances in mineral processing technologies suitable for critical metal recovery from mining and processing wastes. *Cleaner Engineering and Technology* 7: 100451.

Xiao, W., Lv, X., Zhao, Y. et al. (2020). Ecological resilience assessment of an arid coal mining area using index of entropy and linear weighted analysis: a case study of Shendong Coalfield, China. *Ecological Indicators* 109: 105843.

Zenker, A., Cicero, M.R., Prestinaci, F. et al. (2014). Bioaccumulation and biomagnification potential of pharmaceuticals with a focus to the aquatic environment. *Journal of Environmental Management* 133: 378–387.

Zhao, S., Fan, J., and Sun, W. (2014). Utilization of iron ore tailings as fine aggregate in ultra-high performance concrete. *Construction and Building Materials* 50: 540–548.

Zhou, X., Xue, Z., and Seydehmet, J. (2021). An empirical study on industrial eco-efficiency in arid resource exploitation region of northwest China. *Environmental Science and Pollution Research* 28 (38): 53394–53411.

2

Restoring Biodiversity and Ecosystem Services on Post-Industrial Land

Challenges and Opportunities

Sameer Shekhar[1], Shubham Abhishek[1,2], Bhanu Pandey[1], and Siddharth Singh[1,2]

[1] *CSIR-Central Institute of Mining and Fuel Research, Barwa Road, Dhanbad, Jharkhand, India*
[2] *Academy of Scientific and Innovative Research (AcSIR), Ghaziabad, India*

2.1 Introduction

Industrialization has brought enormous benefits to human society, including improved living standards, economic growth, and technological advancements. However, the expansion of industrial activities has also resulted in severe environmental degradation and the loss of biodiversity, particularly in post-industrial areas. Many industrial facilities, which were established during the industrialization age, have lost their original functions but remain noteworthy examples of industrial heritage (Elhan 2009). Additionally, as a result of economic growth, various facilities such as factories, transportation infrastructure, and buildings were constructed on extensive tracts of land, but now these structures have fallen into disuse, leading to adverse environmental impacts. The consequences of this transformation can be dire, inflicting substantial harm on the environment and the species that depend on it. Over time, these abandoned and neglected spaces have not only diminished the social and economic vitality of surrounding areas but have also detrimentally affected biodiversity, ecosystem functions, and ecosystem services (Gallagher et al. 2018).

Restoring biodiversity and ecosystem services on post-industrial land is critical for contemporary restoration ecology (Prach and Tolvanen 2016). Human exploitation and ecosystem conversion have led to extensive loss of biodiversity, impacting human health, well-being, and global sustainability (Bullock et al. 2011). Post-industrial lands constitute about one percent of the global land surface, including waste deposits, ash, and other residues, characterized by economic disadvantage, environmental degradation, and social distress (Walker 1999; Prach and Tolvanen 2016; Salgueiro et al. 2020). However, they offer potential environmental benefits (Ling et al. 2007) through redevelopment as green spaces, improving quality of life, achieving environmental and social benefits, enhancing access to green spaces, and reducing urban heat island effects (Stubbs 2008; Bardos et al. 2016; Kristiánová et al. 2016; Pandey 2020c). Transforming post-industrial lands into healthy ecosystems provides

Biodiversity and Ecosystem Services on Post-Industrial Land, First Edition.
Edited by Vimal Chandra Pandey.

essential services for human well-being (Pandey 2017; Pandey et al. 2019; Pandey 2020a; Pandey et al. 2023). Despite research supporting the importance of biodiversity to ecosystem health, little attention has been given to the direct relationship between biodiversity and ecosystem services (VIDICAN et al. 2022). The profound negative impacts of human activities on global ecosystems emphasize the imperative need to utilize novel ecosystems for the restoration and management of these modified ecosystems (Guo 2019).

Post-industrial land restoration efforts have been ongoing for decades in countries like the United States, the United Kingdom, and Australia, but success varies due to specific challenges (Bullock et al. 2011; Prach and Tolvanen 2016). Contamination, limited finances, and conflicting land-use demands pose significant obstacles. Industrial activities leave behind hazardous waste, making restoration challenging and costly. Remediation methods like phytoremediation, bioremediation, and chemical treatment have mixed results. Limited financial resources hinder complete restoration, affecting ecological benefits. Conflicting land-use demands near urban areas create difficulties in prioritizing restoration. Despite challenges, restoring biodiversity and ecosystem services on post-industrial land is crucial for sustainable land use and mitigating industrial impacts (Bardos et al. 2020). It benefits the environment, and local communities' quality of life, and promotes ecotourism while enhancing ecosystem resilience against global environmental challenges (Pandey et al. 2022).

2.2 Post-Industrial Landscapes: A Brief Overview

Post-industrial landscapes refer to areas that have been abandoned or transformed after the cessation of industrial activities, resulting in degraded landscapes and the loss of ecological functions (LeClerc 2015). These areas include former factories, mines, power plants, and other industrial sites. The community composition in the post-industrial areas, i.e., organism composition and relationships developing on soil/soil substratum, is different from pre-industrial habitats. The most vital point for understanding the ecosystem function of industrial areas is the biodiversity, ecosystem function, and ecosystem service relationship. Post-industrial landscapes have unique features that distinguish them from other landscapes. First, these landscapes often include edge-lands, which are areas lacking institutional programs and regulations, resulting in waste accumulation (Humphris and Rauws 2021). Second, they carry moral significance, acting as a counterbalance to the complex industrial society that preceded them. Third, they contribute to a regional identity, giving individuals a sense of belonging to a broader context (Höfer 1998). Last, these landscapes often exhibit an imposed authenticity, influenced by the nature of industrial cities and the demand for post-industrial environments (Loures 2015).

2.2.1 Types of Post-Industrial Landscapes

The development of post-industrial landscapes involves transforming the land for other uses, such as housing, parks, or commercial development, while also preserving the site's industrial heritage and addressing any environmental contamination (Sierka et al. 2018).

Various studies have examined the benefits and barriers of post-industrial land transformation. For better understanding, this landscape can be classified into different categories (Luís Loures 2015). One type is brownfields, which are abandoned or under-used industrial or commercial sites that have potential environmental contamination (Jacek et al. 2022). Brownfields are often located in urban areas and require environmental remediation before they can be redeveloped (Green 2018). Another type is former industrial sites that have been repurposed for new uses such as parks, cultural centers, and residential areas (Pandey 2020a, 2020b). These sites often retain some of their industrial characters and may include preserved industrial artifacts or architecture (Höfer 1998). Post-mining landscapes are another type of post-industrial landscape, resulting from the abandonment of coal or mineral mines (Gerwin et al. 2023; Pandey et al. 2023). These landscapes often present significant environmental challenges, including land subsidence, soil erosion, and acid mine drainage (Louis Loures 2014). Post-industrial rural landscapes are the result of the decline of traditional agricultural practices and the expansion of industrial agriculture. These landscapes often include abandoned farms, rural factories, and other forms of rural infrastructure (Coulibaly and Li 2020).

2.2.2 Drivers of Post-Industrialization

The drivers of post-industrialization are multifaceted and complex. Some of the drivers of this phenomenon include technological advancements, globalization, shifts in consumer preferences, and changes in the nature of work. Technological advancements have played a significant role in driving post-industrialization (Włodarczyk-Marciniak et al. 2020). The development of new technologies has enabled the automation of many manufacturing processes, which has led to increased efficiency and productivity (Javaid et al. 2021). This, in turn, has led to a decline in the number of manufacturing jobs available, while increasing the demand for service-based jobs that require higher levels of education and skill.

Globalization is another driver of post-industrialization (Hamerla 2021). The growth of international trade and the rise of global supply chains have made it easier for companies to outsource manufacturing and other lower-skilled jobs to countries with lower labor costs (Nicita et al. 2013). This has resulted in a decline in manufacturing employment in countries that were previously industrialized.

Shifts in consumer preferences have also played a role in driving post-industrialization (Fine and Leopold 1990). As societies become more affluent, people tend to place a greater emphasis on services and experiences rather than material goods (Hüppauf 2004). This has led to a growing demand for service-based jobs in sectors such as healthcare, education, and entertainment.

Changes in the nature of work have also contributed to post-industrialization (Julia Kovalchuk 2021). As technology has made it easier to automate many routine and repetitive tasks, there has been a shift toward jobs that require higher levels of creativity, critical thinking, and problem-solving skills (Dogan et al. 2023). This has created a demand for workers with higher levels of education and training, which has contributed to the growth of service-based industries.

2.3 Biodiversity on Post-Industrial Land

2.3.1 Current Status and Trends of Biodiversity on Post-Industrial Land

The present condition and patterns of biodiversity on post-industrial land exhibit complex and diverse characteristics, based upon a variety of factors including the specific geographical position, earlier land utilization, initiatives toward restoration, and the regional background (Raven and Wagner 2021). Post-industrial land frequently commences its existence in a state of degradation, mostly due to the extensive industrial activities that have transpired. The primary consequence of industrialization on biodiversity frequently manifests as the depletion of indigenous flora and fauna populations, primarily attributed to the degradation and fragmentation of their natural habitats. Industrial activities have the potential to induce significant ecological repercussions, leading to the degradation and loss of vital ecosystems such as forests, wetlands, and other natural habitats. Consequently, this can disrupt the delicate balance of biodiversity, resulting in the displacement or outright elimination of species that previously thrived within these areas (Shivanna 2022). Industrial activities have the potential to result in the deposition and buildup of various contaminants within the soil matrix, including but not limited to heavy metals, chemical compounds, and hazardous substances (Brown et al. 2022). The presence of these contaminants may persist over extended durations, thereby presenting a significant hazard to the ecological well-being of both flora and fauna. The presence of soil contamination poses an important challenge to the restoration of biodiversity on post-industrial land, as it interferes with the successful establishment and growth of plants while also disrupting crucial soil-based ecological processes (Han and Xia 2021).

Heavy metals, chemical compounds, and dangerous chemicals, among other pollutants, have the ability to deposit and accumulate in the soil matrix as a consequence of industrial activity (Pandey and Bajpai 2019). These pollutants may be persistent for long periods of time, posing a serious risk to the ecological health of both flora and fauna (Yan and Liu 2021; Pandey and Bajpai 2019). On post-industrial land, soil pollution poses a significant obstacle to the restoration of biodiversity because it hinders plant establishment and development and disturbs essential soil-based biological processes (Sharma et al. 2021).

There are several examples of post-industrial land being successfully restored, which resulted in the restoration of biodiversity. For instance, extensive restoration work has turned abandoned coal mines in Germany's once-heavily industrialized Ruhr area into flourishing natural reserves. Similar success stories that show the possibility for biodiversity recovery on post-industrial land may be found in various regions of the globe (Bradu et al. 2022). Despite the notable progress made, there remain several obstacles impeding the restoration of biodiversity on post-industrial landscapes. The presence of soil contamination poses a substantial challenge, potentially impeding the effectiveness of restoration efforts (Wuana and Okieimen 2011). The establishment and maintenance of diverse and resilient ecosystems can face challenges related to the availability of sufficient plant propagules, the lack of connectivity between restored habitats, and the proliferation of invasive species (Koch et al. 2021).

2.3.2 Factors Affecting Biodiversity on Post-Industrial Land

Factors affecting biodiversity on post-industrial land are manifold. It mainly includes factors such as (i) Air and water pollution: air pollution harms biodiversity and draws species near extinction, affecting wildlife through direct or indirect exposure (Sonwani and Maurya 2019) while on the other hand, water pollution also has the tendency to cause long-term modifications of biodiversity (Gebretsadik 2016). (ii) Habitat destruction and Fragmentation: Habitat destruction is the primary cause of species extinction worldwide (Gebretsadik 2016). Fragmentation is one of the most critical threats to biodiversity and ecosystem services throughout the world, such as pollination, seed dispersal, herbivore, and carbon sequestrate (Brooks et al. 2002; Gebretsadik 2016). (iii) Global Climate Change: Global climate change is occurring at an unprecedented rate today mainly due to an increased emission of greenhouse gases such as atmospheric CO_2 (Hui 2013). Global temperatures have increased by about 0.6–0.76 °C since the mid-1800s and are predicted to further increase by 1.4–5.8 °C by 2100 (Hui 2013). Since climate is the fundamental factor that determines organism life-stages such as plant germination and flowering, it can severely alter habitats and food sources for animals, and ultimately, could have significant impacts on the biodiversity of species and ecosystems around the world (Hui 2013). Effects of climate change on biodiversity and ecosystems mainly include species extinction and biodiversity loss, phenology changes, shifts in geographic change, and ecosystem functioning and service changes. (iv) Invasive Species: The introduction of invasive species is a tremendous threat to biodiversity crisis. The species, which is not native to the ecosystem, arrives or is introduced mostly via humans in the new ecosystem and starts to pullulate is called invasive species (Singh et al. 2021). The invasive species wrack up the native ecosystem in many ways such as modifying the habitat, importing pathogens, being herbivorous on plants in the native ecosystem, leading to the decline of genetic diversity by hybridizing with natives, for the resources they directly compete with and prey on the native species (Doherty et al. 2016; Singh et al. 2021). (v) Natural Disasters: Natural catastrophes, for instance, volcanos, wildfires, floods, hurricanes, draughts, epidemics, tsunamis, etc., cause a heavy loss of biodiversity (Rondeau et al. 2020; Singh et al. 2021).

2.3.3 Importance of Biodiversity for Ecosystem Services Provision

Biodiversity plays a pivotal role in the complex structure of our planet's ecosystems, encompassing the complex array of life forms, genetic diversity, and the interplay of various ecosystems present on Earth (Costanza 2020). It serves as an essential component in facilitating a diverse array of ecosystem services that are crucial for the overall welfare of human beings and the effective functioning of our biosphere (Hasan et al. 2020). In recent years, extensive scientific research has shed light on the profound interconnection between biodiversity and the provision of ecosystem services. One noteworthy illustration is the scientific investigation carried out by Cardinale et al. (2011). A comprehensive meta-analysis was undertaken, encompassing a total of 92 studies, revealing a significant correlation between increased biodiversity and the enhancement of primary productivity, nutrient cycling, and overall ecosystem stability. This observation suggests that the decline in biodiversity can result in noteworthy adverse effects on the provision of ecosystem services (Zhao et al. 2020).

Ecosystems that exhibit high levels of biodiversity, such as forests and wetlands, have the inherent ability to capture and retain substantial quantities of carbon dioxide, therefore serving as a means of reducing the adverse effects of climate change. Pan et al. (2011) conducted thorough research to investigate the correlation between biodiversity and carbon sequestration in various ecosystems. The findings of the research indicate a positive correlation between biodiversity levels and carbon sequestration in various terrestrial ecosystems, including forests and grasslands. The diminishment of biodiversity poses a significant threat to the capacity of ecosystems to effectively control the Earth's temperature and buffer the adverse effects of climate change (Weiskopf et al. 2020).

Pollinators, such as bees, butterflies, and other insects, are essential agents in the process of pollination for blooming plants. Their crucial function involves aiding crop production and maintaining the integrity of natural plant communities. The decrease in populations of pollinators has generated apprehensions about the preservation of food security and the stability of ecosystems. The significance of pollinator variety for crop pollination was examined in research conducted by Klein et al. (2007). The researchers discovered that the presence of varied pollinator groups resulted in increased fruit set and seed output in both wild and cultivated plant species. This research highlights the significance of maintaining a wide range of pollinator populations in order to ensure long-term agricultural production and the preservation of natural habitats (Beillouin et al. 2021).

Healthy ecosystems, particularly wetlands and riparian zones, play a critical role in purifying water and regulating nutrient cycling. A study by Mitsch and Gosselink (2015) explored the relationship between wetland biodiversity and water quality improvement. The researchers found that wetlands with higher species diversity exhibited enhanced water purification capacities, effectively removing pollutants and excess nutrients. This emphasizes the importance of preserving wetland biodiversity for the provision of clean water and the maintenance of healthy aquatic ecosystems (Díaz et al. 2019).

2.4 Ecosystem Services on Post-Industrial Land

2.4.1 Types of Ecosystem Services Provided by Post-Industrial Land

Ecosystem services are defined as all of the physical, social, and economic advantages that people may obtain from natural ecosystems (Costanza et al. 1997; Millennium Ecosystem Assessment 2005). Ecosystems provide several benefits through supporting, provisioning, regulating, and cultural services (Otsuka et al. 2021). Ecosystem services on restored fly ash dumpsites are well described in detail in the book "Phytomanagement of Fly Ash" (Pandey 2020b). Similarly, the ecosystem services on restored mine land are defined in the book "Eco-Restoration of Mine Land," which offers an ecosystem-based solution for remediating mine spoil and re-establishing vegetation in disused mine lands (Pandey et al. 2023). These ecosystem services may be provided by post-industrial land, which consists of regions that were formerly used for industrial purposes but have subsequently been recovered or restored (Merwin et al. 2022). These services help ensure the environmental, social, and economic viability of the surroundings.

2.4.1.1 Provisioning Ecosystem Services

Provisioning services are tangible products that may be derived from ecosystems, including food, fiber, raw materials, water, genetic resources, minerals, and medicinal resources (Hasan et al. 2020).

Restoration of post-industrial land often involves re-establishing native vegetation, managing soil health, and enhancing habitat quality (Teague and kreuter 2020). This process positively impacts provisioning services like crop yield improvement, enhanced water filtration, and increased availability of resources (Rigolon et al. 2022).

2.4.1.2 Supporting Services

Supporting services are the essential ecosystem components that make it possible for other services to function properly. These functions include nutrient cycling, soil formation, primary production (plant photosynthesis), and maintaining biodiversity (Hasan et al. 2020).

Ecological restoration of post-industrial land is common, establishing new habitats and aiding biodiversity conservation (Simonson et al. 2021). These habitats may sustain a wide range of plant and animal species, including rare and endangered species. These locations offered critical habitat for a variety of bird species, aiding conservation efforts. Plant variety and habitat for various species may also be supported by post-industrial environments (Enssle and Kabisch 2020).

2.4.1.3 Regulating Services

Regulating services encompass a variety of vital processes that play a crucial role in safeguarding the long-term functionality of ecosystems (Kabisch et al. 2017). These processes work diligently to uphold the stability and integrity of ecosystem characteristics, thereby ensuring the sustained functioning of the ecosystem over time. They encompass ecological processes and dynamics, pest and disease regulation, water and air purification, waste management, and other related aspects (Wu et al. 2015).

Post-industrial land may help to mitigate climate change by absorbing carbon dioxide from the atmosphere (Eric et al. 2008). Carbon absorption and storage are improved by restoring vegetation and establishing forests on such land (Dangwal et al. 2022).

Water regulation services may be provided by post-industrial land by regulating stormwater runoff and enhancing water quality (Zhang et al. 2020). Areas that have been restored or re-vegetated may absorb and filter rainfall, lowering the danger of floods and increasing water infiltration rates (Kosanic and Petzold 2020).

2.4.1.4 Cultural Services

Cultural services encompass intangible advantages that contribute to the enrichment of recreational activities, spiritual thought, cognitive growth (particularly in the realm of education), and aesthetic interactions (Jin et al. 2017).

Reclaimed post-industrial land may be used for recreational and cultural experiences. These places may be turned into parks, green spaces, or cultural landscapes, providing aesthetic and recreational advantages to the surrounding community (Semeraro et al. 2021).

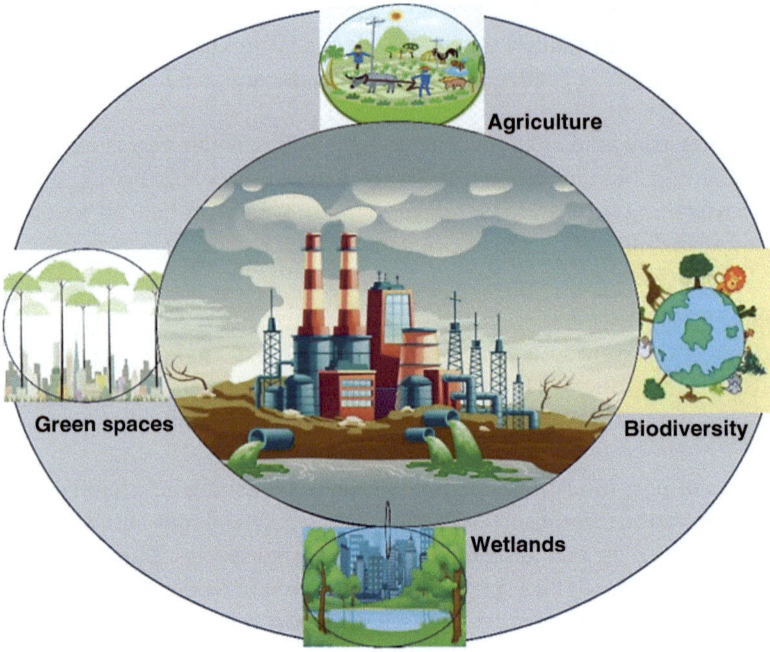

Figure 2.1 Utilization of ecosystem services on post-industrial land.

Post-industrial lands may be used as living laboratories for environmental research and teaching. These sites provide possibilities to investigate ecological processes, ecosystem restoration approaches, and species colonization patterns (Geng et al. 2014) (Figure 2.1).

2.4.2 Quantification and Valuation of Ecosystem Services

Ecosystem services are the benefits that ecosystems provide to humans, including provisioning, regulating, cultural, and supporting services (Mengist et al. 2020). Ecosystem services play a vital role in supporting human well-being and maintaining ecological integrity (Comberti et al. 2015). Quantifying and valuing ecosystem services allows policymakers, economists, and ecologists to assess their importance and incorporate them into decision-making processes. Understanding the value of ecosystem services is crucial for effective environmental management. There are various approaches available to quantify and value ecosystem services.

2.4.2.1 Quantification of Ecosystem Services

1) **InVEST model:**

   The InVEST (Integrated Valuation of Ecosystem Services and Trade-offs) model, developed by Guerry et al. (2012), is a widely used tool for quantifying and mapping ecosystem services. It integrates spatial data and models to estimate the supply, demand, and value of services such as carbon sequestration, water purification, and habitat provision.

2) **Ecological production function (EPF):**

The EPF approach, proposed by Boyd and Banzhaf (2007), quantifies ecosystem services by establishing the relationship between ecological inputs (e.g., habitat area, biodiversity) and the resulting services (e.g., pollination, pest control). This approach allows for the estimation of service outputs based on ecological parameters.

2.4.2.2 Valuation of Ecosystem Services

Market-based approaches: Market-based valuation methods use market prices or economic transactions to estimate the value of ecosystem services. For instance, Costanza et al. (1997) conducted a ground-breaking study that estimated the global value of ecosystem services at $33 trillion per year using market prices for various services.

2.4.2.3 Non-market Valuation Techniques

Non-market valuation methods aim to estimate the value of ecosystem services that are not traded in traditional markets. Contingent valuation (CV) and choice experiments (CEs) are commonly used non-market valuation techniques. For example, Hanley et al. (1998) applied a CV approach to estimate the value of improvements in water quality resulting from changes in land management practices.

2.4.3 Impacts of Post-Industrialization on Ecosystem Services Provision

The transformation of landscapes worldwide has raised global concerns increasing the need to rethink landscapes and protect the environment (Luis Loures 2014). The impacts of post-industrialization on ecosystem services provision are complex and multifaceted. However, since positive and negative impacts may vary from one landscape to another, it remains a challenge and a task for those aiming to transform these landscapes to develop a more systematic and theoretically fruitful methodology that incorporates both economic, socio-cultural, and environmental aspects. The implementation of restoration or habitat restoration initiatives on post-industrial landscapes has the potential to significantly enhance the provision of ecosystem services. Conversely, land use change (LUC) and urbanization often exert detrimental impacts on the ecosystem services offered by these landscapes.

As heavy industries decline, there is an increased likelihood of abandoned industrial sites being repurposed for ecological restoration (Mills et al. 2017). These sites can be transformed into urban parks, wildlife reserves, or green spaces, contributing to biodiversity conservation, recreational opportunities, and aesthetic values (Rodríguez-Loinaz et al. 2015). The severe consequences of post-industrialization on the provision of ecosystem services can be attributed to the modification of land use and the emergence of urbanization patterns (Qiu et al. 2021). LUC is intimately associated with ecosystem services through the changes in the interactions between humans and the environment and can greatly affect services ranging from biodiversity to climate (Hasan et al. 2019). Moreover, LUC modifies ecosystems, thereby changing their functions and structure. Tolessa et al. (2017) concluded that different anthropogenic activities and climate change affected negatively ecosystem services. Urbanization and infrastructure development often lead to habitat destruction, fragmentation, and loss, which in turn negatively impacts ecosystem services such as wildlife habitat, pollination, and water filtration (Ouyang et al. 2021).

Achieving a balance between the positive and negative impacts of post-industrialization on ecosystem services requires proactive and holistic approaches (Opoku and Boachie 2020). Urban planning that incorporates green infrastructure, conservation strategies, and sustainable land use practices is crucial (Bliege Bird and Nimmo 2018). Investing in research and technology to mitigate the environmental footprint of new industries and urban expansion can help safeguard ecosystem services.

2.5 Restoring Biodiversity and Ecosystem Services on Post-Industrial Land

2.5.1 Strategies for Restoration and Rehabilitation of Post-Industrial Landscapes

Restoration and rehabilitation efforts are significantly hampered by post-industrial landscapes, which are marked by closed industries and mines as well as deteriorated ecosystems (Gallagher 2021). The emphasis is on finding fundamental ideas, strategies, and methods that have produced ecological recovery and sustainable landscapes. Soil contamination from industrial activity frequently necessitates repair and rehabilitation (Foster 2022). Engineered solutions as well as phytoremediation and bioremediation have all been studied by scientists. Phytoremediation, which cleans up contaminated soil by using plants, has demonstrated its potential. Due to their adaptability and capacity to improve the health of the soil, native plant species are frequently favored. One of the main goals of post-industrial landscape restoration is to create or improve habitats. Increased biodiversity and ecosystem resilience result from including a variety of habitat types, such as wetlands, woods, grasslands, and aquatic systems. In urban post-industrial settings, the installation of green infrastructure, such as green roofs, can help create habitat (Starczewski et al. 2023). Affected hydrological systems and water pollution are common problems in post-industrial areas. Successful restoration depends on regulating stormwater runoff, reducing contaminants, and restoring natural water flows. Restoring hydrological functions and enhancing water quality can be accomplished with the use of strategies like built wetlands, riparian buffer zones, and rainwater harvesting systems (LeClerc and Wiersma 2017).

2.5.2 Challenges and Opportunities for Restoration

Restoration is a critical process for reversing environmental degradation and promoting the recovery of ecosystems. It involves the active manipulation of ecosystems to enhance their ecological functioning and resilience. While restoration presents numerous challenges, it also offers significant opportunities for conserving biodiversity, improving ecosystem services, and mitigating the impacts of climate change.

2.5.2.1 Challenges for Restoration

The process of restoration is essential for stopping environmental degradation and fostering ecosystem recovery. Significant obstacles stand in the way of restoration efforts, including habitat fragmentation and the resulting loss of biodiversity (Raven and Wagner 2021).

Reduced connection caused by fragmented landscapes restricts species mobility and hinders the natural processes required for ecosystem repair. In order to enhance the efficacy of restoration efforts, a study by Fahrig et al. (2019) underlines the significance of taking landscape context and interconnectedness into consideration.

For restored ecosystems to be resilient and long-lived, genetic diversity is essential. Genetic diversity is decreased as a result of the fact that many restoration initiatives rely on a small number of people or cultivars (Kolar et al. 2023).

Invasive species can negatively impact ecosystem function and make restoration difficult. They frequently displace native species, interfere with ecological processes, and change the nutrient cycle. To improve ecosystem recovery, Simberloff et al.'s (Simberloff et al. 2013) study emphasizes the significance of managing invasive species in restoration operations.

The resilience and long-term survival of restored ecosystems depend on genetic diversity. However, a lot of restoration initiatives only use a few people or cultivars, which results in a lack of genetic variety (Sun et al. 2022). To increase the adaptability and resilience of ecosystems, a study by Young et al. (2005) emphasizes the significance of incorporating genetic concerns into restoration operations.

2.5.2.2 Opportunities for Restoration

Enhancing ecosystem services including the provision of clean water, carbon sequestration, and climate management may be possible through restoration. For sustainable growth and human well-being, these ecosystem services are essential. Bullock et al. (2011) demonstrate the capacity of restoration to provide a variety of ecosystem services and emphasize the significance of taking socioeconomic advantages into account when planning restoration.

By capturing carbon dioxide and improving carbon storage in soils and vegetation, restoration is essential in reducing climate change. Ecosystem restoration can help with climate change mitigation by serving as a sink for greenhouse gases (Singh et al. 2021). The significance of including restoration in measures for adapting to climate change is emphasized by Chazdon et al. (2016). By creating adequate habitats and encouraging population recovery, restoration can aid in the conservation of endangered species. Restoration initiatives can encourage species mobility and gene flow by establishing or improving habitat connectivity, lowering the danger of population isolation. Chimura (2022) goes over the value of restoration in preserving biodiversity and averting the extinction of species. Getting local communities involved in restoration projects might be a great way to promote conservation. It may assist in reviving traditional knowledge and cultural practices related to ecosystem management. In order to achieve long-term success and community support, a study by Miyagawa et al. (2023) emphasizes the significance of incorporating social factors and local viewpoints into restoration initiatives.

2.5.3 Case Studies of Successful Restoration Projects

Ecological restoration plays a crucial role in conserving and rehabilitating degraded ecosystems. Over the years, numerous restoration projects have been implemented worldwide, aiming to restore ecosystems to their original or improved states. A list of case studies of successful restoration projects is listed below highlighting the diverse approaches and outcomes, emphasizing the importance of evidence-based practices for achieving successful restoration.

2.5.3.1 Case Study 1: "The Loess Plateau Restoration Project"

The Loess Plateau Restoration Project in China is one of the world's largest and most successful ecological restoration efforts. This project aimed to combat soil erosion and restore degraded land through the implementation of comprehensive measures, including reforestation, terracing, and sustainable land management practices. The project spanned over two decades, leading to significant improvements in soil conservation, water availability, and biodiversity. The study by Zhang et al. (2015) demonstrated the positive impact of the restoration efforts on soil quality, vegetation cover, and hydrological processes.

2.5.3.2 Case Study 2: "The Florida Everglades Restoration Plan"

The Florida Everglades, a unique wetland ecosystem in the United States, has been severely impacted by human activities, including drainage, water diversion, and pollution. To restore and protect this ecosystem, the Comprehensive Everglades Restoration Plan (CERP) was initiated. Szantoi et al. (2015) analyzed the effectiveness of several restoration projects within the CERP. They highlighted the positive outcomes of the restoration efforts, such as improved water flow, enhanced wildlife habitat, and increased populations of endangered species. The study emphasized the need for adaptive management and long-term monitoring to ensure the success of large-scale restoration projects.

2.5.3.3 Case Study 3: "The Oostvaardersplassen Rewilding Project"

The Oostvaardersplassen in the Netherlands is a renowned example of a successful rewilding project. Originally a polder, the area was transformed into a diverse wetland habitat, promoting the establishment of self-regulating ecosystems. The project aimed to mimic natural processes and allow natural selection to shape the ecosystem. Knijn et al. (2020) assessed the ecological outcomes of the rewilding project and found that it resulted in increased biodiversity, including the return of iconic large herbivores and carnivores. The study highlighted the importance of large-scale restoration projects that focus on ecological processes rather than specific target species.

2.5.3.4 Case Study 4: "The Great Barrier Reef Restoration Initiative"

The Great Barrier Reef, a UNESCO World Heritage site, has faced significant threats from climate change, pollution, and coral bleaching. In response, the Great Barrier Reef Restoration Initiative was launched to restore and protect this iconic ecosystem. Harrison et al. (2021) discussed various restoration approaches, including coral larval reseeding, assisted evolution, and water quality improvement. The study highlighted the potential of these interventions to enhance coral reef resilience and promote ecosystem recovery. However, it also emphasized the need for continued efforts in reducing greenhouse gas emissions to address the underlying causes of coral reef decline.

2.5.3.5 Case Study 5: "The Rhône River Restoration Project"

The Rhône River in Europe has undergone significant alterations due to dam construction, channelization, and floodplain modification. The Rhône River Restoration Project aimed to enhance ecological connectivity, restore natural habitats, and improve the river's overall health.

2.5.3.6 Case Study 6: Renaturation of Coal-Mining Sites in the Lusatian Lakeland, Germany

The Lusatian Lakeland region in Germany has undergone significant ecological restoration following the cessation of coal mining activities. The restoration of former open-pit mining sites into water bodies and recreational areas is a prominent feature of the landscape transformation. A research paper by Sun (2021) examines the restoration process, focusing on the ecological aspects and the establishment of habitats in the renaturated mining sites. The study highlights the successful recovery of plant species diversity, the creation of new wetland habitats, and the re-establishment of endangered flora and fauna. It also discusses the challenges faced during the restoration process and provides valuable insights for future restoration projects.

2.5.3.7 Case Study 7: Kurasia Opencast Mines in Chirimiri Area of SECL

The reclamation strategies adopted in the Kurasia opencast mines in Chirimiri area of SECL were studied during July 2005 and April 2006 (Maiti 2007). The overburden dumps and mined-out areas were reclaimed by the plantation of tree species. The plantation was carried out by Chhattisgarh Van Vikas Nigam at the rate of Rs. 50/tree sapling, which includes digging of pits and overall maintenance for three years. To assess the biological reclamation aspects, different aged reclaimed overburden dumps were selected based on age of the plantation, and spoil samples were also taken from the rhizosphere of different tree species to study the amelioration of dump physicochemical and notional properties.

- Tree species composition in young vegetation stand (two years old, plantation year: 2003–2004).
- Tree species composition in intermediate age of four years old, plantation year 2001.
- Tree species composition in old reclaimed dumps (six years, plantation carried out in 1999).

2.6 Policy and Management for Post-Industrial Landscapes

2.6.1 International and National Policies and Frameworks for Post-Industrial Landscapes

There are several strategies that are adapted at the international and national levels for conservation of post-industrial landscapes. Some of these are as follows:

2.6.1.1 International Conservation Strategies

Conserving post-industrial landscapes is not an issue confined to any one country or community. It is a crucial global concern. Several international treaties and agreements are in place in an attempt to strengthen international participation and commitment toward conserving post-industrial landscapes (Table 2.1). Some of these are:

2.6.1.2 National Conservation Strategies

Formal policies and programs for conservation and sustainable utilization of biodiversity resources date back to several decades. The concept of environmental protection is enshrined in the Indian constitution in articles 48A and 51A(g) (Table 2.2).

Table 2.1 List of various international conservation strategies.

Serial number	Convention/treaty	Year	Description
1	Convention on Biological Diversity (CBD)	1992	Biodiversity conservation, sustainable resource use, equitable benefit sharing
2	Convention on International Trade in Endangered Species of Wild Flora and Fauna (CITES)	1975	Protection of endangered wildlife from international trade
3	Convention on Wetlands of International Importance (Ramsar Convention)	1975	Conservation of designated wetlands of international importance
4	European Landscape Convention (ELC)	2000	Landscape importance, holistic approach to social, cultural, economic, and ecological dimensions
5	United Nations Sustainable Development Goals (SDGs)	Ongoing	Addressing sustainability in urban development and environmental protection, including post-industrial areas

Table 2.2 List of various national conservation strategies.

Serial number	Policy/Act/Regulation	Year	Description
1	Indian Forest Act	1927	Consolidates laws relating to forests and forest produce.
2	Indian Forest Service (IFS)	1966	Manages and protects forests in India, and implements National Forest Policy.
3	Wild Life (Protection) Act	1972	Protects wildlife, and controls poaching and illegal trade.
4	Water (Prevention and Control of Pollution) Act	1974	Prevention and control of water pollution, maintaining water wholesomeness.
5	Water (Prevention and Control of Pollution) Cess Act	1977	Levies a cess on water consumption for pollution control.
6	Forest Conservation Act	1980	Regulates and restricts de-reservation of forests or non-forest land use.
7	Environment (Protection) Act	1986	Protection and improvement of the environment, prevention of environmental pollution.
8	Public Liability Insurance	1991	Provides for compensation for accidents involving hazardous substances.
9	National Environment Tribunal	1995	Provides strict liability for environmental damage from hazardous substances.
10	Biological Diversity Act	2002	Conservation of biological resources and associated knowledge, the establishment of National Biodiversity Authority.
11	Scheduled Tribes and Other Traditional Forest Dwellers (Recognition of Forest Rights) Act	2006	Recognizes forest rights of forest-dwelling Scheduled Tribes and traditional forest dwellers.
12	National Green Tribunal (NGT)	2010	Provides expedited resolution of environmental cases, and specializes in environmental disputes.

2.6.2 Management Approaches and Best Practices

Management approaches and best practices for post-industrial landscapes involve a wide range of strategies aimed at revitalizing and transforming abandoned or underutilized industrial areas into vibrant and sustainable spaces (Edler et al. 2019). These approaches typically include elements of urban planning, environmental remediation, community engagement, and economic development. One effective strategy for post-industrial landscapes is the adaptive reuse of existing structures. This approach involves repurposing old industrial buildings for new uses, such as residential, commercial, cultural, or recreational purposes. It helps preserve the historical and architectural value of these structures while promoting economic activity (Betts et al. 2021). Many post-industrial landscapes suffer from contamination or environmental degradation. Successful management approaches often include remediation efforts to address pollution and restore ecological health. Bengtsson et al. (2019) discuss various ecological restoration techniques, including phytoremediation, bioremediation, and soil amendment, and highlight their applicability in post-industrial contexts.

Brownfield sites, often found in post-industrial landscapes, are abandoned or underutilized properties with potential environmental contamination. Redevelopment of brownfield sites can stimulate economic growth and create new opportunities. Integrating sustainable infrastructure and design principles is crucial for transforming post-industrial landscapes into environmentally friendly and resilient spaces. A study by Sheppard et al. (2018) emphasizes the importance of landscape architecture in promoting sustainable design practices, such as green roofs, rainwater harvesting, and urban agriculture, to enhance the ecological and social value of post-industrial sites. Involving local communities in the decision-making process is essential for successful post-industrial landscape management (Lazdinis et al. 2019). Community engagement fosters a sense of ownership, ensures diverse perspectives, and increases the chances of long-term success. A study by Rydin and Holman (2019) examines different forms of community participation and highlights their significance in achieving sustainable and inclusive post-industrial redevelopment.

Revitalizing post-industrial landscapes often involves strategies for economic development and job creation. Stathakis and Fakiolas (2020) explore the role of innovation and entrepreneurship in stimulating economic growth in post-industrial areas, emphasizing the importance of fostering innovation ecosystems and supporting small and medium-sized enterprises. Regular monitoring and evaluation of post-industrial landscape projects are crucial to ensure progress, identify challenges, and make necessary adjustments. A study by Colantonio et al. (2019) provides insights into monitoring and evaluation frameworks, including the use of performance indicators and stakeholder engagement, to assess the effectiveness of post-industrial regeneration initiatives (Vanbergen et al. 2020).

2.6.3 Stakeholder Engagement and Participatory Approaches

Stakeholder engagement and participatory approaches are crucial when it comes to the revitalization and management of post-industrial landscapes (Pólvora et al. 2020). These landscapes often face complex challenges such as environmental degradation, economic decline, and social disconnection. Engaging stakeholders and employing participatory approaches can help address these challenges, foster community involvement, and ensure the long-term success and sustainability of post-industrial landscape projects. The first step in stakeholder engagement is identifying and understanding the various individuals, groups, and organizations that have a vested interest in the post-industrial landscape (Solman et al. 2021). This may include residents, businesses, community organizations, environmental groups, government agencies, and academic institutions. Each stakeholder group may have different perspectives, concerns, and goals that need to be considered.

Developing positive and trust-based relationships with stakeholders is essential. It is important to establish open lines of communication, listen to their concerns and ideas, and involve them in the decision-making process (Marques et al. 2020). Regular meetings, workshops, and public forums can provide opportunities for stakeholders to share their knowledge, experiences, and aspirations for the post-industrial landscape. Engaging stakeholders in the planning process helps ensure that diverse perspectives are taken into account and that decisions are made collectively. This participatory approach can involve collaborative workshops, charrettes, or design studios, where stakeholders work together with planners, architects, and landscape designers to develop shared visions, goals, and strategies for the post-industrial landscape. This process can lead to more inclusive and innovative solutions that address the needs and aspirations of the community (Erfani and Roe 2020).

Post-industrial landscapes are often located in communities that have experienced economic decline and social challenges. Engaging local communities in the revitalization process can help empower them and build local capacity. This can be done through training programs, skill-building workshops, and the creation of local employment opportunities. By involving community members in the implementation and maintenance of the landscape, a sense of ownership and pride can be fostered, leading to increased social cohesion and well-being (Ferreira et al. 2020).

Many post-industrial landscapes suffer from environmental degradation resulting from industrial activities. Stakeholder engagement can help ensure that ecological considerations are integrated into the design and management of the landscape. Involving environmental experts and organizations can provide valuable insights into habitat restoration, biodiversity conservation, and sustainable land use practices (Vitálišová et al. 2021). This can help create a resilient and ecologically valuable landscape that benefits both people and the environment. Effective stakeholder engagement requires transparency and accountability throughout the entire process. This includes providing stakeholders with access to information, sharing project updates, and being responsive to their feedback and concerns. Regular communication and feedback loops help build trust and ensure that stakeholders are actively involved in the decision-making and implementation phases (Paddeu and Aditjandra 2020).

Post-industrial landscape projects are often long-term endeavors that require ongoing monitoring and adaptation. Stakeholder engagement should not be limited to the initial planning stages but should continue throughout the lifecycle of the project. Regular evaluations, feedback mechanisms, and adaptive management strategies ensure that the project remains responsive to changing needs, priorities, and circumstances (Kovács et al. 2021).

2.7 Key Findings and Implications

Post-industrial landscapes refer to areas that have undergone a significant transformation from industrial to post-industrial activities, typically characterized by the decline or cessation of heavy manufacturing and the emergence of new economic sectors. The study of post-industrial landscapes is crucial for understanding the complex processes of urban and regional development in the contemporary world. One of the key findings regarding post-industrial landscapes is the potential for urban regeneration and revitalization. Studies have shown that former industrial sites can be transformed into vibrant cultural and recreational spaces, promoting economic development and enhancing the quality of life for residents. For instance, the research by Evans et al. (2016) examines the conversion of a former steelworks site into a mixed-use development with a focus on arts, culture, and heritage. The findings suggest that such projects can attract new investments, create job opportunities, and foster social cohesion. This highlights the importance of adaptive reuse and creative urban planning in post-industrial areas.

Furthermore, the ecological impact of industrial activities and the subsequent remediation efforts are critical considerations in post-industrial landscapes. Scientific research has demonstrated the long-lasting effects of industrial pollution on the environment, including soil and water contamination, air pollution, and habitat degradation. Studies by Lasat (2002) and Moscovici et al. (2017) highlight the importance of environmental remediation and restoration in post-industrial regions. The findings emphasize the need for sustainable practices, green infrastructure, and the incorporation of ecological considerations in urban planning processes to mitigate the ecological consequences of industrialization.

Additionally, the emergence of the knowledge economy and the role of innovation in post-industrial landscapes have been the focus of extensive research. The work of Saxenian (1996) explores the importance of regional innovation systems and the formation of technology clusters in post-industrial regions such as Silicon Valley. The findings suggest that the concentration of high-tech industries, research institutions, and entrepreneurial activities can drive economic growth and foster innovation. This highlights the significance of fostering a culture of innovation, supporting research and development, and promoting collaboration between academia, industry, and government in post-industrial landscapes.

Social and cultural dimensions are also crucial considerations in understanding post-industrial landscapes. Research by Zukin (2009) examines the transformation of industrial neighborhoods into creative and cultural hubs, often known as creative clusters or cultural quarters. The findings emphasize the role of the creative class, artists, and cultural institutions in revitalizing post-industrial areas. This highlights the need to support cultural activities, creative industries, and community engagement to enhance the social fabric and cultural vitality of post-industrial landscapes.

2.8 Challenges and Future Research Directions

Post-industrial landscapes, often characterized by abandoned factories, mines, and other industrial sites, present unique challenges and opportunities for researchers and policy-makers. These landscapes often pose environmental and socio-economic challenges, requiring careful management and redevelopment strategies. As societies transition from industrial to post-industrial economies, understanding and effectively managing these landscapes becomes crucial for sustainable development, environmental remediation, and social revitalization. Remediation of post-industrial landscapes involves addressing pollution, hazardous materials, and ecological degradation resulting from past industrial activities. Future research can focus on developing innovative and cost-effective remediation techniques. For instance, a study by Pivetz (2001) provides insights into the potential of using plants for removing contaminants from soils and groundwater.

Quantifying and valuing ecosystem services often face challenges related to data availability, uncertainty in ecosystem processes, and the complex interactions between ecological and socio-economic systems (Dushkova and Haase 2020). Addressing these challenges requires further study and the development of standardized protocols. While provisioning and regulating services have been relatively well-studied, cultural ecosystem services (e.g., recreational and spiritual values) have received less attention (Daniel et al. 2012). Future studies should focus on developing methods to quantify and value these services, considering the diverse cultural perspectives and preferences of different communities.

Research can explore strategies for repurposing industrial structures and integrating them into new urban landscapes. A study by Gurran and Blakely (2007) offers valuable insights into the social, economic, and environmental benefits of adaptive reuse in post-industrial areas. Understanding the social and cultural dynamics of post-industrial landscapes is crucial for successful revitalization efforts. Research can delve into the impacts of industrial decline on local communities, social inequalities, and cultural heritage preservation. Andersson et al. (2019) explore the social dimensions of urban green spaces in post-industrial settings and highlight the importance of inclusive planning processes. Research can focus on sustainable design principles, green infrastructure, and urban resilience in post-industrial areas. Gasparatos et al. (2017) present a comprehensive framework for sustainable brownfield redevelopment, integrating ecological, economic, and social dimensions. A study by Evans et al. (2016) provides an overview of urban innovation and identifies potential pathways for economic diversification in post-industrial areas.

Brownfield sites, which are abandoned or underutilized industrial areas, offer great potential for redevelopment. Research in this area emphasizes the need for sustainable land-use planning, incorporating concepts such as mixed land-use, smart growth, and green infrastructure. Salisbury et al. (2021) discuss the role of brownfield redevelopment in urban revitalization and highlight the importance of community involvement, environmental justice, and economic viability in achieving successful outcomes.

Efficient management of resources and infrastructure in post-industrial landscapes is critical for their sustainable development. Research in this domain includes studies on renewable energy systems, water management, waste recycling, and green infrastructure planning. A comprehensive review by Shu et al. (2023) examines the sustainable infrastructure approaches and technologies for the redevelopment of post-industrial landscapes, emphasizing the importance of circular economy principles and the integration of nature-based solutions.

The integration of digital technologies and smart city concepts can enhance the redevelopment of post-industrial landscapes. Research focuses on the application of the Internet of Things (IoT), artificial intelligence, and data-driven decision-making processes. Vosloo (2018) discusses the potential of digital technologies in post-industrial landscapes, including smart energy systems, intelligent transportation, and sensor-based environmental monitoring.

2.9 Conclusion

The chapter illuminates the potential of post-industrial landscapes for sustainable development, environmental remediation, and social revitalization. It underscores the importance of informed policymaking and effective strategies to transform these landscapes into vibrant and sustainable spaces. The interplay of environmental, social, and economic factors in post-industrial transformations requires holistic approaches and innovative solutions. Successful restoration projects exemplify the positive outcomes achievable through adaptive reuse, ecological restoration, and community engagement. The chapter advocates for continued research in environmental remediation techniques, adaptive reuse strategies, social and cultural dynamics, sustainable urban design, economic transformation, and participatory planning to address the challenges and unlock the opportunities presented by post-industrial landscapes. By navigating these research areas, policymakers and stakeholders can navigate the complexities of post-industrial redevelopment, ensuring a balance between environmental conservation, social well-being, and economic growth.

References

Andersson, E., Barthel, S., and Ahrné, K. (2019). Postindustrial urban green space: a social-ecological perspective. *Urban Forestry & Urban Greening* 40: 1–9.

Assessment, M.E. (2005). *Ecosystems and human well-being: wetlands and water.* World Resources Institute.

Bardos, R.P., Jones, S., Stephenson, I. et al. (2016). Optimising value from the soft re-use of brownfield sites. *Science of the Total Environment* 563: 769–782.

Bardos, P., Spencer, K.L., Ward, R.D. et al. (2020). Integrated and sustainable management of post-industrial coasts. *Frontiers in Environmental Science* 8: 86.

Beillouin, D., Ben-Ari, T., Malézieux, E. et al. (2021). Positive but variable effects of crop diversification on biodiversity and ecosystem services. *Global Change Biology* 27 (19): 4697–4710.

Bengtsson, J., Nilsson, M., and Woldeamanuel, T. (2019). Ecological restoration of industrial sites: from theory to practice. *Sustainability* 11 (18): 5081.

Betts, M.G., Phalan, B.T., Wolf, C. et al. (2021). Producing wood at least cost to biodiversity: integrating T riad and sharing–sparing approaches to inform forest landscape management. *Biological Reviews* 96 (4): 1301–1317.

Bliege Bird, R. and Nimmo, D. (2018). Restore the lost ecological functions of people. *Nature Ecology & Evolution* 2 (7): 1050–1052.

Boyd, J. and Banzhaf, S. (2007). What are ecosystem services? The need for standardized environmental accounting units. *Ecological Economics* 63 (2-3): 616–626.

Bradu, P., Biswas, A., Nair, C. et al. (2022). Recent advances in green technology and Industrial Revolution 4.0 for a sustainable future. *Environmental Science and Pollution Research* 1–32.

Brooks, T.M., Mittermeier, R.A., Mittermeier, C.G. et al. (2002). Habitat loss and extinction in the hotspots of biodiversity. *Conservation Biology* 16 (4): 909–923.

Brown, S.C., Mellin, C., García Molinos, J. et al. (2022). Faster ocean warming threatens richest areas of marine biodiversity. *Global Change Biology* 28 (19): 5849–5858.

Bullock, J.M., Aronson, J., Newton, A.C. et al. (2011). Restoration of ecosystem services and biodiversity: conflicts and opportunities. *Trends in Ecology & Evolution* 26 (10): 541–549.

Cardinale, B.J., Matulich, K.L., Hooper, D.U. et al. (2011). The functional role of producer diversity in ecosystems. *American Journal of Botany* 98 (3): 572–592.

Chazdon, R.L., Broadbent, E.N., Rozendaal, D.M. et al. (2016). Carbon sequestration potential of second-growth forest regeneration in the Latin American tropics. *Science Advances* 2 (5): e1501639.

Chimura, R. (2022). *Intermediate Community Use during Brownfield Remediation Phytoremediation's Applicability Analysis in the Post-Industrial City of Malmö*, 10–65. Swedish University of Agricultural Sciences.

Colantonio, A., Manzi, T., and Van der Borgh, C. (2019). Monitoring urban regeneration projects: a review of indicators and evaluation frameworks. *Sustainability* 11 (4): 1093.

Comberti, C., Thornton, T.F., De Echeverria, V.W., and Patterson, T. (2015). Ecosystem services or services to ecosystems? Valuing cultivation and reciprocal relationships between humans and ecosystems. *Global Environmental Change* 34: 247–262.

Costanza, R. (2020). Valuing natural capital and ecosystem services toward the goals of efficiency, fairness, and sustainability. *Ecosystem Services* 43: 101096.

Costanza, R., d'Arge, R., De Groot, R. et al. (1997). The value of the world's ecosystem services and natural capital. *Nature* 387 (6630): 253–260.

Coulibaly, B. and Li, S. (2020). Impact of agricultural land loss on rural livelihoods in peri-urban areas: empirical evidence from Sebougou, Mali. *Land* 9 (12): 470.

Dangwal, B., Rana, S.K., Negi, V.S., and Bhatt, I.D. (2022). Forest restoration enhances plant diversity and carbon stock in the sub-tropical forests of western Himalaya. *Trees, Forests and People* 7: 100201.

Daniel, T.C., Muhar, A., Arnberger, A. et al. (2012). Contributions of cultural services to the ecosystem services agenda. *Proceedings of the National Academy of Sciences* 109 (23): 8812–8819.

Díaz, S., Settele, J., Brondízio, E.S. et al. (2019). Pervasive human-driven decline of life on Earth points to the need for transformative change. *Science* 366 (6471): eaax3100.

Dogan, E., Cuomo, F., and Battisti, L. (2023). Reviving urban greening in post-industrial landscapes: the case of Turin. *Sustainability* 15 (17): 12760.

Doherty, T.S., Glen, A.S., Nimmo, D.G. et al. (2016). Invasive predators and global biodiversity loss. *Proceedings of the National Academy of Sciences* 113 (40): 11261–11265.

Dushkova, D. and Haase, D. (2020). Not simply green: nature-based solutions as a concept and practical approach for sustainability studies and planning agendas in cities. *Land* 9 (1): 19.

Edler, D., Keil, J., Wiedenlübbert, T. et al. (2019). Immersive VR experience of redeveloped post-industrial sites: the example of "Zeche Holland" in Bochum-Wattenscheid. *KN Journal of Cartography and Geographic Information* 69: 267–284.

Elhan, S. (2009). Kentsel bellek bağlamında sanayi mirasının değerlendirilmesi: İstanbul Haliç Örneği. Doctoral dissertation. Fen Bilimleri Enstitüsü.

Enssle, F. and Kabisch, N. (2020). Urban green spaces for the social interaction, health and well-being of older people—an integrated view of urban ecosystem services and socio-environmental justice. *Environmental Science & Policy* 109: 36–44.

Erfani, G. and Roe, M. (2020). Institutional stakeholder participation in urban redevelopment in Tehran: an evaluation of decisions and actions. *Land Use Policy* 91: 104367.

Evans, J., Karvonen, A., and Raven, R. (2016). The experimental city: new modes and prospects of urban transformation. In: *The Experimental City*, 1–12. Routledge.

Fahrig, L., Arroyo-Rodríguez, V., Bennett, J.R. et al. (2019). Is habitat fragmentation bad for biodiversity? *Biological Conservation* 230: 179–186.

Ferreira, V., Barreira, A.P., Loures, L. et al. (2020). Stakeholders' engagement on nature-based solutions: a systematic literature review. *Sustainability* 12 (2): 640.

Fine, B. and Leopold, E. (1990). Consumerism and the industrial revolution. *Social History* 15 (2): 151–179.

Eric, S., Robert, B., Stephen, F. et al. (2008). *Carbon sequestration to mitigate climate change*, 3097. US Geological Survey, Fact Sheet.

Foster, J. (2022). *Post-industrial Urban Greenspace Ecology, Aesthetics and Justice*. Taylor & Francis.

Gallagher, F.J. (2021). Natural transformation of post-industrial lands: Liberty State Park in Jersey City, NJ (USA). In: *Urban Wastelands: A Form of Urban Nature?* (ed. F. Di Pietro and A. Robert), 19–38. Springer.

Gallagher, F., Goodey, N.M., Hagmann, D. et al. (2018). Urban re-greening: a case study in multi-trophic biodiversity and ecosystem functioning in a post-industrial landscape. *Diversity* 10 (4): 119.

Gasparatos, A., von Maltitz, G.P., Johnson, F.X. et al. (2017). Design strategies for sustainable brownfield development. *Journal of Cleaner Production* 155: 115–123.

Gebretsadik, T. (2016). Causes for biodiversity loss in Ethiopia: a review from conservation perspective. *Journal of Natural Sciences Research* 6 (11): 32–40.

Geng, X., Wang, X., Yan, H. et al. (2014). Land use/land cover change induced impacts on water supply service in the upper reach of Heihe River Basin. *Sustainability* 7 (1): 366–383.

Gerwin, W., Raab, T., Birkhofer, K. et al. (2023). Perspectives of lignite post-mining landscapes under changing environmental conditions: what can we learn from a comparison between the Rhenish and Lusatian region in Germany? *Environmental Sciences Europe* 35 (1): 1–24.

Green, T.L. (2018). Evaluating predictors for brownfield redevelopment. *Land Use Policy* 73: 299–319.

Guerry, A.D., Ruckelshaus, M.H., Arkema, K.K. et al. (2012). Modeling benefits from nature: using ecosystem services to inform coastal and marine spatial planning. *International Journal of Biodiversity Science, Ecosystem Services & Management* 8 (1–2): 107–121.

Guo, C. (2019). *New Natural Landscape: Restoring biodiversity in post-industrial region*. Parkstad-Limburg.

Gurran, N. and Blakely, E.J. (2007). The role of adaptive reuse in sustainable historic preservation. *Journal of Planning Literature* 21 (3): 279–300.

Hamerla, A. (2021). Development of e-services in the process of post-mining and post-industrial area management in the Silesia region. *Zeszyty Naukowe. Organizacja i Zarządzanie/Politechnika Śląska*.

Han, H. and Xia, S. (2021). An agro-based society after post-industrial society: from a perspective of economic growth paradigm. *Social Sciences* 10 (12): 455.

Hanley, N., Wright, R.E., and Adamowicz, V. (1998). Using choice experiments to value the environment. *Environmental and Resource Economics* 11: 413–428.

Harrison, P.L., Babcock, R.C., Bull, G.D. et al. (2021). The Great Barrier Reef restoration initiative: a cross-sectoral collaboration to promote ecosystem resilience at multiple scales. *Frontiers in Marine Science* 8: 671010.

Hasan, S., Shi, W., Zhu, X., and Abbas, S. (2019). Monitoring of land use/land cover and socioeconomic changes in south china over the last three decades using landsat and nighttime light data. *Remote Sensing* 11 (14): 1658.

Hasan, S.S., Zhen, L., Miah, M.G. et al. (2020). Impact of land use change on ecosystem services: a review. *Environmental Development* 34: 100527.

Höfer, W. (1998). Post-industrial landscape. *Urban Ecology* 671–675.

Hui, D. (2013). Global climate change and biodiversity: issues and future research. *Journal of Biodiversity and Endangered Species* 1 (1): 1–2.

Humphris, I. and Rauws, W. (2021). Edgelands of practice: post-industrial landscapes and the conditions of informal spatial appropriation. *Landscape Research* 46 (5): 589–604.

Hüppauf, B. (2004). Globalization-threats and opportunities. In: *Globalization and the Future of German* (ed. A. Gardt and B. Hüppauf), 3–24.

Jacek, G., Rozan, A., Desrousseaux, M., and Combroux, I. (2022). Brownfields over the years: from definition to sustainable reuse. *Environmental Reviews* 30 (1): 50–60.

Javaid, M., Haleem, A., Singh, R.P., and Suman, R. (2021). Substantial capabilities of robotics in enhancing industry 4.0 implementation. *Cognitive Robotics* 1: 58–75.

Jin, G., Deng, X., Chu, X. et al. (2017). Optimization of land-use management for ecosystem service improvement: a review. *Physics and Chemistry of the Earth, Parts A/B/C* 101: 70–77.

Kabisch, N., van den Bosch, M., and Lafortezza, R. (2017). The health benefits of nature-based solutions to urbanization challenges for children and the elderly–a systematic review. *Environmental Research* 159: 362–373.

Klein, A.M., Vaissière, B.E., Cane, J.H. et al. (2007). Importance of pollinators in changing landscapes for world crops. *Proceedings of the Royal Society B: Biological Sciences* 274 (1608): 303–313.

Knijn, R., Kuiters, L., Slager, D., and Swart, D. (2020). Ecological outcomes of rewilding: lessons for the Netherlands from other European initiatives. *Ambio* 49 (2): 697–711.

Koch, J., Pearson, D.E., Huebner, C.D. et al. (2021). Restoration of landscapes and habitats affected by established invasive species. In: *Invasive Species in Forests and Rangelands of the United States* (ed. T.M. Poland, T. Patel-Weynand, and D.M. Finch), 185. Springer Nature Switzerland AG.

Koch, J., Pearson, D.E., Huebner, C.D., et al. (2021). Restoration of landscapes and habitats affected by established invasive species. *Invasive species in forests and rangelands of the United States*, 185.

Kolar, V., Chmelová, E., Bílková, M. et al. (2023). Muddying the unexplored post-industrial waters: biodiversity and conservation potential of freshwater habitats in fly ash sedimentation lagoons. *Science of the Total Environment* 900: 165803.

Kosanic, A. and Petzold, J. (2020). A systematic review of cultural ecosystem services and human wellbeing. *Ecosystem Services* 45: 101168.

Kovács, E., Mile, O., Fabók, V. et al. (2021). Fostering adaptive co-management with stakeholder participation in the surroundings of soda pans in Kiskunság, Hungary–an assessment. *Land Use Policy* 100: 104894.

Kovalchuk, J. (ed.) (2021). *Post-industrial Society: The Choice Between Innovation and Tradition*. Springer Nature.

Kristiánová, K., Gécová, K., and Putrová, E. (2016). Old industrial sites–conversion to parks: potential of Bratislava. *Procedia Engineering* 161: 1858–1862.

Lasat, M.M. (2002). Phytoextraction of toxic metals: a review of biological mechanisms. *Journal of Environmental Quality* 31 (1): 109–120.

Lazdinis, M., Angelstam, P., and Pülzl, H. (2019). Towards sustainable forest management in the European Union through polycentric forest governance and an integrated landscape approach. *Landscape Ecology* 34: 1737–1749.

LeClerc, E. (2015). Assessing the legacy of the pine point mine: post-industrial land cover and land use. Doctoral dissertation. Memorial University of Newfoundland.

LeClerc, E. and Wiersma, Y.F. (2017). Assessing post-industrial land cover change at the Pine Point Mine, NWT, Canada using multi-temporal landsat analysis and landscape metrics. *Environmental Monitoring and Assessment* 189: 1–19.

Ling, C., Handley, J., and Rodwell, J. (2007). Restructuring the post-industrial landscape: a multifunctional approach. *Landscape Research* 32 (3): 285–309.

Loures, L. (2015). Post-industrial landscapes as drivers for urban redevelopment: public versus expert perspectives towards the benefits and barriers of the reuse of post-industrial sites in urban areas. *Habitat International* 45: 72–81.

Maiti, S.K. (2007). Minesoil properties of different aged reclaimed coal mine overburden dumps of Korba, Gevra and Kusmunda area of SECL, India. *MINETECH* 28 (2&3): 93–98.

Marques, M., Oliveira, M., and Borges, J.G. (2020). An approach to assess actors' preferences and social learning to enhance participatory forest management planning. *Trees, Forests and People* 2: 100026.

Mengist, W., Soromessa, T., and Feyisa, G.L. (2020). A global view of regulatory ecosystem services: existed knowledge, trends, and research gaps. *Ecological Processes* 9: 1–14.

Merwin, L., Umek, L., and Anastasio, A.E. (2022). Urban post-industrial landscapes have unrealized ecological potential. *Restoration Ecology* 30 (8): e13643.

Mills, J.G., Weinstein, P., Gellie, N.J. et al. (2017). Urban habitat restoration provides a human health benefit through microbiome rewilding: the Microbiome Rewilding Hypothesis. *Restoration Ecology* 25 (6): 866–872.

Mitsch, W.J. and Gosselink, J.G. (2015). *Wetlands*. John Wiley & Sons.

Miyagawa, T., Olver, C., Otsuka, N., and Abe, H. (2023). Partnership-based policies and plans for open space management of case studies in post-industrial landscapes. *GEOMATE Journal* 25 (108): 97–105.

Moscovici, A.M., Banescu, O.A., and Vaduva, R. (2017). Integrating brownfield sites into city redevelopment strategies. *International Multidisciplinary Scientific GeoConference: SGEM* 17: 675–682.

Nicita, A., Ognivtsev, V., and Shirotori, M. (2013). *Global supply chains: Trade and economic policies for developing countries*, vol. 55. UN.

Opoku, E.E.O. and Boachie, M.K. (2020). The environmental impact of industrialization and foreign direct investment. *Energy Policy* 137: 111178.

Otsuka, N., Abe, H., Isehara, Y., and Miyagawa, T. (2021). The potential use of green infrastructure in the regeneration of brownfield sites: three case studies from Japan's Osaka Bay Area. *Local Environment* 26 (11): 1346–1363.

Ouyang, X., Tang, L., Wei, X., and Li, Y. (2021). Spatial interaction between urbanization and ecosystem services in Chinese urban agglomerations. *Land Use Policy* 109: 105587.

Paddeu, D. and Aditjandra, P. (2020). Shaping urban freight systems via a participatory approach to inform policy-making. *Sustainability* 12 (1): 441.

Pan, Y., Birdsey, R.A., Fang, J. et al. (2011). A large and persistent carbon sink in the world's forests. *Science* 333 (6045): 988–993.

Pandey, V.C. (2017). Managing waste dumpsites through energy plantations. In: *Phytoremediation Potential of Bioenergy Plants* (ed. K. Bauddh, B. Singh, and J. Korstad), 371–386. Singapore: Springer https://doi.org/10.1007/978-981-10-3084-0_15.

Pandey, V.C. (2020a). Fly ash ecosystem services. In: *Phytomanagement of Fly Ash*, 257–288. Amsterdam: Elsevier https://doi.org/10.1016/B978-0-12-818544-5.00009-2.

Pandey, V.C. (2020b). *Phytomanagement of Fly Ash*. Elsevier https://doi.org/10.1016/C2018-0-01318-3.

Pandey, V.C. (2020c). Fly ash deposits–a potential sink for carbon sequestration. In: *Phytomanagement of Fly Ash*, 235–255. Amsterdam: Elsevier https://doi.org/10.1016/B978-0-12-818544-5.00008-0.

Pandey, V.C., Ahirwal, J., Roychowdhury, R., and Chaturvedi, R. (2023). *Eco-Restoration of Mine Land*. Wiley.

Pandey, V.C. and Bajpai, O. (2019). Phytoremediation: from theory towards practice. In: *Phytomanagement of Polluted Sites* (ed. V.C. Pandey and K. Bauddh), 1–49. Amsterdam: Elsevier https://doi.org/10.1016/B978-0-12-813912-7.00001-6.

Pandey, V.C., Gajic, G., Sharma, P., and Roy, M. (2022). *Adaptive Phytoremediation Practices: Resilience to Climate Change*. Elsevier https://doi.org/10.1016/C2020-0-00560-8.

Pandey, V.C., Rai, A., and Korstad, J. (2019). Aromatic crops in phytoremediation: from contaminated to waste dump sites. In: *Phytomanagement of Polluted Sites* (ed. V.C. Pandey and K. Bauddh), 255–275. Amsterdam: Elsevier https://doi.org/10.1016/B978-0-12-813912-7.00009-0.

Pivetz, B.E. (2001). *Phytoremediation of Contaminated Soil and Ground Water at Hazardous Waste Sites*. US Environmental Protection Agency, Office of Research and Development, Office of Solid Waste and Emergency Response.

Pólvora, A., Nascimento, S., Lourenço, J.S., and Scapolo, F. (2020). Blockchain for industrial transformations: a forward-looking approach with multi-stakeholder engagement for policy advice. *Technological Forecasting and Social Change* 157: 120091.

Prach, K. and Tolvanen, A. (2016). How can we restore biodiversity and ecosystem services in mining and industrial sites. *Environmental Science and Pollution Research* 23: 13587–13590.

Qiu, H., Hu, B., and Zhang, Z. (2021). Impacts of land use change on ecosystem service value based on SDGs report—Taking Guangxi as an example. *Ecological Indicators* 133: 108366.

Raven, P.H. and Wagner, D.L. (2021). Agricultural intensification and climate change are rapidly decreasing insect biodiversity. *Proceedings of the National Academy of Sciences* 118 (2): e2002548117.

Rigolon, A., Fernandez, M., Harris, B., and Stewart, W. (2022). An ecological model of environmental justice for recreation. *Leisure Sciences* 44 (6): 655–676.

Rodríguez-Loinaz, G., Alday, J.G., and Onaindia, M. (2015). Multiple ecosystem services landscape index: a tool for multifunctional landscapes conservation. *Journal of Environmental Management* 147: 152–163.

Rondeau, D., Perry, B., and Grimard, F. (2020). The consequences of COVID-19 and other disasters for wildlife and biodiversity. *Environmental and Resource Economics* 76: 945–961.

Rydin, Y. and Holman, N. (2019). Meanings of participation in planning. *Sustainability* 11 (15): 4155.

Salgueiro, P.A., Prach, K., Branquinho, C., and Mira, A. (2020). Enhancing biodiversity and ecosystem services in quarry restoration–challenges, strategies, and practice. *Restoration Ecology* 28 (3): 655–660.

Salisbury, A., Gallagher, F.J., Parag, H.A. et al. (2021). Plant diversity increases in an urban wildland after four decades of unaided vegetation development in a post-industrial site. *Urban Ecosystems* 24: 95–111.

Saxenian, A. (1996). *Regional Advantage: Culture and Competition in Silicon Valley and Route 128, with a New Preface by the Author*. Harvard University Press.

Scott, A. and Storper, M. (2003). Regions, globalization, development. *Regional Studies* 37 (6-7): 579–593.

Semeraro, T., Scarano, A., Buccolieri, R. et al. (2021). Planning of urban green spaces: an ecological perspective on human benefits. *Land* 10 (2): 105.

Sharma, D., Rao, K., and Ramanathan, A.L. (2021). A systematic review on the impact of urbanization and industrialization on Indian coastal mangrove ecosystem. In: *Coastal Ecosystems: Environmental Importance, Current Challenges and Conservation Measures* (ed. S. Madhav, S. Nazneen, and P. Singh), 175–199. Springer.

Sheppard, S.R., Vignola, R., and Klaiber, H.A. (2018). The role of landscape architecture in post-industrial urban regeneration. *Sustainability* 10 (10): 3471.

Shivanna, K.R. (2022). Climate change and its impact on biodiversity and human welfare. *Proceedings of the Indian National Science Academy* 88 (2): 160–171.

Shu, C., Ruda, M., Dzhumelia, E. et al. (2023). Restoring soil cover and plant communities with Arbuscular Mycorrhizal fungi as an essential component of DSS for environmental safety management in post-industrial landscapes. *Agronomy* 13 (5): 1346.

Sierka, E., Wheeler, A., and Woźniak, G. (2018). Urban and industrial habitats: how important they are for ecosystem services. In: *Ecosystem Services and Global Ecology* (ed. L. Hufnagel). IntechOpen https://doi.org/10.5772/intechopen.75723.

Simberloff, D., Martin, J.L., Genovesi, P. et al. (2013). Impacts of biological invasions: what's what and the way forward. *Trends in Ecology & Evolution* 28 (1): 58–66.

Simonson, W.D., Miller, E., Jones, A. et al. (2021). Enhancing climate change resilience of ecological restoration—a framework for action. *Perspectives in Ecology and Conservation* 19 (3): 300–310.

Singh, V., Shukla, S., and Singh, A. (2021). The principal factors responsible for biodiversity loss. *Open Journal of Plant Science* 6 (1): 011–014.

Solman, H., Smits, M., van Vliet, B., and Bush, S. (2021). Co-production in the wind energy sector: a systematic literature review of public engagement beyond invited stakeholder participation. *Energy Research & Social Science* 72: 101876.

Sonwani, S. and Maurya, V. (2019). Impact of air pollution on the environment and economy. In: *Air Pollution: Sources, Impacts and Controls*, 113–134. Wallingford: CAB International.

Starczewski, T., Rogatka, K., Kukulska-Kozieł, A. et al. (2023). Urban green resilience: experience from post-industrial cities in Poland. *Geoscience Frontiers* 14 (4): 101560.

Stathakis, E. and Fakiolas, A. (2020). Economic development strategies in post-industrial cities: the role of innovation and entrepreneurship. *Sustainability* 12 (1): 395.

Stubbs, M. (2008). Natural green space and planning policy: devising a model for its delivery in regional spatial strategies. *Landscape Research* 33 (1): 119–139.

Sun, J. (2021). Connectivity restoration for fishes in post-industrial rivers of north east England. Doctoral dissertation. Durham University.

Sun, J., Tummers, J.S., Galib, S.M., and Lucas, M.C. (2022). Fish community and abundance response to improved connectivity and more natural hydromorphology in a post-industrial subcatchment. *Science of the Total Environment* 802: 149720.

Szantoi, Z., Escobedo, F.J., Abd-Elrahman, A. et al. (2015). Classifying spatially heterogeneous wetland communities using machine learning algorithms and spectral and textural features. *Environmental Monitoring and Assessment* 187: 1–15.

Teague, R. and Kreuter, U. (2020). Managing grazing to restore soil health, ecosystem function, and ecosystem services. *Frontiers in Sustainable Food Systems* 157.

Tolessa, T., Senbeta, F., and Kidane, M. (2017). The impact of land use/land cover change on ecosystem services in the central highlands of Ethiopia. *Ecosystem Services* 23: 47–54.

Vanbergen, A.J., Aizen, M.A., Cordeau, S. et al. (2020). Transformation of agricultural landscapes in the Anthropocene: nature's contributions to people, agriculture and food security. In: *Advances in Ecological Research*, vol. 63 (ed. D.A. Bohan and A.J. Vanbergen), 193–253. Academic Press.

Vidican, R., Pleşa, A., Mihăiescu, T. et al. (2022). The need to restore post-industrial ecosystems: importance and functions of riparian brownfields. *Bulletin of the University of Agricultural Sciences & Veterinary Medicine Cluj-Napoca. Agriculture* 79 (2): https://doi.org/10.15835/buasvmcn-agr:2022.0033.

Vitálišová, K., Murray-Svidroňová, M., and Jakuš-Muthová, N. (2021). Stakeholder participation in local governance as a key to local strategic development. *Cities* 118: 103363.

Vosloo, P. (2018). Post-industrial urban quarries as places of recreation and the new wilderness–a South African perspective. *Town and Regional Planning* 72: 43–57.

Walker, L.R. (1999). *Ecosystems of Disturbed Ground*. Elsevier.

Weiskopf, S.R., Rubenstein, M.A., Crozier, L.G. et al. (2020). Climate change effects on biodiversity, ecosystems, ecosystem services, and natural resource management in the United States. *Science of the Total Environment* 733: 137782.

Włodarczyk-Marciniak, R., Sikorska, D., and Krauze, K. (2020). Residents' awareness of the role of informal green spaces in a post-industrial city, with a focus on regulating services and urban adaptation potential. *Sustainable Cities and Society* 59: 102236.

Wu, F., Zhan, J., Su, H. et al. (2015). Scenario-based impact assessment of land use/cover and climate changes on watershed hydrology in Heihe River Basin of northwest China. *Advances in Meteorology* 2015: 1–11.

Wuana, R.A. and Okieimen, F.E. (2011). Heavy metals in contaminated soils: a review of sources, chemistry, risks and best available strategies for remediation. *International Scholarly Research Notices* 2011: 402647.

Yan, H. and Liu, G. (2021). Fire's effects on grassland restoration and biodiversity conservation. *Sustainability* 13 (21): 12016.

Young, T., Petersen, D.A., and Clary, J.J. (2005). The ecology of restoration: historical links, emerging issues and unexplored realms. *Ecology Letters* 8 (6): 662–673.

Zhang, M., Wang, K., Liu, H. et al. (2015). How ecological restoration alters ecosystem services: an analysis of vegetation carbon sequestration in the karst area of northwest Guangxi, China. *Environmental Earth Sciences* 74: 5307–5317.

Zhang, Y., Zhao, W., Chen, X. et al. (2020). Assessment on the effectiveness of urban stormwater management. *Water* 13 (1): 4.

Zhao, Y., Liu, Z., and Wu, J. (2020). Grassland ecosystem services: a systematic review of research advances and future directions. *Landscape Ecology* 35: 793–814.

Zukin, S. (2009). *Naked City: The Death and Life of Authentic Urban Places*. Oxford University Press.

3

Spontaneous Flora on Post-industrial Metalliferous Sites

Ksenija Jakovljević[1], Tomica Mišljenović[2], and Dragana Ranđelović[3]

[1] *Department of Ecology, Institute for Biological Research 'Siniša Stanković' - National Institute of the Republic of Serbia, University of Belgrade, Serbia*
[2] *Institute of Botany and Botanical Garden, Faculty of Biology, University of Belgrade, Serbia*
[3] *Institute for Technology of Nuclear and Other Mineral Raw Materials, Serbia*

3.1 Introduction

Intensive industrialization has driven the development of human society, but it has also led to severe pollution and depletion of natural resources. During these processes, many necessary facilities, infrastructures, waste products, and landfills have taken up significant areas of surrounding land, contributing to environmental pressure and landscape fragmentation. In recent decades, industrial globalization, transformation, and relocation of companies and industrial sectors have resulted in rising number of abandoned and degraded post-industrial sites that have shifted from natural and semi-natural to peri-urban and urban areas. Increasing demand for additional resources has led to the spread of industrial facilities into new areas where natural values are often still preserved, paving the way for further environmental degradation. The dependence of humanity on raw materials extracted from nature is a pervasive phenomenon. The mining industry and subsequent man-made structures cause serious long-term environmental impacts. Multiple land rehabilitation measures should be an integral part of the mining life cycle, and environmental impacts over time should be adequately assessed for the various environmental media in each case. Similarly, the post-industrial legacy must be reintegrated into the surrounding urban or natural landscape with appropriate multifunctional solutions that include the remediation of contaminated sites.

Post-industrial sites represent very specific and patchy habitats of degraded land that are often targeted by state legislatures for future restoration. The United Nations Decade for Ecosystem Restoration (2021–2030) calls for greater attention to this issue and the development of various approaches to halt ecosystem loss, minimize impacts, and reverse the degradation of global ecosystems. In this sense, ecological restoration of post-industrial sites based on native species and their potential to provide ecosystem services and halt

Biodiversity and Ecosystem Services on Post-Industrial Land, First Edition.
Edited by Vimal Chandra Pandey.

biodiversity loss should be further advocated and emphasized among scientists, stakeholders, and practitioners from different disciplines dealing with these types of sites.

In addition, the spontaneous flora of post-industrial sites could find its purpose among restoration practices that support most of the UN sustainable development goals (SDG), such as SGD 1 (no poverty – by using an economically oriented phytoremediation approach), 7 (affordable and clean energy – by using appropriate native species as energy crops), 11 (sustainable cities and communities – by addressing urban post-industrial sites), 13 (climate change mitigation – by providing increase of green areas, climate adaptation, and resilience), 14 (life under water – by reducing negative impacts on surface and groundwater and their biota), and 15 (life on land – by restoring and sustainably using terrestrial ecosystems and their services).

3.2 Definition and Types of Post-industrial Sites

Mining, which involves the extraction and processing of ores, is one of the main causes of soil degradation. It not only destroys the soil and underlying bedrock but also generates large amounts of waste material and tailings, which due to their quantity and instability (susceptibility to erosion) are a long-term source of pollution for the surrounding terrestrial and aquatic ecosystems (Figure 3.1; Jakovljević et al. 2021). Currently, more than 100 km^2 of various surface mining structures and mining waste disposal sites have been mapped worldwide (Maus et al. 2022), although the actual number is likely much higher. The unfavorable physical and chemical properties of such substrates limit plant growth and are the main reason that these areas are often bare for long periods. In addition to unfavorable, very low, or very high pH values and a general lack of nutrients, they are also characterized by high metal(loid) contents, whose total concentration and availability determine their toxicity. Phytoavailability depends not only on the concentrations and properties of the elements, i.e., the chemical form in which they occur, but also on the properties of the substrate (low pH, organic matter and clay as promoters of availability) (Corzo Remigio et al. 2020).

Products derived from metallic ores are one of the backbones of the technological progress of the 21st century society. These are limited resources whose exploitation and demand for raw materials are increasing every year. Due to the growing global demand for mineral resources, old deposits are rapidly being depleted and new deposits (very often of lower quality and on a larger spatial scale) are being discovered and mined. During the mining and processing of ores, the natural environment is impacted and severely damaged. As a result of the mining and processing of mineral resources, existing ecosystems are covered with the waste. Mining waste represents a significant portion of the total processed mass, which must be properly stored and managed. Globally, there are thousands of active mining waste disposal sites in the form of tailings and waste rock dumps, while the estimated total amount of solid waste generated from the primary production of mineral and metal products reaches 100 billion tons per year and is expected to increase in the future (Tayebi-Khorami et al. 2019).

Mining waste represents one of the largest waste streams in the world (Bian et al. 2012). Mining and processing of various metallic ores (such as copper, silver, gold, nickel, rare earth, platinum, lead, and zinc ores) generally produce three types of solid mining

Figure 3.1 Different types of post-industrial mining sites. (a) Sb-As-Cr tailings in Lojane, North Macedonia; (b) Flotation tailings deposition site at Pb-Zn mine Veliki Majdan, Serbia; (c) Smelter slag deposition site at Bor copper mine basin, Serbia; (d) Acid mine drainage lake at Bor copper mine waste sites, Serbia.

wastes: – mine rock overburden wastes, flotation tailings, and slags (Manca et al. 2021). The geochemical composition of these wastes is complex and depends on the geological setting, the present mineralization, and the mining and processing methods used (Lottermoser 2011).

In the opening of an ore deposit, rock that contains no ore or only a small amount of ore (usually less than 1%) is called barren rock and is transported to the overburden. The overburden is usually deposited in the form of cup-shaped piles that occupy larger areas and affect the surrounding microclimate and environment. The overburden piles consist of larger pieces of crushed rocks that have a coarse-grained composition, and their substrate, formed by the deposition of rocks from different depths, is often very heterogeneous. It usually contains increased amounts of metals released by the oxidation of rocks from deeper layers that have reached the surface of the mine waste piles. Weathering of mine waste rock can be a significant source of metals to the environment. Particular attention is given to the weathering of reactive minerals, especially sulfides, which are associated with the formation of acid mine drainage (AMD). This phenomenon is related to the oxidation of sulfide-rich minerals upon contact with the atmosphere, moisture, and certain types of bacteria, producing acidic solutions rich in metals. The formation of AMD is considered a serious environmental problem in the mining sector because it can severely contaminate surface and groundwaters and affect soils (Akcil and Koldas 2006). The potential for acid mine water generation depends on the amount and type of acid-forming and acid-neutralizing minerals in the overburden. Oxidation of acid-forming minerals (especially pyrite) leads to a decrease in pH and an increase in metal concentration (e.g., Fe, Mn, Cu, Zn, Cd, Cr, Pb, etc.) in the mine wastewater, especially at pH values $\cong 2$ (Saria et al. 2006). Increased potential for AMD formation in mine wastes tends to undermine natural revitalization processes (Lottermoser 2011). Similarly, neutral or alkaline drainage waters can cause the mobilization of metal(loid)s from mine wastes, resulting in the release of oxyanions and leaching of elements such as As, Sb, Se, and Mo (Vriens et al. 2019a).

Depending on rock characteristics, waste geochemistry, and atmospheric conditions, metals may remain in solution or be deposited in secondary phases during weathering of waste rock. Trace elements may precipitate as oxides, hydroxides, or carbonates, adsorbed to oxyhydroxides (Vriens et al. 2019b), or co-precipitate with other solid phases (Ranđelović et al. 2020a). The total metal concentration in a mine waste is usually not representative of its release potential. The chemical phase in which trace metals are present determines how available they are for release to the environment and uptake by biota.

Each mining waste site is a special case in terms of its characteristics and potential impact on the environment, and specific studies are needed to determine which elements pose the greatest risk to the environment and which factors determine biogeochemical processes. The studies by Lin et al. (2022) on tailings at Dexing, China, the largest copper mine in Asia, found that Cu, Cr, and As are the major elements released from the rock material. While As was predominantly placed in the residual fraction (over 96%), Cu and Cr were predominantly in the acid-soluble and oxidizable fraction (up to 29 and 11%, respectively). Consequently, the main factors affecting the mobility of Cu and Cr were pH and the presence of acidophilic bacteria. Analysis of the mobility of metals in tailings from the Pb-Zn mine deposits of Sidi Driss-Tamra, Tunisia, revealed that an increased concentration of Zn (18.6%) and Mn (11.6%) were released into water-soluble fractions, while Cu and Pb were mainly present in the mobile phase extracted with the chelating agent Na_2-EDTA (21 and 51.6%, respectively) (Trifi et al. 2018). The complex mineralogical paragenesis as well as the severe climatic conditions were recognized as important factors affecting metal speciation and mobility in the studied medium.

After the ore is separated and extracted by the froth flotation method (used to separate the mineral from the gangue), the remaining material is transported in the form of a suspension or paste to the specially designed basins. These features are known as flotation residues and pose an increased risk to the environment during or after the completion of the disposal process. Under normal operating conditions, they threaten the land on which they are located, the surface and groundwater in the watershed, and the air through the effects of aeolian erosion. When deposited in the active tailings pile, the flotation material with the potential to form acid mines should be water-saturated to minimize oxidation of the sulfide minerals. Dold (2014) study showed that pH decreased over time from alkaline to acidic conditions and that a transition from reducing to oxidizing conditions occurred. In addition, important cations and anions present in the flotation debris accumulated on the surface of the tailings due to capillary transport and evaporation processes. Studies on the bioavailability of metals from the flotation residues of the Almagrera sulfides (pyrite belt of the Iberian Peninsula, Spain) revealed an increased content of Mn, Pb, and As in the mobile fractions (81, 74, and 43%, respectively), as noted by Álvarez-Valero et al. (2009), indicating an increased potential for mobilization of the detected contaminants in the environment. During the generation of acid mine drainage, these elements are readily released, while part of the mobile fraction of soluble secondary minerals precipitates, and becomes mobile again during rainy periods.

The economically valuable ore obtained by the flotation process is melted down in a suitable smelter and then further processed. As a by-product of the metallurgical processes, metallurgical slag is generated from the melted ore. It is usually deposited in specially designed slag dumps, while in certain cases it may be co-deposited with the overburden to reduce the volume and surface area of the landfill. Metallurgical slags vary in chemical and phase composition depending on the type of ore processed, the type of furnaces used, and the cooling method. The composition of ferrous slags (produced during the extraction of Fe from ores) is dominated by Ca and Si, while the composition of non-ferrous slag (produced during the extraction of non-ferrous metals from ores) is usually dominated by Fe and Si with some amounts of Al and Ca (Piatak et al. 2015). Metal concentrations are lower in ferrous slags than in non-ferrous ones, which produce acidic leachates and may release larger amounts of metals during weathering.

Coal mining continues to be a growing extraction industry, producing over eight billion tons in 2022. The production of 1 ton of coal generates about 0.4 tons of waste and in the leading coal mining the United States, Russia, China, India, Germany, and others mining waste accounts for about 46% of annual waste production (Fecko et al. 2013). Depending on the operating stage of the coal mine, there are four main types of generated waste: mine waste coal spoils, coal tailings, coal sludge, and fly ash deposits (Figure 3.2). The waste coal spoils consist of various igneous and sedimentary rocks and can sometimes have high carbon content, often exceeding 20% (Yong et al. 2022). These types of deposits are susceptible to metal leaching and the generation of AMD associated with the distribution of sulfate-bearing minerals in the rocks, similar to metal-bearing tailings. Coal tailings consist of processed rock, with most particles less than 0.25 mm in size and representing a fine-grained, heterogeneous mixture of various solids (20–35% w/w) suspended in contaminated water (Wang et al. 2014). Depending on their properties, certain tailing materials can be used as a component of energy fuels or for the recovery of selected metals. The addition of alkaline, loamy coal tailings containing 3% macronutrients, high carbon content, and low metal concentration is beneficial for soil improvement and had a positive effect on tomato plant germination (Yong et al. 2022). Coal sludge is produced during coal washing and contains coal dust, chemicals, and metals such as As, Cr, Cd, B, Hg, and Ni (Śliwka et al. 2022). The introduction of new technologies for sludge treatment has reduced the production of coal sludge. Existing coal sludge has been studied as a potential incineration material, as a component in the production of artificial aggregates and mixtures for the reclamation of degraded land (Li et al. 2021; Adamczyk and Pomykała 2022). After coal combustion in power plants, fly ash and bottom ash are produced as non-combustible residues. They are transported to specially constructed ash ponds and deposited in the form of sludge. These deposits are known as fly ash and are the most common byproduct of the

Figure 3.2 Kostolac coal basin in Serbia. (a) Coal mining and waste deposition sites; (b) Coal sludge deposition site with isolation barrier; (c) Fly ash deposit site.

coal combustion process. Their generation is estimated at 600 million tons per year worldwide (Pandey 2020a). Fly ashes are usually alkaline and fine-grained and consist mainly of oxides such as SiO_2, Al_2O_3, Fe_2O_3, CaO, MgO, TiO_2 (Vilakazi et al. 2022). The chemical composition of fly ash depends on the mineralogy of coal, combustion method, combustion temperature, etc. In addition, fly ash is enriched with pollutants such as metals (e.g. As, Cd, Hg, Pb, Fe, Mn, Ni, Mo, Cu, Zr, Co), radioactive elements (U, Ra, Th), and organic impurities (polycyclic aromatic hydrocarbons, polychlorinated dibenzo-p-dioxins, polychlorinated biphenyls, and others) (Jambhulkar et al. 2018; Kursun Unver and Terzi 2018). The varying content of metals in fly ash is a consequence of their varying concentrations in coal. The average values reach up to 100 ppm in most coal types worldwide, except for Fe and Mn, whose concentrations are higher (Czech et al. 2020). About 25% of fly ash from coal is used worldwide in processes such as aggregate production, mine backfilling, adsorption of wastewater pollutants, and soil improvement (Pandey 2020a). The remaining fly ash remains unused in constructed deposit sites and represents a potential environmental risk, but also a potential habitat for a variety of tolerant organisms.

Various types of mining waste pose a direct threat to the environment. They are a source of dust to surrounding areas, while atmospheric water runoff from their surfaces contaminates surrounding soils and waterways, introducing toxic substances into ecosystems and food chains. Other chemical and physical pollution that may result from mining operations include the continuous generation of contaminated mine water, various types of mine facilities collapsing (including the failure of tailings dams), reduction in water and soil quality due to ground subsidence, etc.

Recent trends in waste recycling, reduction of waste generation and waste streams, and the transition to a circular economy and low-carbon technologies require a reduction in the global production of mining waste, as well as the recycling, reuse, and recovery of useful components from the generated waste. Because the amount of waste generated in the mining industry is large and it contains a significant amount of ores and useful elements, research and technology efforts are focused on developing efficient technologies for its use and reduction. In addition, future volumes of mining waste should be lower and less toxic, increasing the prospects for permanent remediation and the successful development of a sustainable plant cover.

Areas used as quarries also experience numerous environmental degradations, such as soil and rock destruction, alteration of terrain morphology, air and groundwater pollution, and increases in erosion tendency (Zuquette et al. 2004), all of which lead to destruction and alteration of habitats, followed by loss of biodiversity (Figure 3.3a). In addition to dust and particulate matter emissions, the environmental impacts of quarrying stone, sand, gravel, or shale can include various geological hazards and metal pollutants (Wang et al. 2018). Certain types of quarries, such as serpentinite or granite quarries, are susceptible to metal contamination due to the presence of elements such as As, Co, Cr, Fe, Mn, and Ni in environmentally soluble fractions from rock material (Lago-Vila et al. 2015; Arenas-Lago et al. 2016). However, the level of contamination is generally lower than at metalliferous mining sites. This is confirmed, for example, by the research of Adabanija and Oladunjoye (2014), which found a lower Integrated Pollution Index and environmental impact for granite, sand, and laterite quarries compared to tantalite, beryl, and tourmaline

Figure 3.3 Post-industrial sites in Serbia. (a) Active quarry site in Serbia; (b) Municipal landfill site, Bor, Serbia.

mining sites in Nigeria. Stone exploitation results in the complete removal of the original surface, which abruptly changes the original ecosystems and the configuration of the site. Clean, rocky surfaces are challenging, xerothermic habitats for future revegetation. Some of the major limiting factors include substrate texture, compaction, unfavorable water-holding capacity, nutrient deficiencies, the presence of metals, and others. Specific methods that allow restoration of such sites include control blasting to create less steep slopes and additional fractures that allow better moisture retention (Jim 2001).

Brownfields, especially those associated with urban and industrial areas, pose a significant environmental problem as the soil is modified by industrial waste of various types and toxicity. Considering the presence of various pollutants, including hazardous substances, the reuse of such wastes is often very demanding, costly, and involves great risks. This is the reason why they usually remain abandoned for a very long period (Song et al. 2019). The characterization of brownfields for remediation and rehabilitation purposes is often based on site-specific conditions (size, use, contamination level) and the contextual environment (land use, market development, urban development, etc.) (Dasgupta and Tam 2009). Contamination, especially with metals, is one of the most important environmental problems of brownfields. As the expansion of urbanization leads to the integration of more and more brownfields from suburban areas into cities, the reuse and remediation of such land, especially through nature-based solutions, is gaining increasing attention (Song et al. 2019). The nature of anthropogenic influence and the disturbance regime of brownfields affect their potential function as habitat. A metal contamination survey of 625 industrial brownfields in China revealed elevated levels of Pb, Cd, Zn, Ni, and As, with As being the dominant contaminant (Peng et al. 2022). In addition, sites influenced by chemical production, electrical machinery manufacturing, and metal product manufacturing had the highest contaminant levels, indicating an increased health risk from metal exposure. Investigation of metal concentrations in brownfield sites in Cleveland, USA, found contamination with Cd, Cr, Cu, and Pb above regulatory cleanup levels, which may be attributed to the industrial past of the area (Jennings et al. 2002). Assessment of metal bioavailability in brownfield sites in Wolverhampton, UK, revealed complex metal speciation related to the historical use of the site. Soils enriched in Zn had its elevated concentrations in environmentally mobile phases, while soils enriched in Cu, Ni, and Pb had their pre-available concentrations in residual phases (Thums et al. 2008).

Landfilling is the most common waste disposal system, despite ever-improving recycling techniques. Depending on the type of waste, three main groups of landfill sites can be distinguished: municipal waste, industrial waste, and both municipal and industrial waste (Kim and Owens 2010; Figure 3.3b). While municipal waste landfills are primarily characterized by organic contamination, the origin of contamination in industrial and combined waste landfills is much more diverse and in some cases even hazardous in nature. This is especially true for illegal landfills without any control mechanisms. Landfills tend to produce leachate containing elevated concentrations of organic and inorganic pollutants as well as metals. The characteristics of leachate depends on the type of waste deposited, the rate of decomposition, hydrological and climatic conditions, etc. Produced landfill leachate is a serious potential source of surface and groundwater pollution in the affected area, especially in the case of illegal/non-sanitary landfills that do not have an insulating layer between the deposited waste and the underlying soil. Landfills pose a constant environmental threat to their surroundings, as waste decomposition, leachate generation, and landfill gas production can continue for several decades after the landfill is officially closed. The main metals detected in leachate from landfills are usually As, Cd, Cr, Hg, Cu, Zn, and Pb. Their concentration tends to be higher in the early stages of leachate formation due to acidification caused by the production of organic acids during hydrolysis and microbial degradation of biodegradable wastes (Naveen et al. 2018; Iravanian and Ravari 2020). The analysis of leachate pollution from municipal solid waste at the global scale showed that the concentration of metals (especially Cd and Pb) in landfill leachates had the highest values in areas with dry tropical climates, except for Fe, which had the highest values in landfills with dry temperate climates (Ma et al. 2022). The study by Chaudhary et al. (2021) on the change in the pollution index of leachate from landfills in Delhi, India, showed an increase in metal content in the time series, indicating the presence of various e-wastes, and an exponential increase in the case of Cr, Zn, and Pb. Soil quality and metal content were investigated at fifty-six sanitary and non-sanitary landfills in China, and Cr was identified as the dominant metallic pollutant with a pollution index of 44.3 (Wang et al. 2022). Significant differences were found in the concentrations of As, Cr, Pb, and Zn in soil samples from sanitary and non-sanitary landfills, with higher values for non-sanitary landfills. It was also found that the type of landfill, age, and precipitation are important factors contributing to the migration of metals in landfill soils.

3.2.1 Impact of Post-industrial Sites on the Environment

Post-industrial sites have a variety of environmental impacts, especially when abandoned or inadequately managed. They affect directly and indirectly the atmosphere, hydrosphere, pedosphere, and biosphere in the surrounding and modify their quality and function. Multidisciplinary efforts are underway to identify the conditions, mechanisms, and most appropriate technologies for remediating and protecting such impacted sites.

One of the biggest environmental disasters in the mining industry are the failure of mine dams, which leave a large ecological footprint and destroy surrounding areas. Even after remediation, erosion, remobilization, and transport of metals and their transfer into food chains continue over the long term, threatening ecosystem services and human health. The collapse of the tailings dam at the Aznalcóllar mine in Spain directly contaminated more than 4000 ha of soils in the valleys of the Guadiamar and Agrio rivers with sulfidic tailings sludge. Initially, severe contamination of soils with elements such as As, Cd, Cu, Pb, Tl, and Zn was reported (Cabrera et al. 1999), while soil quality studies after sludge removal revealed elevated concentrations of Cu, Pb, Zn, As, Co, Cd, Hg, and Sb above background levels, exceeding those suitable for cultivation in the case of As (Galán et al. 2002). Additionally, As, Cd, Cu, and Pb concentrations were found above toxic levels in *Cynodon dactylon* and *Sorghum halepense* from contaminated soils covered with sludge, while Cd concentration reached toxic levels in plant tissues on remediated sites (Madejón et al. 2002). Moreover, the increased concentration of As, Cd, Cu, Tl, Sn, and Pb in the population of the lizard *Psammodromus algirus* from contaminated sites confirmed the persistence of the main pollutants in the trophic network almost ten years after the accident (Márquez-Ferrando et al. 2009). Similarly, the study by Ranđelović et al. (2020a) on the mobility of elements in the affected sediments and soils three years after the tailing spill at the Stolice antimony mine site (Western Serbia) revealed that the content of As, Cd, Pb, and Sb as the main pollutants was strongly related to the content of the sand fraction, while Fe-Mn oxides were the main sinks for the toxic elements. The predominant native plant colonizing the Stolice tailings, *Equisetum ramosissimum*, showed the ability to uptake and phytostabilize the available fractions of Fe, Mn, Zn, Pb, and Sb in its roots (Ranđelović et al. 2019).

The phenomenon of AMD is also an environmental threat with potentially large and long-term impacts on the environment and its components. Efforts are constantly being made to mitigate these impacts through diverse treatment methods and technologies, with varying degrees of success. The impact of AMD on soils and sediments near the polymetallic Dabaoshan mine in China showed changes in pH and metal content, with increased potential ecological risk for Cd, Cu, Pb, As, and Zn. The mobilization potential of Cd in soils affected by acid mine water was high, as ~40% of the total Cd content was found in the extractable fraction (Luo et al. 2020). In addition, Cd in this area showed an increased transfer rate from soil to edible plants (0.12–0.25), representing a risk to human health (Liao et al. 2016). Furthermore, the affected soils were found to be unsuitable for growing edible crops. The median total cancer risk exceeded USEPA thresholds, particularly for Cd and As, indicating an elevated carcinogenic risk to area residents.

Quarrying activities can also result in the release of metals into the environment by contaminating dust and soil and polluting runoff water and existing vegetation. Peña González et al. (2006) found a strong correlation between suspended soils and mobilized As, Cd, Cr, Fe, Mn, and Pb in effluents from granite quarries in Galicia, Spain. Similarly, Rodríguez-Seijo and Andrade (2017) found elevated levels of Co, Cr, and Ni in soils from serpentine quarries in Spain. Tree plants in the vicinity of the Ishiagu granite quarry in Nigeria showed elevated content of Cd, Pb, and Zn in soils and plant leaves, whereas soil Cd concentrations were significantly higher compared to average concentrations in soils around the world (Ogbonna et al. 2011).

Landfills are inevitably linked to urban and industrial development and growing waste production on a global scale. They are also considered a significant environmental risk, affecting the soil, air, and water in the surrounding areas. Many studies have been conducted to quantify their impact on the environment. Metal pollution in the vicinity of open dumpsites in Sylhet, Bangladesh, was studied by Alam et al. (2020). The metal content in the surrounding soils showed a trend: Fe > Mn > Pb > Zn > Cu > Cd and moderate loading of Pb and Cd. On the other hand, the metal content in the plants was within the FAO permissible limits, with the highest transfer rate from soil to plant recorded for Cu (120) and the lowest for Fe (0.77). Ecological risk assessment of contaminants in soils in the vicinity of the landfill in Shannan, Tibet, revealed a low-risk level for As, Cu, Ni, Pb, and Zn, and an elevated ecological risk level for Cr and Hg, which may have been related to the presence of e-waste (Zhou et al. 2023). The investigation of water, sediment, and soil quality after the closure of a landfill in northern Lithuania showed that the effects of the landfill were still present, as the levels of As, Cu, Cr, Ni, and Pb were elevated in the surrounding waters. Although the soil contamination coefficient did not indicate contamination with any single element, the total soil contamination factor reached dangerous levels (Baziene et al. 2020), indicating the need for long-term monitoring of metal dispersion. The study of metal content in *Tanacetum vulgare* growing in an urban landfill in Štěpánovice, Czech Republic, revealed the highest transfer from soil to plant tissue for Cd (between 1.4–11.08 in roots and 1.3–14.9 in leaves; Adamcová et al. (2017)). The total content of metals in the tissues of *T. vulgare* was in the normal range for Cd, Pb, and Mn, and was slightly increased in the case of Cu, Ni, and Zn (Kabata-Pendias 2010).

3.3 Spontaneous Flora: Characteristics and Benefits

3.3.1 Definition and Types of Spontaneous Flora

An important aspect of the restoration of anthropogenically disturbed sites, including post-industrial sites, is the occurrence of spontaneous flora, whose composition, ecological characteristics, and rate of succession indicate the overall success of the revitalization processes (Prach et al. 2001). This type of flora can be synanthropic or non-synanthropic, depending on how it is associated with man-made habitats. The dominance of non-synanthropic flora is mostly related to the undisturbed areas characterized by autochthonous flora, with low participation of species of alien origin (Kotanska et al. 2015). On the other hand, synanthropic flora is strongly associated with anthropogenically influenced habitats (Wittig 2004). Depending on their origin, two main groups can be distinguished: apophytes (native species) and anthropophytes (anthropochorous or introduced species, i.e., species of foreign origin), which can be archaeophytes (introduced before 1492, i.e., the discovery of America) and neophytes (or kenophytes; introduced after 1492), depending on the period of introduction (Essl et al. 2021). While the occurrence of neophytes is usually accurately documented due to their recent arrival, archaeophytes are often difficult to distinguish from native flora (Wittig 2004). Previous studies show that the general trend is a decrease in the number of archaeophytes (Pyšek et al. 2005; Eichenberg et al. 2021), while the number of neophytes either remains constant if they do not form stable communities after invasion (Pyšek et al. 2004), or increases, which is more common, especially if a somewhat longer period is observed (Chocholoušková and Pyšek 2003; Eichenberg et al. 2021).

3.3.2 Adaptability and Resilience of Spontaneous Flora

Numerous plant characteristics can influence the success of the restoration process, and life form is one of the most important variables. A particularly high participation of therophytes has been observed at severely disturbed sites. Due to numerous adaptations, species with this life form are able to maintain their populations at disturbed sites despite numerous habitat and condition constraints that often change significantly over time. The root of the therophyte strategy is to invest in the production of a large number of small seeds rather than a perennial character. On the other hand, a decline has been observed in chamaephytes, phanerophytes, and geophytes. Different types of responses have been found in hemicryptophytes. Namely, the erect forms are particularly sensitive to site disturbance, while the number of plants with flat rosettes has not changed significantly (McIntyre et al. 1995).

The study of the vegetation that developed on municipal waste and its succession over 13 years shows that the number of therophytes is significantly higher in the early stages than in the later stages (Huber-Humer and Klug-Pümpel 2004). A similar decline in therophyte participation has been observed in coal mine waste pile vegetation in China (Li et al. 2008). Due to the demanding conditions required for the development of phanerophytes, this life form is considered ecologically luxurious, but significant tree involvement has been noted on the metalliferous spoil heaps, which are among the most extreme of the types studied. Thus, numerous phanerophytes have been intensively studied for metal (hyper)accumulation, especially those from the genera *Salix*, *Populus*, *Ulmus*, etc. (Chang et al. 2005; Mleczek et al. 2018; Suo et al. 2021).

In addition to producing large numbers of viable seeds, the advantageous characteristics of pioneer species include anemochory as a dispersal mechanism and high tolerance of plants to harsh substrate conditions that allow them to take root undisturbed (Giannini et al. 2017). Anemochory allows species to be independent of other species in seed dispersal (Cusser and Goodell 2013), which has been particularly important in recent years when climate change has caused disruptions in the life cycle stages of many organisms (Memmott et al. 2007). One of the beneficial characteristics of spontaneous flora is also clonal growth. This allows the plants to form dense stands and outcompete other species in forming stable communities, as has been found with *Solidago gigantea* (Jakobs et al. 2004) and *Calamagrostis epigejos* (Ranđelović et al. 2020b).

Of particular importance is the ability of spontaneous flora to adapt their life cycles and flower earlier in response to stressful conditions (Grime 1993). Their great ecological plasticity not only allows them to cope with the harsh characteristics of these sites, which is why they are the first and only species to appear in severely degraded areas, but also gives them a competitive advantage over native species. Considering that these species take advantage of opportunities where there is less competition from environmental stressors (various types of disturbances), they emerge and generally complete their development quickly after seeding. Because of this strategy, these species are also referred to as "opportunists" (Trueman et al. 2022). Although these species may pose a threat to native flora due to their competitive nature, they not only increase the richness and diversity of the flora, but also initiate the process of succession in biologically empty spaces, allowing the entry and survival of native flora and creating more stable coenotic relationships (Ranđelović and Jovanović 2023).

There are also numerous physiological adaptations that allow spontaneous flora to colonize post-industrial sites. Although adaptations vary by habitat, plants most often cope with nutrient deficiencies, a lack of water, an unfavorable thermal regime, as well as elevated levels of metals and metalloids. Mycorrhiza has been shown to be of great importance for plants to mitigate the stress caused by habitat conditions (Carrenho et al. 2007). Mycorrhizal fungi have been shown to interact with the majority of plant species (more than 80%; Leung et al. 2013). The fungal hyphae not only bind metals and metalloids, reducing their availability to plants (Bąba et al. 2016), but also absorb water, reducing the water deficit common in such habitats, increasing uptake of N, P, K, and micronutrients, resistance to pathogens, etc. (Entry et al. 2002; Smith and Read 2010; Leung et al. 2013; Bąba et al. 2016). Due to their ability to survive in extremely unfavorable habitats, including excessive metal and metalloid concentrations, mycorrhizal fungi make an important contribution to the phytoremediation process. They increase plant tolerance to soil toxicity while allowing hyperaccumulating species to grow faster and produce greater biomass (Leung et al. 2013).

3.4 Spontaneous Flora in Post-industrial Sites: Current Findings

The flora of post-industrial sites, although differing to some extent according to the type of habitat, has some common features, especially in the dominance of species of certain families, life forms, and distribution patterns. The floristic analysis of coal mine sedimentation basins in Poland showed the dominance of several families: mainly Asteraceae, but also families of grasses in a broader sense (Poaceae, Cyperaceae, Juncaceae) and Fabaceae (Kompała-Bąba et al. 2022). The same is true for the spontaneous flora in ultramafic post-mining sites in Lower Silesia, where, in addition to the Asteraceae as the most dominant family, species from the Poaceae, Fabaceae, and Caryophyllaceae families occur in large numbers, with *Achillea millefolium*, *Artemisia vulgaris*, *Calamagrostis epigejos*, *Arrhenatherum elatius*, *Lotus corniculatus*, etc. as the most dominant species (Koszelnik-Leszek et al. 2013; Kasowska and Koszelnik-Leszek 2014). The significant contribution of Asteraceae to the taxonomic spectrum of spontaneous flora in degraded areas is mainly due to the large proportion of the family in general, which includes more than 23 000 species within about 1600 genera (Simpson 2010), as well as a large number of non-native species that easily adapt to different habitat conditions. A high proportion of grasses is also characteristic of habitats that have been modified by human activities, such as post-industrial sites. This is due to the high adaptability of these species, as well as their exceptional diversity resulting from the easy formation of hybrids and new forms (Rostanski and Wozniak 2001). One of the advantageous characteristics of grasses in overgrowing post-industrial sites is a well-developed root system that covers the substrate and prevents erosion. Soils at post-industrial sites tend to have a rough texture and have discontinuities in the soil layers, and the roots of plants, especially grasses, help to stabilize them. Although species composition varies depending on the type of post-industrial site, some species with a particularly large ecological amplitude were found on each of these sites. These include, for example, *Agropyron repens*, *A elatius*, *C. epigejos*, *Dactylis glomerata*, *Lolium perenne*, *Poa annua*, etc. (Rostanski and Wozniak 2001; Ranđelović et al. 2020b). Species from the Fabaceae family also play a particularly important role in the overgrowth of post-industrial substrates with respect to their pronounced nutrient deficiency. Indeed, legumes are known to bind these nutrients to the substrate in reaction with nitrogen-fixing bacteria, leading to their reaccumulation and improvement of N content and overall substrate characteristics over time. The species most commonly found on post-industrial sites are *Medicago lupulina*, *L. corniculatus*, *Trifolium pratense*, *T. medium*, etc. (Vidican et al. 2022).

3.4.1 Spontaneous Flora on Mine Waste Sites

The toxicity of metals and metalloids, usually present in high concentrations, makes mine waste deposits as one of the most extreme types of post-industrial sites. The persistence of metals and metalloids accumulating in the substrate poses the greatest problem to all living organisms associated with the contaminated areas, as well as a potential health risk to humans, who are the last step in the food chain. Conventional remediation of these sites based on physico-chemical processes proved to be costly as well as time and space-consuming, making phytoremediation an effective and environmentally friendly way to clean and neutralize the contaminants. Although the process of phytoremediation involves the controlled planting of contaminated areas, the effects of spontaneously occurring plants on such areas should not be neglected, as this flora is the more appropriate way of remediation in many respects (Figure 3.4). For example, the species *Baccharis trimera*, *Senecio leptolobus*, and *S. brasiliensis* growing spontaneously at a polymetallic site in Brazil were shown to efficiently absorb Cd, Cu, Zn, and Pb (Boechat et al. 2016). *Calamagrostis epigejos*, a clonal grass commonly found as a pioneer plant on overburdened mine waste, has been shown to have some potential for phytoremediation. This species takes up a large portion of extractable metal concentrations from the soil, but stores them in the root, making it a good candidate for phytostabilization (Ranđelović et al. 2020b). *Phragmites australis*, a cosmopolitan plant species that also belongs to the Poaceae family, grows spontaneously in mine tailings ponds. Similar to *C. epigejos*, this species has good potential for phytostabilization, but also for rhizofiltration, containing a significant amount of accumulated Cu

Figure 3.4 Spontaneous flora in various post-industrial metalliferous sites in Serbia [(a and b) Fly ash disposal site in Kostolac, Serbia, and (c and d) Overburden mine waste site, Bor, Serbia].

and Pb (Prica et al. 2019). In search of previously unknown candidates for phytoremediation, an intensive field study was conducted in three different mining sites in Morocco. Of the total 46 taxa discovered, eight took up the elements Cd, Cu, Pb, and Zn, which were present in excessive concentrations in the substrate, at concentrations that make them potentially useful for phytostabilization. These species are: *Citrullus vulgaris*, *Forsskaolea tenacissima*, *Hammada scoparia*, *Hirschfeldia incana*, *Lactuca viminea*, *Lycium intricatum*, *Portulaca oleracea*, and *Stipa capensis* (Midhat et al. 2019). The study of flora in six extremely contaminated spoil heaps in Spain, belonging to the former Pb/Zn and Hg/As mines, revealed a total of 79 taxa, most of which (63 taxa) were detected on alkaline substrate in the Hg-As spoil heap. The study on the Pb/Zn-rich tailings revealed that 22 taxa occur at these sites, with most of them preferring acidic substrates, while only six species were found in both types of spoil heaps – *Erica arborea*, *E. cinerea*, *Festuca rubra*, *Pteridium aquilinum*, *Rubus ulmifolius*, and *Ulex cantabricus* (Fernández et al. 2017).

Because of their association with sites rich in metals or metalloids, some of these species are considered indicators of metalliferous sites. This is especially true for the species *Minuartia verna*, one of the best-studied metallophytes in the genus *Minuartia*, mainly because of its tolerance to elevated concentrations of various elements and, at the same time, the wide distribution of the species (Baker and Proctor 1990; Jakovljević et al. 2022a). *Noccaea* species, especially *N. caerulescens*, are species known by their association with metalliferous sites, which are mainly rich in Ni, Zn, and Cd (Gonneau et al. 2014; Jacobs et al. 2018; Mišljenović et al. 2018, 2020; Kozhevnikova et al. 2020). *Noccaea caerulescens*, a hyperaccumulator of Ni, Zn, and Cd, occurs on post-industrial (post-mining) sites, especially on calamine sites rich in Zn, Pb, and Cd. In addition, this species inhabits ultramafic sites (rich in Ni), but also those non-metalliferous sites (Reeves et al. 2001), both in post-industrial and "natural" or "undisturbed" areas.

Of particular interest are the species that spontaneously overgrow abandoned mine sites rich in extremely toxic metals and metalloids such as Tl and As. An example of this is Allchar, the abandoned mine in North Macedonia, which is unique in the world due to its extremely high As and Tl concentrations (Boev et al. 2012). Studies have shown that despite the extremely high concentrations of these two elements in the substrate (142 $g\,kg^{-1}$ As and 18 $g\,kg^{-1}$ Tl; Đorđević et al. 2021) this area is inhabited by a considerable number of species, some of which are endemic (e.g. *Centaurea kavadarcensis*, *Thymus allchariensis*, *Onobrychis degenii*, *Viola allchariensis*, *V. arsenica*, etc.; Bačeva Andonovska et al. 2021). Given the highly specific substrate characteristics, the presence of edaphic endemics is certainly not surprising. Some of these taxa, such as three members of the genus *Viola* (two endemic – *V. allchariensis* and *V. arsenica* and one widespread *V. tricolor* subsp. *macedonica*), not only successfully overgrow this area spontaneously, but also take up Tl in concentrations exceeding the notional hyperaccumulation threshold (Jakovljević et al. 2023). At the same time, high concentrations of As and Tl in the substrate were also detected in an As tailings pile in Yunnan (SW China), where the species *Pteris vittata*, a known hyperaccumulator of As that grew spontaneously in this area, also hyperaccumulated thallium (Wei et al. 2020).

The spontaneous flora of post-mining sites often consists of seedlings of woody species that have been shown in many studies to tolerate elevated concentrations of metals and metalloids in the substrate very well. Such species belong to the genera *Salix*, *Populus*, and *Betula* (Chang et al. 2005; Mleczek et al. 2018; Mataruga et al. 2020). *Betula pendula* is one of the pioneer species that is highly tolerant to various environmental stresses such as metal excess, nutrient deficiency, and low temperatures, which is why it is often found on highly contaminated sites and in temperate and boreal forests (Sitko et al. 2021). Because of its plasticity, the species can often be used to restore degraded sites.

Studies of parasitic plants occurring spontaneously on post-industrial sites provide important information for studying metal(oid) uptake from this specific substrate. Previous studies have shown that species of the genus *Orobanche*, the most numerous holoparasitic genus of the family Orobanchaceae, when growing on sites with excessive metal concentrations, reduce metal concentrations in host plants by taking up metals into their tissues. For example, the polyphagous *O. laxissima* from the Caucasus, when parasitizing *Fraxinus angustifolia* and *Punica granatum*, reduces concentrations of these metals in host plants thanks to its potential to take up Ni, Cd, and Zn (Piwowarczyk et al. 2021). A similar finding was made for *O. lutea* at sites with excessive metal concentrations (Pb, Zn, and Cd), where parasitism led to reduced uptake by *Medicago falcata* as a host plant and this metal distribution resulted in lower toxic effects in both species (Turnau et al. 2018).

One of the well-studied overburden mine waste sites are the coal mine spoil piles, which are colonized by different plant species depending on the conditions. The greatest differences have been found in mine spoil/substrate pH, which is often extreme, ranging from very acidic to very alkaline, greatly affecting the type of flora that develops. At highly acidic sites, the flora is represented by cryptogamous species, such as lichens (e.g., *Cladonia* spp.). At less acidic sites, cryptogams are replaced by vascular plants, e.g., communities with acidophilic species such as *Galium saxatile*, *Calluna* spp. *Erica* spp., and with calciphilic *Linum catharticum* and *Carlina vulgaris* (Trueman et al. 2022). Coal mine spoil heaps are often colonized by various grass species, such as *C. epigejos*, *P. australis*, *Festuca arundinacea*, *F. rubra*, etc. Their common feature is high abundance, and often these species form almost monodominant stands that do not allow the invasion of other species, which negatively affects plant richness and diversity (Błońska et al. 2019).

Fly ash disposal sites pose a major environmental challenge, mainly because of the possible release of metals and metalloids into the environment, so revegetation of such areas could be of great benefit. One of the species that grows successfully in fly ash basins is *Typha latifolia*, a species that has already proven to be very successful in wastewater treatment due to its high accumulation potential (Ciria et al. 2005). At fly ash sites, this species can also be successfully used in ecological engineering, a process where a sustainable system of the natural environment and human society is formed, benefiting both (Pandey et al. 2014). The grasses such as *A. elatius*, *C. epigejos*, *C. dactylon*, *D. glomerata*, *F. rubra*, *Phleum pratense*, *Triticum aestivum*, etc. also proved to be very important in revegetation process, as shown also for legumes (*T. pratense*, *T. repens*, *Medicago sativa*, *Melilotus officinalis*, etc.), species particularly important in respect of nitrogen fixation (Gajić et al. 2019). All these species are not only able to tolerate the toxicity of the substrate, but also successfully cope with other challenges of such habitats, such as nutrient deficiency, high-temperature fluctuations, water deficit, intense radiation, etc.

3.4.2 Spontaneous Flora on (Metalliferous) Quarries

Studies of spoil heaps of active and abandoned marble quarries in the Apuan Alps of Tuscany (Italy) revealed the presence of 125 taxa (Gentili et al. 2011). Due to extremely unfavorable abiotic characteristics, the plant cover is patchily distributed, and almost half of the species were detected only once or twice. The most frequently detected species include the thermophilic species *Brachypodium genuense* and *Festuca ovina* s.l. from the Poaceae family and the woody *Ostrya carpinifolia*. Although the studied area is located in a humid climatic zone with only one dry month, species from dry habitats dominate due to site characteristics (water deficit). The predominant life form is therophyte, which is consistent with the fact that annual plants are most common in degraded areas, especially in heavily modified quarries.

Considering the constraints in quarries, often physical rather than related to metal excess, Lago-Vila et al. (2017) investigated differences in responses of plants grown spontaneously in quarries and abandoned Pb/Zn mines. The tolerance of *F. ovina* and *Cytisus scoparius* as spontaneously grown species was analyzed and additionally compared with *Sinapis alba* and *Brassica juncea*, the species with previously confirmed tolerance to metalliferous sites. As expected, soil from abandoned mine sites proved to be more toxic for the studied species, especially for *C. scoparius*, and *F. ovina*, whereas *S. alba* and *B. juncea* showed significantly weaker responses to the increased toxicity of the substrate. The response of plants to the metalliferous substrate depends not only on the metal concentration itself, but also on substrate properties such as reaction and organic matter content.

One of the problems faced by plant species in quarries is related to dust exposure, which is mainly determined by the intensity of activities and concentration of particles, their size and chemical properties, microclimatic conditions, etc. To evaluate the response of plants to quarry activities, the spontaneously growing flora in Lokpaukwu in Nigeria was analyzed, namely the species *Baphia pubescens*, *Vitex doniana*, *Alchornea cordifolia*, *Nauclea latifolia* and *Napoleona imperialis* (Ogbonna et al. 2020). It was shown that dust accumulation is significantly higher in the quarry compared to the control and that it is influenced more by leaf structure and the presence of hairs than by leaf size. Plant leaves in quarry samples are characterized by a higher presence of metals originating both from the activities in the quarries themselves and from the machinery used in their production, i.e., from the combustion of fuels in them.

3.4.3 Spontaneous Flora on Brownfield Sites

Brownfields are sites that have been polluted by industrial activities, which is why they are underutilised or even abandoned. Due to the large number of pollutants and the high cost of possible reclamation, these habitats often remain in their original state for a long period. Despite the numerous challenges faced by plants at such sites (lack or absence of organic matter and nutrients in the soil, low water holding capacity, extreme pH and temperature, presence of toxic substances, etc.), a prolonged period without additional contamination allows spontaneous flora to emerge. The specificity of habitat conditions also affects the plant species found here, which are often different from those found on less disturbed sites. These sites are often characterized by "ecological recombination," - the appearance of new communities of native and alien species.

Changes in plant composition were studied on a post-industrial brownfield site in New Jersey, USA, 50 years after its closure to observe the relationship between plant cover and soil properties (Salisbury et al. 2021). Results obtained in 2017 showed the presence of 72 plant taxa, with a trend toward increasing diversity as plant community succession progressed (48 taxa were recorded in 2008 and 62 in 2016). However, the secondary succession process was found to be slower compared to abandoned agricultural lands due to legacy and limited availability of propagation material due to isolation from natural sites.

Spontaneous flora was also studied in Nitrastur, one of the largest brownfields in Spain, focusing on its potential for phytoremediation (Matanzas et al. 2021). A total of six species from this site, heavily contaminated with As and Pb, were analyzed as potentially suitable for use in phytoremediation processes. The results show that only *Lotus hispidus* and *M. lupulina* are able to extract elements, while most of the studied taxa are only suitable for stabilization. Studies on the phytoremediation potential of *Acer pseudoplatanus*, *Betula celtiberica*, and *Buddleja davidii* in an urban wasteland in the same area showed that the accumulation strategy of the plants depends on the level of metal contamination (Fernández-Braña et al. 2023). When these plants grew in a moderately contaminated area, they expressed a phytoextraction strategy, whereas in the heavily metal(loid)-contaminated area, most of the accumulated concentrations remained in the root and only a small portion was translocated to the shoot, indicating the phytostabilization potential of the species.

3.4.4 Spontaneous Flora on Landfills

Landfills represent a very specific environment, considering the wide variation of abiotic factors, due to which these habitats are considered unstable and difficult to form permanent stands. The stability of vegetation cover is a very reliable indicator of spatial changes in contamination. Due to the diversity of conditions that prevail in these types of habitats, the number of plant species that overgrow them is not insignificant. For example, a total of 195 taxa were found in one of the landfills in the Czech Republic with a size of 0.07 km^2 (Vaverková et al. 2019). The number of taxa, as well as their origin (native or allochthonous species), allergen production, importance for agriculture, and rooting depth, were monitored over several years. The analysis showed that the number of species increased over time, both overall and by type of origin. The least change was found in apophytes (native species), while the number of introduced and invasive species increased the most. These species are an important source of pathogens, pest and weed infestations, and also allergenic pollen, over both short and long periods. As expected, the number of trees and shrubs changed the least, while the number of annual herbs increased the most. This is consistent with the high participation of annuals on disturbed sites and their strategy of investing in the production of large numbers of small seeds rather than a perennial character. A constant presence throughout the monitoring period was observed for species such as: *A. vulgaris*, *Acer platanoides*, *A. pseudoplatanus*, *Bromus erectus*, *Cichorium intybus*, *Convolvulus arvensis*, *Picris hieracioides*, *Plantago lanceolata*, *P. major*, *T. pratense*, while species such as *Amaranthus retroflexus*, *Brassica oleracea*, *Echium vulgare*, *Geum urbanum*, *Lactuca serriola*, *M. lupulina*, *P. oleracea*, *Reynoutria japonica*, and *Robinia pseudoacacia* were recorded only in the last years of research.

The study of plant diversity in four different types of landfills (municipal solid waste landfill, serpentine dump, post-metallurgical waste dump, and slag dump) in Poland revealed the presence of 259 taxa (Koszelnik-Leszek et al. 2013). Although belong to the same group, site specificity is reflected in the fact that only five species (*A. elatius*, *Cirsium arvense*, *Daucus carota*, *E. vulgare*, and *L. corniculatus*) occur at all sites studied, but the dominance of the

Asteraceae, Poaceae, and Fabaceae families and hemicryptophytes as a life form stands out as a common feature. These peculiarities of the habitat are caused by the differences in the material of the depositional surfaces, as well as by the different successive phases in which the plant cover is found. The values of the ecological indicators also proved to be similar, whereas significant differences were found in the species composition and the relative participation of certain groups to the origin of the species and the period of their introduction. Thus, 100% apophytes (native species) were found in the serpentine dump, whereas in the post-metallurgical waste dump, slag dump, and municipal waste dump this percentage was 33, 49, and 61%, respectively.

A considerable number of species (244 taxa) were also found at the Batumi landfill in Georgia, a 20-hectare area that has been used as a landfill for the last 60 years. The taxonomic spectrum follows the pattern already observed at similar sites, with a dominance of Asteraceae, Poaceae, and Fabaceae (Koszelnik-Leszek et al. 2013; Kompała-Bąba et al. 2022). However, in contrast to the study of landfills in Poland (Koszelnik-Leszek et al. 2013), the proportion of native species is twice as low as that of allochthonous species, which have been shown to originate mainly from East Asia, the European region, and the Mediterranean, which is consistent with the characteristics of the flora of the wider area.

3.4.5 Allochthonous and Invasive Species on Post-industrial Metalliferous Sites

The conquest of new areas by invasive species is particularly pronounced in fragmented habitats, and habitat fragmentation is known to be one of the most important factors negatively affecting the biodiversity of native plant species (Cardinale et al. 2012). However, the relationship between allochthonous, especially invasive, and native species in anthropogenically modified habitats is still poorly understood. This is of particular interest in highly disturbed areas such as post-mining landscapes, where site characteristics are so altered that they have enormous impacts on ecosystem functioning (Anibaba et al. 2023).

Although previous analyses show a general decline in the number of archaeophytes while neophytes are increasing (Pyšek et al. 2005), studies on post-coal mine heaps in Poland (Anibaba et al. 2023) show that archaeophytes still outnumber neophytes. This was also confirmed by the studies of Simonová and Lososová (2008) at anthropogenically modified sites in the Czech Republic. The invasive species that occur most frequently on post-coal mine heaps in Poland are *Solidago canadensis*, *S. gigantea*, *Erigeron annuus*, and *E. canadensis*. *Erigeron* spp. are particularly important for spontaneous overgrowth on post-industrial sites because of their abundant seed production (up to 50 000 seeds per year; Dauer et al. 2007). However, due to their low competitive ability, these species are more common in early successional stages (Anibaba et al. 2023).

An increase in the number of allochthonous and invasive species during the 8-year monitoring period was also noted at one of the landfills in the Czech Republic, where invasive species accounted for 7.2% of all recorded species in the last year of monitoring (compared to 1.5% at baseline). Some of the invasive species recorded are *Erigeron canadensis*, *E. annuus*, *R. japonica*, *R. pseudoacacia*, *Datura stramonium*, *P. oleracea*, etc. (Vaverková et al. 2019).

Ambrosia artemisiifolia, a common ragweed, is a noxious weed capable of invading a wide range of disturbed habitats due to a broad range of ecological tolerance (Makra et al. 2022). Not only is this species introduced in most parts of the world (with the exception of North America, where it is considered native), but it is also considered invasive in most countries and has been found in various types of disturbed sites, from landfills to mining dumps. In Serbia, this species was found in the post-flotation tailing site of one of the largest Sb deposits in Europe, in a fly ash deposit in a lignite mine, and near a borate-rich coal mine (Ranđelović et al. 2020c). Although this species has been shown to take up and transport significant concentrations of metals to the shoot, due to its dominance in disturbed sites where it occurs as a pioneer species and its high competitiveness compared to native species, this species should be actively suppressed and its spread in natural ecosystems, where it can lead to imbalance and loss of biodiversity of native species, should be prevented in time.

3.4.6 Rare, Endemic, and Relict Species on Post-industrial Metalliferous Sites

The floristic richness of post-industrial sites is reflected not only in the absolute number of species inhabiting these habitats, but also in the character of the present species themselves, as many of them are rare (often endangered), endemic, and/or relict species. Some of them have even disappeared from their natural habitats and have found refuge in post-industrial areas for the survival of their populations (Woźniak et al. 2018).

More than 150 endangered and protected species have been found in the post-industrial areas of the Silesian Upland in Poland. Among them, the largest number of species belongs to the Orchidaceae family, which accounts for almost 50% of all orchid species known so far in Poland (Bacler-Żbikowska and Nowak 2022). In this study, quarries were found to be the post-industrial sites with the highest number of endangered species, while sand and gravel pits, as one of the most common types of degraded sites, had slightly fewer species. However, it is interesting to note that the analyses showed that some of the species present in (almost) all habitat types studied are wetland species (e.g., *Schoenoplectus tabernaemontani*) or even aquatic species (e.g., *Najas marina* or *Utricularia australis*), which is atypical for post-industrial sites with water deficit. In addition, a considerable number of species were found in marsh and fen vegetation (class *Scheuchzerio-Caricetea nigrae*) and vegetation of freshwater macrophytes (class *Potametea*) (Bacler-Żbikowska and Nowak 2022). The presence of rare species in the quarries in this area was also highlighted by Czylok et al. (2008), which noted the wetland species *Liparis loeselii* and *Pinguicula vulgaris*, but also the terrestrial orchid *Malaxis monophyllos*, all of which are listed in the Polish Red Data Book (Zarzycki and Szelag 2006).

Dump deposits of marble quarries in Tuscany (Italy) also proved to be of great conservation importance, given the number of endemic taxa found. A total of six endemic species were found: *Globularia incanescens*, *Cerastium apuanum*, *Carum apuanum*, *Biscutella apuana*, *Asperula purpurea* subsp. *apuana*, and *Leontodon anomalus* (Gentili et al. 2011).

3.4.7 Medicinal and Aromatic Plants on Post-industrial Metalliferous Sites

The use of medicinal plants has increased greatly in recent decades. However, in addition to their use for medicinal purposes, these plants can be important in other ways, including the revegetation of sites damaged by various anthropogenic activities. *Tussilago farfara* is one of the medicinal plants that occurs as a pioneer species on bare slopes or various degraded sites, including those with excessive metal concentrations. Habitats of the species include post-mining coal dumps, where *T. farfara* has been shown to enhance microbial community characteristics (Stefanowicz et al. 2015), as well as active and abandoned spoil heaps rich in Pb, Zn, Sb, and As (Jakovljević et al. 2020; Figure 3.5a). When growing on metalliferous substrates, this species showed two different strategies: either accumulation and storage of metals in the root or translocation of metals to the shoot as part of the detoxification strategy through defoliation (Jakovljević et al. 2020). Metal concentrations in the shoot below hyperaccumulation thresholds for all elements analyzed do not support using the species in phytoextraction, and the species should be used with caution if collected from metal-rich sites. However, there are numerous positive side effects of this species in technogenic substrate. They are primarily reflected in the stabilization of the substrate by binding particles, preventing erosion and the spread of potentially toxic elements to surrounding habitats (Stefanowicz et al. 2015).

Species of the genus *Plantago*, often used in traditional medicine, very successfully overgrow post-industrial sites that are often heavily polluted with metals and metalloids (Figure 3.5b). *Plantago major* and *P. lanceolata*, two of the best-known *Plantago* species, have been found in tailings piles from Pb/Zn mines and near the abandoned metal smelting plant in Poland. In addition to the usual benefits of overgrowing degraded sites, these two species showed phytostabilization potential due to their significant metal tolerance and thus can be successfully incorporated into the process of reclamation (Gucwa-Przepióra et al. 2016). Similar results were shown for *P. major* on the spoil heaps in Romania (Levei et al. 2018).

(a) (b)

Figure 3.5 Medicinal plants in abandoned antimony mine Stolice, Serbia. (a) *Tussilago farfara*; (b) *Plantago lanceolata*.

The studies of medicinal and aromatic plants were also conducted in the periphery of the coal-fired power plant in Romania, where the presence of species such as *Hypericum perforatum*, *Artemisia absinthum*, *Chelidonium majus*, and *Crataegus monogyna*, which are known to be used for medicinal purposes, was detected. However, low concentrations of Cd, Hg, and Pb in the plant parts indicate the safe use of these species for medicinal purposes (Badea 2015). The presence of medicinal plant species has also been noted at landfills, where they were either present prior to the change in use of the site and the creation of the landfills or emerged along with other species during the overgrowth process. Thus, a considerable number of medicinal plant species have been found at landfills in the Czech Republic (*Primula veris*, *Cornus mas*, *T. farfara*, *P. major*, *P. lanceolata*, *Reseda luteola*; Vaverková et al. 2019).

Aromatic plant species are often recorded as pioneer plants in the colonization of various industrial waste sites, showing tolerance to various metals (Ranđelović 2023). Since their essential oils - volatile secondary metabolites generally remain free of contaminants, their increased use for metal remediation at contaminated sites is feasible and beneficial (Pandey et al. 2023a). Vetiver plants (e.g., *Vetiveria zizanioides*) were found to be ecologically plastic and particularly tolerant of a wide range of pH levels and the presence of metals in the substrate (Chen et al. 2004). Similar findings were made for species from the genera *Salvia*, *Matricaria*, *Rosmarinus*, *Ocimum*, *Geranium*, etc. (Pandey et al. 2023a; Ranđelović 2023).

3.4.8 Pioneer Flora and Vegetation on Post-industrial Metalliferous Sites

The establishment of spontaneous vegetation in biologically vacant areas, such as post-industrial sites, after the cessation of the anthropogenic activities that caused the site alteration is certainly very important not only for the improvement of site characteristics, but also for laying the foundation for further successional phases. However, these initial phases are extremely sensitive to invasion by alien plant species, so the establishment of native species should be encouraged during this particular phase (Anibaba et al. 2023). The development of pioneer species on bare metalliferous substrates derived from mining, mine waste disposal, or ore processing is extremely important in creating the basic conditions for colonization by other plants after the physico-chemical properties of the substrate have been improved and it has been enriched with nutrients. Rhizosphere microorganisms, which can stimulate plant growth, also play an important role in colonizing bare metalliferous substrates. In mining areas in arid and semiarid regions, the resistance of pioneer plants to drought and tolerance to high concentrations of metal(oid)s in the soil is of particular importance. At the same time, tolerance to soil salinity may be a significant advantage for plants in the initial colonization of substrates containing metals or metalloids in high concentrations.

In the early stages of succession, grasses and low shrubs play a special role. For example, in the tailings of copper and iron mines in Liguria, Italy, low-growing grasses and herbs such as *Deschampsia flexuosa*, *Minuartia laricifolia* subsp. *ophiolitica*, *Sesamoides interrupta*, and *Festuca robustifolia* were present in pioneer communities that spread discontinuously on the bare substrate, while *Asplenium adiantum-nigrum* and *Cerastium ligusticum* were found in parts with well-developed substrate (Roccotiello et al. 2015).

Grasses are particularly important in the primary colonization of tailings piles, as confirmed by Shu et al. (2005) in a study of Pb/Zn tailings in southern China. Due to the very high concentrations of toxic metals, natural plant colonization was very limited and localized, but a total of 54 species from 51 genera and 21 families were detected in the spontaneous vegetation patches, of which Poaceae dominated with 13 species. Numerous factors influence the success of colonization of bare patches. Of particular importance is the ability for seed dispersal, with anemochory being the most advantageous dispersal mechanism, as it enables the plant to be independent of other species in completing life cycle stages for many organisms (Memmott et al. 2007). Substrate pH response is also important for site colonization. In Ontario (Canada), spoil heaps with high concentrations of Cu and Ni remain largely unvegetated without the addition of lime or any form of fertilization, except patches of metal-tolerant grass *Deschampsia caespitosa* (Bagatto and Shorthouse 1999) that have developed spontaneously. Variations in substrate response on mine waste in England, together with available concentrations of Zn and Pb, were found to be the main reason for variations in floristic composition on mine waste in England (Morrey et al. 1988). In addition, substrate moisture content was found to be an important factor limiting the occurrence and development of spontaneous vegetation most often on biologically vacant sites, as was the case with *D. caespitosa* at the sites studied. Species composition is largely determined by the gradient of metal(loid)s contamination, as noted in dry grassland communities where, under conditions of excessive elemental concentrations in the substrate, species sensitive to metals retreated, while species tolerant of metals became more prominent (Woch et al. 2016). The flora of post-industrial sites is strongly influenced by the flora of the immediate surroundings, as migration is difficult to avoid. Here, dense forestation could be very beneficial because it has been shown that when it surrounds different types of post-industrial sites, it can significantly reduce the introduction of dispersal products from anthropogenically modified habitats, which has an impact on the conservation of natural flora (Kopeć et al. 2011). Another important predictor was the depth of the surface soil, as shallow soils favored the establishment of grasslands rich in metals. Plant community success and successional tendencies also depend on the structure of the post-industrial site, as a mosaic-like structure allows the presence of species with different characteristics and different life strategies, all of which influence vegetation success (Woźniak et al. 2015). Habitat heterogeneity has been recognized as one of the most important determinants of plant community diversity in industrial areas, particularly in mining activities (Rich et al. 2015). Despite the diversity of conditions observed in post-industrial sites, the dominance of a single species was common, due to their abundance and the influence on shaping habitat conditions (water regime, nutrient availability, etc.) (Błońska et al. 2019).

Although therophytes often dominate the spectrum of life forms in these types of habitats due to their adaptability to altered conditions, the study of the vegetation of the Les Malines mining district in southern France, which is highly contaminated with Zn, Pb, Cd, As, and Tl, showed that annual species were not found on bare soil and a clear dominance of hemicryptophytes was observed at all sites (Escarré et al. 2011). Similarly, the role of therophytes appears to be much smaller in the initial stages of colonization of metalliferous mining sites and some coal waste dumps (Piekarska-Stachowiak et al. 2014; Ranđelović et al. 2014). Although several mining sites had been abandoned for a long time (hundreds of years), no trees from the surrounding chestnut forest colonized the dumps. On the contrary, it was found that trees played the most important role in the formation of more fertile patches, also called "fertile islands" (Parraga-Aguado et al. 2014). Munford et al. (2022) suggest that *Betula papyrifera*, *Populus tremuloides*, and *Pohlia nutans* as metal/acid tolerant pioneer species facilitate colonization of less tolerant species on Ni-Cu tailings piles in Ontario, Canada. Several woody species have also been noted as pioneer species in the colonization of abandoned old mining sites in New Caledonia (Jaffré et al. 1994). In the early stages of colonization, pioneer species colonize bare areas that are usually protected from strong winds and intense sunlight. They are generally slow-growing, except for some woody species such as *Casuarina collina*, *Alphifonia neocaledonica*, *Dodonaea viscosa*, *Acacia spirorbis*, and *Carpolepis laurifolia*. Regarding the origin of the plants, the main trend in the natural revegetation of severely degraded areas is towards quality reduction and increasing participation of allochthonous species with a high percentage of species considered invasive (Gentili et al. 2011).

Tolerance to a combination of edaphic factors, including nutrient deficiency and inadequate soil structure, is important for the successful colonization of bare metalliferous substrates. Several halophytes from the Chenopodiaceae family, as well as Tamaricaceae, Plumbaginaceae, and Asteraceae, have been found in various stages of succession, including early pioneer stages at mining sites in SE Spain (Parraga-Aguado et al. 2013).

At the Balangero asbestos mine in Italy, which was in operation until the 1990s, different dynamic phases of bare soil colonization have been studied (Favero-Longo et al. 2006). Vegetation studies have shown that, in addition to vascular plants, lichens play an important role in colonization, among which the crustose lichens *Candelariella vitellina* and *Scoliciosporum umbrinum* and the foliose lichen *Xanthoparmelia tinctina* stand out. Two hyperaccumulator plant species, *Noccaea sylvium* and *M. laricifolia* with low cover characterize the early colonization phase, while they are followed by stands of *Noccaea alpestris*, which, thanks to their large cover, prevent the spread of toxic asbestos fibers.

In addition to species with broader ecological preferences, there are also pioneer species associated exclusively with bare metalliferous soils. These include *Gypsophila perfoliata*, a West Asian species considered rare in Central Europe and most likely introduced with iron ore. In Hungary, this species was found exclusively on highly modified anthroposols, bare metalliferous soils, or mine dumps and has a strong pioneer character (Schmotzer et al. 2021). Similar findings were made for *Crepidorhopalon tenuis*, a small plant endemic to Zambezy with high tolerance to excessive concentration of Cu in the substrate (Faucon et al. 2011).

Metal-tolerant pioneer plants are often associated with mycorrhizal fungi that increase plant tolerance to metals, although some mechanisms by which this occurs are not fully understood. For example, ectomycorrhizal filamentous fungi of the genus *Suillus* prevent the transfer of large amounts of toxic metals to the plant-fungus interface due to the metal-specific efflux system, but do not affect the transfer of nutrients to the host plant (Colpaert et al. 2011).

3.5 Ecosystem Services of Spontaneous Flora on Post-industrial Metalliferous Sites

The concept of ecosystem services (ES) stands for ". . .those aspects of ecosystems that are used (actively or passively) to produce human well-being" (Fisher et al. 2009) or as "natural capital that provides life-sustaining services of enormous value" (Daily and Matson 2008). According to Millennium Ecosystem Assessment (MEA) (2005), there are four different

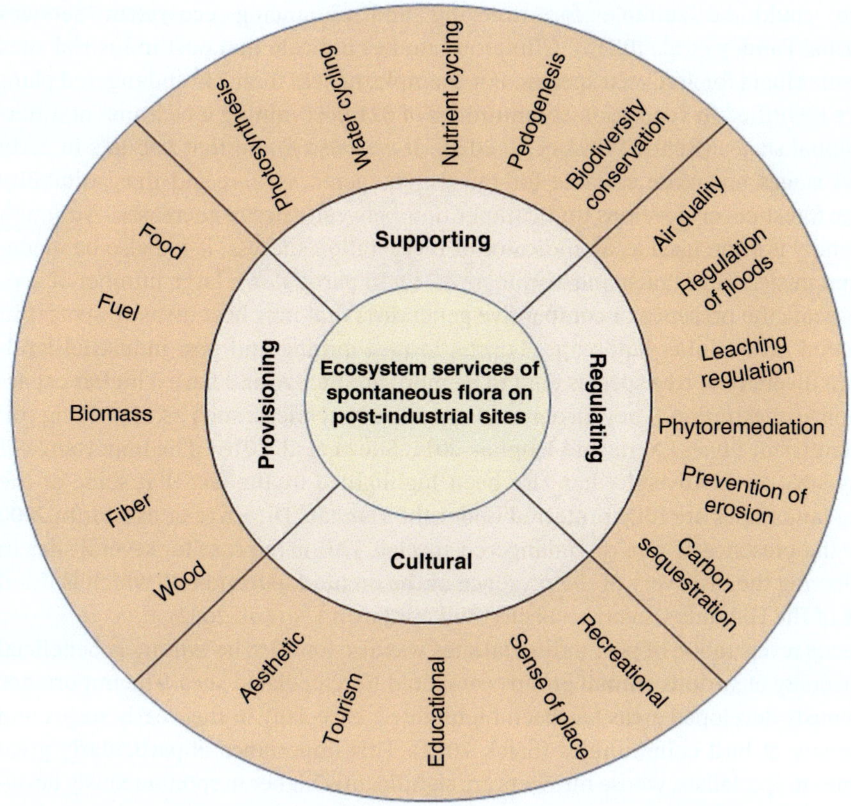

Figure 3.6 Ecosystem services of spontaneous flora on post-industrial sites.

service categories, depending on the area in which they provide certain goods: (i) provisioning services including drinking water, food, fuel, wood, and other commodities, (ii) regulatory services provided by various groups of organisms, such as erosion control, crop pollination, water purification, and flood control; (iii) supporting services that include biodiversity conservation, pedogenesis, nutrient cycling etc.; while (iv) cultural services include recreation, aesthetic values, and well-being (Figure 3.6). Although there is increasing interest in the study and implementation of ecosystem services, this concept is still not widely used in practical planning and decision making, including rehabilitation of post-industrial sites. In this regard, ecosystem services assessment should be incorporated into post-industrial site reclamation planning and could demonstrate the potential for environmental and economic benefits following the reclamation process.

3.5.1 Ecosystem Services and Biodiversity of Spontaneously Vegetated Post-industrial Sites

Post-industrial sites, especially metalliferous ones, are usually considered unattractive and ecologically problematic, especially if they are located in urban or peri-urban areas. However, when covered with spontaneous vegetation, they contribute significantly to biodiversity, and if surrounding ecosystems can provide sufficient plants, spontaneous revegetation could be suitable for restoring and enhancing ecosystem services (Pandey 2020a; Pandey et al. 2023b). Numerous studies indicate that post-industrial sites are important refugia for analyzed species. For example, no less than 235 endangered plant species were identified in vegetation communities of 321 post-mining landscapes at different successional stages (Řehounková et al. 2020). It was also found that habitats in early successional stages are more suitable for most endangered species and that suitability decreases as forest cover develops and competition between species increases. Although species diversity is often used as an indicator of revegetation success, it can also be inadequate or even misleading (Prach and Tolvanen 2016). In particular, a large number of species may indicate the presence of competitive generalists that may host invasive species in the revegetated post-mining landscape. Forests in post-mining and post-industrial landscapes with a diversity of tree species tend to be more productive and have a higher capacity for carbon sequestration. They also provide regulatory services, such as improving soil quality and nutrient fluxes (Aerts and Honnay 2011; Souza et al. 2016). The importance of post-mining sites for biodiversity has also been highlighted by the fact that some of the post-mining landscapes are fully protected under the Habitats Directive or as Natura 2000 sites due to the presence of rare or endangered species. This is the case for several sites in Poland, following the discovery of the presence of the orchid *Liparis loselii*, which is listed in Annex II of the Habitats Directive (Bacler-Żbikowska and Nowak 2022).

Spontaneous revegetation of post-industrial sites was also found to be extremely beneficial to the biodiversity of various animal groups compared to remediated sites. The importance of spontaneously developed areas has been highlighted, especially in their early stages and for the diversity of bird communities (Šálek 2012). This importance is particularly great when it comes to specialists, whose numbers are significantly higher in spontaneously developed areas than in reclaimed ones, and it was found that the conservation value of these communities also decreases over time, suggesting that the early stages are the most valuable for this group. A similar finding was made for several arthropod groups colonizing engineered mining sites in the Czech Republic (Tropek et al. 2012), where significantly higher numbers of species of concern or special conservation interest were found on spontaneously revegetated sites compared to engineered sites. Therefore, official support for natural revegetation of mining sites rather than engineered reclamation by legislators and decision-makers has been proposed. In addition, ecosystems in the early stages of succession in post-mining landscapes, such as meadows and young forests and legumes, can promote bee diversity by providing forage or suitable nesting sites, which is particularly important for pollinator conservation (Łowicki and Fagiewicz 2021). Considering that animals pollinate almost 90% of all plants, the central importance of this process for biodiversity conservation becomes clear. However, results indicate the importance of a mosaic distribution of different successional stages for restoration (Řehounková et al. 2020). This is based on the fact that this type of distribution provides different habitats for species of special conservation interest.

3.5.2 Ecosystem Services of Special Value in Post-industrial Sites

In post-mining and post-industrial landscapes, several types of ecosystem services are particularly important. These include preventing soil erosion, purifying water, and improving climate. All of these factors contribute to the successful establishment of vegetation cover.

Ecosystem services provided by rehabilitated mining sites also include forage for livestock, provision of clean water and air, sequestration of carbon, habitat for various organisms, and maintenance of biodiversity, timber, and natural resources used to produce traditional products or in folk medicine (Garris et al. 2016).

The importance of pioneer plants in the revegetation of spoil heaps and the subsequent improvement of soil quality and functions was highlighted by Álvarez-Rogel et al. (2021). The authors also noted that after the development of the plant cover, the subsequent improvement of soil quality allows, to a certain extent, the restoration of soil functions regardless of the concentrations of metals or metalloids.

Revegetation and improvement of soil properties of abandoned mine overburden dumps benefit several important ecosystem services, including reducing the spread of polluted soils, providing habitat for diverse organisms, and improving soil nutrient status (Conesa et al. 2007). Patches of spontaneously developed vegetation can also compensate for fluctuations in soil conditions between seasons and increase biological activity, suggesting greater resilience to climate change compared to unvegetated mine tailings (Álvarez-Rogel et al. 2022). Two native, spontaneously growing plant species on lead-zinc smelter slags in northwestern Guizhou, *Chenopodium ambrosioides* and *Neyraudia reynaudiana*, were studied to assess their potential for phytoremediation, effects on nutrient status, and enzyme activity of the slag matrix (Salisbury et al. 2017). In addition to indicating the phytostabilization potential of spontaneously growing plants in urban brownfields enriched with high concentrations of As, Cu, Cr, and Zn, the plants were found to significantly increase nutrient concentrations (N, P, K, and organic matter), stimulate enzyme activity, and enhance rhizosphere microbial respiration compared to unvegetated substrate (Qiu et al. 2019).

3.5.3 Variability of Ecosystem Services in Post-industrial Sites

Different approaches to mine rehabilitation result in significant differences in the provision of ecosystem services at rehabilitated post-mining and post-industrial metalliferous sites. Assessment of ecosystem services allows comparison between approaches and selection of those that provide the widest range of ecosystem services. Post-mining/industrial restoration options can be evaluated using the concept of ecosystem services to improve the overall sustainability of the landscape, especially in cases where it is not practical or possible to restore previously present ecosystems (Larondelle and Haase 2012). Although there are certain limitations when considering the general environmental conditions of post-industrial habitats, their heterogeneity enables consideration and incorporation of different ecosystem services. For example, significant spatial variability was also found within rehabilitated mining sites, which can be explained by the heterogeneity of numerous environmental factors such as topography, aspect, slope, soil type, and quality. It was found that the areas with gentle slopes provided more ecosystem services when converted to agricultural land compared to the zones with steeper slopes, which provided more ecosystem services when converted to forest (Wang et al. 2017). Another important factor contributing to the variability of ecosystem services on the reclaimed mining sites could be the microbiota of the rehabilitated soil (Garris et al. 2016). The study by Xiao et al. (2019) showed that certain root-associated bacteria in an abandoned antimony dump were controlled by soil properties and contributed significantly to nutrient enrichment and metal cycling. These factors contributed to the successful colonization of the tailings pile by *Bidens bipinnata*. The importance of key microbial taxa in the revegetation of mine tailings was also shown for stibnite (Sb_2S_3) tailings, where Sun et al. (2020) highlighted the crucial role of key taxa in providing phosphorus and iron and promoting plant growth. Rehabilitated fly ash sites, both naturally and man-assisted, proved to be able to provide various ecosystem services from all four categories: provisioning, regulating, supporting, and cultural services (Pandey 2020a,b,c). Regulating services are among the most important, considering the variety of beneficial processes, and spontaneously growing plant species might have a crucial role in many of them, such as phytoremediation, prevention of erosion, regulation of metal leaching, regulation of flood, etc. Thus, it has been shown that *T. latifolia* could be beneficial in the remediation of fly ash basins (Pandey et al. 2014), *D. glomerata*, *C. epigejos*, *F. rubra*, etc. play a major role in phytostabilization processes (Gajić et al. 2016, 2019), *Vetiveria zizanioides* in air pollution control (Verma et al. 2014), etc.

3.5.4 Economic Value and Potential Applications of Native Plants on Post-industrial Sites

The study of the added economic value of native plants that have spontaneously colonized abandoned mine dumps in Chile revealed that a large number of these plant species have at least one or more known uses, including in traditional medicine, as ornamental or melliferous plants, and for the production of raw materials (Orchard et al. 2009). Seed dispersal is also important for the revegetation of post-mining sites, and grazing plays an important role in this regard (Hindersah et al. 2019). However, grazing should be kept away from areas where concentrations of metals taken up by roots and accumulated in plant tissue have been shown to be low, but metal-rich dust particles adhering to plant surfaces or direct ingestion of soil can pose a serious health problem for cattle feeding in these areas (Conesa and Faz 2011).

Spontaneously vegetated mining sites represent a source of extremely valuable genetic material of species that can be used for phytoremediation, including phytostabilization and phytoextraction (Salas-Luévano et al. 2009). Plants that develop spontaneously on metalliferous substrates, including abandoned mine waste and metalliferous post-industrial sites, and are referred to as metallophytes, have several tolerance mechanisms that allow them to survive on extreme substrates with extremely high concentrations of metals or metalloids. According to Baker (1987), metallophytes can be divided into three basic groups depending on their tolerance strategy: excluders, indicators, or accumulators. A special group of accumulators is the metal(oid) hyperaccumulator plants, which accumulate toxic elements in their leaves at concentrations several hundred times higher than other plants in a given habitat without symptoms of toxicity. To date, only slightly more than 700 hyperaccumulators of various elements are known, of which over 2/3 hyperaccumulate nickel (Jakovljević et al. 2022b). The plants that fall into the group of excluders or indicator plants can be used for phytostabilization of abandoned mining sites, while certain hyperaccumulators that produce a significant amount of biomass and are resistant to unfavorable soil properties and various plant diseases have significant potential for use in phytoextraction and/or phytomining. In phytostabilization, the contaminated substrate is prevented from spreading and its physical and chemical properties are improved, while in phytoextraction, metals or metalloids accumulate in high concentrations in the aboveground parts of hyperaccumulator plants. In addition to hyperaccumulators, phytoextraction can also be made by plants that are accumulators or indicators of certain elements but produce a large amount of biomass, such as species of the genus *Salix*. The biomass of hyperaccumulating plants can be used to leach pure metals or to produce metal salts in phytomining. The application of this technology offers particular potential for the remediation of lateritic nickel mines, such as those operated in Indonesia or Malaysia. The process is effective in preventing erosion while generating additional revenue through the recovery of residual nickel (van der Ent et al. 2013). One of the plants with great potential for Ni phytomining in lateritic Ni mines in Sabah, Malaysia, is *Phyllanthus rufuschaneyi*, an endangered taxa with extremely high nickel concentrations in the aboveground organs. The use of this species in phytomining also contributes to its conservation and demonstrates the importance of preserving the habitat and genetic material of rare species that can be used in phytoremediation or other biotechnologies (van der Ent et al. 2013).

The Ni hyperaccumulator *Odontarrhena muralis* s.l. has been successfully used in field trials for nickel agromining on ultramafic soils in Europe (Kidd et al. 2018), but this species is also a common spontaneous colonizer of mine tailings with high Ni concentrations. Spontaneously growing patches of *O. muralis* s.l. with nickel concentrations in the aerial parts exceeding $10\,000\,\mathrm{mg\,kg^{-1}}$ were found on asbestos waste in central and W Serbia (unpublished, Figure 3.7). Due to the large root biomass, this species has a special importance for erosion control and phytostabilization of metalliferous sites, including those created after mining and industry.

In addition to *Odontarrhena* spp., numerous *Noccaea* species that hyperaccumulate Ni, Zn, Cd, or Pb have been spontaneously discovered on abandoned mine waste. One of the best-studied species is *N. caerulescens* and its various ecotypes that hypertolerate and/or

Figure 3.7 *Odontarrhena muralis* s.l. growing spontaneously on an asbestos dump in Serbia.

hyperaccumulate Zn, Cd, or Ni (Assunção et al. 2003). This species was also found on spontaneously vegetated ultramafic mine spoils in northwestern Spain, showing considerable potential for application in Ni extraction, i.e., the phytomining process (Cerdeira-Pérez et al. 2019). Spontaneously growing *Noccaea rotundifolia* was found on mine tailings from lead-zinc mines in northern Italy, with lead and zinc concentrations exceeding hyperaccumulation limits (Reeves and Brooks 1983). High concentrations of Pb, Zn, and Cd were also found in shoots of *N. praecox* growing spontaneously at a polluted site contaminated with metals near a lead mine and smelter in Slovenia (Vogel-Mikuš et al. 2005). *Noccaea praecox* is also a Ni hyperaccumulator, with Ni uptake and hyperaccumulation varying greatly depending on various factors, including microedaphic factors (Mišljenović et al. 2018, 2020). Although most *Noccaea* species do not produce sufficient biomass for use in agromining, several species of this genus could be used in the revegetation of post-industrial metalliferous sites as part of phytostabilization, but with careful selection of the appropriate ecotype due to wide variation in metal tolerance and accumulation. In addition, seeds of *Noccaea* accessions growing spontaneously on post-industrial metalliferous sites could be used as a solid foundation for the revegetation of these sites.

In recent years, special attention has been paid to a group of plants that are not only important for phytoremediation, but also represent important bioenergy resources. These include species from the genera *Miscanthus*, *Populus*, *Ricinus*, *Jatropha*, etc., as well as grasses such as *Festuca arundinacea*, *Panicum virgatum*, *Phalaris arundinacea*, etc. (Sanderson and Adler 2008; Pandey et al. 2016). Using species that have other beneficial properties in addition to phytoremediation potential allows us to get more as a result while reducing human health impacts. This represents the basis of sustainable phytoremediation, a concept proposed by Pandey et al. (2015). Highly productive species from the genus *Miscanthus* are useful for heat production in addition to their demonstrated ability for phytostabilization due to the high heating value of the dry biomass, as well as for the production of bioethanol from lignocellulose as a building material (Han et al. 2011). Carbon sequestration is also one of the beneficial properties of *Miscanthus*, which has been observed particularly in *M.* × *giganteus* (Pidlisnyuk et al. 2014). Carbon sequestration ability has also been observed in *Jatropha curcas* (Pandey et al. 2012), a species whose seed oil is used for biodiesel production. *Populus* species also represent an energy crop. Their lignin and hemicellulose can be used to produce bioethanol, and because of their high calorific value, poplars can be used to produce thermal energy (Pandey et al. 2016).

3.6 Conclusions

The expansion of post-industrial sites in the future, as a result of industrial development, also leads to the development of their spontaneous flora. Although to some extent dependent on the type of the post-industrial sites, there are some common characteristics of the flora analyzed. Asteraceae, Poaceae, and Fabaceae proved to be the most dominant families, whereas therophytes and hemicryptophytes are the predominant life forms. Despite the expectation that most species in post-industrial sites are allochthonous, native species are shown to dominate in a significant number. The biological value of these habitats is further highlighted by the significant participation of endemic, rare, relict, medicinal, and aromatic plants. In addition to increasing biodiversity, the flora of post-industrial sites participate in a variety of ecosystem services from all four major groups. They also represent an important resource of species that could be used in the remediation of post-industrial sites, while simultaneously supporting several UN SDGs. Interest in the study and implementation of ecosystem services is steadily increasing. However, they are still not considered in many aspects of industrial activity planning and post-industrial site remediation, and therefore their implementation in this area should be one of the main objectives in the future.

Acknowledgments

The Ministry of Science, Technological Development and Innovation of the Republic of Serbia supported this research (grant numbers 451-03-66/2024-03/ 200007, 451-03-65/2024-03/ 200178 and 451-03-66/2024-03/200023).

References

Adabanija, M.A. and Oladunjoye, M.A. (2014). Geoenvironmental assessment of abandoned mines and quarries in South-western Nigeria. *Journal of Geochemical Exploration* 145: 148–168.

Adamcová, D., Radziemska, M., Ridošková, A. et al. (2017). Environmental assessment of the effects of a municipal landfill on the content and distribution of heavy metals in *Tanacetum vulgare* L. *Chemosphere* 185: 1011–1018.

Adamczyk, J. and Pomykała, R. (2022). Coal sludge permeability assessment based on Rowe cell consolidation, and filtration investigations. *Minerals* 12 (2): 212.

Aerts, R. and Honnay, O. (2011). Forest restoration, biodiversity and ecosystem functioning. *BMC Ecology* 11 (1): 1–10.

Akcil, A. and Koldas, S. (2006). Acid mine drainage (AMD): causes, treatment and case studies. *Journal of Cleaner Production* 14 (12–13): 1139–1145.

Alam, R., Ahmed, Z., and Howladar, M.F. (2020). Evaluation of heavy metal contamination in water, soil and plant around the open landfill site Mogla Bazar in Sylhet, Bangladesh. *Groundwater for Sustainable Development* 10: 100311.

Álvarez-Rogel, J., Peñalver-Alcalá, A., and González-Alcaraz, M.N. (2022). Spontaneous vegetation colonizing abandoned metal (loid) mine tailings consistently modulates climatic, chemical and biological soil conditions throughout seasons. *Science of the Total Environment* 838: 155945.

Álvarez-Rogel, J., Peñalver-Alcalá, A., Jiménez-Cárceles, F.J. et al. (2021). Evidence supporting the value of spontaneous vegetation for phytomanagement of soil ecosystem functions in abandoned metal (loid) mine tailings. *Catena* 201: 105191.

Álvarez-Valero, A.M., Sáez, R., Pérez-López, R. et al. (2009). Evaluation of heavy metal bio-availability from Almagrera pyrite-rich tailings dam (Iberian Pyrite Belt, SW Spain) based on a sequential extraction procedure. *Journal of Geochemical Exploration* 102 (2): 87–94.

Anibaba, Q.A., Dyderski, M.K., Woźniak, G., and Jagodziński, A.M. (2023). Native plant community characteristics explain alien species success in post-industrial vegetation. *NeoBiota* 85: 1–22.

Arenas-Lago, D., Andrade, M.L., Vega, F.A., and Singh, B.R. (2016). TOF-SIMS and FE-SEM/EDS to verify the heavy metal fractionation in serpentinite quarry soils. *Catena* 136: 30–43.

Assunção, A.G., Bookum, W.M., Nelissen, H.J. et al. (2003). Differential metal-specific tolerance and accumulation patterns among *Thlaspi caerulescens* populations originating from different soil types. *New Phytologist* 159 (2): 411–419.

Bąba, W., Błońska, A., Kompała-Bąba, A. et al. (2016). Arbuscular mycorrhizal fungi (AMF) root colonization dynamics of *Molinia caerulea* (L.) Moench. in grasslands and post-industrial sites. *Ecological Engineering* 95: 817–827.

Bacler-Żbikowska, B. and Nowak, T. (2022). Role of post-industrial sites in maintaining species diversity of rare, endangered and protected vascular plant species on the example of the urban-industrial landscapes. In: *Green Scenarios: Mining Industry Responses to Environmental Challenges of the Anthropocene Epoch*, 245–264. CRC Press.

Bačeva Andonovska, K., Stafilov, T., and Matevski, V. (2021). Accumulation abilities of endemic plant species from the vicinity of an As-Sb-Tl abandoned mine, Allchar, Kožuf Mountain. In: *Contaminant Levels and Ecological Effects: Understanding and Predicting with Chemometric Methods*, 375–402. Cham: Springer.

Badea, D.N. (2015). Determination of potentially toxic heavy metals (Pb, Hg, Cd) in popular medicinal herbs in the coal power plant area. *Revista de Chimie* 66 (8): 1132–1136.

Bagatto, G. and Shorthouse, J.D. (1999). Biotic and abiotic characteristics of ecosystems on acid metalliferous mine tailings near Sudbury, Ontario. *Canadian Journal of Botany* 77 (3): 410–425.

Baker, A.J.M. (1987). Metal tolerance. *New phytologist* 106: 93–111.

Baker, A.J.M. and Proctor, J. (1990). The influence of cadmium, copper, lead, and zinc on the distribution and evolution of metallophytes in the British Isles. *Plant Systematics and Evolution* 173: 91–108.

Baziene, K., Tetsman, I., and Albrektiene, R. (2020). Level of pollution on surrounding environment from landfill aftercare. *International Journal of Environmental Research and Public Health* 17 (6): 2007.

Bian, Z., Miao, X., Lei, S. et al. (2012). The challenges of reusing mining and mineral-processing wastes. *Science* 337 (6095): 702–703.

Błońska, A., Kompała-Bąba, A., Sierka, E. et al. (2019). Diversity of vegetation dominated by selected grass species on coal-mine spoil heaps in terms of reclamation of post-industrial areas. *Journal of Ecological Engineering* 20: 209–217.

Boechat, C.L., Pistóia, V.C., Gianelo, C., and Camargo, F.A.D.O. (2016). Accumulation and translocation of heavy metal by spontaneous plants growing on multi-metal-contaminated site in the Southeast of Rio Grande do Sul state, Brazil. *Environmental Science and Pollution Research* 23: 2371–2380.

Boev, B., Jovanovski, G., and Makreski, P. (2012). Geology and mineralogy of Allchar Sb-As-Ti-Au deposit. *Geologica Macedonica* 26: 215–232. (2012).

Cabrera, F., Clemente, L., Barrientos, E.D. et al. (1999). Heavy metal pollution of soils affected by the Guadiamar toxic flood. *Science of the Total Environment* 242 (1–3): 117–129.

Cardinale, B.J., Duffy, J.E., Gonzalez, A. et al. (2012). Biodiversity loss and its impact on humanity. *Nature* 486 (7401): 59–67.

Carrenho, R., Trufem, S.F.B., Bononi, V.L.R., and Silva, E.S. (2007). The effect of different soil properties on arbuscular mycorrhizal colonization of peanuts, sorghum and maize. *Acta Botanica Brasilica* 21: 723–730.

Cerdeira-Pérez, A., Monterroso, C., Rodríguez-Garrido, B. et al. (2019). Implementing nickel phytomining in a serpentine quarry in NW Spain. *Journal of Geochemical Exploration* 197: 1–13.

Chang, P., Kim, J.Y., and Kim, K.W. (2005). Concentrations of arsenic and heavy metals in vegetation at two abandoned mine tailings in South Korea. *Environmental Geochemistry and Health* 27: 109–119.

Chaudhary, R., Nain, P., and Kumar, A. (2021). Temporal variation of leachate pollution index of Indian landfill sites and associated human health risk. *Environmental Science and Pollution Research* 28: 28391–28406.

Chen, Y., Shen, Z., and Li, X. (2004). The use of vetiver grass (*Vetiveria zizanioides*) in the phytoremediation of soils contaminated with heavy metals. *Applied Geochemistry* 19 (10): 1553–1565.

Chocholoušková, Z. and Pyšek, P. (2003). Changes in composition and structure of urban flora over 120 years: a case study of the city of Plzeň. *Flora-Morphology, Distribution, Functional Ecology of Plants* 198 (5): 366–376.

Ciria, M.P., Solano, M.L., and Soriano, P. (2005). Role of macrophyte *Typha latifolia* in a constructed wetland for wastewater treatment and assessment of its potential as a biomass fuel. *Biosystems Engineering* 92 (4): 535–544.

Colpaert, J.V., Wevers, J.H., Krznaric, E., and Adriaensen, K. (2011). How metal-tolerant ecotypes of ectomycorrhizal fungi protect plants from heavy metal pollution. *Annals of Forest Science* 68 (1): 17–24.

Conesa, H.M. and Faz, Á. (2011). Metal uptake by spontaneous vegetation in acidic mine tailings from a semiarid area in south Spain: implications for revegetation and land management. *Water, Air, & Soil Pollution* 215: 221–227.

Conesa, H.M., García, G., Faz, Á., and Arnaldos, R. (2007). Dynamics of metal tolerant plant communities' development in mine tailings from the Cartagena-La Unión Mining District (SE Spain) and their interest for further revegetation purposes. *Chemosphere* 68 (6): 1180–1185.

Corzo Remigio, A., Chaney, R.L., Baker, A.J. et al. (2020). Phytoextraction of high value elements and contaminants from mining and mineral wastes: opportunities and limitations. *Plant and Soil* 449: 11–37.

Cusser, S. and Goodell, K. (2013). Diversity and distribution of floral resources influence the restoration of plant–pollinator networks on a reclaimed strip mine. *Restoration Ecology* 21 (6): 713–721.

Czech, T., Marchewicz, A., Sobczyk, A.T. et al. (2020). Heavy metals partitioning in fly ashes between various stages of electrostatic precipitator after combustion of different types of coal. *Process Safety and Environmental Protection* 133: 18–31.

Czylok, A., Rahmonov, O., and Szymczyk, A. (2008). Biological diversity in the area of quarries after sand exploitation in the eastern part of Silesian Upland. *Teka Kom. Ochr. Kszt. Środ. Przyr. OL PAN A* 5: 15–22.

Daily, G.C. and Matson, P.A. (2008). Ecosystem services: from theory to implementation. *Proceedings of the National Academy of Sciences* 105 (28): 9455–9456.

Dasgupta, S. and Tam, E.K.L. (2009). Environmental review: a comprehensive review of existing classification systems of brownfield sites. *Environmental Practice* 11 (4): 285–300.

Dauer, J.T., Mortensen, D.A., and Vangessel, M.J. (2007). Temporal and spatial dynamics of long-distance Conyza canadensis seed dispersal. *Journal of Applied Ecology* 44 (1): 105–114.

Dold, B. (2014). Evolution of acid mine drainage formation in sulphidic mine tailings. *Minerals* 4 (3): 621–641.

Đorđević, T., Drahota, P., Kolitsch, U. et al. (2021). Synergetic Tl and As retention in secondary minerals: an example of extreme arsenic and thallium pollution. *Applied Geochemistry* 135: 105114.

Eichenberg, D., Bowler, D.E., Bonn, A. et al. (2021). Widespread decline in Central European plant diversity across six decades. *Global Change Biology* 27 (5): 1097–1110.

Entry, J.A., Rygiewicz, P.T., Watrud, L.S., and Donnelly, P.K. (2002). Influence of adverse soil conditions on the formation and function of arbuscular mycorrhizas. *Advances in Environmental Research* 7 (1): 123–138.

Escarré, J., Lefèbvre, C., Raboyeau, S. et al. (2011). Heavy metal concentration survey in soils and plants of the Les Malines mining district (Southern France): implications for soil restoration. *Water, Air, & Soil Pollution* 216: 485–504.

Essl, F., Glaser, M., and Schertler, A. (2021). New and old invaders in forests in eastern Austria: the role of species attributes and invasion history. *Flora* 283: 151922.

Faucon, M.P., Parmentier, I., Colinet, G. et al. (2011). May rare metallophytes benefit from disturbed soils following mining activity? The case of the *Crepidorhopalon tenuis* in Katanga (DR Congo). *Restoration Ecology* 19 (3): 333–343.

Favero-Longo, S.E., Siniscalco, C., and Piervittori, R. (2006). Plant and lichen colonization in an asbestos mine: spontaneous bioattenuation limits air dispersion of fibres. *Plant Biosystems* 140 (2): 190–205.

Fecko, P., Tora, B., and Tod, M. (2013). Coal waste: handling, pollution impacts and utilization. In: *The Coal Handbook: Towards Cleaner Production*, 63–84. Woodhead Publishing Limited, Elsevier.

Fernández, S., Poschenrieder, C., Marcenò, C. et al. (2017). Phytoremediation capability of native plant species living on Pb-Zn and Hg-As mining wastes in the Cantabrian range, north of Spain. *Journal of Geochemical Exploration* 174: 10–20.

Fernández-Braña, A., Salgado, L., Gallego, J.L.R. et al. (2023). Phytoremediation potential depends on the degree of soil pollution: a case study in an urban brownfield. *Environmental Science and Pollution Research* 30: 1–12.

Fisher, B., Turner, R.K., and Morling, P. (2009). Defining and classifying ecosystem services for decision making. *Ecological Economics* 68 (3): 643–653.

Gajić, G., Djurdjević, L., Kostić, O. et al. (2016). Assessment of the phytoremediation potential and an adaptive response of *Festuca rubra* L. sown on fly ash deposits: native grass has a pivotal role in ecorestoration management. *Ecological Engineering* 93: 250–261.

Gajić, G., Mitrović, M., and Pavlović, P. (2019). Ecorestoration of fly ash deposits by native plant species at thermal power stations in Serbia. In: *Phytomanagement of Polluted Sites*, 113–177. Elsevier.

Galán, E., González, I., and Fernández-Caliani, J.C. (2002). Residual pollution load of soils impacted by the Aznalcóllar (Spain) mining spill after clean-up operations. *Science of the Total Environment* 286 (1–3): 167–179.

Garris, H.W., Baldwin, S.A., Van Hamme, J.D. et al. (2016). Genomics to assist mine reclamation: a review. *Restoration Ecology* 24 (2): 165–173.

Gentili, R., Sgorbati, S., and Baroni, C. (2011). Plant species patterns and restoration perspectives in the highly disturbed environment of the Carrara marble quarries (Apuan Alps, Italy). *Restoration Ecology* 19 (101): 32–42.

Giannini, T.C., Giulietti, A.M., Harley, R.M. et al. (2017). Selecting plant species for practical restoration of degraded lands using a multiple-trait approach. *Austral Ecology* 42 (5): 510–521.

Gonneau, C., Genevois, N., Frérot, H. et al. (2014). Variation of trace metal accumulation, major nutrient uptake and growth parameters and their correlations in 22 populations of *Noccaea caerulescens*. *Plant and Soil* 384: 271–287.

Grime, J.P. (1993). Vegetation functional classification systems as approaches to predicting and quantifying global vegetation change. In: *Vegetation Dynamics & Global Change*, 293–305. Boston, MA: Springer US.

Gucwa-Przepióra, E., Nadgórska-Socha, A., Fojcik, B., and Chmura, D. (2016). Enzymatic activities and arbuscular mycorrhizal colonization of *Plantago lanceolata* and *Plantago major* in a soil root zone under heavy metal stress. *Environmental Science and Pollution Research* 23: 4742–4755.

Han, M., Kim, Y., Koo, B.C., and Choi, G.W. (2011). Bioethanol production by *Miscanthus* as a lignocellulosic biomass: Focus on high efficiency conversion to glucose and ethanol. *Bioresources* 6 (2): 1939–1953.

Hindersah, R., Kalay, A.M., Komarya, A., and Kamaluddin, N.N. (2019). Natural revegetation of tailing deposited on agricultural area in Buru, Maluku. *IOP Conference Series: Earth and Environmental Science* 308 (1): 012054. IOP Publishing.

Huber-Humer, M. and Klug-Pümpel, B. (2004). The vegetation on different top covers of an abandoned solid waste landfill. *Die Bodenkultur* 55 (4): 155–163.

Iravanian, A. and Ravari, S.O., 2020. Types of contamination in landfills and effects on the environment: a review study. In *IOP Conference Series: Earth and Environmental Science* (Vol. 614, No. 1, p. 012083). IOP Publishing.

Jacobs, A., Drouet, T., and Noret, N. (2018). Field evaluation of cultural cycles for improved cadmium and zinc phytoextraction with *Noccaea caerulescens*. *Plant and Soil* 430: 381–394.

Jaffré, T., Rigault, F., and Sarrailh, J.M. (1994). Revegetation of former mining sites. *Bois et Forets des Tropiques* 242: 387179.

Jakobs, G., Weber, E., and Edwards, P.J. (2004). Introduced plants of the invasive *Solidago gigantea* (*Asteraceae*) are larger and grow denser than conspecifics in the native range. *Diversity and Distributions* 10 (1): 11–19.

Jakovljević, K., Bani, A., Pavlova, D. et al. (2022b). Hyperaccumulator plant discoveries in the Balkans: accumulation, distribution, and practical applications. *Botanica Serbica* 46 (2): 161–178.

Jakovljević, K., Mišljenović, T., Bačeva Andonovska, K. et al. (2023). Thallium hyperaccumulation status of the violets of the Allchar arsenic–thallium deposit (North Macedonia) confirmed through synchrotron μXRF imaging. *Metallomics* 15 (11): mfad063.

Jakovljević, K., Mišljenović, T., Savović, J. et al. (2020). Accumulation of trace elements in *Tussilago farfara* colonizing post-flotation tailing sites in Serbia. *Environmental Science and Pollution Research* 27: 4089–4103.

Jakovljević, K., Ranđelović, D., and Mišljenović, T. (2021). Phytoremediation of mine waste disposal sites: current state of knowledge and examples of good practice. In: *Biotechnology for Sustainable Environment*, 223–250. Springer.

Jakovljević, K., Tomović, G., Baker, A.J.M. et al. (2022a). Strategies of accumulation of potentially toxic elements in *Minuartia recurva* and *M. bulgarica*. *Environmental Science and Pollution Research* 29 (28): 43421–43434.

Jambhulkar, H.P., Shaikh, S.M.S., and Kumar, M.S. (2018). Fly ash toxicity, emerging issues and possible implications for its exploitation in agriculture; Indian scenario: a review. *Chemosphere* 213: 333–344.

Jennings, A.A., Cox, A.N., Hise, S.J., and Petersen, E.J. (2002). Heavy metal contamination in the brownfield soils of Cleveland. *Soil and Sediment Contamination* 11 (5): 719–750.

Jim, C.Y. (2001). Ecological and landscape rehabilitation of a quarry site in Hong Kong. *Restoration Ecology* 9 (1): 85–94.

Kabata-Pendias, A. (2010). *Trace Elements in Soils and Plants*. Boca Raton: CRC press.

Kasowska, D. and Koszelnik-Leszek, A. (2014). Ecological features of spontaneous vascular flora of serpentine post-mining sites in Lower Silesia. *Archives of Environmental Protection* 40 (2): 33–52.

Kidd, P.S., Bani, A., Benizri, E. et al. (2018). Developing sustainable agromining systems in agricultural ultramafic soils for nickel recovery. *Frontiers in Environmental Science* 6: 44.

Kim, K.R. and Owens, G. (2010). Potential for enhanced phytoremediation of landfills using biosolids–a review. *Journal of Environmental Management* 91 (4): 791–797.

Kompała-Bąba, A., Hanczaruk, R., and Bąba, W. (2022). Post-industrial habitats as centres of biodiversity–a case study on a sample of sedimentation pools. In: *Green Scenarios: Mining Industry Responses to Environmental Challenges of the Anthropocene Epoch*, 115–130. CRC Press.

Kopeć, D., Zając, I., and Halladin-Dąbrowska, A. (2011). The influence of surrounding vegetation on the flora of post-mining area. *Biodiversity Research and Conservation* 24 (2011): 29–38.

Koszelnik-Leszek, A., Podlaska, M., and Tomaszewska, K. (2013). Diversity of vascular flora of waste dumps and dumping grounds in Lower Silesia. *Archives of Environmental Protection* 39 (2): 81–105.

Kotanska, M., Buziak-Chmielowiec, E., Dabrowska, A. et al. (2015). Human impact on the plant cover of four villages in SE Poland. *Steciana* 19 (2): 115–121.

Kozhevnikova, A.D., Seregin, I.V., Aarts, M.G., and Schat, H. (2020). Intra-specific variation in zinc, cadmium and nickel hypertolerance and hyperaccumulation capacities in *Noccaea caerulescens*. *Plant and Soil* 452: 479–498.

Kursun Unver, I. and Terzi, M. (2018). Distribution of trace elements in coal and coal fly ash and their recovery with mineral processing practices: a review. *Journal of Mining and Environment* 9 (3): 641–655.

Lago-Vila, M., Arenas-Lago, D., Rodríguez-Seijo, A. et al. (2015). Cobalt, chromium and nickel contents in soils and plants from a serpentinite quarry. *Solid Earth* 6 (1): 323–335.

Lago-Vila, M., Rodríguez-Seijo, A., Arenas-Lago, D. et al. (2017). Heavy metal content and toxicity of mine and quarry soils. *Journal of Soils and Sediments* 17: 1331–1348.

Larondelle, N. and Haase, D. (2012). Valuing post-mining landscapes using an ecosystem services approach—an example from Germany. *Ecological Indicators* 18: 567–574.

Leung, H.M., Zhen-Wen, W.A.N.G., Zhi-Hong, Y.E. et al. (2013). Interactions between arbuscular mycorrhizae and plants in phytoremediation of metal-contaminated soils: a review. *Pedosphere* 23 (5): 549–563.

Levei, L., Kovacs, E., Hoaghia, M.A., and Ozunu, A. (2018). Accumulation of heavy metals in *Plantago major* grown in urban and post-industrial areas. *Studia Universitatis Babeş-Bolyai. Chemia* 63 (1): 87–98.

Li, D., Yin, N., Xu, R. et al. (2021). Sludge amendment accelerating reclamation process of reconstructed mining substrates. *Scientific Reports* 11 (1): 2905.

Li, S., Yang, B., and Wu, D. (2008). Community succession analysis of naturally colonized plants on coal gob piles in Shanxi mining areas, China. *Water, Air, and Soil Pollution* 193: 211–228.

Liao, J., Wen, Z., Ru, X. et al. (2016). Distribution and migration of heavy metals in soil and crops affected by acid mine drainage: public health implications in Guangdong Province, China. *Ecotoxicology and Environmental Safety* 124: 460–469.

Lin, H., Tang, Y., Dong, Y. et al. (2022). Characterization of heavy metal migration, the microbial community, and potential bioremediating genera in a waste-rock pile field of the largest copper mine in Asia. *Journal of Cleaner Production* 351: 131569.

Lottermoser, B.G. (2011). Recycling, reuse and rehabilitation of mine wastes. *Elements* 7 (6): 405–410.

Łowicki, D. and Fagiewicz, K. (2021). A new model of pollination services potential using a landscape approach: a case study of post-mining area in Poland. *Ecosystem Services* 52: 101370.

Luo, C., Routh, J., Dario, M. et al. (2020). Distribution and mobilization of heavy metals at an acid mine drainage affected region in South China, a post-remediation study. *Science of the Total Environment* 724: 138122.

Ma, S., Zhou, C., Pan, J. et al. (2022). Leachate from municipal solid waste landfills in a global perspective: characteristics, influential factors and environmental risks. *Journal of Cleaner Production* 333: 130234.

Madejón, P., Murillo, J.M., Marañón, T. et al. (2002). Bioaccumulation of As, Cd, Cu, Fe and Pb in wild grasses affected by the Aznalcóllar mine spill (SW Spain). *Science of the Total Environment* 290 (1–3): 105–120.

Makra, L., Matyasovszky, I., and Deák, Á.J. (2022). Ragweed in Eastern Europe. In: *Invasive Species and Global Climate Change*, 141–157. GB: CABI.

Manca, P.P., Massacci, G., Pintus, D., and Sogos, G. (2021). The flotation of sphalerite mine tailings as a remediation method. *Minerals Engineering* 165: 106862.

Márquez-Ferrando, R., Santos, X., Pleguezuelos, J.M., and Ontiveros, D. (2009). Bioaccumulation of heavy metals in the lizard *Psammodromus algirus* after a tailing-dam collapse in Aznalcóllar (Southwest Spain). *Archives of Environmental Contamination and Toxicology* 56: 276–285.

Matanzas, N., Afif, E., Díaz, T.E., and Gallego, J.R. (2021). Phytoremediation potential of native herbaceous plant species growing on a paradigmatic brownfield site. *Water, Air, & Soil Pollution* 232: 1–14.

Mataruga, Z., Jarić, S., Marković, M. et al. (2020). Evaluation of *Salix alba, Juglans regia* and *Populus nigra* as biomonitors of PTEs in the riparian soils of the Sava River. *Environmental Monitoring and Assessment* 192: 131.

Maus, V., Giljum, S., da Silva, D.M. et al. (2022). An update on global mining land use. *Scientific Data* 9 (1): 1–11.

McIntyre, S., Lavorel, S., and Tremont, R.M. (1995). Plant life-history attributes: their relationship to disturbance response in herbaceous vegetation. *Journal of Ecology* 83: 31–44.

Memmott, J., Craze, P.G., Waser, N.M., and Price, M.V. (2007). Global warming and the disruption of plant–pollinator interactions. *Ecology Letters* 10 (8): 710–717.

Midhat, L., Ouazzani, N., Hejjaj, A. et al. (2019). Accumulation of heavy metals in metallophytes from three mining sites (Southern Centre Morocco) and evaluation of their phytoremediation potential. *Ecotoxicology and Environmental Safety* 169: 150–160.

Millennium Ecosystem Assessment (MEA) (2005). *Ecosystems and Human Well-Being: Current State and Trends*, vol. 1. Washington, DC: Island Press.

Mišljenović, T., Jakovljević, K., Jovanović, S. et al. (2018). Micro-edaphic factors affect intra-specific variations in trace element profiles of *Noccaea praecox* on ultramafic soils. *Environmental Science and Pollution Research* 25: 31737–31751.

Mišljenović, T., Jovanović, S., Mihailović, N. et al. (2020). Natural variation of nickel, zinc and cadmium (hyper) accumulation in facultative serpentinophytes *Noccaea kovatsii* and *N. praecox*. *Plant and Soil* 447: 475–495.

Mleczek, M., Gąsecka, M., Waliszewska, B. et al. (2018). *Salix viminalis* L.-a highly effective plant in phytoextraction of elements. *Chemosphere* 212: 67–78.

Morrey, D.R., Baker, A.J.M., and Cooke, J.A. (1988). Floristic variation in plant communities on metalliferous mining residues in the northern and southern Pennines, England. *Environmental Geochemistry and Health* 10: 11–20.

Munford, K.E., Asemaninejad, A., Basiliko, N. et al. (2022). Native plants facilitate vegetation succession on amended and unamended mine tailings. *International Journal of Phytoremediation* 24 (9): 963–974.

Naveen, B.P., Sumalatha, J., and Malik, R.K. (2018). A study on contamination of ground and surface water bodies by leachate leakage from a landfill in Bangalore, India. *International Journal of Geo-Engineering* 9: 1–20.

Ogbonna, P.C., Emea, R., and Teixeira da Silva, J.A. (2011). Heavy metal concentration in soil and woody plants in a quarry. *Toxicological & Environmental Chemistry* 93 (5): 895–903.

Ogbonna, C.E., Nwafor, F.I., and Ogbonnaya, O.O. (2020). Dust accumulation, heavy metal content and stomata morphology of some medicinal plants at Rock Quarrying locations at Lokpaukwu, Nigeria. *International Journal of Environment and Climate Change* 10 (12): 540–549.

Orchard, C., Lobos, P.L., and Ginocchio, R. (2009). Phytostabilization of massive mine wastes with native phytogenetic resources: potential for sustainable use and conservation of the native flora in north-central Chile. *Ciencia e investigación agraria: revista latinoamericana de ciencias de la agricultura* 36 (3): 329–352.

Pandey, V.C. (2020a). *Phytomanagement of Fly Ash*. Elsevier.

Pandey, V.C. (2020b). Fly ash ecosystem services. In: *Phytomanagement of Fly Ash*, 257–288. Amsterdam: Elsevier https://doi.org/10.1016/B978-0-12-818544-5.00009-2.

Pandey, V.C. (2020c). Fly ash deposits–a potential sink for carbon sequestration. In: *Phytomanagement of Fly Ash*, 235–255. Amsterdam: Elsevier https://doi.org/10.1016/B978-0-12-818544-5.00008-0.

Pandey, V.C., Ahirwal, J., Roychowdhury, R., and Chaturvedi, R. (2023b). *Eco-Restoration of Mine Land*. Wiley (Authored Book), ISBN: 9781119872252.

Pandey, V.C., Ancona, V., Roy, M., and Ranđelović, D. (2023a). *Aromatic Plant-Based Phytoremediation*. Elsevier (Authored Book), ISBN: 9780443190827: https://doi.org/10.1016/C2022-0-00341-X.

Pandey, V.C., Bajpai, O., and Singh, N. (2016). Energy crops in sustainable phytoremediation. *Renewable and Sustainable Energy Reviews* 54: 58–73.

Pandey, V.C., Pandey, D.N., and Singh, N. (2015). Sustainable phytoremediation based on naturally colonizing and economically valuable plants. *Journal of Cleaner Production* 86: 37–39.

Pandey, V.C., Singh, K., Singh, J.S. et al. (2012). *Jatropha curcas*: a potential biofuel plant for sustainable environmental development. *Renewable and Sustainable Energy Reviews* 16 (5): 2870–2883.

Pandey, V.C., Singh, N., Singh, R.P., and Singh, D.P. (2014). Rhizoremediation potential of spontaneously grown *Typha latifolia* on fly ash basins: study from the field. *Ecological Engineering* 71: 722–727.

Parraga-Aguado, I., Gonzalez-Alcaraz, M.N., Alvarez-Rogel, J. et al. (2013). The importance of edaphic niches and pioneer plant species succession for the phytomanagement of mine tailings. *Environmental Pollution* 176: 134–143.

Parraga-Aguado, I., Querejeta, J.I., González-Alcaraz, M.N. et al. (2014). Usefulness of pioneer vegetation for the phytomanagement of metal (loid) s enriched tailings: grasses vs. shrubs vs. trees. *Journal of Environmental Management* 133: 51–58.

Peña González, E., Suárez López, J., Delgado Martín, J. et al. (2006). Analysis of the mobilization of solid loads and heavy metals in runoff waters from granite quarries. *Environmental Geology* 50: 823–834.

Peng, J.Y., Zhang, S., Han, Y. et al. (2022). Soil heavy metal pollution of industrial legacies in China and health risk assessment. *Science of the Total Environment* 816: 151632.

Piatak, N.M., Parsons, M.B., and Seal, R.R. II (2015). Characteristics and environmental aspects of slag: a review. *Applied Geochemistry* 57: 236–266.

Pidlisnyuk, V., Stefanovska, T., Lewis, E.E. et al. (2014). *Miscanthus* as a productive biofuel crop for phytoremediation. *Critical Reviews in Plant Sciences* 33 (1): 1–19.

Piekarska-Stachowiak, A., Szary, M., Ziemer, B. et al. (2014). An application of the plant functional group concept to restoration practice on coal mine spoil heaps. *Ecological Research* 29: 843–853.

Piwowarczyk, R., Ochmian, I., Lachowicz, S. et al. (2021). Correlational nutritional relationships and interactions between expansive holoparasite *Orobanche laxissima* and woody hosts on metal-rich soils. *Phytochemistry* 190: 112844.

Prach, K., Pyšek, P., and Bastl, M. (2001). Spontaneous vegetation succession in human-disturbed habitats: a pattern across seres. *Applied Vegetation Science* 4 (1): 83–88.

Prach, K. and Tolvanen, A. (2016). How can we restore biodiversity and ecosystem services in mining and industrial sites? *Environmental Science and Pollution Research* 23: 13587–13590.

Prica, M., Andrejić, G., Šinžar-Sekulić, J. et al. (2019). Bioaccumulation of heavy metals in common reed (*Phragmites australis*) growing spontaneously on highly contaminated mine tailing ponds in Serbia and potential use of this species in phytoremediation. *Botanica Serbica* 43 (1): 85–95.

Pyšek, P., Chocholousková, Z., Pyšek, A. et al. (2004). Trends in species diversity and composition of urban vegetation over three decades. *Journal of Vegetation Science* 15 (6): 781–788.

Pyšek, P., Jarošík, V., Chytrý, M. et al. (2005). Alien plants in temperate weed communities: prehistoric and recent invaders occupy different habitats. *Ecology* 86 (3): 772–785.

Qiu, J., Wu, Y., Luo, Y. et al. (2019). Effects of two pioneer plants on habitat improvement and heavy metal migration of lead-zinc smelting slags. *Journal of Agro-Environment Science* 38 (4): 798–806.

Ranđelović, D. (2023). Aromatic herbs and shrubs in phytoremediation. In: *Aromatic Plant-Based Phytoremediation, Socio-Economic and Agricultural Sustainability*, 75–124. Elsevier.

Ranđelović, D., Cvetković, V., Mihailović, N., and Jovanović, S. (2014). Relation between edaphic factors and vegetation development on copper mine wastes: a case study from Bor (Serbia, SE Europe). *Environmental Management* 53: 800–812.

Ranđelović, D., Jakovljević, K., and Jovanović, S. (2020b). The application of *Calamagrostis epigejos* (L.) Roth. in phytoremediation technologies. In: *Phytoremediation Potential of Perennial Grasses*, 259–282. Elsevier.

Ranđelović, D., Jakovljević, K., Mišljenović, T. et al. (2020c). Accumulation of potentially toxic elements in invasive *Ambrosia artemisiifolia* on sites with different levels of anthropogenic pollution. *Water, Air, & Soil Pollution* 231: 1–14.

Ranđelović, D. and Jovanović, S. (2023). Understanding the role of ruderal plant species in restoration of degraded lands. In: *Bio-Inspired Land Remediation*, 31–67. Cham: Springer International Publishing.

Ranđelović, D., Mihailović, N., and Jovanović, S. (2019). Potential of *Equisetum ramosissimum* Desf. for remediation of antimony flotation tailings: a case study. *International Journal of Phytoremediation* 21 (7): 707–713.

Ranđelović, D., Mutić, J., Marjanović, P. et al. (2020a). Geochemical distribution of selected elements in flotation tailings and soils/sediments from the dam spill at the abandoned antimony mine Stolice, Serbia. *Environmental Science and Pollution Research* 27: 6253–6268.

Reeves, R.D. and Brooks, R.R. (1983). Hyperaccumulation of lead and zinc by two metallophytes from mining areas of Central Europe. *Environmental Pollution Series A, Ecological and Biological* 31 (4): 277–285.

Reeves, R.D., Schwartz, C., Morel, J.L., and Edmondson, J. (2001). Distribution and metal-accumulating behavior of *Thlaspi caerulescens* and associated metallophytes in France. *International Journal of Phytoremediation* 3 (2): 145–172.

Řehounková, K., Vítovcová, K., and Prach, K. (2020). Threatened vascular plant species in spontaneously revegetated post-mining sites. *Restoration Ecology* 28 (3): 679–686.

Rich, K.J., Ridealgh, M., West, S.E. et al. (2015). Exploring the links between post-industrial landscape history and ecology through participatory methods. *PLoS One* 10 (8): e0136522.

Roccotiello, E., Marescotti, P., Di Piazza, S. et al. (2015). Biodiversity in metal-contaminated sites–problem and perspective–a case study. In: *Biodiversity in Ecosystems-Linking Structure and Function*. IntechOpen.

Rodríguez-Seijo, A. and Andrade, M.L. (2017). Characterization of soil physico-chemical parameters and limitations for revegetation in serpentine quarry soils (NW Spain). *Journal of Soils and Sediments* 17: 1321–1330.

Rostanski, A. and Wozniak, G. (2001). Grasses in the spontaneous vegetation of the post-industrial waste sites. In: *Studies on Grasses in Poland*, 313–327. W. Szafer Institute of Botany, Polish Academy of Science.

Salas-Luévano, M.A., Manzanares-Acuña, E., Letechipía-de León, C., and Vega-Carrillo, H.R. (2009). Tolerant and hyperaccumulators autochthonous plant species from mine tailing disposal sites. *Asian Journal of Experimental Sciences* 23 (1): 27–32.

Salisbury, A., Gallagher, F.J., Parag, H.A. et al. (2021). Plant diversity increases in an urban wildland after four decades of unaided vegetation development in a post-industrial site. *Urban Ecosystems* 24: 95–111.

Salisbury, A.B., Reinfelder, J.R., Gallagher, F.J., and Grabosky, J.C. (2017). Long-term stability of trace element concentrations in a spontaneously vegetated urban brownfield with anthropogenic soils. *Soil Science* 182 (2): 69–81.

Sanderson, M.A. and Adler, P.R. (2008). Perennial forages as second generation bioenergy crops. *International Journal of Molecular Sciences* 9 (5): 768–788.

Saria, L., Shimaoka, T., and Miyawaki, K. (2006). Leaching of heavy metals in acid mine drainage. *Waste Management & Research* 24 (2): 134–140.

Schmotzer, A., Takács, A., and Koscsó, J. (2021). A newcomer from the east: naturalisation of *Gypsophila perfoliata* L. around the city of Miskolc (Northeast Hungary). *Thaiszia* 31 (2): 171–194.

Shu, W.S., Ye, Z.H., Zhang, Z.Q. et al. (2005). Natural colonization of plants on five lead/zinc mine tailings in Southern China. *Restoration Ecology* 13 (1): 49–60.

Simonová, D. and Lososová, Z. (2008). Which factors determine plant invasions in man-made habitats in the Czech Republic? *Perspectives in Plant Ecology, Evolution and Systematics* 10 (2): 89–100.

Simpson, M.G., 2010. Plant Systematics (2a Edição) Cap. 2: Phylogenetic Systematics (pp. 17–52). Elsevier.

Sitko, K., Opała-Owczarek, M., Jemioła, G. et al. (2021). Effect of drought and heavy metal contamination on growth and photosynthesis of silver birch trees growing on post-industrial heaps. *Cells* 11 (1): 53.

Śliwka, M., Kępys, W., and Pawul, M. (2022). Analysis of the properties of coal sludge in the context of the possibility of using it in biological reclamation. *Gospodarka Surowcami Mineralnymi* 38: 173–189.

Smith, S.E. and Read, D.J. (2010). *Mycorrhizal Symbiosis*. Academic press.

Song, Y., Kirkwood, N., Maksimović, Č. et al. (2019). Nature based solutions for contaminated land remediation and brownfield redevelopment in cities: a review. *Science of the Total Environment* 663: 568–579.

Souza, C.A., Gallardo, A.L.C.F., Silva, É.D.D. et al. (2016). Environmental services associated with the reclamation of areas degraded by mining: potential for payments for environmental services. *Ambiente & Sociedade* 19: 137–168.

Stefanowicz, A.M., Kapusta, P., Błońska, A. et al. (2015). Effects of *Calamagrostis epigejos*, *Chamaenerion palustre* and *Tussilago farfara* on nutrient availability and microbial activity in the surface layer of spoil heaps after hard coal mining. *Ecological Engineering* 83: 328–337.

Sun, X., Xu, R., Dong, Y. et al. (2020). Investigation of the ecological roles of putative keystone taxa during tailing revegetation. *Environmental Science & Technology* 54 (18): 11258–11270.

Suo, Y., Tang, N., Li, H. et al. (2021). Long-term effects of phytoextraction by a poplar clone on the concentration, fractionation, and transportation of heavy metals in mine tailings. *Environmental Science and Pollution Research* 28 (34): 47528–47539.

Šálek, M. (2012). Spontaneous succession on opencast mining sites: implications for bird biodiversity. *Journal of Applied Ecology* 49 (6): 1417–1425.

Tayebi-Khorami, M., Edraki, M., Corder, G., and Golev, A. (2019). Re-thinking mining waste through an integrative approach led by circular economy aspirations. *Minerals* 9 (5): 286.

Thums, C.R., Farago, M.E., and Thornton, I. (2008). Bioavailability of trace metals in brownfield soils in an urban area in the UK. *Environmental Geochemistry and Health* 30: 549–563.

Trifi, M., Dermech, M., Abdelkrim, C. et al. (2018). Extraction procedures of toxic and mobile heavy metal fraction from complex mineralogical tailings affected by acid mine drainage. *Arabian Journal of Geosciences* 11: 1–12.

Tropek, R., Kadlec, T., Hejda, M. et al. (2012). Technical reclamations are wasting the conservation potential of post-mining sites. A case study of black coal spoil dumps. *Ecological Engineering* 43: 13–18.

Trueman, I.C., Tokarska-Guzik, B., and Cohn, E.V. (2022). Brownfield sites as hot spots of plant diversity. In: *Green Scenarios: Mining Industry Responses to Environmental Challenges of the Anthropocene Epoch*, 87–102. CRC Press.

Turnau, K., Jędrzejczyk, R., Domka, A. et al. (2018). Expansion of a holoparasitic plant, *Orobanche lutea* (Orobanchaceae), in post-industrial areas-a possible Zn effect. *Science of the Total Environment* 639: 714–724.

van der Ent, A., Baker, A.J.M., van Balgooy, M.M.J., and Tjoa, A. (2013). Ultramafic nickel laterites in Indonesia (Sulawesi, Halmahera): mining, nickel hyperaccumulators and opportunities for phytomining. *Journal of Geochemical Exploration* 128: 72–79.

Vaverková, M.D., Winkler, J., Adamcová, D. et al. (2019). Municipal solid waste landfill–vegetation succession in an area transformed by human impact. *Ecological Engineering* 129: 109–114.

Verma, S.K., Singh, K., Gupta, A.K. et al. (2014). Aromatic grasses for phytomanagement of coal fly ash hazards. *Ecological Engineering* 73: 425–428.

Vidican, R., Pleşa, A., Mihăiescu, T. et al. (2022). The need to restore post-industrial ecosystems: importance and functions of riparian brownfields. *Bulletin of University of Agricultural Sciences and Veterinary Medicine Cluj-Napoca. Agriculture* 79 (2): 59–70.

Vilakazi, A.Q., Ndlovu, S., Chipise, L., and Shemi, A. (2022). The recycling of coal fly ash: a review on sustainable developments and economic considerations. *Sustainability* 14 (4): 1958.

Vogel-Mikuš, K., Drobne, D., and Regvar, M. (2005). Zn, Cd and Pb accumulation and arbuscular mycorrhizal colonisation of pennycress *Thlaspi praecox* Wulf. (Brassicaceae) from the vicinity of a lead mine and smelter in Slovenia. *Environmental Pollution* 133 (2): 233–242.

Vriens, B., Peterson, H., Laurenzi, L. et al. (2019b). Long-term monitoring of waste-rock weathering at the Antamina mine, Peru. *Chemosphere* 215: 858–869.

Vriens, B., Skierszkan, E.K., St-Arnault, M. et al. (2019a). Mobilization of metal (oid) oxyanions through circumneutral mine waste-rock drainage. *ACS Omega* 4 (6): 10205–10215.

Wang, C., Harbottle, D., Liu, Q., and Xu, Z. (2014). Current state of fine mineral tailings treatment: a critical review on theory and practice. *Minerals Engineering* 58: 113–131.

Wang, H., Zhang, B., Bai, X., and Shi, L. (2018). A novel environmental restoration method for an abandoned limestone quarry with a deep open pit and steep palisades: a case study. *Royal Society Open Science* 5 (5): 180365.

Wang, J., Zhao, F., Yang, J., and Li, X. (2017). Mining site reclamation planning based on land suitability analysis and ecosystem services evaluation: a case study in Liaoning province, China. *Sustainability* 9 (6): 890.

Wang, S., Han, Z., Wang, J. et al. (2022). Environmental risk assessment and factors influencing heavy metal concentrations in the soil of municipal solid waste landfills. *Waste Management* 139: 330–340.

Wei, X., Zhou, Y., Tsang, D.C. et al. (2020). Hyperaccumulation and transport mechanism of thallium and arsenic in brake ferns (*Pteris vittata* L.): a case study from mining area. *Journal of Hazardous Materials* 388: 121756.

Wittig, R. (2004). The origin and development of the urban flora of Central Europe. *Urban Ecosystems* 7: 323–329.

Woch, M.W., Kapusta, P., and Stefanowicz, A.M. (2016). Variation in dry grassland communities along a heavy metals gradient. *Ecotoxicology* 25: 80–90.

Woźniak, G., Markowicz, A., Borymski, S. et al. (2015). The relationship between successional vascular plant assemblages and associated microbial communities on coal mine spoil heaps. *Community Ecology* 16: 23–32.

Woźniak, G., Sierka, E., and Wheeler, A. (2018). *Urban and Industrial Habitats: How Important They are for Ecosystem Services*, 169–194. London, UK: IntechOpen.

Xiao, E., Ning, Z., Xiao, T. et al. (2019). Variation in rhizosphere microbiota correlates with edaphic factor in an abandoned antimony tailing dump. *Environmental Pollution* 253: 141–151.

Yong, M.T., Babla, M., Karan, S. et al. (2022). Coal tailings as a soil conditioner: evaluation of tailing properties and effect on tomato plants. *Plant Growth Regulation* 98 (3): 439–450.

Zarzycki, K. and Szeląg, Z. (2006). Czerwona lista roślin naczyniowych w Polsce. In: *Czerwona lista roślin i grzybów Polski*, 11–20. Kraków: Inst. Bot. im. W. Szafera PAN.

Zhou, W., Dan, Z., Meng, D. et al. (2023). Distribution characteristics and potential ecological risk assessment of heavy metals in soils around Shannan landfill site, Tibet. *Environmental Geochemistry and Health* 45 (2): 393–407.

Zuquette, L.V., Pejon, O.J., and dos Santos Collares, J.Q. (2004). Land degradation assessment based on environmental geoindicators in the Fortaleza metropolitan region, state of Ceará, Brazil. *Environmental Geology* 45 (3): 408–425.

4

Restoration Ecosystem Toward Spontaneous Succession on Reclaimed Mining Sites

Marcin Pietrzykowski[1], Bartłomiej Woś[1], Justyna Likus-Cieślik[1], Marek Pająk[1], Edyta Sierka[2], and Barbara Stalmachová[3]

[1] *Department of Ecological Engineering and Forest Hydrology, Faculty of Forestry, University of Agriculture in Krakow, Krakow, Poland*
[2] *Institute of Biology, Biotechnology and Environmental Protection, University of Silesia in Katowice, Katowice, Poland*
[3] *Department of Environmental Engineering, Technical University of Ostrava, Ostrava-Poruba, Czech Republic*

4.1 Introduction

Spoil heaps and open-cast pits are two of the consequences of mining for fossil fuels and mineral resources that, despite the measures taken to transform economies and obtain energy from renewable sources, constitute an environmental problem. In addition, these by-products of mining generate an ongoing need for land reclamation (Pietrzykowski and Krzaklewski 2018). The geomorphological effects of both active mining operations and abandoned mines are diverse and include waste rock spoil heaps, contaminated sites, brownfields, and underutilized fallow land. Both surface and underground mining affect all components of the ecosystem (Bell and Donnelly 2006). Moreover, they despoil the pre-mining landscape and transform it into a new technogenic environment – referred to as the post-mining landscape or novel ecosystem — which is often far from equilibrium and, thus, sensitive to failure (Martín-Duque et al. 2010; Hobbs et al. 2013). Globally, spoil heaps as by-products of the Anthropocene geological age cover over 1% of the current landmass (Walker 1992). Novel ecosystems that develop on spoil heaps are still alien communities in the landscape and a serious environmental challenge in many mining regions around the world (Hobbs et al. 2013). An understanding of the technical handling of spoil heaps and their natural significance in urban and industrial areas will enable individuals and communities to support and restore ecological functions, including plant community succession, dynamic soil processes, soil microbiological activity, afforestation efficiency and biomass, and carbon sequestration (Pietrzykowski and Daniels 2014; Pietrzykowski and Krzaklewski 2018). Moreover, post-mining sites and their vegetation form key types of terrestrial ecosystems, and when aided by reclamation treatments, they perform ecological functions that are important for increasing the human environment's resilience to climate change, as well as providing ecosystem services that include crucial carbon sequestration

Biodiversity and Ecosystem Services on Post-Industrial Land, First Edition.
Edited by Vimal Chandra Pandey.

and water retention (Pietrzykowski and Krzaklewski 2018; Misebo et al. 2022; Pandey et al. 2023). These factors are critical for maintaining biological diversity, which is essential for sustaining ecological processes at both regional and global scales (Scherer-Lorenzen 2008). Hence, the need to mitigate the impact of coal mining on the environment and human life as part of a minimally invasive transformative process on the one hand, and the fundamental changes in the condition and functioning of the Earth's systems on the other (Steffen et al. 2015), by optimizing spoil heap reclamation treatments. Essential solutions include facilitating carbon storage in the vegetation, and soil and ecosystem services that underpin climate regulation (Zedler and Kercher 2005; Conti and Díaz 2013). Therefore, it is important to comprehensively understand the correlations between functional elements within reclaimed ecosystems and spoil heaps where spontaneous succession occurs.

Exposed substrates after mining exploitation usually exhibit physical and chemical properties that are unfavorable for plants, including extremely coarse texture, lack of soil organic matter (SOM), nutrient deficiency, unfavorable water–air properties, and, frequently, high salinity. These substrates are also characterized by poor water retention (i.e., low water capacity and high permeability), which varies depending on the type of vegetation cover (Pietrzykowski 2014). To support the environmental functions of these sites, for the past fifty years, they have been reclaimed by introducing afforestation or sowing grass and perennial species seeds on the sites (Macdonald et al. 2015). If spoil heaps remain untreated, they become subject to spontaneous vegetation growth and subsequent colonization by groups of organisms, such as bacteria, that decompose dead plant matter. Succession, based on rewilding or passive restoration, is used to reclaim post-industrial areas at a reduced cost. The use of succession increases the species diversity of an ecosystem and increases the chances of restoring fully functional ecosystems that are resistant to disease and other stresses (Macdonald et al. 2015; Woś et al. 2023a). The optimal strategies of reclamation include leaving strips of land for the spontaneous development of vegetation to diversify the habitat conditions in uniform post-mining areas. In both cases (reclamation and succession), such ecosystems may play an important function in mitigating climate change via carbon sequestration and water retention on a local and global scale.

This chapter presents an overview of issues related to the use of succession for reclaiming post-mining sites. The dynamics of ecosystem restoration under reclamation and succession, and the potential importance of succession for increasing the biodiversity of post-mining sites and its impact on soil formation processes, are discussed.

4.2 Biodiversity of Succession Communities on Mine Sites

The natural regeneration and restoration of forest ecosystems and tree stands typically constitute a long-term process (Espigares et al. 2011; Ahirwal and Pandey 2020). In contrast, typically reclamation is costly and may lead to significant biodiversity loss (Prach and Walker 2011). The biodiversity of plant communities that emerge through natural succession compared to afforestation is often similar in the first regeneration stage (Pietrzykowski 2008; Mudrák et al. 2010; Šebelíková et al. 2016). However, spontaneously revegetated post-mining sites may act as important secondary habitats for many threatened vascular plant species (Prach et al. 2011; Šebelíková et al. 2016; Řehounková et al. 2020). For instance, a well-documented phenomenon in post-industrial sites is the appearance of rare plant species, such as orchids, in biocenoses formed during succession (Pietrzykowski 2008; Rahmonov et al. 2019). Numerous rare and protected species have established niches in such environments, including *Lycopodium clavatum*, *Listera ovata*, *Dactylorhiza majalis*, *Epipactis atrorubens*, *Epipactis. helleborine*, *Epipactis palustris*, *Ornithogalum umbellatum*, and *Aquilegia vulgaris*. In total, 11 such species in Poland have been found at reclaimed sites. According to published data, 619 species of vascular plants have been recorded on heaps of the Silesian Voivodeship in Poland, including 14 species that are obligate halophytes (Sierka 2013). Tropek et al. (2010) recorded the occurrence of 93 species of vascular plants on selected mine spoil disposal sites in the Czech Republic while conducting faunal studies. In addition, Stalmachová and Frnka (2003) recorded 614 plant species on the spoil heaps of the Upper Silesia in the Czech Republic, and Rostański and Zhukov (2001) found over 260 species on selected mine spoil heaps in Ukraine.

The spontaneous development of vegetation on spoil heaps that are not subject to reclamation is a long-term process. This is because the excavated material is usually transported with heavy trucks that exert significant pressure on the ground (approximately $5\,kg\,cm^{-2}$ or more) and cause significant compaction, which inhibits root growth and hinders seed germination (Kirmer and Mahn 2001). The succession of plant communities on spoil heaps demonstrates primary plant succession characteristics. This means that the initiation of the development of the vegetation cover starts on rudimentary anthropogenic soils with devoid of initial of diasporas in the substrate. At the time of initiation, one of the major factors of community development on spoil heaps is the transport of diasporas from the surrounding area. The plants and plant communities colonize anthropogenic substrates in an approximate order and, thus, may be considered indicators of the condition of the soils and their development (Stalmachová and Sierka 2014).

The species that occur in the vicinity of the mining site are referred to as the local species pool (Zobel 1997). In other words, species richness on a smaller scale primarily depends on the number of species available in the surrounding area as a microcosm of the larger scale. This hypothesis is corroborated by findings reported by Ash et al. (1994), which show that spoil heaps are inhabited by the 20 species that produce very small seeds, have specialized requirements, and can travel up to 40 km to reach suitable places within post-industrial areas. In addition, these results show that the composition of spontaneously formed brownfield flora is dominated by anemochorous (Figure 4.1), heliophilous, and expansive species.

Furthermore, the species composition and diversity of spontaneous vegetation on brownfields do not depend on the size of the heaps. They are, however, closely linked with the local sources of diasporas and, therefore, depend on the diversity of flora in the vicinity. It was also found that, in addition to the advancement of the succession process, the number of species shared by the surrounding areas (Stalmachová 1992; Jochimsen 1996) and significant proportions of random and rare species increased (Figure 4.2).

The plants that colonize spoil heaps during the initial stages of succession on various types of heaps consist mainly of pioneer species, such as *Conyza canadensis*, *Senecio viscossus*, *Oenothera biennis*, and *Chamaenerion dodonaei*, as well as perennials such as *Tussilago farfara* and *Calamagrostis epigejos*. The latter species, usually dominate over other species while spreading and preventing succession from progressing further, as in degenerating forest communities (Sierka and Chmura 2005).

Figure 4.1 Spontaneous succession on hard coal mine spoil heap built mainly with carboniferous rocks (shells and sandstones) – anemochoric species, Zabrze, Poland. *Source:* Photo by E. Sierka, September 2021.

Figure 4.2 The process of spontaneous succession on hard coal mine spoil heap, Ostrava, Czech Republic. *Source:* Photo by B. Stalmachová, July 2018.

The initial stage on the northern to eastern slopes of fine-grained heaps is dominated by the bryophytes *Ceratodon purpureus*, *Bryum argenteum*, and *Funaria hygrometrica*, into which, over a period of three to five years, penetrate *Poa compressa*, *Poa annua*, and anemochorous phanerophytes, including *Betula pendula* and *Populus tremula*, among others (Stalmachová and Frnka, 2003).

On older heaps (i.e., 10–60 years old), decreasing stress from species competition leads to the formation of single-species-dominant aggregations that feature, for example, *Calamagrostis epigejos* and *Solidago* sp. or, on wet substrates, *Chamaerenion dodonaei* and *Myricaria germanica* (Stalmachová 1992), species that prefer the gravel areas of riverbanks. In heap locations, Rostański (2006) distinguished three groups of species that dominate numerically during spontaneous succession (Figure 4.3).

Anemochory and other light-seed-producing species appear first on the heaps (group A). Subsequently, the quantitative proportion of annual and perennial herbaceous plants (group B) and those that occur in all stages of vegetation development (group C) increases. In the next stage, perennials dominate the open space with a significant proportion of meadow species (group D), followed by perennial species that grow in both open and shaded areas (group E). Woody species (group F) appear in one of the last stages.

Long-term research conducted in the field has provided exciting results about the mechanisms of spontaneous vegetation, ecosystem development, and biodiversity enrichment within a broad spatiotemporal context (Woźniak 2010; Kompała-Bąba 2013). Industrial areas function as local biocenters and ecological corridors that participate in the coherent functioning of the overall natural environment (Rosenberg et al. 1997). These areas can also be considered alternatives to protecting native biodiversity (Oberndorfer et al. 2007). However, over a long period of time, only a relatively small number of native species in such areas have been recorded (Ratcliffe 1974). Anthropophytes accounted for

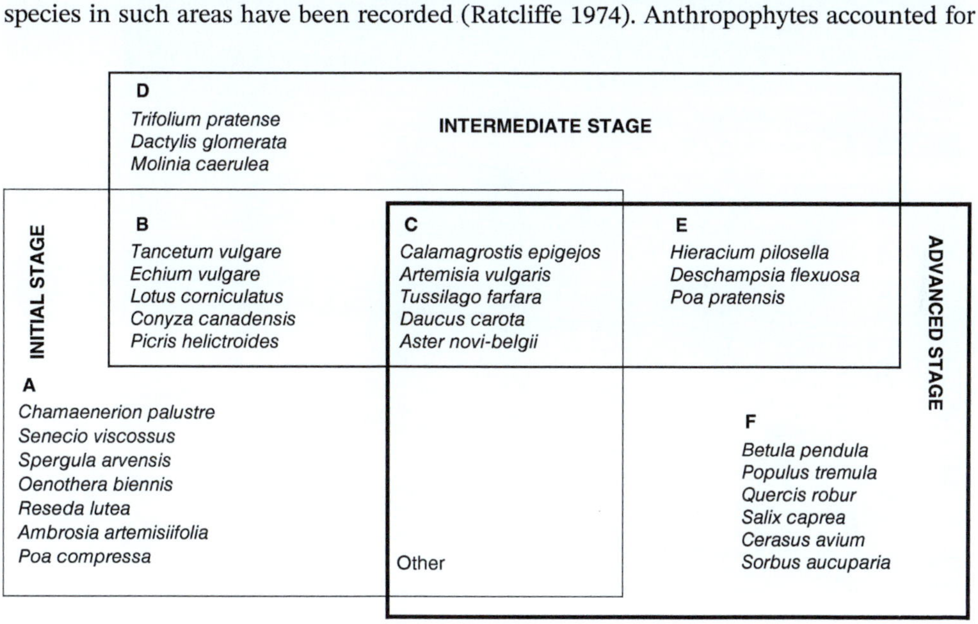

Figure 4.3 The stages of spontaneous succession on hard coal mine spoil heaps. *Source:* Adapted from Rostański (2006).

about 30% of species, and their number increased with the age of the heap (Rostański 2006; Woźniak 2010).

Ash et al. (1994) reported that native species thrive in post-industrial ecosystems, which provide them with propagules and microhabitats for their development. Heap ecosystems are threatened by expansive native species, however, which may threaten biological diversity (e.g., the common *Calamagrostis epigejos*, as well as alien species, such as *Solidago gigantea* and *Aster novi-belgii*). Both of these can spontaneously appear or be sown during inadequately managed renovation projects (Prach and Pyšek 2001).

Animal communities develop alongside vegetation succession (Sierka and Sierka 2008), many of which are often protected by law or included on Red Lists of threatened species (Tropek et al. 2010). Vojar (2006) noted that post-mining areas in the Czech Republic have been crucial for the survival of amphibians. The dominant species in patches of vegetation on spoil heaps typically vary, and these species, as described by Whittaker (1965), are the most competitive due to their biomass, biological activity, and use of resources. However, less abundant species typically define the scale of biodiversity in the patches.

This phenomenon is favored by the diverse microhabitat conditions of these areas, depending on their lithology, the method of overburden management, and the diaspora bank (Figures 4.4 and 4.5). Often, due to the non-selective deposition of diversified geological formations or the planned or accidental disruption of water availability in soils at post-mining sites, a mosaic of habitats and, consequently, a mosaic of plant communities forms (Pietrzykowski 2008; Rahmonov et al. 2019). Hendrychová et al. (2012) determined

Figure 4.4 Water as a key succession factor in an area of poor sandy substrates at a sand pit excavation influences the rate of plant settlement in the area (Szczakowa sand mine, southern Poland). *Source:* Photo by M. Pietrzykowski, June 2007.

Figure 4.5 Habitat diversity in a post-mining area under primary succession. Lusatian Lignite District, eastern Germany. *Source:* Photo by M. Pietrzykowski, May 2018.

that 47% of identified invertebrate species occurred on both reclaimed and succession sites in post-mining areas after lignite mining, although many were found exclusively on succession (37%) or reclamation (16%) sites. The authors concluded that reclamation and succession should be combined in practice to support invertebrate colonization (Hendrychová et al. 2012). In contrast, Tropek et al. (2012) reported a high proportion of endangered arthropod species on successional plots that were nearly absent from reclaimed plots. They suggested that technical reclamation is ineffective for biodiversity conservation and should be reserved for well-justified cases of public concern (Tropek et al. 2012). In addition, species richness and the abundance of amphibians were higher at succession sites compared to reclamation sites (Vojar et al. 2016). However, not all unreclaimed sites result in ecosystems with enhanced biodiversity (Moradi et al. 2018). Biodiversity at unreclaimed sites is often concentrated in small areas with specific conditions that constitute rather small sections of the whole area (Hendrychová and Bogusch 2016).

Spontaneous succession is the unmediated development of plant and animal communities on recently cleared sites in the landscape. With land reclamation, the facilitation model of ecological succession is especially particularly in practice. Succession begins with the initiation of simple biocenoses of R- and S-strategists (e.g., therophytes, chamaephytes, or phanerophytes), depending on the site conditions. The ruderal and segetal species from the early succession stages support development in the soil substrate and decrease and balance environmental extremes, including high irradiance, high substrate temperature, and a lack of nutrients (especially nitrogen), among others (Kompała-Bąba et al. 2020). The gradual shading of the ground surface, formation of litterfall, and gradual changes in soil properties create conditions that permit the spread of typical species for more mature succession stages. Seeding of the reserves or the distance of the site from diaspora sources, however, may be limiting factors (Stalmachová and Sierka 2014).

4.3 Mine Soil Development Under Reclamation and Successional Communities

Ecosystem development at post-mining sites involves not only with the colonization of plant and animal species but also with soil development processes (Pietrzykowski and Krzaklewski 2018; Uzarowicz et al. 2020; Mukhopadhyay and Masto 2022). The evidence regarding the effects of succession and afforestation on the dynamics of chemical and biological development in initial mining or post-mine soils, however, is contradictory. Chodak et al. (2009) observed that the reclamation of sand pits not only improved the mine soil's microbial properties more than that of succession but also promoted the faster development of metabolic capabilities characteristic of natural forest soil communities. Conversely, Šnajdr et al. (2013) noted that in clay-heavy soils on spoil heaps after lignite mining near Sokolov (Czech Republic), the microbial biomass and enzyme activities in sites under spontaneous succession were similar to that of afforested sites with different tree species. Under these habitat conditions, only soils under afforested stands of black and grey alder exhibited increased microbiological properties compared to those under succession communities (Helingerová et al. 2010; Šnajdr et al. 2013). Moreover, the soil macrofauna density was higher on reclaimed and afforested land with alders than on unreclaimed sites on spoil heaps from lignite mining (Frouz et al. 2007). However, the impact of reclamation on soil compaction may slow the development of the soil profile and deeper soil biota communities (Moradi et al. 2020).

At post-mining sites, the soil is formed either from overburden (spoils) as a parent material, a process that is often facilitated by succession (Macdonald et al. 2015; Woś et al. 2018), or by the application of topsoil (Čížková et al. 2018). Overburden materials typically contain low or no recent soil organic carbon (SOC) from litterfall (Misebo et al. 2022), and reclaimed topsoil also has lower SOC than undisturbed soils due to mineralization losses during storage and manipulation (Akala and Lal 2001; Shrestha and Lal 2011). Among soil properties, SOC and total nitrogen (TN) serve as suitable indicators of reclaimed mine soil quality and the rate of pedogenesis (Bandyopadhyay et al. 2020; Woś et al. 2023b). SOC plays a crucial role in improving soil properties, enhancing soil structure, and increasing soil porosity, aeration, and water-holding capacity (Yan et al. 2020). As a result of the increased carbon input from litter and belowground biomass, the revegetation of spoil heaps increases SOC accumulation (Józefowska et al. 2017). To further enhance SOC accumulation, identification of vegetation types and reclamation techniques with high storage capacity is vital for mitigating global climate change (Čížková et al. 2018; Yan et al. 2020).

The vegetation types selected for reclamation and appearing in succession play a significant role in carbon cycling due to their different means of storing carbon (C) in biomass, SOM, and litter (Misebo et al. 2022). Different types of vegetation and reclamation technology at mining sites affect the SOC and TN by varying organic matter input rates, decomposition rates, and organic matter distribution throughout the soil's depth. A critical review by Vindušková and Frouz (2013) reported findings on significantly lower soil C storage in coniferous forests than in grasslands and deciduous forests at reclaimed post-mining sites in the Northern Hemisphere. Zhang et al. (2020) also noted a higher accumulation of SOC and TN under grassland than under woodland in reclaimed mine soils on the Loess Plateau in China. Sroka et al. (2018) reported the significant role that black alder (*Alnus glutinosa*), a nitrogen-fixing species, plays in the accumulation of SOC and TN in sandy mining soils. Woś et al. (2022) investigated the effect of tree species distribution on carbon stock (Cstock) in the organic and mineral horizons and found that soil under pine had higher Cstock in the Oi + Oe horizons, whereas alder resulted in higher Cstock values in the 0–5 cm soil horizons compared to that of non-nitrogen-fixing tree species. In terms of reclamation, significantly higher SOC accumulation was observed at sites reclaimed with topsoil compared to that of bare rock under spontaneous succession (Akala and Lal 2001; Čížková et al. 2018). Comparative studies conducted in the Szczakowa sand pit (southern Poland) on the formation of soils during succession and full reclamation showed that reclamation increased the accumulation of humus in soils. After 25 years, approximately 9.0 $Mg\,ha^{-1}$ of organic carbon had accumulated in the soil of the sites left to succession, and approximately 14.5 $Mg\,ha^{-1}$ C was present at the reclaimed sites (Pietrzykowski and Krzaklewski 2007). Frouz et al. (2009) observed a similar pattern in the clay-like soils of post-lignite-mining areas. High carbon accumulation was observed in post-mining soils under 22–32-year-old stands of different tree species compared to sites left to natural regeneration and succession (Table 4.1; Frouz et al. 2009). Compared to succession, reclamation increased humus accumulation, humification, and humus transformation at higher rates on spoil heaps post-lignite mining. However, the differences

Table 4.1 The effects of succession and reclamation on soil organic carbon (SOC) accumulation in uppermost mineral soil initial horizons, as shown by selected European cases.

Vegetation	Age (years)	SOC accumulation ($kg\,ha^{-1}$ per year)	Substrate	Reference and type of mine extraction
Succession, forest	5	79	Quaternary sands	Pietrzykowski and Krzaklewski 2007; sand mining
	17	143		
	20	141		
	25	185		
Reclamation, forest	5	782		
	17	318		
	20	334		
	25	297		
Succession, forest	28	150	Neogene clays	Frouz et al. 2009; lignite mining
Reclamation, forest	22–33	250–1280		
Succession, grassland	50	Insignificant change in SOC content	Neogene clays	Čížková et al. 2018; lignite mining
Reclamation, grassland	50	1600	Neogene clays with topsoiling	

between reclaimed and unclaimed sites decreased with site age and were very small at 40-year-old sites (Abakumov et al. 2013). Increasing trends of SOC accumulation over time were observed on mine sites left to succession in the North Dakota Coal Basin in the United States, and the rate of SOC accumulation was 131 kg ha^{-1} per year (Wali 1999). Research conducted in the Szczakowa sand pit found that the SOC accumulation in the endohumus horizons was on average 186 kg ha^{-1} per year under natural regeneration (succession) and 296 kg ha^{-1} per year under the stands introduced during reclamation (Pietrzykowski and Krzaklewski 2007). Assuming that SOC accumulation remains at these levels, the estimated SOC storage in soils under succession will reach a state similar to that of the SOC in the podzols of undisturbed and natural sites in approximately 350 years, whereas that of soils in reclaimed areas will do so in approximately 250 years. It is worth noting, however, that it is difficult to predict the dynamics of changes in SOC accumulation in the future, which depend on many variables, although similar estimation results were obtained by modeling SOC accumulation in soils under succession communities in post-lignite-mining areas in prairie regions of the United States. These studies estimated that post-mining soils would reach values of SOC storage similar to that of the natural soils of the adjacent areas after about 250–300 years (West and Wali 2002).

Nitrogen (N) accumulation is also related to the development of organic horizons and the accumulation of organic matter during pedogenesis in post-mining sites. N is a scarce nutrient at post-mining sites and the rate of its accumulation may vary substantially (Woś et al. 2022, 2023b). The rate of N accumulation in post-mining soils formed under succession communities on heaps from post-lignite mining in the prairie region of the United States was estimated at approximately 24 kg ha^{-1} per year in the first 17 years and up to 36 kg ha^{-1} per year by year 30. Subsequently, however, a decrease in the rate of N accumulation of approximately 16.2 kg ha^{-1} per year was observed (Wali 1999). Studies conducted in Poland on sand pit excavations (Pietrzykowski and Krzaklewski 2007) estimated the N accumulation in the organic and mineral soil horizons under succession communities at about 17.6 kg ha^{-1} per year (on average over 25 years), and in the case of reclaimed areas, this value was much higher at approximately 40 kg ha^{-1} per year. Pietrzykowski et al. (2017) emphasized the beneficial effects of green manuring with lupine (*Lupinus luteus*) cultivation in the temporal climatic zone (central and Eastern Europe conditions) and mineral fertilization during biological reclamation on sites with high nitrogen accumulation in the early years.

4.4 Biomass and Wood Productivity Potential of Reclaimed and Successional Plant Communities

The benefits of reclamation and artificial afforestation of post-mining sites include higher wood and biomass production in reclaimed areas than in those left to succession (Skousen et al. 1994; Pietrzykowski and Krzaklewski 2009; Vacek et al. 2021). Conversely, Frouz et al. (2015) reported higher woody biomass only in younger stands (<20 years) of reclaimed sites compared to sites left to natural regeneration, although the difference disappeared in older stands (Frouz et al. 2015). Plant communities from succession are often heterogeneous and significantly different from each other even over a small area (Figure 4.6; Pietrzykowski and Krzaklewski 2009). This observation may explain the differences reported in the

Figure 4.6 Irregular surface of post-mine pit excavation coverage by succession communities, an area left for succession (Szczakowa sand mine, southern Poland). *Source:* Photo by B. Woś, May 2017.

literature regarding the growth parameters achieved by stands from succession and afforestation and their influence on soil properties (Groninger et al. 2017). In North America, when succession occurs under favorable conditions, some fast-growing timber trees may grow to harvestable size as early as 30–40 years after disturbance, while slower-growing hardwoods may require 50–60 years or longer. Other sites may remain in the grass-herb-shrub stage, with only scattered trees, for several decades after a disturbance, because the soil conditions are unsuitable or the understory vegetation is too competitive for tree growth (Groninger et al. 2017). On the one hand, as most post-mining sites are intended for forestry use, large areas cannot be left untended for decades, if doing so fails to result in woody vegetation (Krümmelbein et al. 2012); on the other, succession communities contribute to improving forest ecosystems' resilience to climate change and environmental stresses (Conradi et al. 2021), along with increasing their biodiversity (Řehounková et al. 2020). Therefore, finding a balance between wood production and environmental protection suggests the need to include the most promising communities identified from succession in the planned afforestation of post-mining sites.

4.5 Plant Communities from Succession as Indicators of Site Conditions

The correct identification of habitat conditions at mining sites, and the planning of species composition for afforestation have foundational impacts on the stability of reclaimed stands and developed forest ecosystems (Heinsdorf 1996; Pietrzykowski et al. 2010).

The assessment and classification of habitats on mining soils reclaimed for forestry are crucial for selecting the optimal composition of species for afforested sites (Pietrzykowski 2014; Krzaklewski 2017). When selecting appropriate tree species, their ecological requirements must be well-suited to the habitat's conditions (i.e., climate, fertility, and soil humidity; Krzaklewski 2017). An important indicator for assessing the initial conditions, the difficulties of biological reclamation, and the site conditions is the vegetation that spontaneously appears in such areas (Krzaklewski 1988). Plants in post-mining sites function as bioindicators because their presence or absence, and the reactions of individual plants in the population, indicate the intensity of the ecological factors involved (Krzaklewski 1977, 2017). Each plant species has specific ecological requirements from the environment. Therefore, plants have long been used as indicators of habitat status (Ellenberg 2009), including in the practice of ecological engineering (Begemann and Schiechtl 1999) and in relation to reclaimed post-mining areas as indicators of reclamation success (Krzaklewski 2017). The spontaneous appearance of vegetation on post-mining sites provides information not only about site conditions but also about the direction of changes in habitat conditions, which are important features at post-mining sites where soil properties often change abruptly (Pietrzykowski 2014).

One of the least expensive methods for establishing an initial diagnosis of reclamation difficulties and assessing habitat conditions is to examine the rate of overgrowth in the area from succession vegetation using the method proposed by Krzaklewski (1982). In this method, the coverage of the area by vegetation (as a percentage) and the area's age after the end of mining exploitation are used to calculate the surface overgrowth index (Wz = coverage in %/age in years). The higher the overgrowth rate, the easier the biological reclamation of the site will be. Based on the overgrowth index, post-industrial areas were divided into three groups (Krzaklewski 1982):

I) sites with very slow overgrowth rates or practically failing to overgrow (i.e., no spontaneous overgrowth for 10 years) means biological reclamation will be very challenging.
II) slowly overgrowing sites (i.e., no spontaneous overgrowth for 5–10 years) means biological reclamation will be difficult.
III) fast overgrowing sites (i.e., no overgrowth for no more than two years) means biological reclamation will be easy.

Another pioneering method developed by Krzaklewski (2017) can be applied to the second and third groups and employs the indicator properties of plant species and soil characteristics. This method includes the degree of coverage of post-mining sites by vegetation, divided into dominant and co-dominant species, which indicate the habitat conditions. For example, in post-lignite-mining areas in Poland, oligotrophic sands are dominated by *Corispermum leptopterum*, mesotrophic sandy loams by *Lepidium densiflorum* and *Equisetum arvense*, and fertile loams and clays by *Tussilago farfara* (Figure 4.7). Thus, the habitat conditions, in conjunction with the soil properties, can be described. By analyzing these factors, the current site fertility can be determined, and appropriate methods of biological reclamation can be selected.

Figure 4.7 An example of *Tussilago farfara* as an indicator of fertile soils at post-mining sites, according to the pioneering assumptions of the Krzaklewski method (1977). Loams and Clays on the external heap of the Szczerców Field, Bełchatów lignite mine, central Poland. *Source:* Photo by B. Woś; June 2018.

4.6 Managed Succession on Post-Mine Sites

Utilizing natural, spontaneous processes that occur during ecosystem renewal represents one of the most effective and inexpensive solutions for restoring landscape functions in post-mining areas. Managed succession to the regeneration of landscape fragments is managed succession. This process employs practitioners' experience and knowledge about the processes of spontaneous succession, and its patterns and rules, to accelerate and direct the reclamation process in a relatively inexpensive and ecologically optimal way.

Managed succession involves a combination of natural succession and human intervention (i.e., control of invasive species and planting of trees typical of later succession stages; Hendrychová et al. 2020). Its main objective is to protect and restore ecological stability and comprehensively protect and regenerate local biodiversity. The method uses a close-to-nature approach to rehabilitating recently damaged elements in open landscapes. In doing so, it relies on the self-regeneration capabilities of regeneration and the natural process of biocenosis renewal.

Technologies for managed succession may be utilized in areas where spontaneous succession would take too long, and the migration routes of biological diasporas have been disrupted. The application of managed succession is grounded in the tolerance model of succession, the principles of which permit the planting and seeding of species typical of more mature stages in the area, as long as they are comparable with the natural communities. It has been established (Stalmachová and Frnka, 2003; Sierka et al. 2012) that, especially in ecotonal and forest communities, multi-species combinations decrease the possibility of the competitive exclusion of more competitive species, even in areas where the initial site conditions somewhat diverge from natural conditions (e.g., through made-up ground or the planting of "talus forests" in the waste rock with no overlay, etc.).

The planting of framework wood species, combined with the spontaneous succession of herbs and complementary shrubs during land reclamation, results in areas progressing toward the renewal of natural and near-natural communities and ecosystems in an open landscape. Forms of managed succession may arise even in places where "normal" reclamation is applied, and where the target is arable land, among other possibilities (Lacková et al. 2012).

Increasing the biodiversity of all ecosystem services (Woźniak et al. 2018), which depend on ecosystem functions, is the main aim of ecological restoration (Whisenant 1999; Farber et al. 2006). In post-mining and post-industrial sites, biodiversity and ecosystem function restoration and enhancement should be related to the broader landscape and various local microhabitats within a broad spatiotemporal context (Sierka et al. 2022).

4.7 Example of a Reclamation by Using Vegetation Communities from Succession

The mosaics of succession communities can be successfully used as elements to increase the biodiversity of communities and the ecosystem (Figure 4.8). In the 1990s, in the pit of the Szczakowa sand mine (southern Poland), a method of forest reclamation that used vegetation from succession was implemented (Krzaklewski and Frączek 1999; Pietrzykowski 2015). First, the succession vegetation appeared in the areas that were left undisturbed for several years after exploitation was mapped. The soil properties and groundwater depths were subsequently determined, and a map of potential habitats with varying degrees of fertility and vegetation cover with the dominant species was developed. Reclamation treatments based on this material were planned for selected areas. For example, a category of areas without vegetation or with herbaceous vegetation cover <50% (e.g., herb communities with *Corynephorus canescens*) was designated. In this category, a full cycle of reclamation treatments was planned, including the addition of organic matter, the introduction of legume plants, mineral fertilization, and afforestation. Then, a category with vegetation coverage of 50–75% was selected. In this category, a partial cycle of reclamation treatment was planned that involved fertilization and green manure without the addition of organic matter.

For areas with vegetation coverage >75% (i.e., communities with predominant populations of *Calamagrostis epigejos* and *Poa compressa*, surface coverage in the herbaceous layer of 75–100%, and *Pinus sylvestris* and *Betula pendula* in the tree layer), only mineral fertilization with NPK was planned. The next category consisted of areas covered mainly with vegetation characteristic of coniferous forests, with coverage in the tree layer >50%, in the lower layer (shrubs) >25%, and in the herbaceous vegetation layer >75%. In this category, the only treatments consisted of regulating the species composition by cutting trees in excessively dense

Figure 4.8 An example of including succession communities in reclaimed and restored forest ecosystems. Szczakowa sand pit, southern Poland. *Source:* Photo by M. Pietrzykowski, spring 2007.

biogroups and planting in coverage gaps. Wetlands were distinguished as a separate category; these occurred along watercourses (e.g., drainage channels) and in areas with higher groundwater levels (approximately >40 cm). Such areas were included as a valuable element for increasing the biodiversity of the restored ecosystems and the rare species habitat. This method was successfully implemented on 257 ha in the Szczakowa sand mine pit (Krzaklewski and Frączek 1999; Pietrzykowski and Krzaklewski 2009).

4.8 Conclusion

The potential for using spontaneous succession processes in the reclamation of post-mining areas remains underestimated. The mosaics of succession communities can be successfully employed to increase the biodiversity of regenerated communities. However, biodiversity of plant communities in the restored post-mine ecosystem by way of succession and afforestation in the long-time perspective can be generally similar in both cases. However, spontaneously revegetated post-mining sites may act as important secondary habitats for many threatened animal and plant species, such as orchids. In the context of biodiversity, restored wetlands along watercourses (e.g., drainage channels) and areas with higher groundwater levels (approximately >40 cm) should be given special attention. Current evidence for the different effects of succession and afforestation on the dynamics of the developing chemical and biological properties of post-mining soils is contradictory. Results from some studies suggest that succession communities have similar effects on soil properties as that of afforestation communities. Other studies have reported that reclamation treatments influence the intensification of the soil-forming process, which particularly manifests in a higher accumulation of carbon and nitrogen than that of soils under succession. The benefits of the reclamation and man-made afforestation of post-mining sites include increased wood and biomass production in reclaimed areas compared to those left to succession. In any case, prior to reclamation treatment, succession vegetation can be employed as an indicator of habitat conditions. The use of succession increases the chances of restoring fully functional ecosystems that are resistant to disease and other stressors. For best practice reclamation solutions, some areas of post-mining sites should be left undisturbed, however, to enable the development of spontaneous vegetation for diversifying habitat conditions in otherwise uniform post-mining areas.

Acknowledgments

This study was financed by the National Science Centre, Poland (Grant Number 2020/39/B/ST10/00862).

References

Abakumov, E.V., Cajthaml, T., Brus, J., and Frouz, J. (2013). Humus accumulation, humification, and humic acid composition in soils of two post-mining chronosequences after coal mining. *Journal of Soils and Sediments* 13: 491–500.

Ahirwal, J. and Pandey, V.C. (2020). Restoration of mine degraded land for sustainable environmental development. *Restoration Ecology* 29 (4): e13268.

Akala, V.A. and Lal, R. (2001). Soil organic carbon pools and sequestration rates in reclaimed minesoils in Ohio. *Journal of Environmental Quality* 30: 2098–2104.

Ash, H.J., Gemmell, R.P., and Bradshaw, A.D. (1994). The introduction of native plant species on industrial waste heaps: a test of immigration and other factors affecting primary succession. *Journal of Applied Ecology* 31: 74–84.

Bandyopadhyay, S., Novo, L.A.B., Pietrzykowski, M., and Maiti, S.K. (2020). Assessment of forest ecosystem development in coal mine degraded land by using integrated mine soil quality index (IMSQI): the evidence from India. *Forests* 11 (12): 1310.

Begemann, W. and Schiechtl, H.M. (1999). *Ecological Engineering in Ground and Water Construction*. Arkady, Warsaw, Poland.

Bell, F.G. and Donnelly, L.J. (2006). *Mining and Its Impact on the Environment*. London and New York: Taylor & Francis Group.

Chodak, M., Pietrzykowski, M., and Niklińska, M. (2009). Development of microbial properties in a chronosequence of sandy mine soils. *Applied Soil Ecology* 41 (3): 259–268.

Čížková, B., Woś, B., Pietrzykowski, M., and Frouz, J. (2018). Development of soil chemical and microbial properties in reclaimed and unreclaimed grasslands in heaps after opencast lignite mining. *Ecological Engineering* 123: 103–111.

Conradi, T., Henriksen, M.V.J., and Svenning, J.-C. (2021). Global change, novel ecosystems and the ecological restoration of post-industrial areas: the case of a former brown coal mine in Søby, Denmark. *Applied Vegetation Science* 24 (3): e12605.

Conti, G. and Díaz, S. (2013). Plant functional diversity and carbon storage – an empirical test in semi-arid forest ecosystems. *Journal of Ecology* 101: 18–28.

Ellenberg, H. (2009). *Vegetation Ecology of Central Europe*, Reprint. Cambridge: Cambridge University Press.

Espigares, T., Moreno-de las Heras, M., and Nicolau, J.M. (2011). Performance of vegetation in reclaimed slopes affected by soil erosion. *Restoration Ecology* 19 (1): 35–44.

Farber, S., Costanza, R., Childers, D.L. et al. (2006). Linking ecology and economics for ecosystem management. *BioScience* 6 (2): 121–133.

Frouz, J., Dvorščík, P., Vávrová, A. et al. (2015). Development of canopy cover and woody vegetation biomass on reclaimed and unreclaimed post-mining sites. *Ecological Engineering* 84: 233–239.

Frouz, J., Pižl, V., Cienciala, E., and Kalčík, J. (2009). Carbon storage in post-mining forest soil, the role of tree biomass and soil bioturbation. *Biogeochemistry* 94: 111–121.

Frouz, J., Pižl, V., and Tajovský, K. (2007). The effect of earthworms and other saprophagous macrofauna on soil microstructure in reclaimed and un-reclaimed post-mining sites in Central Europe. *European Journal of Soil Biology* 43 (Supplement 1): S184–S189.

Groninger, J., Skousen, J., Angel, P. et al. (2017). Mine reclamation practices to enhance forest development through natural succession. In: *The Forestry Reclamation Approach: Guide to Successful Reforestation of Mined Lands*. Gen. Tech. Rep. NRS-169. (ed. M.B. Adams), 8-1–8-7. Newtown Square, PA: U.S. Department of Agriculture, Forest Service, Northern Research Station.

Heinsdorf, D. (1996). Development of forest stands in the Lusatian lignite mining district after mineral fertilization adopted to site and tree species. *Water, Air & Soil Pollution* 91: 33–42.

Helingerová, M., Frouz, J., and Šantrůčková, H. (2010). Microbial activity in reclaimed and unreclaimed post-mining sites near Sokolov (Czech Republic). *Ecological Engineering* 36 (6): 768–776.

Hendrychová, M., Šálek, M., Tajovský, K., and Řehoř, M. (2012). Soil properties and species richness of invertebrates on afforested sites after brown coal mining. *Restoration Ecology* 20 (5).

Hendrychová, M. and Bogusch, P. (2016). Combination of reclaimed and unreclaimed sites is the bestpractice for protection of aculeate Hymenoptera species on browncoal spoil heaps. *Journal of Insect Conservation* 20: 807–820.

Hendrychová, M., Svobodova, K., and Kabrna, M. (2020). Mine reclamation planning and management: integrating natural habitats into post-mining land use. *Resources Policy* 69: 101882.

Hobbs, R.J., Higgs, E.S., and Hall, C. (2013). *Novel Ecosystems: Intervening in the New Ecological World Order*, 384. Wiley-Blackwell.

Jochimsen, M.E.A. (1996). Reclamation of colliery mine spoil founded on natural succession. *Water, Air, and Soil Pollution* 91 (1–2): 99–108.

Józefowska, A., Pietrzykowski, M., Woś, B. et al. (2017). The effects of tree species and substrate on carbon sequestration and chemical and biological properties in reforested post-mining soils. *Geoderma* 292: 9–16.

Kirmer, A. and Mahn, E.G. (2001). Spontaneous and initiated succession on unvegetated slopes in the abandoned lignite-mining area of Goitsche, Germany. *Applied Vegetation Science* 4 (1): 19–27.

Kompała-Bąba, A. (2013). *Abiotic and Biotic Factors Affecting the Diversity of Ruderal Vegetation (Silesian Uplands Poland)*. Sorus, Poznań, s. 160.

Kompała-Bąba, A., Sierka, E., Dyderski, M.K. et al. (2020). Do the dominant plant species impact the substrate and vegetation composition of post-coal mining spoil heaps? *Ecological Engineering* 143: 105685.

Krümmelbein, J., Bens, O., Raab, T., and Naeth, M.A. (2012). A history of lignite coal mining and reclamation practices in Lusatia, eastern Germany. *Canadian Journal of Soil Science* 92: 53–66.

Krzaklewski, W. (1977). The spontaneous plants as a site conditions indicator and beckground to reforestation of spoil banks exemplified by Lignite Mine "Adamów" (in Polish). Ph.D. Thesis, University of Agriculture in Kraków, Poland, pp. 74.

Krzaklewski, W. (1982). Die Möglichkeiten der Ausnutzung der natürlichen Vegetation in der Recultivierung der Nachabbaugelände (Abraum oder Berghalden). *Internationale Haldenfachtagung*, 79–84, Essen (7–10 September 1982). Kommunalverband Ruhrgebit.

Krzaklewski, W. (1988). Some methodological aspects of planning and implementation of forest reclamation on the example of open cast mining. *Academy Mineral Metallurgy Krakow Science Paper AGH* 1222: 331–338.

Krzaklewski, W. (2017). *The Basics of Forest Reclamation*. Krakow: University of Agriculture in Krakow Publishing Hous.

Krzaklewski, W. and Frączek, M. (1999). The method of reclamation to forestry of old sand pits with the use of vegetation from spontaneous succession. Proceedings International Conference, Open-pit mining – Environment – Reclamation with particular emphasis on KWB "Bełchatów", pp. 111–127, Krakow (8–9 June 1999). Bełchatow: SC Drukpol Press.

Lacková, E., Žampachová, D., Čmielová, L., and Stalmachová, B. (2012). Effect of forest reclamation on the Křivý Důl spoil heap (Upper Silesia, Czech Republic). *Carpathian Journal of Earth and Environmental Sciences*, Baia Mare, Romania 7 (1): 205–212.

Macdonald, S.E., Landhäusser, S.M., Skousen, J. et al. (2015). Forest restoration following surface mining disturbance: challenges and solutions. *New Forests* 46: 703–732.

Martín-Duque, J.F., Sanz, M.A., Bodoque, J.M. et al. (2010). Restoring earth surface processes through landform design. A 13-year monitoring of a geomorphic reclamation model for quarries on slopes. *Earth Surface Processes and Landforms* 35 (5): 531–548.

Misebo, A.M., Pietrzykowski, M., and Woś, B. (2022). Soil carbon sequestration in novel ecosystems at post-mine sites—a new insight into the determination of key factors in the restoration of terrestrial ecosystems. *Forests* 13 (1): 63.

Moradi, J., Potocký, P., Kočárek, P. et al. (2018). Influence of surface flattening on biodiversity of terrestrial arthropods during early stages of brown coal spoil heap restoration. *Journal of Environmental Management* 220: 1–7.

Moradi, J., John, K., Vicentini, F. et al. (2020). Vertical distribution of soil fauna and microbial community under two contrasting post mining chronosequences: sites reclaimed by alder plantation and unreclaimed regrowth. *Global Ecology and Conservation* 23: e01165.

Mudrák, O., Frouz, J., and Velichová, V. (2010). Understory vegetation in reclaimed and unreclaimed post-mining forest stands. *Ecological Engineering* 36 (6): 783–790.

Mukhopadhyay, S. and Masto, R.E. (2022). Comparative evaluation of *Cassia siamea* and *Albizia lebbeck* for their potential to recover carbon and nutrient stocks in a chronosequence post-mining site. *CATENA* 208: 105726.

Oberndorfer, E., Lundholm, J., Bass, B. et al. (2007). Green roofs as urban ecosystems: ecological structures, functions, and services. *BioScience* 57 (10): 823–833.

Pandey, V.C., Ahirwal, J., Roychowdhury, R., and Chaturvedi, R. (2023). *Eco-Restoration of Mine Land*. Wiley.

Pietrzykowski, M. (2008). Soil and plant communities development and ecological effectiveness of reclamation on a sand mine cast. *Journal of Forest Science* 54 (12): 554–565.

Pietrzykowski, M. (2014). Soil quality index as a tool for Scots pine (*Pinus sylvestris*) monoculture conversion planning on afforested, reclaimed mine land. *Journal of Forestry Research* 25: 63–74.

Pietrzykowski, M. (2015). Reclamation and reconstruction of terrestrial ecosystems on mine sites – ecological effectiveness assessment. In: *Energy Science and Technology Series*, vol. 2 (ed. S. Sri, S.U. Chandra, and P. Ram), 121–151, 120-151. Houston, New Delhi: Studium Press LLC.

Pietrzykowski, M. and Daniels, W.L. (2014). Estimation of carbon sequestration by pine (*Pinus sylvestris* L.) ecosystems developed on reforested post-mining sites in Poland on differing mine soil substrates. *Ecological Engineering* 73: 209–218.

Pietrzykowski, M., Gruba, P., and Sproull, G. (2017). The effectiveness of Yellow lupine (*Lupinus luteus* L.) green manure cropping in sand mine cast reclamation. *Ecological Engineering* 102: 72–79. https://doi.org/10.1016/j.ecoleng.2017.01.026.

Pietrzykowski, M. and Krzaklewski, W. (2007). Soil organic matter, C and N accumulation during natural succession and reclamation in an opencast sand quarry (southern Poland). *Archives of Agronomy and Soil Science* 53 (5): 473–483.

Pietrzykowski, M. and Krzaklewski, W. (2009). *Reclamation to Forest in Sand-filing Mining Exemplified on Szczakowa Sand Mine Cast. Monography*. Krakow: Department of Forest Ecology, University of Agriculture in Krakow and Monos.

Pietrzykowski, M. and Krzaklewski, W. (2018). Reclamation of mine lands in Poland. *Bio-Geotechnologies for Mine Site Rehabilitation* 493–513.

Pietrzykowski, M., Krzaklewski, W., Pająk, M. et al. (2010). *Analysis and Optimization of Site Classification and Criteria of Forest Reclamation Assessment in Selected Mining Areas in Poland*. Krakow: University of Agriculture in Krakow Publishing Hous.

Prach, K. and Pyšek, P. (2001). Using spontaneous succession for restoration of human disturbed habitats: experience from Central Europe. *Ecological Engineering* 17: 55–62.

Prach, K., Řehounková, K., Řehounek, J., and Konvalinková, P. (2011). Ecological restoration of central European mining sites: a summary of a multi-site analysis. *Landscape Research* 36 (2): 263–268.

Prach, K. and Walker, L.R. (2011). Four opportunities for studies of ecological succession. *Trends in Ecology & Evolution* 26 (3): 119–123.

Ratcliffe, D. A. (1974). Ecological effects of mineral exploitation in the United Kingdom and their significance to nature conservation. *Proceedings of the Royal Society of London A* 339: 355–372. https://doi.org/10.1098/rspa.1974.0126.

Rahmonov, O., Rozkowski, J., and Szymczyk, A. (2019). Is the vegetation succession a threat for rare and protected species in a sand quarry? Case study of the Kuźnica Warężyńska sand quarry (Southern Poland). *IOP Conference Series: Earth and Environmental Science* 362: 012010.

Rosenberg, D.K., Noon, B.R., and Meslow, E.C. (1997). Biological corridors: form, function, and efficacy. *BioScience* 47 (10): 667–687.

Rostański, A. (2006). *Spontaneous Succession of Vegetation Cover on Dumps the Coal Mining Industry in the Upper Silesia*. (In Polish). Katowice: University of Silesia Publisher.

Rostański, A. and Zhukov, S. (2001). Comparison of the flora of mining spoil heaps of Upper Silesia (Poland) and Donetsk coal district (Ukraine). *Natura Silesiae Superioris* (Supplement): 67–77.

Scherer-Lorenzen, M. (2008). Functional diversity affects decomposition processses in experimental grasslands. *Functional Ecology* 22: 547–555.

Šebelíková, L., Řehounková, K., and Prach, K. (2016). Spontaneous revegetation vs. forestry reclamation in post-mining sand pits. *Environmental Science and Pollution Research* 23: 13598–13605.

Shrestha, R.K. and Lal, R. (2011). Changes in physical and chemical properties of soil after surface mining and reclamation. *Geoderma* 3–4: 168–176.

Sierka, E. (2013). Biodiversity of the Silesian Voivodship postindustrial areas-resources, protection, threats. Strategia ochrony przyrody województwa śląskiego do roku 2030. *Raporty Opinie* 6: 7–58.

Sierka, E. and Chmura, D. (2005). Role of *Calamagrostis epigeios* in forest communities of Silesian Upland (S Poland). In: *Biology of Grasses* (ed. L. Frey), 343–352. Kraków: W. Szafer Institute of Botany, Polish Academy of Sciences.

Sierka, E., Radosz, Ł., Ryś, K., and Woźniak, G. (2022). Ecosystem services and post-industrial areas. In: *Green Scenarios: Mining Industry Response to Environmental Challenges of the Anthropocene Epoch: International Mining Forum (2021, 2022)* (ed. A. Dyczko, A. Jagodziński, and G. Woźniak), 265–274. Leiden: CRP Press/Balkema.

Sierka, W. and Sierka, E. (2008). The effect of flooded mine subsidence on thrips and forest biodiversity in the Silesian upland of Southern Poland – a case study. *Acta Phytopathologica et Entomologica Hungarica* 43 (1): 345–353.

Sierka, E., Stalmachová, B., Molenda, T. et al. (2012). *Environmental and Socio-economic Importance of Mining Subsidence Reservoirs*, 112. BEN - Technicka Literatura, Praha.

Skousen, J., Johnson, C., and Garbutt, K. (1994). Natural revegetation of 15 abandoned mine land sites in West Virginia. *Journal of Environmental Quality* 23: 1224–1230.

Šnajdr, J., Dobiášová, P., Urbanová, M. et al. (2013). Dominant trees affect microbial community composition and activity in post-mining afforested soils. *Soil Biology and Biochemistry* 56: 105–115.

Sroka, K., Chodak, M., Klimek, B., and Pietrzykowski, M. (2018). Effect of black alder (*Alnus glutinosa*) admixture to Scots pine (*Pinus sylvestris*) plantations on chemical and microbial properties of sandy mine soils. *Applied Soil Ecology* 124: 62–68.

Stalmachová, B. (1992). Succession of plant communities on the OKR heaps. (In Czech). Dissertation. Ústav pro životní prostředí UK Praha. 332 s.

Stalmachová, B. and Frnka, T. (2003). Managed succession – principles of mining landscape regeneration. (In Czech). *Proceedings of Conference 'Strategie obnovy hornické krajiny'*, Ostrava (25–26 September 2003), p. 150–158.

Stalmachová, B. and Sierka, E. (2014). Managed succession in reclamation of postmining landscape. Technická univerzita v Košiciach, 1. Vyd., Košice, s.85.

Steffen, W., Richardson, K., Rockström, J. et al. (2015). Planetary boundaries: guiding human development on a changing planet. *Science* 347: 6223.

Tropek, R., Kadlec, T., Hejda, M. et al. (2012). Technical reclamations are wasting the conservation potential of post-mining sites. A case study of black coal spoil dumps. *Ecological Engineering* 43: 13–18.

Tropek, R., Kadlec, T., Karesova, P. et al. (2010). Spontaneous succession in limestone quarries as an effective restoration tool for endangered arthropods and plants. *Journal of Applied Ecology* 47: 139–147.

Uzarowicz, Ł., Wolińska, A., Błońska, E. et al. (2020). Technogenic soils (Technosols) developed from mine spoils containing Fe sulphides: microbiological activity as an indicator of soil development following land reclamation. *Applied Soil Ecology* 156: 103699.

Vacek, Z., Linda, R., Cukor, J., Vacek, S., Šimůnek, V., Gallo, J., Vančura, K. (2021). Scots pine (*Pinus sylvestris* L.), the suitable pioneer species for afforestation of reclamation sites? *Forest Ecology and Management* 485, 118951.

Vindušková, O. and Frouz, J. (2013). Soil carbon accumulation after open-cast coal and oil shale mining in Northern Hemisphere: a quantitative review. *Environmental Earth Sciences* 69: 1685–1698.

Vojar, J. (2006). Colonization of post-mining landscapes by amphibians: a review. *Scientia Agriculturae Bohemica* 37: 35–40.

Vojar, J., Doležalová, J., Solský, M. et al. (2016). Spontaneous succession on spoil banks supports amphibian diversity and abundance. *Ecological Engineering* 90: 278–284.

Wali, M.K. (1999). Ecological succession and the rehabilitation of disturbed terrestial ecosystems. *Plant and Soil* 213: 195–220.

Walker, B.H. (1992). Biodiversity and ecological redundancy. *Conservation Biology* 6 (1): 18–23.

West, T.O. and Wali, M.K. (2002). Modeling regional carbon dynamics and soil erosion in disturbed and rehabilitated ecosystems as affected by land use and climate. *Water, Air & Soil Pollution* 138: 41–163.

Whittaker, R.H. (1965). Dominance and diversity in land plant communities. *Science* 147: 250–260.

Whisenant, S. (1999). *Repairing Damaged Wildlands: A Process-Orientated, Landscape-Scale Approach*, vol. 1. New York: Cambridge University Press.

Woś, B., Chodak, M., Józefowska, A., and Pietrzykowski, M. (2022). Influence of tree species on carbon, nitrogen, and phosphorus stocks and stoichiometry under different soil regeneration scenarios on reclaimed and afforested mine and post-fire forest sites. *Geoderma* 415: 115782.

Woś, B., Nezhad, M.T.K., Mustafa, A. et al. (2023a). Soil carbon storage in unreclaimed post mining sites estimated by a chronosequence approach and comparison with historical data. *CATENA* 220: 106664.

Woś, B., Józefowska, A., Chodak, M., and Pietrzykowski, M. (2023b). Recovering of soil organic matter and associated C and N pools on regenerated forest ecosystems at different tree species influence on post-fire and reclaimed mine sites. *Geoderma Regional* 33: 1–12.

Woś, B., Pietrzykowski, M., and Józefowska, A. (2018). Reclaimed mine soil substrates and tree stands vs. successional forest floor vegetation: a case study of developing ecosystems on afforested mine sites. *Ecological Engineering* 120: 504–512.

Woźniak, G. (2010). *Diversity of Vegetation on coalmine heaps of the Upper Silesia (Poland).* (In Polish), 319. Kraków: W. Szafer Institute of Botany, Polish Academy of Science.

Woźniak, G., Sierka, E., and Wheeler, A. (2018). Urban and industrial habitats: how important they are for ecosystem services. In: *Ecosystem Services and Global Ecology* (ed. L. Hufnagel), 169–194. IntechOpen.

Yan, M., Fan, L., and Wang, L. (2020). Restoration of soil carbon with different tree species in a post-mining land in eastern Loess Plateau, China. *Ecological Engineering* 158: 106025.

Zedler, J.B. and Kercher, S. (2005). Wetland resources: status, trends, ecosystem services, and restorability. *Annual Review of Environment and Resources* 30: 39–74.

Zhang, P.P., Zhang, Y.L., Jia, J.C. et al. (2020). Revegetation pattern affecting accumulation of organic carbon and total nitrogen in reclaimed mine soils. *Peer J* 8: e8563.

Zobel, M. (1997). The relative role of species pools in determining plant species richness: an alternative explanation of species coexistence? *Trends in Ecology and Evolution* 12: 266–269.

5

Plant Diversity on Post-Industrial Land

Resilience and Restoration

Shivali Sharma[1], *Jonnada Likhita*[1], *Sunny Sharma*[2], *Gaurav Sharma*[1], *Amit Kumar*[3], *Rupesh Kumar*[4], *and Vimal Chandra Pandey*[5]

[1] *Department of Fruit Science, Rani Lakshmi Bai Central Agricultural University, Jhansi, Uttar Pradesh, India*
[2] *Department of Horticulture, School of Agriculture, Lovely Professional University, Phagwara, Punjab, India*
[3] *School of Hydrology and Water Resources, Nanjing University of Information Science and Technology, Nanjing, China*
[4] *Jindal Global Business School, O.P. Jindal Global University, Sonipat, Haryana, India*
[5] *CSIR-National Botanical Research Institute, Lucknow, Uttar Pradesh, India*

5.1 Introduction

Industries are starting to be acknowledged as a significant new global change factor that may have an impact on terrestrial ecosystems (de Souza Machado et al. 2018; Rillig 2012). Industrial pollution affects all aspects of our environment, including the food and the air. Due to its indistinctness, it might present a significant hazard and potentially remain unnoticed. Soil pollution is the condition when the soil contains excessive amounts of compounds, agrochemicals, and other substances that surpass a specific limit and disrupt the growth and survival of plants and animals. Soil contamination causes a variety of pressures that impede the growth of the plants. Additionally, industrial pollutants raise the soil's salinity, making it insufficient for plant development. Consequently, the soil becomes deficient in minerals and becomes unusable. Although the global population is growing daily, agricultural yield is not rising at the same rate. Industrial pollution affects the biodiversity and abundance of organisms cohabiting in the soil, resulting in significant detrimental effects on soil functionality. A steady decrease in the activity of soil microbes leads to a decrease in the process of nitrogen cycling (Jacoby et al. 2017). This, in turn, has an impact on the structure and fertility of the soil, ultimately affecting crop yield. The generation of microplastics can adversely affect plant communities through many mechanisms. A significant amount of microplastics will be deposited in both terrestrial and marine ecosystems as a result of several factors, such as the wearing down of synthetic materials and big plastic objects during production or in their natural habitats, and the erosion of tires while driving. Microplastics can find their way into soil through soil additions, irrigation, diffuse

Biodiversity and Ecosystem Services on Post-Industrial Land, First Edition.
Edited by Vimal Chandra Pandey.

urban runoff, floods, air fallout (Qi et al. 2018; Zhang and Liu 2018), and plastic mulching. (Rillig 2012; Mahon et al. 2017) (Hurley et al. 2018; Dris et al. 2015). As a result, microplastics can show up in soil as fibers, films, or granules with a range of shapes, compositions, and abundances, and at concentrations as high as 7% in the vicinity of industrial sites. (Fuller and Gautam 2016) For example, because plastic mulching is heavily employed to increase plant production, agricultural areas would be greatly impacted by microplastic films (Brodhagen et al. 2017), but microplastic pieces or microfibers would typically have a greater impact on waterways or roadsides. According to (Sommer et al. 2018), plant communities are significantly harmed by industrial waste.

However, creating and preserving urban woods is a crucial aspect of communal life. Promoting sustainable development, in particular in urban areas, may have a major influence on a city's long-term sustainability and total environmental impact (Voget and Fischer 2012). In addition to lowering the urban heat island effect, urban trees may purify the air. They have the power to sequester carbon from the atmosphere and even replenish it below. Tree growth may be negatively impacted by urbanization in several ways. This is a frequent and widespread occurrence in the Amazon Rainforest, where urban growth has devastated thousands of acres of natural forests. Local governments remove once-thick and impenetrable native flora and till the earth underneath it. Not just cities, but humans, in general, are to blame for this. Large tracts of tropical forests have been drastically converted to agricultural and grazing fields, mining sites, and other uses as a result of the Amazon's development into a key supplier of goods sold globally, including food and minerals. This shift has disastrous effects on the ecosystem (Richards and VanWey 2015). Cities pose a threat to forests unless they are managed carefully. Trees are important to cities for various reasons. They offer entertainment places, timber, and shade, as was previously said. Trees are an essential part of the ecology, both within and outside of cities. They offer ecological services as well.

Plant conservation is therefore essential. Due to the relatively lower popularity of plants compared to animals, the maintenance of plant diversity has received much less focus than the preservation of animal species (Goettsch et al. 2015). Plant conservation is allocated significantly fewer resources in comparison to animal conservation (Havens et al. 2014). Nevertheless, we hold plants in higher regard than they do. Although plant products are essential for human survival and health, animals provide flesh, leather, and other materials, none of which is indispensable. In addition to a wide range of other commodities, including fibers and timber, plants also provide essential resources like clean water and erosion control. Furthermore, they serve as a source of nourishment for both humans and animals. Urban dwellers employ a diverse range of plant species for various purposes, while rural residents typically make use of even greater numbers. This is in contrast to the fact that most commercial plant products come from a limited selection of plant species. Depending entirely on such organisms would be detrimental to one's health and monotonous. Various additional species hold importance in traditional medicine, while wild plant foods play a crucial role in providing nourishment and ensuring food security. Furthermore, plants provide the three-dimensional framework for animal life and movement as well as the food that the majority of animals eat, making them the foundation of all terrestrial ecosystems. The present state of the diversity of terrestrial plants worldwide, the main risks to their survival, and the most important conservation measures are the main topics of this assessment. It focuses mostly on the tropics where the majority of plant species occur; yet little is understood about them.

5.2 Numerous Elements that Promote Industrialization

5.2.1 Industrialization

The major reason why forest areas are being depleted is because emerging nations are clearing them to make room for their rapidly expanding populations and to meet their industrial and infrastructure demands. There are primarily two ways that humans utilize forestry resources. First, the human population directly uses forest resources – trees, plants, herbs, grasses, etc. – to meet the increasing need for fuel, cow feed needed for milk production, medicine, etc., without removing the forest area. Second, forest stands are chopped down in big sections to develop infrastructure. These sections are then used to build industrial facilities, housing complexes, farmhouses, livestock farms, and other structures, as well as to use the land for agriculture. As population density rises, more and more development-related projects are being undertaken, which furthers industrialization. According to Augmented Industrialization (2009), wood-based industries primarily employ forest trees by cutting them down to produce timber and other related products. Government data reveals that encroachments and industrial projects have resulted in the loss of forests equivalent to approximately two-thirds of the territory of Haryana during the previous 30 years. The encroachments account for an area of 15 000 km^2, while the industrial projects have caused the loss of 14 000 km^2. The government has recently accepted that artificial woods are irreplaceable (Himadri Ghosh IndiaSpend.com 2016).

5.2.2 Farming Practices

According to recent research, industrial agriculture causes deforestation on more than five million hectares of forest area annually (Curtis et al. 2018). The extent of deforestation caused by the cultivation of economically significant crops, such as palm oil trees, has remained unchanged from 2001 to 2015, despite various efforts to reduce it. The corporates are the primary actors driving the pattern of deforestation.

5.2.3 Urbanization Process

One of the main features of an economy like India's that is expanding quickly is urbanization. Nevertheless, in addition to the benefits, uncontrolled urbanization frequently has negative effects that significantly worsen the environment. 4.5% of the reduction in tree cover in India from 2001 to 2019 occurred in areas where deforestation was primarily caused by urbanization, commodity-driven deforestation, and shifting agriculture, according to a report by The Sustainability Consortium, World Resources Institute, and the University of Maryland titled "Tree Cover Loss by Driver."

5.3 Industrial Impact on Plant Communities

5.3.1 Effect of Different Pollutions Caused by Industries

5.3.1.1 Soil Pollution

The effects of soil pollution include radioactive fallout, microbial disease, salinity, and metal contamination, among other forms of industrial pollutants (FAO 2020). Other factors contributing to the problem include improper drainage system maintenance, runoff of sewage and gasoline from vehicles, leaching of waste from garbage dumps, underground storage tank failures, infiltration of contaminated surface water, the release of industrial waste into the soil, and inappropriate irrigation methods. Precipitation effectively removes all industrial pollutants, which subsequently infiltrate the soil and present a significant health hazard to both flora and fauna. Thus, there is a great need to take action to remove these contaminants from agricultural land.

5.3.1.2 Air Pollution

Mohammadi (2015) specifies air pollution as the presence of one or more pollutants in the atmosphere for a prolonged period and in significant amounts, which can potentially harm people, life on earth, or their natural functions, and disrupt the overall quality of life. Industry is the main contributor to air pollution. PMs refer to an intricate mixture of microscopic solid fragments and liquid droplets (except pure water) dispersed throughout the atmosphere. They are one of the most destructive and severe types of air pollution. They may travel great distances (by wind) after being formed as a result of numerous industrial operations, and they can have detrimental short- and long-term impacts (Sakellaris et al. 2017; Tang et al. 2012). The levels of particulate matter (PM), specifically PM10 fragments, can increase substantially as a result of various factors, including industrial operations, population expansion, and the combustion of fossil fuels, especially during peak traffic periods in urban areas. The release of PM into the air has experienced a partial reduction and has been effectively managed in recent years due to the implementation of legislation and the rigorous enforcement of both domestic and global regulations. Nevertheless, achieving complete eradication of these chemicals would entail significant costs and pose considerable technical obstacles (Biglari et al. 2017). This hazardous pollutant has detrimental effects on plants, such as severe leaf damage, decreased growth, and decreased fruiting. NO_2, which has a reddish-brown hue and an unpleasant stench and decreases the quality of both indoor and outdoor air, is another significant pollutant created by a variety of natural and human activities. This substance stimulates the lower respiratory tract physiologically and contributes to global warming (Khaniabadi et al. 2017). Sulfur dioxide (SO_2) is a prominent air pollutant that greatly adds to pollution in the air and considerably hampers visibility. Human activities emit SO_2 into the atmosphere. The combustion of coal and petroleum fuels, which are abundant in sulfur, predominantly leads to the emission of sulfur into the atmosphere. This has a deleterious impact on the ecosystem and the well-being of living organisms. Through synergistic effects, SO_2 exacerbates air pollution whenever it interacts with suspended particles and moisture droplets, enhancing the risk of respiratory disorders in the community (Ware et al. 1981).

5.3.1.3 Radioactive Pollution

Smičiklas and Šljivić-Ivanović (2016) specify that radiation damage refers to the existence of hazardous elements or compounds in living organisms, which is considered harmful. Since the inception and utilization of nuclear reactors in 1945, there has been a substantial introduction of radionuclides into the environment. According to Miura (2016), soils that have been significantly polluted with radioactive materials become less productive and are thus considered deteriorated. For instance, severe deterioration of agricultural land interferes with crop yield due to uranium ore extraction and processing, nuclear fuel overproduction and recycling, and radioactive waste processing and disposal. The primary sources for such radioactive particles include nuclear power generation, various industries, pharmaceuticals, agriculture, and research activities. The presence of toxins in the soil inhibits the growth of plants. The primary means of translocation for these hazardous radioactive substances within plants is through the roots, causing significant disruption to metabolism (Golmakani et al. 2008; Singh et al. 2019).

5.3.1.4 Microbiological Contamination

The waste materials emitted from municipal and industrial wastewater contain a variety of bacteria, protozoa, and viruses (Oguis et al. 2019). Therefore, either through direct touch or ingestion of soil, or indirectly by consuming infected crops, this water serves as a medium for causing intestinal infections (Pandey et al. 2014). Agricultural workers are the most impacted group among all of them because they are more frequently exposed to polluted soils. Due to their sessile nature, crops are quickly contaminated by irrigation-related wastewater contact (Brega Filho and Mancuso 2003). *Escherichia coli* can be transmitted from the soil to the lettuce leaves either through the root system or by entering through infections in the plant's tissues. Additionally, injuries allowed *Salmonella* and *E. coli to* enter tomato and lettuce plants (Liu et al. 2018a,b). While the pathogenic strain of *E. coli* O_{157}:H_7 may remain in spinach leaves for 14 days. The bacterium *E. coli* can survive in soil for one month, as demonstrated by Zhang et al. (2009). The incorporation of dangerous bacteria within plant tissues can enhance their survival (Lim et al. 2014).

5.3.1.5 Heavy Metals

Heavy metals are a significant category of industrial pollutants due to their intricate nature and the extended time required for natural remediation. In addition, they possess metallic properties such as malleability, conductivity, ductility, ligand selectivity, and cation stability. The elements mentioned by Raskin et al. (1994) possess a comparatively elevated atomic weight and an atomic number above 20. Even though certain metals – like Co, Fe, Mn, Cu, Mo, Ni, Zn, and V – are not as essential to life as other metals, geological processes and human activities – like burning fossil fuels, mining and smelting metals, applying pesticides and fertilizers in agriculture, manufacturing batteries and other metal products in industries, and getting rid of sludge, sewage, and municipal waste – make them more dangerous to plants and animals (Figure 5.1). The components present in soil solutions consist of metals bound to silicate minerals, as well as free metal ions, soluble metal compounds, and insoluble inorganic metal compounds such as carbonates and phosphates. In addition, they encompass both convertible and non-exchangeable ions that are deposited onto the surfaces of inorganic substances (Marques et al. 2009).

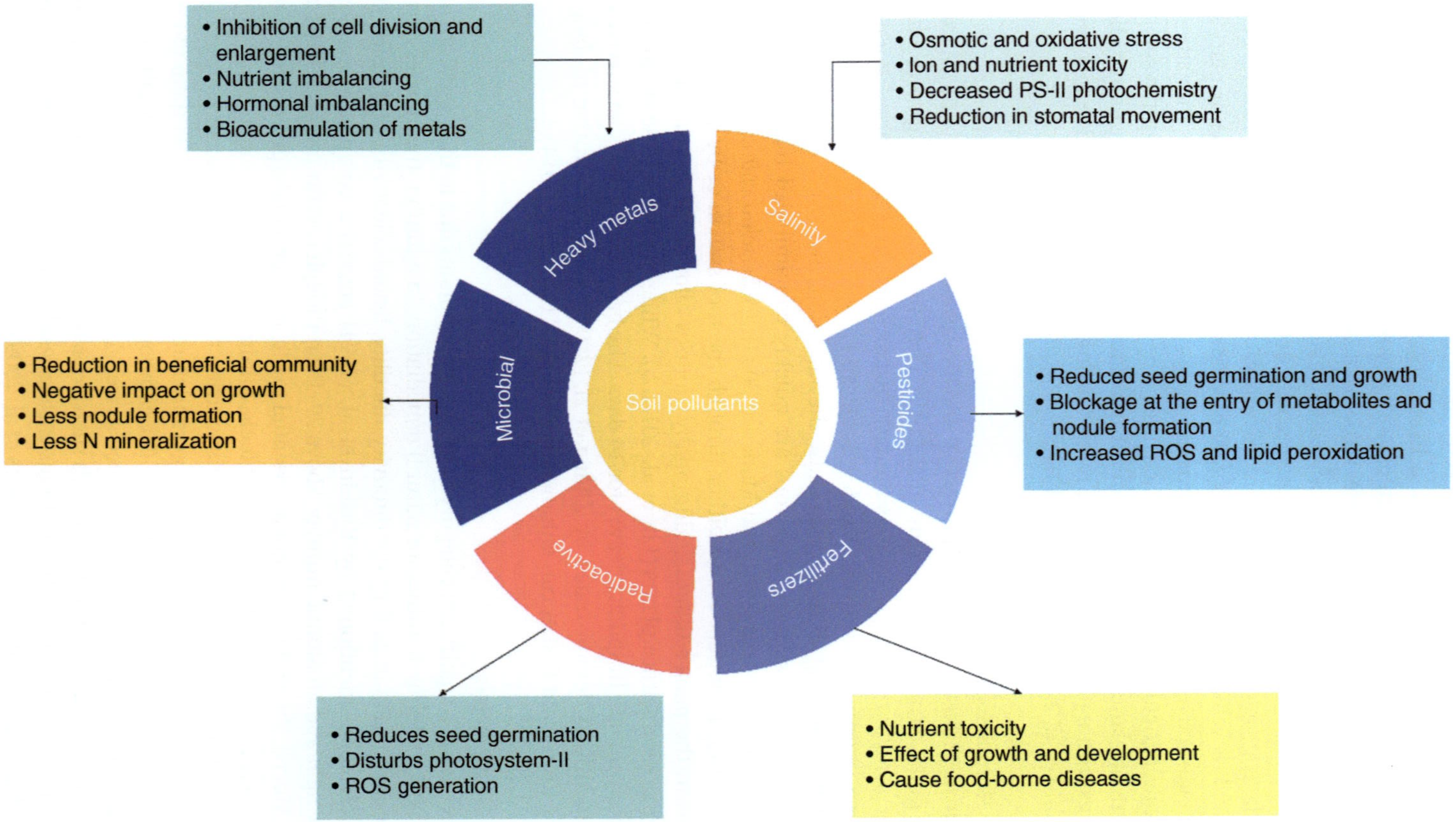

Figure 5.1 Effects of soil pollutants on plants.

5.4 Effects of Toxins Released by Various Industries

Toxic chemicals are chemicals that possess the ability to be harmful or have detrimental effects on health due to their resistance to rapid decomposition in the environment. These compounds become incorporated into the tissue of small creatures and can be subsequently transferred up the food chain. Exposure to toxic waste, whether it is deposited in soil, water bodies, or the atmosphere, can have detrimental effects on humans, animals, and plants. Certain poisons, such as lead and mercury, exhibit long-lasting persistence in the environment and gradually build over an extended period. Historically, there was insufficient regulation of hazardous wastes, resulting in significant contamination of both populations and the environment. In India, the Ministry of Environment, Forest and Climate Change (MoEFCC) and state agencies of environmental protection have been responsible for supervising toxic waste. Currently, there is a requirement for hazardous waste to be managed with specific procedures and to be disposed of in specifically designated facilities.

5.4.1 Effects of Salinity

Salinity disrupts the ecological equilibrium of the region by not only reducing yields of crops but also significantly affecting the chemical and physical properties of the soil, mostly leading to soil erosion (Rahneshan et al. 2018). Salt accumulation in the soil can have various negative effects on plants, including hindering seed germination, nutrient and water uptake, plant growth, and reproductive development. This can lead to physiological damages such as ion toxicity; nutrient deficiency; and osmotic, metabolic, and oxidative stress. When the salt concentration in the soil exceeds that inside the root cells, water will be taken from the roots by the soil, causing the plant to wilt and eventually die. Additionally, the presence of high salinity levels can lead to imbalances in soil pH, resulting in both acidity and alkalinity. This imbalance, caused by an excess of sodium, might subsequently lead to a scarcity of micronutrients (Rahneshan et al. 2018).

Regarding salinity, different plant species exhibit varying degrees of sensitivity. For instance, tomatoes, onions, and lettuce belong to the resistant variety, while halophytes are considered the tolerant kind. The absorption of phosphorus significantly decreased due to the precipitation of phosphate ions with calcium ions (Bano and Fatima 2009). Photosynthesis, a crucial physiological parameter, is significantly impaired by salinity. This impairment is characterized by a decrease in leaf area, chlorophyll content, transpiration rate, and stomatal conductance. Additionally, salinity has a moderate impact on the PSII photosynthesis and electron transport chain (Parihar et al. 2015). Ashraf and Harris (2004) and Netondo et al. (2004), it has been shown that it leads to a decrease in the availability of photosynthetic assimilation and the associated hormones.

5.5 Mechanisms of Plant Resilience

There are many conceptual models for characterizing plant growth and survival strategies. In ecology, plant strategies were defined to account for leading dimensions of variation between species.

5.5.1 Response of Plants to Industrial Waste

Various sectors and organizations in the area of phytoremediation have implemented multiple phytoremediation strategies. The earliest idea of utilizing green plants, such as green belts, for phytoremediation, originated from three countries: Kenya, Britain, and the United States (Andy 1991). The green belt is a designated area consisting of a certain arrangement of trees, chosen for their species and shape, which when planted around a pollution source, effectively reduce air pollution by capturing and storing hazardous substances sustainably. Green belts are sometimes referred to as trees, as they mostly consist of trees along with shrubs and ground species. The aim of the green belt could differ across different industries. It is crucial to consider multiple criteria when selecting plant species for environmental improvement, including the agro-climatic conditions of the province, the specific pollution to be addressed, the ability of plant species to withstand the pollutant, and the overall landscape and topography of the region (Khan and Abbasi 2000). Plants remove harmful components from the environment through three methods: absorption and deposition of substances through leaves, and the collection of PM on the downwind side due to the sluggish passage of air (Pal et al. 2000). Green belt plantations have the potential to mitigate air pollution and modify urban environments by reducing temperatures. Additionally, evapotranspiration can contribute to enhancing human well-being (Tzoulas et al. 2007). The presence of vegetation in landfills appears to be the most effective technique for in situ remediation. This technique offers several benefits, including stabilizing against wind and water erosion, decontamination of toxic heavy metals through plant absorption, and the creation of an aesthetically pleasing landscape that can serve as a shelter and habitat for wildlife.

5.5.2 Phytoremediation of Heavy Metal-Contaminated Soil

Soil remediation is an emerging technology that has garnered significant interest due to its environmentally benign nature and cost-effectiveness. Phytoextraction is a specific process of phytoremediation where green plants absorb metal and pesticides in the soil and retain them in their various components. The fly ash-grown species recommended by various researchers are *Calotropis procera*, *Cassia tora*, *Chenopodium album*, *Sida cardifolia*, *Blumea lacera* (Gupta and Sinha 2008); *Prosopis juliflora*; *Acacia auriculiformis*, *Leucaena leucocephala*; *Cicer arietinum* (Gupta et al. 2004); *Sesbania cannabina*; *Cassia siamea* (Tripathi et al. 2004); *Agropyron elongatum*, *Festuca arundinacea*, *Melilotus officinalis* (Mulhern et al. 1989); *Liquidambar styraciflua*, *Platanus occidentalis* (Carlson and Adriano 1991).

5.5.3 Role of Plants in Phytoremediation of Industrial Waste

The plant species mentioned are *Melilotus alba* (Gutenmann et al. 1976), *Mesembryanthemum nodiflorum*, *Enchylaena tomentosa*, *Halosarcia halocnemoides*, and *Halosarcia pergranulata* (Jusaitis and Pillman 1997). Through the implementation of the phytoremediation technique, one can effectively track the development of versatile tree species and the release of metals from fly ash dykes into these soils. Although recent progress in phytoremediation has enabled the genetic modification of suitable plants to enhance their tolerance, uptake, and ability to remediate contaminated sites, further research and advancements are required to establish on-site phytoremediation technologies that can function in real-time. *Jatropha curcas* was subjected to testing to assess its capacity for phytoremediation of fly ash, taking into consideration all of these criteria. *J. curcas* is a member of the *Euphorbiaceae* family and is a substantial shrub or little tree. Due to its exceptional resilience, minimal water needs, and ability to endure pressure, it is seen as a prospective crop for future biofuel production (Kaushik et al. 2007). The plant can attain a height ranging from 3 to 8 m and exhibits rapid growth with minimal maintenance (Singh 2007). Plant diversity on anthropogenic sites all over the world is given in Table 5.1. These anthropogenic sites are fly ash dumpsites, coal mine spoil sites, industrial estates, and red mud dumpsites.

Various methods for remediation, such as stabilization, solidification, and containment, have been proposed for the purification of on-site soils contaminated with arsenic. However, all of these methods require proper monitoring and continuous observation to ensure the proper behavior of arsenic throughout the soil column (Chang et al. 2011). According to studies conducted by Francesconi et al. (2002), hyperaccumulators that thrive in soil contaminated with arsenic had the ability to amass significant quantities of this toxic element. Nevertheless, due to their limited biomass and slow growth rate, most of these plants possess minimal prospects for practical application in the field (Garbisu et al. 2002). Rhizoremediation is a very efficient technique for degrading contaminants, using the synergistic interaction between plants and rhizospheric microorganisms. Microorganisms have been discovered to possess diverse strategies for tolerating different metals. These processes include exclusion, active removal, biosorption, precipitation, and bioaccumulation, which can occur both inside and outside the cells (Haferburg and Kothe 2007). Some bacterial strains can have a notable influence on the metabolic cycle of arsenic by transforming it into different species with different solubility, mobility, bioavailability, and toxicity (Jiang et al. 2008). Various technologies exist for the treatment of stormwater, including chemical, physical, and biological approaches. These technologies can be expensive and have different levels of effectiveness for different metals when used to treat water with low levels of metal contamination. Alternatively, phytofiltration has been proposed as an effective and environmentally friendly method to decrease the concentration of heavy metals in contaminated water. High metal-accumulating plants are utilized as biofilters in phytofiltration, a very effective method for extracting metals from polluted rivers (Xie et al. 2009). Due to their ability to tolerate and thrive in environments with elevated levels of heavy metals, numerous aquatic plants are commonly present in polluted water sources and have been suggested as a potential solution for wastewater treatment. Most of the species discovered so far are not suitable for onsite phytofiltration since they have slow growth rates and small root and shoot biomass. Conversely, plants that develop well usually have low levels of metal accumulation and a restricted ability to withstand heavy metals (Srivastava et al. 2005).

The significance of several microbial groups (bacteria, fungi, and microalgae) in converting Cr(VI) to Cr(III) is believed to be crucial for the remediation of chromate-contaminated soil and water. Undoubtedly, there is a strong desire to develop a very effective biological system to address the environmental problems associated with hexavalent chromium. The ability of a bacterial strain to decrease hexavalent chromium is a favorable characteristic for

Table 5.1 Plant diversity on anthropogenic sites over the world.

Serial number	Plants	Amendments	Remark	Country	References
	Fly ash dumpsites				
1	*Liquidambar styraciflua*, *Platanus occidentalis*, *Pinus palustris*, *Pinus taeda*		Hardwood species likely, *P. occidentalis* and *L. styraciflua* survived and grew on fly ash deposit.	United States	McMinn et al. (1982)
2	*Acer rubrum*, *L. styraciflua*, *P. occidentalis*, *P. taeda*, *Populus* spp. N-fixing shrub *Myrica cerifera* dominates the understory.		This case study revealed that fly ash deposits may provide conditions more favorable for tree growth than local soils. Trees have the ability to grow on fly ash deposit without using topsoil. *P. occidentalis* and *L. styraciflua* grew best on the wet fly ash deposit.	United States	Carlson and Adriano (1991)
3	*Betula* spp., *Salix* spp.		Results showed that soil development on fly ash deposits correlated more closely with site age than it did with the plant community composition.	England	Shaw (1992)
4	Herbaceous legumes (*Rhynchosia minima*, *Macroptilium atropurpureum*, *Crotalaria juncea*), Stoloniferous perennial grasses (*Cynodon dactylon*, *Cynodon nlemfuensis*), indigenous nitrogen-fixing leguminous trees (*Acacia gerrardii*, *Sesbania sesban*)	Soil, nitrogen fertilizers, herbicide	The proposed low input approach was not successful on the fly ash deposit due to the greater toxicity problems and constraints imposed by the active status of the fly ash deposit. In the longer term, however, after the ash deposit is no longer active and weathering can reduce the toxicities, it is projected that this tactic may yet attain the ultimate goal of a self- sustaining ecosystem.	Zimbabwe	Piha et al. (1995)

5	*Atriplex holocarpa, A. lindleyi, A. vesicaria, Enchylaena tomentosa, Halosarcia halocnemoides, H. pergranulata, Scaevola colloris, Nitraria billardierei, Mesembryanthemum aitonis*	A range of surface amelioration treatments for fly ash deposits were designed and used.	The soil or compost layer offered a seedbed for initial germination against fly ash toxicity, while the stabilization treatment barred wind-blown abolition of both seed and seedbed, though also conferring certain safety to seedlings. Compost layer stabilized with mulch was the ideal amelioration toward biomass production from seeds or transplants.	Australia	Jusaitis and Pillman (1997)
6	*Acacia auriculiformis, Leucaena leucocephala*	LA-VC-N fertilizer, LA-SSC-N fertilizer, LA-VC-RI-N fertilizer, LA-SSC-RI-N fertilizer	Both ameliorants enhanced the physical properties of fly ash deposit by increasing the porosity and aeration for root development, but only sewage sludge compost (SSC) increased nutrient levels (N, P, K) of the infertile fly ash substrate. Even though both species showed potential to establish on amended fly ash deposit, *A. auriculiformis* being the best adapted.	China	Cheung et al. (2000)
7	*A. auriculiformis, Casuarina equisetifolia, Eucalyptus globulus*		Plants can be grown directly on the fly ash deposit without amendment with fertilizers.	India	Zevenbergem et al. (2000)
8	*Chloris gayana, C. dactylon, Conyza bonariensis, Eragrostis sp., Schkuhria pinnata, Pogonarthria squarrosa, Pseudognaphalium undulatum, Pennisetum clandestinum, Hyparrhenia hirta, Chamaecrista biensis, Plantago lanceolata, Walafrida densiflora, Medicago sativa*	Topsoil cover	Flora composition recognizes the species that are adapted to the limiting factors posed by the FA deposits. Thus, the composition outcomes can be used to lead the next rehabilitation programs.	South Africa	Van Rensburg et al. (2003)
9	*Annona squamosa, Azadirachta indica, Cassia siamea, Dalbergia sissoo, Dendrocalamus strictus, Emblica officinalis, Eucalyptus hybrid, Pongamia pinnata, Tectona grandis*	Soil, lime slurry (2.2 t ha^{-1}), FYM (50 t ha^{-1}), *Rhizobium* (leguminous), *Azatobacter* (non-leguminous), VAM	The FYM and biofertilizer are better supportive materials for establishing vegetation on fly ash deposits because both amendments assisted 15 times higher growth in profuse root system in *D. strictus* than the control.	India	Juwarkar and Jambhulkar (2008)

(*Continued*)

Table 5.1 (Continued)

Serial number	Plants	Amendments	Remark	Country	References
10	*Quercus robur*, *Quercus rubra*, *Pinus sylvestris*, *Betula verrucosa*, *Populus tremula*, *Populus nigra*, *Salix fragilis*, *Salix alba*, *Alnus glutinosa*, *Q. robur* and *P. sylvestris*	*Stabilize* – a mixture of fly ash treated with 1–2% CaO, 25% water, slag, and FDG gypsum. Physicochemical properties contribute to the origination of an adverse soil environment (high alkalinity, cementing effects).	Mainly *Robinia pseudoacacia*, *B. verrucosa*, *S. alba*, *S. fragilis*, *P. tremula*, *P. nigra*, *Populus alba*, and *P. sylvestris* can be considered the tree species taxa of very high reclamation significance characterized by good performance already in the first vegetation series of primary succession on fly ash deposit only.	Czech Republic	Cermák (2008)
11	*Casuarina equsitifolia*, *Dalbergia sisso*, *Bauhnia bauhni*, *Delonix regia*, *Michelia champaca*, *Hibiscus rosa-sinensis*, *Bougainvillea glabra*, *Chamaedorea elegans*, *Thuja occidentalis*, *Lawsonia inermis*	Soil + cow manure (2:1 ratio), chemical fertilizers (di-ammonium phosphate; 25 grams (g) pit^{-1}), rhizobium and phosphobacterium biofertilizers (10 g pit^{-1}), humic acid (5 g pit^{-1}).	The microbial activity of the fly ash deposit could be successfully upgraded through the plantation of *D. sisso*, *B. glabra*, *C. equsitifolia*, *D. regia*, and *T. occidentalis* to a large extent and well reclaimed for social forestry purposes.	India	Ram et al. (2008)
12	*Azadirecta indica*, *C. siamea*, *Delbergia sissoo*, *D. strictus*, *E. officinalis*, *Eucalyptus hybrid*, *P. pinnata*, *T. grandis*,	Soil, lime slurry (2.2 t ha^{-1}), FYM (50 t ha^{-1}), *Rhizobium* (leguminous), *Azatobacter* (non-leguminous), VAM	*C. siamea* can be used as a hyper-accumulator species for fly ash deposit remediation, because it is noted that *C. siamea* accumulated all metals at higher concentrations than other plants.	India	Jambhulkar and Juwarkar (2009)
13	*Festuca arundinacea*, *Onobrychis viciifolia*	Incorporating fertilizer and amendments, as composts and modified indigenous volcanic tuff.	The strategy chosen is very significant toward obtaining green armour that rapidly and capably covers the FA catena, as well as, to permit the wildlife habitat development. The fertilizer and modified indigenous volcanic tuff provided establishing conditions for the plant supplying the nutrients. Limiting the transfer of heavy metals in plant tissues is a healthy habitat for wildlife.	Romania	Anca et al. (2011)
14	*Amorpha fruticosa*, *P. alba*, *R. pseudoacacia*, *Tamarix tentandra*		Spontaneously grown woody species can be used in the first step to stabilize the fly ash deposit.	Serbia	Kostic et al. (2012)

15	*A. glutinosa* *Alnus incana*	CFA (control), CFA + MS, CFA + L. Sewage sludge ($4\,Mg\,ha^{-1}$) mixed with grass seedlings ($200\,kg\,ha^{-1}$) and NPK mineral fertilization (N–60, P–36, and K–$36\,kg\,ha^{-1}$) were applied by hydroseeding	The best amendment was lignite culm for fly ash substrate improvement. Especially when the introduced *A. glutinosa* has mainly protective and phyto-amelioration functions and thus offers the improved substrate for the afforestation and next generation of target species.	Poland	Krzaklewski et al. (2012)
16	Leguminous species *Onobrichys viciifolia*, Bird's-foot Trefoil (*Lotus corniculatus*), yellow sweet (*Melilotus officinalis*)	Untreated FA, FA amended with modified/unmodified volcanic tuff, FA treated with modified/unmodified volcanic tuff mixed with organic fertilizer, anaerobically stabilized municipal sludge.	The suggested approach for vegetating the FA deposit was by seeding leguminous crop species *O. viciifolia* followed by the installation of some invasive more resistant species, which allow the ecological restoration of the fly ash deposit with an effective stabilization against the wind, erosion, and leaching.	Romania	Morariu et al. (2013)
17	*Albizia procera, Bombex ceiba, C. equisetifolia, D. sissoo, D. strictus, Eucalyptus tereticornis, Gmelina arborea, Melia azadirachta, Parkinsonia aculeate, Populus deltoides, Populus euphratica, Shorea robusta, T. grandis*	FYM, Mycorrhizal bio-fertilizer	Marked changes in microbial population and their activities as well as in nutritional status were noticed in reclaimed fly ash deposit. So, Mycorrhizal technology has great potential to develop resilient plants as green cover.	India	Das et al. (2013)
18	*Alnus viridis*	FA (control), FA + L, Sewage sludge ($4\,Mg\,ha^{-1}$), Grass seeds (*Dactylis glomerata* and *Lolium multiflorum*) ($200\,kg\,ha^{-1}$), NPK mineral fertilizer (N:60, P:36, K:$36\,kg\,ha^{-1}$)	Application of lignite culm in planting pits, preliminary surface preparation by hydroseeding, and mineral fertilization had the most positive effect on the seedling parameters *of A. viridis*. Outcomes reveal that *A. viridis* can be used as a potential biological stabilizer on fly ash deposits.	Poland	Pietrzykowski et al. (2015)

(*Continued*)

Table 5.1 (Continued)

Serial number	Plants	Amendments	Remark	Country	References
19	*Ailanthus excelsa*, *Albizia lebbeck*, *A. procera*, *D. sissoo*, *E. tereticornis*, *Holoptelea integrifolia*, *L. leucocephala*, *Pithecellobium dulce*, *P. pinnata*, *Prosopis juliflora*, *Ziziphus mauritiana*	Soil layer	Based on the importance value index (IVI) *P. juliflora*, *P. dulce*, and *P. pinnata* have been noticed as dominant species in the fly ash ecosystem. In this regeneration status study, the above three tree species (68.91%) have been found in the good regeneration category which can be used for revegetation of new fly ash deposit.	India	Pandey et al. (2016)
20	Alder species (*A. glutinosa, A. incana, and A. viridis*)	CCW, CCW + L, Start-up mineral fertilization (N: 60, P: 36, K: 36 kg ha^{-1}) and initial stabilization by hydroseeding with sewage sludge (4 Mg ha^{-1} dry mass) and a mixture of grasses (*D. glomerata* and *L. multiflorum*) (200 kg ha^{-1}).	The obtained results supported the findings of a previous study (Pietrzykowski et al. 2015) on alder growth on combustion waste disposal sites, showed the effectiveness of alders, mainly *A. glutinosa*, for biological reclamation of this kind of barrens.	Poland	Pietrzykowski et al. (2018a)
21	*A. glutinosa* (Black alder), *A. incana* (Grey alder)	CCW, CCW + L, CCW + MS Hydro-seeding with sewage sludge (4 Mg dry mass ha^{-1}) mixed with the seeds (200 kg ha^{-1}) of *D. glomerata* and *L. multiflorum*, mineral fertilization (N: 60, P: 36, K 36 kg ha^{-1}).	Despite the good growth attentions, the N:P and N:K ratios in alder leaves largely differed from the optimal values showing poor P and K supply at fly ash deposit. This may pose a threat to further development of the introduced tree plantings. The introduction of alders along with the lignite addition into the planting holes seems to be an effective method of fly ash deposit revegetation.	Poland	Pietrzykowski et al. (2018b)

22	Dominant plant species *R. pseudoacacia*, *Fraxinus* spp.		30 years of pedogenesis changed the original properties of the fly ash substrate as the decrease of pH and salinity, the increase of total nitrogen and total organic carbon contents due to accumulation of organic matter as well as the formation of pedogenic carbonates	Poland	Uzarowicz et al. (2018)
23	*Arrhenatherum elatius*, *D. glomerata*, *Festuca rubra*, *L. multiflorum*, *L. corniculatus*, *M. sativa*, *Populus euramericana*, *R. pseudoacacia*, *S. alba*, *Secale cereale*, *Tamarix tetrandra*, *Vicia villosa*	No use topsoil, application of agronomic measures (800 kg ha^{-1} of 15N:15P:15K mineral fertilizer and the wetting of the seeded area until establishing plant cover), grass-legume mixture (sowing density: 270–300 kg ha^{-1})	Study revealed the effects of weathering and the development of vegetation directly on fly ash deposits. The changes in the fly ash physicochemical properties as significant indicators of pedogenesis on fly ash substrate, with these processes being of special importance for the effective ecological reclamation of fly ash deposits worldwide.	Serbia	Kostić et al. (2018)
24	*Alder* ssp. [*A. viridis* (Green alder), *A. glutinosa* (Black alder), *A. incana* (Grey alder)]	Hydro-seeding with sewage sludge (4 Mg ha^{-1} dry mass) mixed with the seeds (200 kg ha^{-1}) of *D. glomerata* and *L. multiflorum*, Mineral fertilizer (NPK 60:36:36 kg ha^{-1})	The results revealed that the black-, grey- and green-alder growing on barren fly ash deposits have similar fine root biomass growth. The study also indicates a significant role of fine roots in the development of the carbon and nutrient pools in fly ash deposits.	Poland	Świątek et al. (2019)
	Coal mine spoil sites				
25	*A. lebbeck*, *A. procera*, *T. grandis*, *D. strictus*		Plantation age, dominant tree species, and species specificity have a significant impact on herbaceous plant development beneath the woody plantation at Singrauli coalmine fields.	India	Kumar et al. (2020)

(*Continued*)

Table 5.1 (Continued)

Serial number	Plants	Amendments	Remark	Country	References
26	Pine (*Pinus* sp.), spruce (*Picea* sp.), alder (*Alnus* sp.), larch (*Larix decidua*), summer oak (*Q. robur*), lime (*Tilia* sp.).		The study recommends combining spontaneous succession with technical reclamation.	Czech Republic	Vachova et al. (2022)
27	Highest IVI of dominated tree species were *Acacia mangium* (37.75%), followed *by C. siamea* (35.08%), *Paraserianthe falcataria* (28.28%), *Dryobalanops aromatic* (18.85%) *and Samanea saman* (15.23%). Herbaceous species were dominated by *Diplazium esculentum* (66.75%), followed by *Blumea balsamifera* (30.01%), *Ageratum conyzoides* (18.75%), *Acmella oleracea* (14.25%) *and Merremia peltata* (12.75%).		The slightly higher diversity of herbaceous vegetation in the reclamation site may be attributed to higher coverage per species at this site.	Indonesia	Komara et al. (2016)
	Industrial Estate				
28	*Carpobrotus edulis, Sedum reflexum*		PTI values ranged from 5 to 12 PMs adsorbed on the adaxial and abaxial leaf surfaces, as well as near the stomata; Adsorbed Cr, Fe, Pb in *S. reflexum*		Hozhabralsadat et al. (2022)
29	*Cerastium dichotomum, Chenopodium murale*		Soil Cd removed roots and BCFs, TFs, and BACs values		Naz et al. (2024)
30	*Mangifera indica, Ficus bengalensis, Psidium gujava*		Highest Air pollution tolerance index and anticipated pollution index in *M. indica,* Highest metal accumulation index and dust capturing capacity in *F. bengalensis,* Highest metal transferring coefficient in *P. gujava*		Roy et al. (2020)

31	*Melia azedarach* growing in Chromium salt – producing factory wasteland		Adsorb Cd-15.61 $mg\,kg^{-1}$) and Cr 925.07 $mg\,kg^{-1}$	China	Yan et al. (2020)
	Red mud dumpsites				
32	*Grevillea crithmifolia, Spyridium globulosum, Acacia cochlearis, Acacia rostellifera, Hardenbergia comptoniana*		This study provides key mechanisms for creating a viable green cover through plants and functional ecosystems on residue sand embankments.	Australia	Phillips and Courtney (2022)
33	*P. pinnata, A. lebbeck, Terminalia arjuna, Bauhinia variegate, Pennisetum pedicellatum, Stylosanthes hamata*	20% gypsum along with 10% vermicompost and acid-producing bacteria emerged	This study reveals standardizing rehabilitation protocol using vegetation cover for bauxite waste (red mud).	India	Chauhan and Ganguly (2011)
34	*P. juliflora, Acacia nilotica,* *P. pinnata,* *M. azedarach* and *Acacia tortilis*	55% red mud + 25% FYM + 15% gypsum + 5% vegetative dry dust + bacteria + mycorrhizae	*P. juliflora* is the best plant species for the rehabilitation of red mud bauxite wasteland.	India	Chauhan and Silori (2010)
35	*Atriplex canescens*	*Pseudomonas simiae* [a plant growth-promoting rhizobacteria (PGPR)]	*P. simiae, a PGPR,* enhances the tolerance to alkaline red mud residue by modulating physiological processes in *A. canescens*.	China	Shi et al. (2022)

Abbreviations: CCW, coal combustion waste; CM, cow-dung manure; FA, fly ash; FDG, flue gas desulfurization; FYM, farmyard manure; L, lignite culm; MS, miocene acidic sand; RI, rhizobium inoculation; SSC, sewage sludge compost; VAM, vesicular arbuscular mycorrhizae; and VC, vermiculite.

developing a biological approach for effective chromate detoxification, even though the method of Cr(VI)-reduction differs among strains (Ma et al. 2001). Nevertheless, the presence of elevated levels of chromate in the environment sometimes hampers microbial activity and inhibits growth. Several bacterial species, such as *Pseudomonas*, *Micrococcus*, *Bacillus*, *Achromobacter*, *Microbacterium*, *Arthrobacter*, and *Corynebacterium*, have been demonstrated to possess the ability to directly reduce Cr(VI) (Ma et al. 2001).

Determine the specific substances that are emitted into the atmosphere as pollutants and determine the origins of these pollutants. The subsequent course of action should involve promptly devising a strategy to mitigate the emission of these particles and gasses (Duncan et al. 2016). Various plant species have a significant impact on the reduction of environmental pollutants, such as NO_2 and O_3. The effectiveness of this reduction depends on factors such as the arrangement and kind of trees planted, the climate conditions, and the distance between the vegetation and the sources of pollutants (Yli-Pelkonen et al. 2017). Plant species can mitigate ambient air pollution, although this might also result in detrimental impacts on the plants themselves. The effects of air pollutants on the morphology of plant species differ based on their biological characteristics. However, in general, these pollutants interfere with plant function and can result in their mortality at high concentrations. Various categories of industrial pollutants possess distinct characteristics and exhibit a wide range of impacts on trees (Akeredolu et al. 1994). The efficacy of urban green infrastructure strategy in mitigating air pollution and enhancing air quality is contingent upon several factors, such as the climatic conditions of the region, the density and extent of plant species, and the capacity of trees to assimilate environmental pollutants (Fantozzi et al. 2015). Through the process of deposition, suspended particles and gaseous pollutants settle on various areas of trees. These pollutants are then absorbed through the pores on the leaves, triggering various biological processes (Simonich and Hites 1994). In plant habitats where pollutants build on tree leaves, the absence of air exchange for plant respiration poses a hazard to plant survival due to subsequent adverse impacts (Viippola et al. 2018). Furthermore, certain studies have indicated that in regions devoid of plants and trees, as opposed to forested areas, the presence of open-air flow mitigates air pollution by augmenting the kinetic energy of particles and generating unstable atmospheric conditions as a result of the quick transfer of pollutants.

5.5.4 Response of Plants to Air Pollution

Air pollution has become a widespread health issue in recent years. One effective solution to mitigate the issues resulting from this phenomenon is to extensively plant trees in various locations. Trees can remove pollution from the atmosphere, leading to cleaner and purified air (Yli-Pelkonen et al. 2017). It is a very destructive force and a concerning menace that disrupts life in major urban areas. Hence, the initial measure to mitigate pollution in the air is to precisely ascertain the contaminants emitted into the environment and their origins. The subsequent course of action should involve promptly devising strategies to mitigate the emission of these particulates and gasses (Duncan et al. 2016). Various plant species have a significant impact on the reduction of environmental pollutants, such as NO_2 and O_3. The effectiveness of this reduction depends on factors such as the arrangement and kind of trees planted, the weather, and the distance that exists between the vegetation and the sources of pollutants (Yli-Pelkonen et al. 2017).

Plant species can mitigate ambient air pollution, although this may have detrimental consequences on the plants themselves. The effects of pollutants in the air on the morphology of plant species differ based on their biological characteristics. However, in general, these chemicals interfere with plant function and can result in their mortality when present in high concentrations. Various categories of industrial pollutants possess distinct characteristics and have a wide range of impacts on plants (Akeredolu et al. 1994). The efficacy of a green infrastructure strategy in mitigating air pollution and enhancing air quality is contingent upon several factors, such as the climatic conditions prevailing in the region, the density and extent of plant species, and the capacity of trees to assimilate pollutants from the environment (Fantozzi et al. 2015). The accumulation of suspended particulates and gaseous contaminants on various portions of trees results in the absorption of these pollutants through the pores in the leaves, triggering various biological processes (Simonich and Hites 1994). In plant habitats where pollutants build on tree leaves, the absence of air exchange for plant respiration poses a hazard to plant survival due to subsequent adverse impacts (Viippola et al. 2018). Furthermore, certain studies have indicated that in regions devoid of plants and trees, as opposed to forested areas, the presence of open-air flow mitigates air pollution by augmenting the kinetic energy of particles and generating unstable atmospheric conditions as a result of the swift transfer of pollutants (Yli-Pelkonen et al. 2017).

5.5.5 Purification or Removal of Industrial Pollutants

Regrettably, the environment is undergoing significant degradation due to the growing number of societies through urbanization and industrialization. Hence, the establishment of tree-planting infrastructures has garnered the focus of urban planners as a means to mitigate the environmental difficulties arising from urbanization and to effectively curb air pollution at small expense (Escobedo et al. 2011). The municipal officials have established natural habitats and artificial forests in several locations. These places consist of a variety of trees and plants that are well-suited to the local climate, soil, and water conditions. This initiative aims to mitigate the risks associated with pollutant emissions and enhance the quality of the air (Hirabayashi 2021). Urban forest areas should be designed based on the specific technical attributes of trees, such as their height, crown shape, root and leaf type, heat and cold resistance, ecological conditions, and practical applications. It is important to consider that these characteristics vary throughout the year due to seasonal weather conditions, including changes in leafing (Escobedo et al. 2011).

An effective method of utilizing plants for purifying air involves the cultivation of evergreen trees, which retain their leaves for multiple years, as opposed to deciduous trees, which shed their leaves during certain periods of the year (Escobedo et al. 2011). While big leaf stomata can capture PM10 and PM2.5, resulting in a decrease in their presence in the environment, the ongoing absorption of these particulates causes significant harm to plants (Chen et al. 2015). Lichens have been utilized for over a hundred years to assess air pollution by examining the presence of algae and fungi and their ability to determine the level of contaminants. Lichens absorb pollutants from the air, making them a valuable tool for determining the air quality index (Davies et al. 2007). Studies have demonstrated that urban green spaces and trees, which are intended to cleanse dirty air in cities, are effective in eliminating both primary and secondary pollutants. However, for different reasons, city officials do not consistently use this approach. Air entrapment can occur among trees grown in severely polluted locations, particularly under tree canopies. This can lead to an increase in the local amount of polycyclic aromatic hydrocarbons (PAHs) relative to the surroundings. Conversely, urban forests exhibit lower levels of O_3 compared to treeless regions, suggesting that trees absorb and reduce the harmful effects of this pollution on public health through deposition and uptake (Yli-Pelkonen et al. 2017).

5.5.6 Response of Plants to Heavy Metals

Certain plants exhibit the ability to withstand the elevated levels of heavy metals present in their environment. Baker (1981) identified three mechanisms that plants use to tolerate heavy metals. The first mechanism is exclusion, which involves preventing the transport of metals and conserving a consistent amount of metal in the shoot across a wide range of soil concentrations. The second mechanism is inclusion, which maintains a linear relationship between the levels of metal in the shoot and the soil solution. The third mechanism is bioaccumulation, which involves the plants accumulating metals in both high and low levels in the soil at a faster rate than they are lost through catabolism and elimination in the shoot and root.

5.6 Case Studies

5.6.1 Successful Examples of Plant Resilience

Plant resilience encompasses the capacity of plants to endure and recuperate from many pressures and disruptions, including a lack of moisture pests, diseases, and alterations in the environment.

5.6.1.1 Resilience to Drought

Desert plants, such as cacti and succulents, have evolved unique adaptations to thrive in arid environments. They can store water in their tissues, have shallow but widespread root systems, and employ strategies like crassulacean acid metabolism (CAM) photosynthesis to minimize water loss during the day (Nobel 2002).

5.6.1.2 Resilience to Pathogens

Arabidopsis thaliana is a small flowering plant often used in research. It has been extensively studied to understand plant defense mechanisms. The plant can activate various defense pathways, such as the jasmonic acid and salicylic acid pathways, to resist attacks from pathogens (Jones and Dangl 2006).

5.6.1.3 Resilience to Environmental Changes

Mangrove Trees (*Rhizophora* spp.) are trees that grow in coastal saline environments. They exhibit resilience to fluctuations in water salinity, tidal variations, and extreme weather events. Their specialized root systems help stabilize the soil, and they can tolerate both saltwater and freshwater conditions (Kathiresan and Bingham 2001).

5.6.1.4 Resilience to Herbivores

Acacias and Ant Mutualism. Certain acacia species have formed mutualistic relationships with ants. These ants defend the acacia plants against herbivores, and in return, the acacias provide food and shelter for the ants. This interaction enhances the plant's resilience against herbivore attacks (Palmer et al. 2008).

5.6.1.5 Resilience to Fire

Fire-Adapted Eucalyptus Trees Eucalyptus trees in Australia are adapted to frequent wildfires. They possess traits such as thick bark, which protects them from intense heat, and the ability to resprout after a fire. These adaptations contribute to their resilience in fire-prone ecosystems (Gill and Moore 1997). These examples showcase the diverse strategies that plants employ to cope with and recover from different stressors, highlighting the remarkable adaptability and resilience of the plant kingdom.

5.6.2 Species Thriving in Polluted Environments

Some plant and animal species have exhibited the capacity to flourish in contaminated habitats by adjusting to the existence of pollutants. This phenomenon is often referred to as pollution tolerance or resistance.

5.6.2.1 Heavy Metal Tolerance in Plants

Thlaspi caerulescens, often known as Alpine Penny Cress, exhibits remarkable resistance to toxic heavy metals like zinc, cadmium, and nickel. It has adapted mechanisms to hyperaccumulate these metals in its tissues without suffering toxicity. This plant has been extensively studied for phytoremediation purposes in contaminated soils (Baker and Brooks 1989).

5.6.2.2 Oil-Degrading Bacteria

Pseudomonas aeruginosa is an adaptable bacteria that can flourish in hydrocarbon-contaminated settings, such as oil spills. The ability of this substance to break down various contaminants has been extensively researched, which makes it an intriguing choice for bioremediation projects in areas contaminated by oil (Das and Chandran 2011).

5.6.2.3 Air Pollution Tolerance in Plants

Tillandsia usneoides (Spanish Moss) is an epiphytic bromeliad that can tolerate air pollution, including SO_2 and heavy metals. It has been used as a bioindicator of air quality in urban areas, as its health and growth are influenced by atmospheric pollutants.

5.6.3 Wastewater Treatment With Macrophytes

Phragmites australis, often known as Common Reed, is a type of plant that thrives in wetland environments. It has the ability to withstand high amounts of nutrients and is suitable for use in built wetlands designed for treating wastewater. It helps in nutrient uptake and promotes microbial activities that contribute to water purification (Kadlec and Knight 1996).

5.6.3.1 Insect Adaptation to Polluted Water

Chironomus riparius (Midge larvae) have been found in polluted aquatic environments. They display modifications to endure in aquatic environments tainted with organic contaminants, heavy metals, and other hazardous compounds. These examples showcase the fascinating ability of certain animals to not only endure but thrive in polluted environments, contributing to the potential for ecological restoration and environmental remediation efforts.

5.6.4 Natural Recolonization After Industrial Abandonment

Natural recolonization by plants after industrial abandonment is a fascinating process that showcases the resilience and adaptability of vegetation. When industrial areas are abandoned, they often leave behind degraded landscapes with soil contamination and altered ecological conditions. Despite these challenges, various plant species manage to establish and thrive in these abandoned sites. Here are a few examples.

5.6.4.1 Zollverein Coal Mine Industrial Complex, Germany

The Zollverein Coal Mine in Essen, Germany, was once a bustling coal mining and coking plant. After its closure in the 1980s, the site underwent natural recolonization. Today, it is a UNESCO World Heritage Site with a diverse array of plant species reclaiming the once-industrial landscape. Grasses, wildflowers, and even trees have established themselves, transforming the area into a green space (UNESCO World Heritage Centre).

5.6.4.2 Bethlehem Steel Plant, the United States

The Bethlehem Steel Plant in Bethlehem, Pennsylvania, was a major steel-producing facility that closed in the 1990s. Over time, the brownfield site has seen natural recolonization by various plant species. Grasses, shrubs, and even some trees have taken root in the disturbed soil, contributing to the ecological restoration of the area (Bliss 1998).

5.6.4.3 Eifel National Park, Germany

The Eifel National Park in Germany is an example of successful natural recolonization following the closure of military training grounds. The area was heavily disturbed, but after abandonment, natural processes took over. Today, a diverse range of plant species, including heathlands, forests, and wetlands, has returned, demonstrating the resilience of the natural ecosystem (Lorz et al. 2012).

5.6.4.4 Post-Industrial Sites in the United Kingdom

Several post-industrial locations in the United Kingdom, including former coal mines and industries, have experienced natural recolonization. Native plants, including grasses, wildflowers, and even some trees, have recolonized the abandoned areas, contributing to the restoration of ecological processes.

5.6.4.5 Ruhr Area, Germany

The Ruhr Area in Germany, once a heavily industrialized region, has witnessed natural recolonization after the decline of coal and steel industries. Brownfields and former industrial areas have become green spaces with a mix of native vegetation, showcasing the resilience of the landscape. These examples demonstrate that abandoned industrial sites can be reclaimed by nature, and through natural recolonization, ecosystems can regain biodiversity and functionality over time. The process underscores the importance of allowing natural processes to unfold in the restoration of disturbed landscapes.

5.7 Challenges and Failures

5.7.1 Persistent Contamination Issues

5.7.1.1 Phytoremediation Success

Some plants exhibit resilience by thriving in contaminated environments through a process called phytoremediation. For instance, several plant species have effectively gathered and eliminated contaminants from the soil through a process known as detoxification (Salt et al. 1998). Willow trees, such as those mentioned in the study by Meers et al. (2005), are recognized for their capacity to absorb and store heavy metals, aiding in soil remediation endeavors.

5.7.1.2 Adaptation to Stress Conditions

Some plant species show resilience by adapting to stress conditions on post-industrial lands. This adaptation may involve changes in physiological processes or the development of specific traits that confer tolerance to contaminants (Bradshaw 2006). Researchers have observed genetic adaptation in plant populations exposed to contaminated environments, suggesting the potential for evolutionary responses to industrial pollution (Figure 5.2) (Franks et al. 2007).

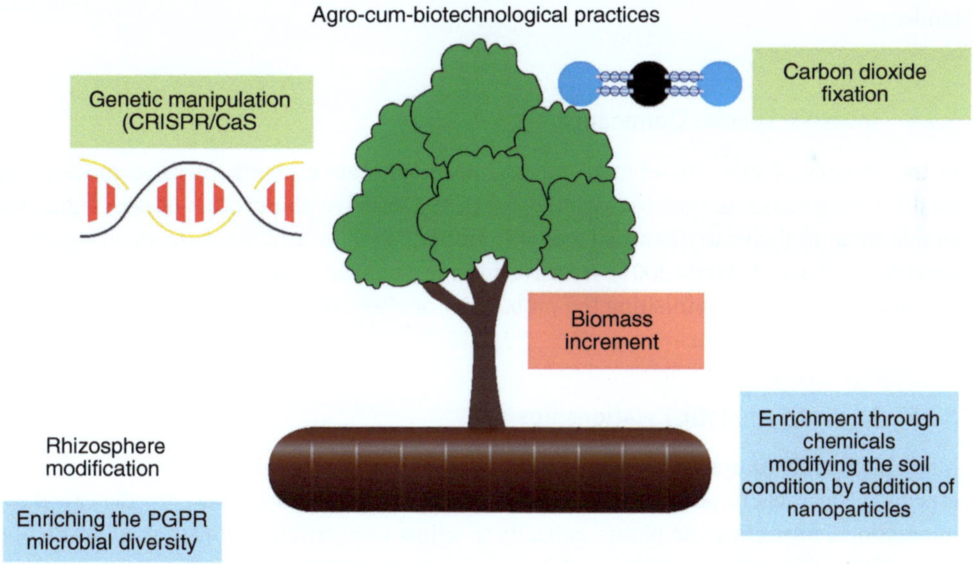

Figure 5.2 Agro-cum-biotechnological approaches for removal of post-industrial pollution.

5.7.1.3 Community-Level Resilience

Successful restoration projects often focus on enhancing overall ecosystem resilience rather than relying solely on individual plant species. Establishing diverse plant communities can contribute to the stability and resilience of post-industrial ecosystems (Walker 1995). Selecting a mix of native plant species with complementary ecological functions can help restore ecosystem services and support long-term resilience.

5.7.2 Lack of Adaptive Strategies in Certain Ecosystems

The lack of adaptive strategies in certain ecosystems on post-industrial land can pose significant challenges to the resilience and restoration of plant diversity. Adaptive strategies pertain to the capacity of plant species and ecosystems to adapt and flourish in the face of altering environmental circumstances. In post-industrial landscapes, where ecosystems may be severely degraded due to pollution, habitat destruction, and altered soil conditions, the absence of effective adaptive strategies can hinder the natural recovery of plant diversity. Here are some key factors contributing to the lack of adaptive strategies and the associated challenges.

5.7.3 Limited Genetic Diversity

Some ecosystems on post-industrial land may suffer from reduced genetic diversity due to contamination or habitat fragmentation (Ellstrand and Elam 1993). Plant populations with limited genetic variety have constraints in their capacity to adjust to fluctuations in environmental conditions. Without a diverse gene pool, plant species may struggle to develop traits that confer resistance to contaminants or improve their overall fitness in the altered landscape.

5.7.4 Invasive Species Dominance

In the absence of effective adaptive strategies among native plants, invasive species may establish dominance in post-industrial ecosystems. Invasive plants often possess traits that enable them to thrive in disturbed and contaminated environments, outcompeting native species (Hobbs and Harris 2001). Invasive species can hinder the regrowth of native plant communities, hence diminishing the probability of effective restoration.

5.7.5 Lack of Symbiotic Relationships

Many plants rely on symbiotic relationships with fungi (mycorrhizae) for nutrient uptake, especially in nutrient-poor or contaminated soils. Industrial activities can disturb these interactions, impacting the plants' capacity to adjust to nutritional constraints (Smith and Read 2008). Without the support of beneficial microorganisms, plants may struggle to establish and grow in nutrient-stressed environments.

5.7.6 Altered Soil Microbial Communities

Industrial activities can alter soil microbial communities, impacting nutrient cycling and soil health. Some plant species depend on specific microbial associations for nutrient acquisition and stress tolerance. Disruption of these microbial communities can limit the adaptive capacity of plants, making it challenging for them to cope with contaminated or degraded soils.

5.7.7 Prolonged Exposure to Contaminants

Certain pollutants can endure in the environment for prolonged durations, impeding the formation of adaptive mechanisms in plant species. Chronic exposure to pollutants can lead to physiological stress and hinder the establishment of resilient plant populations (Baker et al. 1997).

5.8 Ecological Restoration

5.8.1 Principles of Restoration Ecology

5.8.1.1 Establishing Clear Restoration Goals

Establishing clear restoration goals is a fundamental principle of restoration ecology. This involves defining specific, measurable, and achievable objectives that guide the restoration process. Clear goals provide a framework for decision-making, help prioritize actions, and facilitate effective monitoring and adaptive management strategies. Restoration goals may include enhancing biodiversity, improving ecosystem services, restoring natural processes, or recovering specific species. The establishment of well-defined goals contributes to the success and sustainability of restoration projects by providing a shared vision for stakeholders and guiding the allocation of resources.

Restoration goals can vary widely depending on the ecological context, project scale, and desired outcomes. These goals may include:

Biodiversity enhancement: Setting targets for the restoration of native species richness, abundance, and diversity to promote a more resilient and ecologically functional community (Hobbs and Harris 2001).

Ecosystem functionality: Restoring natural processes and ecological functions, such as nutrient cycling, water purification, and pollination, to enhance overall ecosystem health.

Habitat restoration: Re-establishing or improving specific habitats for particular species, with a focus on creating suitable conditions for native flora and fauna (Hobbs and Suding 2009).

Community engagement and education: Engaging local people in restoration endeavors and imparting knowledge about the significance of ecological restoration, cultivating a sense of responsibility and care (Clewell and Aronson 2006).

Climate change resilience: Enhancing ecosystem resilience to climate change impacts by selecting and restoring species and habitats that are adaptive and resilient to changing environmental conditions (Hobbs et al. 2018).

5.8.2 Selecting Appropriate Plant Species

Selecting appropriate plant species is a critical component of successful ecological restoration. The choice of plants should align with the goals of restoration, the characteristics of the degraded site, and the ecological functions of the target ecosystem. Here are key considerations in selecting plant species for restoration ecology.

5.8.2.1 Site-Specific Adaptation

Select plant species that are well-suited to the particular environmental circumstances of the restoration site, such as soil composition, levels of moisture, and seasonality (McKay et al. 2005).

5.8.2.2 Native Species Emphasis

Prioritize native plant species to enhance ecological integrity and support local biodiversity (Hobbs and Harris 2001).

5.8.2.3 Functional Traits and Niches

Consider the functional traits of plant species, such as root structure and growth form, to match specific ecological niches and roles in the ecosystem (Lavorel and Garnier 2002).

5.8.2.4 Successional Dynamics

Ensure that the selection of plant species is in line with the successional stage of the environment to guarantee compatibility with the natural evolution of vegetation (Suding et al. 2015).

5.8.2.5 Diversity and Resilience

Foster plant species diversity to enhance ecosystem resilience, as diverse plant communities are often more resistant to pests, diseases, and environmental changes (Tilman 1999).

5.8.2.6 Invasive Species Management

Be cautious of invasive plant species and implement measures to control or exclude them to prevent competition with native vegetation (Pysek et al. 2020).

5.8.2.7 Pioneer Species for Soil Stabilization

Consider the use of pioneer species with traits that promote soil stabilization and prevent erosion, especially in the early stages of restoration (Walker et al. 2010).

5.8.2.8 Climate Change Adaptation

Factor in climate change considerations by selecting plant species that are resilient to anticipated shifts in temperature, precipitation, and other climate variables.

5.8.2.9 Community and Stakeholder Involvement

Engage local communities and stakeholders in the selection process to incorporate traditional knowledge and community preferences, fostering a sense of ownership and sustainability (Clewell and Aronson 2006).

5.8.3 Monitoring and Adaptive Environment

Implement continuous monitoring of the restoration site to assess the performance of selected plant species and adapt management strategies based on feedback (Hobbs and Harris 2001).

The process of selecting plant species for restoration ecology requires a holistic understanding of ecological processes, local conditions, and the desired outcomes of restoration efforts. Continuous learning and adaptation based on monitoring and evaluation contribute to the success of ecological restoration projects.

5.8.4 Techniques for Restoration

5.8.4.1 Soil Remediation

It has been reported that the pH of the soil in the natural environment has a significant impact on soil biogeochemical processes. Soil pH is considered the primary factor that impacts various soil properties and processes related to plant growth and biomass production. It influences the biological and physicochemical aspects of the soil environment, as well as the interconnected processes that are affected by human activities and can cause changes in soil pH. This research diverges from conventional talks of the multiple factors influencing soil pH, namely soil acidification, by emphasizing the connections and implications of soil biogeochemistry. The text first examines the impact of soil pH on the availability, movement, and biological processes of substances in the soil, and then explores how soil pH is regulated by living organisms. Soil pH has two main applications: nutrient cycling and plant nutrition, as well as soil remediation through biological and physicochemical remediation (Neina 2019).

5.8.4.2 Habitat Reconstruction

Restoring damaged agricultural land is crucial, especially when soil erosion is caused by insufficient vegetation cover on the soil surface. Wind erosion is the primary force responsible, although water erosion can also contribute considerably to degradation. A variety of agricultural and biological strategies are essential in resolving this problem. These techniques include activities such as rotational farming, agro-forestry, reduced tillage, cover crops, vegetative filter strips, residue management, canopy cover regulation, and no-till systems (Lamb et al. 2005). The processes utilized in these biological techniques differ, but together, they strive to protect the soil from erosion. Restoring damaged agricultural land encompasses two primary groups of strategies: agriculture conservation methods and managing the soil techniques.

5.8.4.3 Agricultural Restoration Techniques

Crop Management Implementing good crop management practices is fundamental to reducing wind and water erosion. Maintaining soil cover, especially by leaving crop residues on the soil surface after harvesting, is a key principle in the restoration of degraded agricultural land.

Inter-Cropping Introducing fast-growing legumes like cowpeas and beans early in the season before the canopy develops for main crops like cotton or maize can reduce the impact of raindrops, shielding the soil and promoting regeneration.

Crop Selection Choosing cover crops strategically, especially during longer gaps between harvesting and sowing, can enhance the stability of conservation agriculture systems. Perennials like sugar cane, fodder grasses, sweet potatoes, and tea are more effective in preventing soil erosion.

Crop Rotation Growing different crops in sequential seasons helps avoid pathogen and pest buildup, balances nutrient demand, replenishes nitrogen using green manure, and improves soil structure and fertility.

Cover Cropping Utilizing close-growing cover crops or green manure crops protects the soil surface, reduces erosion by regulating soil heat and temperature, and serves as a source of organic matter as the decomposed vegetation enriches the soil.

Shelterbelts Employed to protect farmlands from creeping sands, shelterbelts reduce wind velocity, enhance the microclimate, and increase livestock yields. These protective barriers, often comprised of species like *Eucalyptus*, *Casuarina* sp., *Populus*, and *Prosopis*, contribute significantly to improved agricultural production.

Strip Cropping Implementing a practice of cultivating crops in alternating strips of row crops and forage/grass can effectively mitigate soil erosion by dividing sloping terrains into sections with varied vegetative cover. This approach successfully intercepts runoff, enhances water penetration, and mitigates soil erosion by integrating rotations and variable strip widths. According to Bravo and Silenzi (2002), it is advisable to have a strip width of around 30 meters, taking into account any necessary modifications for steeper slopes.

5.9 Introduction of Important Species

The reintroduction of keystone species, which play a disproportionately significant role in an ecosystem compared to their abundance, is a key rationale behind many global wildlife restoration initiatives. Plains bison (*Bison* L.) has been identified as a keystone species in the tallgrass prairie ecoregion after restoration endeavors. Nevertheless, there is a dearth of studies regarding comparable impacts in the mixed-grass grasslands, a region where restoration endeavors are currently under progress. The objective of this study is to examine if plant communities in the mixed-grass prairie of the Northern Great Plains (NGP) display traits that align with the four primary keystone effects observed in bison populations in the tallgrass prairie. The study entailed gathering data on the species composition, diversity, abundance, bare ground cover, and plant height in three different scenarios: areas where Bos taurus L. livestock was consistently grazed, areas where livestock was absent for a decade, and areas where bison have been introduced and have resided for a decade. The study yielded inconclusive evidence about the role of bison as an essential species in the NGP mixed-grass prairie. Ultimately, the findings indicate that, following a decade of restoration endeavors, bison in the mixed-grass prairie partially fulfill their function as a keystone species. The study highlights the importance of ongoing, extended data gathering to evaluate the enduring impact of bison in the mixed-grass prairie ecosystem of the NGP (McMillan et al. 2019). According to a study conducted by Siegenthaler et al. (2014), it was shown that Northern peatlands are mostly controlled by bryophytes. Therefore, precise measurements of moss development are essential for comprehending its reactions to global changes. Several techniques have been utilized to quantify moss growth, although the ability to make meaningful comparisons among research is hindered by the lack of clarity regarding the comparable performance of these methods under varied conditions. We conducted a field experiment to evaluate the development and productivity of *Sphagnum fallax* (Sphagnum) and *Polytrichum strictum* (Polytrichum) using two different methods of handling: cutting the plants and leaving them uncut. We measured three growth variables: height growth, length growth, and mass growth. We aimed to compare six different methodological alternatives using the same set of factorial experiments that involved atmospheric carbon dioxide enrichment and nitrogen addition. Our objective was to establish a baseline for these options.

The two techniques of handling produced partially divergent outcomes: in 50% of the instances, one method identified a substantial treatment impact that the other method failed to discover. Remarkably, substantial detrimental impacts on growth were exclusively detected in intact plants exposed to increased levels of carbon dioxide, and in plants that were cut and received more nitrogen. Furthermore, the correlation between measurements obtained using different growth indicators varied depending on the species and, to a lesser degree, the treatments. The growth of both Sphagnum and Polytrichum was suppressed in the presence of increased levels of carbon dioxide, which was correlated with higher concentrations of ammonium. Contrary to Polytrichum, Sphagnum exhibited lower sensitivity, resulting in a reduction in the disparity in height between the two species. Nitrogen addition led to a reduced P/N ratio, likely inducing phosphorus-limiting conditions.

5.10 Success Stories in Restoration

5.10.1 Examples of Successive Rehabilitation Projects

Many post-industrial cities in North America face the dual challenges of having numerous vacant lots and a surprising lack of well-managed green spaces. Previous to the foreclosure crisis of 2008, many cities had been dealing with problems of desertion and abandonment in their urban centers as a result of substantial decreases in industrial and manufacturing sectors, as well as the expansion of suburban development. Metropolitan areas such as Cleveland (Ohio, United States), New Orleans (Louisiana, United States), Detroit (Michigan, United States), and Windsor (Ontario, Canada) have undergone significant population declines, with some surpassing 40%, since reaching their highest point in the mid-20th century.

The ramifications of this demographic transition encompass reduced tax revenues, deteriorating infrastructure, and a deterioration in the standard of municipal services. Services related to the management of abandoned properties, including property auctions, demolitions, and investments in fundamental infrastructure, are currently facing significant pressure. As a result, neighborhoods display inconsistent patterns of empty properties, inadequately carried out demolitions (including buried basements and debris that obstruct redevelopment efforts), declining infrastructure (such as outdated and overloaded sewer lines), and an absence of efforts to establish or sustain ecosystem services. Even when green techniques could provide more cost-effective methods for delivering essential municipal services, such as green infrastructure, these solutions are frequently disregarded. Increasing and broadening the availability of green areas is identified as a vital approach for incorporating the social and ecological aspects of resilience in urban systems. By converting unused land into urban green spaces, it is feasible to counteract the negative effects of social and environmental deterioration, while also promoting a range of ecosystem services.

5.10.2 Lessons Learned from Restoration Failures

Restoration efforts can provide valuable insights even when they result in failures. Examining these failures helps identify challenges and refine strategies for future projects. Several lessons learned from restoration failures include:

Inadequate planning and research: Lack of thorough site assessment, understanding of ecosystem dynamics, and comprehensive planning contribute to restoration failures (Hobbs and Harris 2001; Suding et al. 2015).

Unrealistic goals and expectations: Setting unrealistic goals or expecting rapid success without considering ecological processes can lead to failure (Hobbs and Suding 2009; Suding et al. 2015).

Limited stakeholder engagement: Inadequate involvement of local communities and stakeholders in the planning and decision-making process can hinder project success (Jordan et al. 2015; Clewell and Aronson 2006).

Invasive species and pest management: Neglecting the control of invasive species or pests can undermine restoration efforts (D'Antonio and Meyerson 2002; Suding et al. 2015).

Insufficient monitoring and adaptive management: Lack of regular monitoring and adaptive management strategies to adjust approaches based on real-time feedback can lead to project failure (Hobbs and Suding 2009; Suding et al. 2015).

Inadequate funding and resources: Limited financial resources and insufficient support for long-term maintenance and monitoring can compromise project success (Higgs 2003; Suding et al. 2015).

Climate change considerations: Failure to account for climate change impacts and variability in restoration planning may result in unsuccessful outcomes (Hobbs and Harris 2001; Suding et al. 2015).

Lack of ecological knowledge: Lack of comprehension regarding the past ecology and natural dynamics of the environment can result in erroneous restoration methods (Hobbs and Harris 2001; Suding et al. 2015).

Unintended consequences: Restoration actions may have unintended consequences, such as altering soil structure or hydrology, which can impede project success (Hobbs and Suding 2009; Suding et al. 2015).

Social and political challenges: Sociopolitical factors, including conflicting interests, policy issues, or legal disputes, can hinder successful restoration outcomes (Jordan et al. 2015; Clewell and Aronson 2006). These lessons underscore the importance of adaptive management, stakeholder involvement, and a holistic understanding of ecological processes in successful restoration initiatives.

5.10.3 Future Prospects

5.10.3.1 Advances in Biotechnology

Biotechnology plays a significant role in addressing the challenges associated with plant diversity on post-industrial land. Here are some advances in biotechnology that contribute to the resilience and restoration of plant diversity in such environments:

5.10.3.2 Genetic Engineering for Pollution Tolerance

Bioremediation is becoming more important as a substitute technology for removing basic toxins from soil and water, requiring appropriate procedures for decontamination. Phytoremediation is a method that utilizes green plants to eliminate, confine, or render harmless environmental contaminants (Pandey et al. 2022). This strategy is notable for its environmentally conscious nature, non-destructive characteristics, and cost-effectiveness. Genetic engineering can be used to change plants to improve their ability to take up, transport, and store metals (Pandey et al. 2022; Pandey et al. 2023a). This has the potential to make phytoremediation more effective. Genes related to metal chelation, metal transport, metallothionein (MT), and phytochelatin (PC) were recently introduced into plants to improve the absorption and storage of metals. As a consequence, transgenic plants have been created that can either detoxify or accumulate elements such as cadmium, lead, mercury, arsenic, and selenium. Enhancing our comprehension of the mechanisms that regulate rhizosphere interactions, metal absorption, transportation, and storage in hyperaccumulator plants will aid in the development of advanced transgenic plants that possess exceptional remediation skills. With the discovery of additional genes associated with metal metabolism, especially via genome-sequencing initiatives, there will be increased possibilities for creating genetically modified plants that are specifically designed for phytoremediation and exhibit exceptional efficiency (Pandey et al. 2023b).

5.10.3.3 Phytoremediation

Metal-tolerant, hyperaccumulating plants occur naturally in locations with high metal content (Baker and Brooks 1989). Nevertheless, these plants frequently prove to be less than ideal for phytoremediation as a result of their diminutive stature and limited capacity for biomass generation. In contrast, plants that have strong growth generally have a restricted ability to accumulate metals and a low tolerance for heavy metals. The most suitable crop for phytoremediation should exhibit the following characteristics: (i) the capacity to accumulate the desired metal(s), preferably in the above-ground portions; (ii) tolerance to high concentrations of accumulated metals; (iii) fast growth with biomass that efficiently accumulates metals; and (iv) ease of harvest (Pandey and Bajpai 2019). Chaney et al. (1997) highlighted that metal resistance is more essential for phytoremediation than achieving high biomass production.

Chaney et al. (1997) state that for phytoremediation to be successful, each element must be individually considered due to its distinct soils and plant chemistry. Nevertheless, metals are rarely discovered in a solitary state, and the ability to adapt to numerous metals concurrently may be required, even if the main objective is the collection of one particular

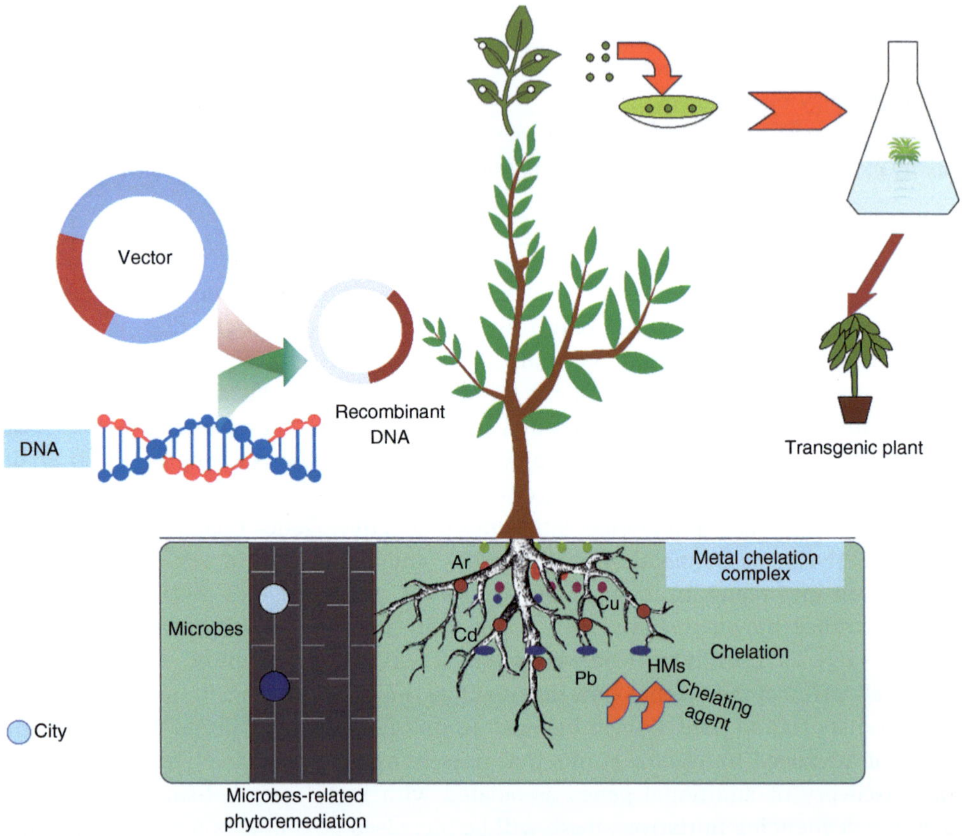

Figure 5.3 Candidate plants for metal tolerance/phytoremediation.

metal through plants. Under certain circumstances, there could be a predilection for extracting many metals concurrently (Figures 5.3 and 5.4).

5.10.3.4 Genetically Modified Plants With Modified Metal Tolerance or Uptake

Mercuric Ion Reduction and Resistance for Phytoremediation Bacteria possess the capacity to transform many heavy metals into less harmful forms. The resistance of Gram-negative bacteria to mercury is mediated by an operon consisting of five to six genes, one of which is the mercuric ion reductase gene (merA). The merA gene encodes a soluble disulfide oxidoreductase that is dependent on NADPH and contains FAD. The enzyme catalyzes the conversion of hazardous Hg^{2+} into less damaging metallic mercury (Hg^0). Experiments conducted with *E. coli* that carried the merA gene showed that it was capable of reducing Hg^{2+}. Additionally, there was a little reduction in activity observed toward Au^{3+} and Ag^+ ions. In addition, the merA gene was discovered to marginally increase the ability of Saccharomyces cerevisiae to tolerate Hg^{2+} (Rensing et al. 1992). These findings indicate that the presence of the merA gene in plants could potentially affect their ability to tolerate metals.

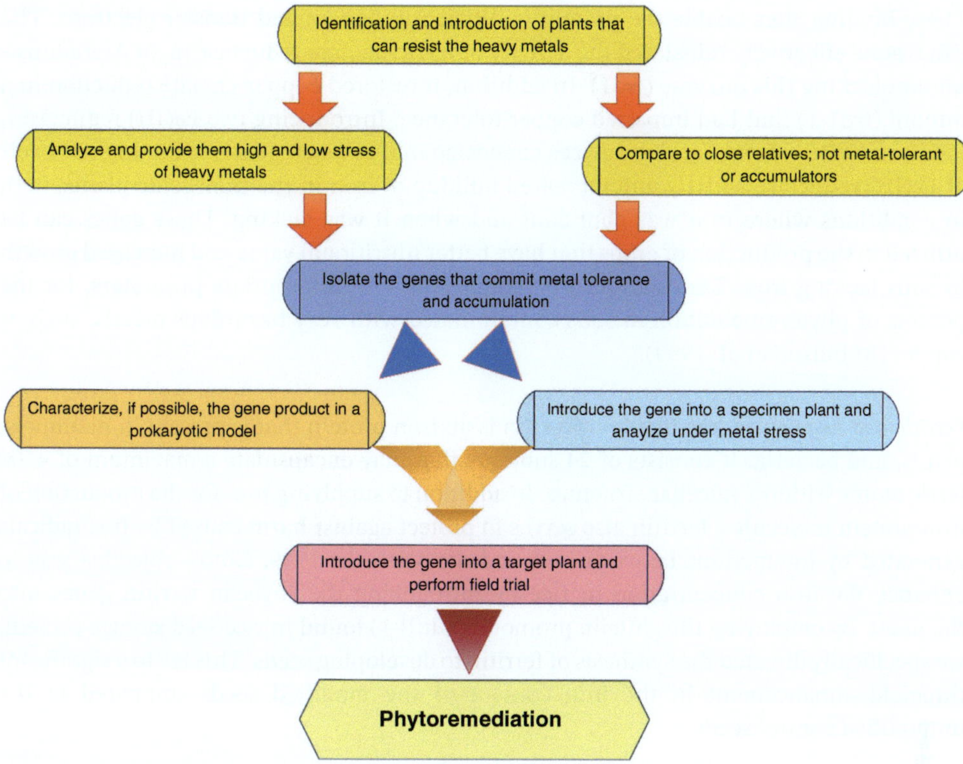

Figure 5.4 Strategies for designing metal-tolerant/accumulator plants using genetic engineering.

MTs and Cd Tolerance MTs and PCs found in plants are rich in cysteine sulfhydryl groups, which form stable complexes by binding and sequestering heavy metal ions. PCs, induced by various metals, exhibit a high affinity for binding Cu and Cd and may contribute to plant Cd tolerance. Arabidopsis mutants, such as cad1-mutants, were isolated by Howden and Cobbett (1992), who displayed increased sensitivity to Cd without significant changes in Cu tolerance. These mutants lacked PC synthesis and exhibited reduced PC synthase activity.

Ferric Reductases and Enhanced Iron Uptake Plants have developed adaptation mechanisms to enhance the accessibility of iron (Fe) for absorption due to the prevalence of insoluble Fe(III) oxides and hydroxides in soil. Proton extrusion, facilitated by an ATPase-driven proton pump, increases the solubility of Fe(III) and promotes its reduction to Fe(II) through reductases located on the plasma membrane. Iron (III) reductases, which are activated when there is a lack of iron, have a vital function in this process. The FRO_2 gene, which encodes a ferric-chelate reductase, was extracted from Fe-deficient Arabidopsis roots. FRO_2 is a member of the flavocytochrome superfamily, which facilitates the transfer of electrons across membranes (Samuelsen et al. 1998). The FRO_2 protein possesses specific locations where heme, NADPH, and FAD cofactors can interact.

These binding sites enable the protein to efficiently donate and transfer electrons. The FRO_2 gene effectively reinstated the activity of ferric-chelate reductase in an Arabidopsis mutant lacking this enzyme (frd1). In addition, it restored copper-chelate reduction in a mutant (frd1-1) that had impaired copper tolerance. Introducing two Fe(III) reductases, FRE1 and FRE2, from Saccharomyces cerevisiae into tobacco plants led to higher levels of Fe(III) reductase activity and increased buildup of iron in the transgenic plants, both in conditions where iron was abundant and when it was lacking. These genes can be utilized in the production of crops that have better nutritional value and increased growth in soils lacking iron. They can also be employed, with appropriate promoters, for the process of phytoremediation in soils contaminated with very hazardous metals, such as copper (Robinson et al. 1999).

Ferritin and Augmented Iron Uptake Ferritin is an iron protein that is present in mammals, plants, and bacteria. It consists of 24 subunits that may encapsulate a maximum of 4500 ferric atoms within a micellar structure. In addition to supplying iron for the production of iron protein molecules, ferritin also serves to protect against harm caused by free radicals generated by interactions between iron and dioxygen. In 1999, Goto's objective was to enhance the iron concentration in rice by introducing the soybean ferritin genes into the plant. By employing the glutelin promoter (GluB-1) found in rice seed storage protein, we specifically directed the synthesis of ferritin to developing seeds. This led to a significant threefold enhancement in the iron content of the modified seeds compared to the unmodified control seeds.

5.10.4 Somatic Cell Hybridization

Somatic cell hybridization, involving the fusion of cells from different plant species, has been explored as a technique to create hybrids with enhanced tolerance to toxic metals. A study was conducted to create somatic cell hybrids, both symmetric and asymmetric, by mixing cells from Brassica juncea (a species renowned for its high biomass and ability to accumulate lead) and *T. caerulescens* (a species known for its ability to hyperaccumulate zinc and nickel). The resulting hybrids exhibited increased resistance to lead (Pb), nickel (Ni), and zinc (Zn). Notably, the total amount of lead phytoextracted was significantly higher in the hybrids due to the substantial biomass they produced. This approach highlights the potential of somatic cell hybridization in creating hybrids with both high biomass and hyperaccumulation capabilities for metals, offering a promising avenue for phytoremediation efforts (Dushenkov et al. 2002; Gleba et al. 1999). Therefore, exploring somatic cell hybridization between candidate plants with high biomass and low biomass metal hyperaccumulators could lead to the development of hybrids with the desired combination of traits for efficient metal remediation.

5.10.5 Potential Genes for Improved Metal Tolerance and Accumulation

The utilization of genetic engineering in phytoremediation faces challenges as a result of the insufficient characterization of plant systems possessing metal resistance or sequestration capabilities. Microorganisms are more comprehensively understood in this aspect, and initial instances of their possible application in the rehabilitation of metal-contaminated soil are beginning to emerge. Nevertheless, the process of transferring genomes between bacteria to plants is intricate, mainly because resistance is usually encoded by extensive plasmids that comprise operons comprising numerous genes. Thus far, only isolated genes from these operons have been effectively transported. The debate presents possible genes and approaches for creating plants that can tolerate and accumulate metals through genetic engineering. It highlights the importance of developing precise molecular pathways that are customized to each metal to achieve effective hyperaccumulation and tolerance. The significance of hyperaccumulator plants, which can accumulate metals in their biomass at levels reaching several percent, as possible sources for phytoremediation genes, is emphasized. Although hyperaccumulator plants have been utilized for phytoremediation without genetic modification, enhancing metal bioavailability could safely improve their effectiveness. Strategies that involve boosting reductase activity or the synthesis of chelating compounds, such as phytosiderophores, can improve the availability of metals without causing harm to the environment. This makes hyperaccumulators utilizing these strategies safer for the aim of phytoremediation. Currently, a multitude of genes with the ability to degrade pesticides have been discovered in different animals. This finding has created opportunities for the development of genetically modified bacteria. The gene atzA encodes atrazine chlorohydrolase, which exhibits significant degrading activity against atrazine, a widely used herbicide that is recognized to potentially endanger species (Neumann et al. 2004).

5.10.6 Mechanisms of Bioremediation by Genetically Engineered Plants

Plants provide an economical and eco-friendly method for purifying soil and water by employing several cellular and molecular pathways to remove toxic heavy metals. Phytoremediation, first described in 1983, is a process that utilizes plants with the ability to accumulate metals to either eliminate heavy metals or transform them into biologically inactive forms. This strategy demonstrates economic viability and a lower level of environmental harm when compared to alternative approaches. Plants have the ability to dissolve metals in the soil, take them up through their roots, and then transport them to the above-ground parts of the plant. Certain plants secrete chelating chemicals in their root areas, which assist in the dissolution and absorption of metals. Hyperaccumulator plants frequently secrete organic acids that aid in the absorption of metals.

The plant roots have a vital function in absorbing metals or radioactive elements from the soil, and these harmful substances can be transported and stored in different areas of the plant. Phytoextraction systems can eliminate heavy metals such as Pb and Cd from polluted soil. Additionally, this technique can also be employed to tackle an excess of selenium (Se). Phytostabilization is an additional method that decreases the presence of dangerous contaminants in the soil by managing components including As, Zn, Cd, and U in contaminated areas.

Numerous genes contribute to the bioremediation capacities of plants by regulating metal uptake, removal, and translocation. Transgenic plants have been developed with MT genes encoding peptides containing cysteine residues, facilitating the removal of heavy metals from contaminated sites. For each metal, specific mechanisms need to be developed for uptake, translocation, and sequestration. Transgenic plants incorporating metal transporter genes, such as ZAT and CAX-2 genes, have been created to enhance the accumulation of metals like Zn, Ca, Cd, and Mn. Overall, the development of transgenic plants with tailored mechanisms holds promise for efficient phytoremediation.

5.10.7 Use of Microbial Communities for Land Rehabilitation

Microorganisms, integral to the broader ecological community, engage in intricate relationships with other organisms. Typically, a bacterium is unable to directly harness the power produced by its metabolic processes. However, the predominant route for pollutant degradation is co-metabolism, as stated by Mekuto et al. (2018). Moreover, microbes frequently demonstrate synergistic interactions in the process of pollution breakdown. A microbe's byproducts can generate conditions that promote the life of another, generating mutually advantageous nutritional interdependence. This complex procedure requires the combined involvement of the bacterial community. Hence, while evaluating the degradation rates of pollutants in the environment, it is crucial to take into account the complete microbial network. Additionally, competitive dynamics prevail among microbes (Dikshitulu et al. 1993). Under competitive circumstances, the sporadic degradation of pollutants is not guaranteed, even if specific pollutant-degrading microorganisms are present. This is because other organisms may have a higher likelihood of acquiring favorable factors, leading to accelerated growth, reproduction, and ultimately establishing dominance in the environment. In such situations, the growth of pollutant-degrading microorganisms may be hindered. Hence, recognizing the composition of microbial communities is essential to forecasting the destiny of contaminants in the environment (Figure 5.5). It facilitates the design of diverse bioremediation strategies based on different pollutants, as well as the cultivation of specialized microbial communities tailored for efficient degradation.

5.10.8 Agrochemicals

Currently, pesticides are of utmost importance in agricultural practices as they efficiently prevent and control crop pests and diseases, hence ensuring superior product quality and minimizing crop losses (Damalas and Eleftherohorinos 2011). Regrettably, the widespread use of pesticides and chemical fertilizers in contemporary agriculture has led to the pollution of diverse natural elements, such as the atmosphere, soil, and water (Craig 2018; Lu et al. 2008). The contamination has detrimental impacts on both the human population and animals through various routes. Organophosphorus insecticides, constituting approximately 45% of global pesticide use, pose a significant threat to ecosystems, as highlighted by the Environmental Protection Agency (Shabbir et al. 2018). Organophosphate poisoning globally causes an estimated 200 000 fatalities annually (Ojha et al. 2013). To combat pesticide pollution, a range of physicochemical procedures have been established to partially or fully restore polluted areas. Nevertheless, the use of these techniques is constrained by their huge expenses and the production of ancillary byproducts during the procedure. Microorganisms, such as bacteria and fungi, that are obtained and tested from various sources, can break down many pesticide residues into harmless environmental substances

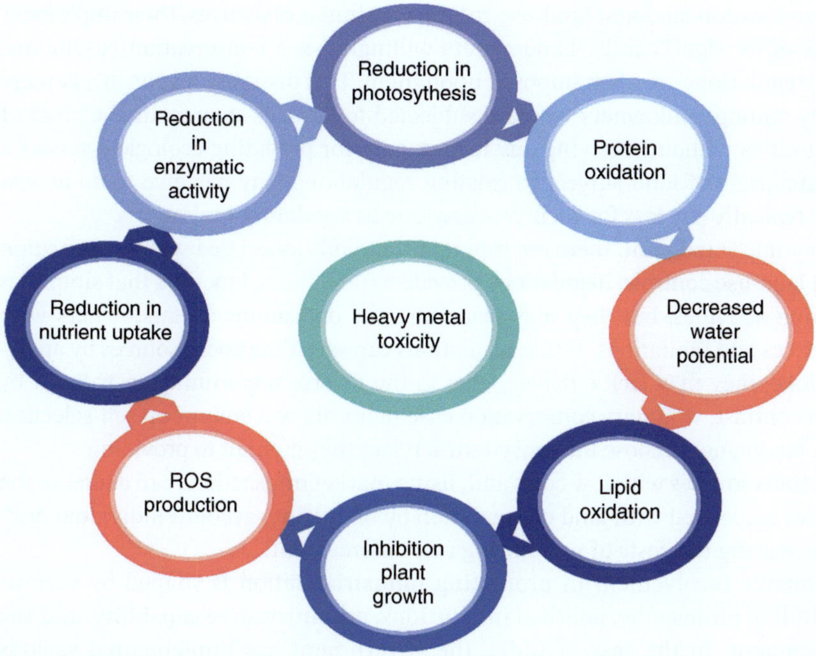

Figure 5.5 Several adverse effects of heavy metals.

(Rayu et al. 2017). The bioremediation method frequently includes biostimulation, which is adding different nutrients to the environment to boost the ability of microbial strains to absorb and break down contaminants (Wang et al. 2017). Nevertheless, the ability of native microbes to break down various pesticide pollutants is usually limited. To address these constraints, the advancement of exceptionally effective and genetically modified bacteria has proved to be a potential resolution. Such engineered bacteria significantly enhance pollutant degradation efficiency and the adaptability of these bacteria to a broader environmental spectrum (Yuanfan et al. 2010).

5.11 Policy Implications

5.11.1 Role of Government Regulations in Industrial Land Management

5.11.1.1 Land-Use Regulation

Land-use laws, related to conservation agreements, function as mechanisms to govern and permit particular uses of privately owned land. Local zoning regulations provide limitations on the division of property and specify the allowable activities inside specified land zones, similar to the restrictions imposed by envelopes for buildings in conservation easements. These restrictions may also apply at the state and federal levels, encompassing wetlands and threatened species regulations that govern land conversion and the preservation of habitats.

Although conservation land and land-use rules have similar objectives, their implementation methods differ significantly. Landowners willingly issue conservation easements, while land-use regulations are often imposed upon them. This disparity results in a perception of inequity among landowners who are subjected to laws, as they incur the costs of missed opportunities without receiving any recompense for providing ecological services. Conversely, purchasers of land subject to existing regulations may perceive them as less unjust, as they typically pay less for their property due to regulatory limitations.

From an economic standpoint, there are inherent trade-offs associated with conservation easements and land-use controls. Regulations provide a standardized method that simplifies land conservation activities, but they may place unequal obligations on landowners with varying preferences and limitations. While regulations can save time and resources by applying a broad stroke, they may lack efficiency due to the diverse opportunity costs faced by landowners. In contrast, voluntary conservation easements allow a bottom-up self-selection process, where landowners choose the ecosystem services they commit to providing.

The method finds sources with low costs and, using market transactions can decrease the societal expenses associated with land conservation by taking into account individual preferences and decreasing the costs of monitoring and enforcement.

The government's involvement in promoting industrialization is shaped by various elements, including philosophy, political institutions, administrative capability, and the level of development. In the case of India, the government has implemented various measures to achieve self-reliance, boost employment, and accelerate economic growth through the development of industries. For Telangana, the 29th state of India, industrialization is seen as a crucial strategy for economic growth and development. The expectations from the Telangana State Government are high, with a focus on job creation for the youth, fostering development in underprivileged areas, maximizing growth opportunities through resource optimization, harnessing the talents and skills of the population, and bringing prosperity to every household. Industrialization is considered the key to realizing these aspirations. The vision for Telangana's industrialization is described as "Research to Innovation; Innovation to Industry; Industry to Prosperity," with a guiding slogan of "In Telangana Innovate, Incubate, Incorporate." The industrial policy framework is designed to create a business regulatory environment that facilitates ease of doing business. The emphasis is on innovation and technology as driving forces for the industries in Telangana.

5.11.2 Incentives for Sustainable Practices

Piñeiro et al. (2020) examined about 18 000 papers to investigate the influence of rewards on farmers' willingness to adopt sustainable agriculture techniques and the consequent achievements. It reveals that programs offering short-term economic benefits yield higher adoption rates than those focused solely on ecological services. Long-term adoption is driven by perceived benefits to farms or the environment. Technical assistance and extension services play a crucial role. Efficient policy formulation necessitates the consideration of the particular characteristics of the group to be targeted population and an in-depth assessment of the pros and cons of economic and social consequences.

Another review performed by Piñeiro et al. (2021) in sustainable agricultural practices are essential for efficient natural resource use, environmental impact mitigation, and climate change adaptation. As these practices often demand significant effort and resources from farmers, effective incentives are crucial for widespread adoption. Despite increasing interest, there has been limited systematic evaluation of the incentives – adoption – outcome chain, determining which incentives best drive adoption and lead to desired sustainability outcomes. This brief presents the findings of a literature review assessing the uptake of agricultural practices under various incentives (market and nonmarket, regulations, and cross-compliance) and their impact on productivity, profitability, and environmental sustainability.

5.12 Conclusion and Ways Forward

Plant-based remediation is a developing technique due to the expensive and environmentally harmful nature of traditional technologies used for ecological remediation. Furthermore, bacteria can alter chemical interactions involving hazardous metals, hence increasing their absorption in the root zone regions for uptake by metallic hyperaccumulating vegetation. Certain molecular approaches can enhance the phytoremediation process by genetically modifying low-biomass metal hyperaccumulator plants to yield high-biomass non-accumulating plants. Vetiver's phytoremediation is a cost-effective solution when compared to traditional phytoremediation methods. Phytoremediation, particularly phytoextraction, is gaining favor as a cost-effective and innovative remediation approach. Phytoextraction is a solar-powered method that can effectively be used to remediate soils contaminated with arsenic.

5.12.1 Recap of Key Findings

A thorough examination of environmental issues in conjunction with modern biotechnology highlights the forefront of environmental biotechnology – constructing genetically modified bacteria for pollutant degradation. The identification of numerous genes associated with pollutant breakdown has enabled the integration of these genes into highly prolific and adaptable recipient bacteria. This has resulted in the creation of multiple genetically modified strains, which substantially enhance the potential and effectiveness of bacteria to cope with heavy metals and other difficult environmental pollutants.

Recognizing the inherent limitations of individual bacteria, the utilization of mixed bacterial consortia – micro-ecological systems where two or more microorganisms cooperate – appears to yield superior results in pollution degradation compared to the use of a single bacterial strain. Furthermore, as industrial activities progress and new pollutants emerge, the primary focus shifts toward discovering genes responsible for degrading these emerging pollutants. There is also potential in screening for mutants with heightened tolerance and/or increased accumulation capabilities. Augmenting the levels of proteins implicated in the containment of metals within cells, such as a group of enzyme PC synthase, and vacuolar transporters, may enhance the capacity of cells to withstand exposure to metals. However, this may not contribute significantly to metal accumulation.

To achieve sufficient accumulation, it is necessary to boost the activity of plasma membrane transporters regulated by non-metal-responsive or positive, instead of negative, metal-responsive promoters. Engineering tolerance and accumulation are crucial for developing appropriate plants for phytoremediation, as both qualities are separate.

5.12.2 Importance of Ongoing Research

The current ongoing research work on plant diversity on post-industrial land is essential for understanding ecological dynamics, promoting resilience, restoring ecosystems, and ensuring the sustainable use of these lands for the benefit of both the environment and the human community. Post-industrial lands often undergo significant ecological changes due to industrial activities. Ongoing research helps scientists and ecologists understand how these changes impact the environment. This involves studying the interactions between different plant species, soil composition, water quality, and the overall ecosystem structure. Understanding these concepts is critical for devising efficient approaches to restore and rehabilitate modified environments. Resilience, denoting an ecosystem's ability to withstand and recover from disturbances, is crucial in post-industrial lands marked by ecological degradation. Research on plant diversity identifies robust species thriving in challenging conditions, enabling the promotion of diverse communities that enhance ecosystems' rebound from stressors like pollution and habitat disruption. This knowledge guides restoration efforts, identifying native plants suited for reintroduction to recreate functional ecosystems and restore lost ecological services. Sustainable use of these lands involves a delicate balance between human activities and ecological preservation, with research identifying plant species suitable for purposes like phytoremediation and sustainable agriculture. Plant diversity, offering ecosystem services such as purification and fertility, contributes to overall environmental health. Research findings inform policymakers, facilitating the development of guidelines promoting sustainable land use, habitat restoration, and biodiversity conservation. Informed decisions at the policy level ensure post-industrial lands align with ecological principles for long-term environmental well-being. In essence, ongoing research on plant diversity on post-industrial land is a multifaceted endeavor that encompasses understanding, promoting resilience, restoring ecosystems, ensuring sustainability, and maximizing benefits for both the environment and human communities. This research contributes valuable knowledge that can guide effective land management practices and policy decisions, fostering a harmonious balance between industrial activities and ecological well-being.

5.12.3 Call to Action for Sustainable Land Management and Conservation

Industrial and organizational psychologists are crucial in supporting the global shift toward environmental sustainability. Today's economic activities and workplaces are undergoing unprecedented changes in their impact on the environment. To clarify, we define key environmental sustainability concepts in work settings and differentiate them from related ideas like social responsibility. Additionally, policies addressing local threats to primate species must adopt a global perspective, considering social and economic factors. Integrated land management skills, extension services, infrastructure development, and credit access are essential for successful agricultural sustainability. Challenges faced by small-scale farmers include land pressure from urbanization, lack of policies, and limited credit availability. The future of agriculture hinges on strategies like sustainable intensification, closing yield gaps, and enhancing resource efficiency to meet growing demand while reducing the industry's global environmental footprint.

References

Akeredolu, F.A., Olaniyi, H.B., Adejumo, J.A. et al. (1994). Determination of elemental composition of TSP from cement industries in Nigeria using EDXRF technique. *Nuclear Instruments and Methods in Physics Research Section A: Accelerators, Spectrometers, Detectors and Associated Equipment* 353 (1–3): 542–545. https://doi.org/10.1016/0168-9002(94)91718-3.

Anca, P., Masu, S., Lixandru, B. et al. (2011). Strategies for covering fly ash dumps with plant species suitable for phytostabilization. *Animal Science and Biotechnologies* 44 (2): 229–234.

Andy, C. (1991). Dictionary of Environment and Development. London: Earth Scan Publications Ltd.

Ashraf, M. and Harris, P.J.C. (2004). Potential bochemical indicators of salinity tolerance in plants. *Plant Science* 166: 3–16.

Baker, A.J.M. (1981). Accumulators and excluders strategies in the response of plants to heavy metals. *Journal of Plant Nutrition* 3: 643–654.

Baker, A.J.M. and Brooks, R.R. (1989). Terrestrial higher plants which hyperaccumulate metallic elements. A review of their distribution, ecology and phytochemistry. *Biorecovery* 1 (2): 81–126.

Baker, A.J.M., McGrath, S.P., Sidoli, C.M.D., and Reeves, R.D. (1997). The possibility of in situ heavy metal decontamination of polluted soils using crops of metal-accumulating plants. *Resources, Conservation and Recycling* 20 (3–4): 221–228.

Bano, A. and Fatima, M. (2009). Salt tolerance in *Zea mays* (L.) following inoculation with rhizobium and pseudomonas. *Biology and Fertility of Soils* 45: 405–413.

Biglari, H., Geravandi, S., Mohammadi, M.J. et al. (2017). Relationship between air particulate matter and meteorological parameters. *Fresenius Environmental Bullet* 26 (6): 4047–4056.

Bliss, L.C. (1998). Rediscovering fire, coal, and biodiversity in Pennsylvania. *Ecological Applications* 8 (3): 821–833.

Bradshaw, A.D. (2006). *Unplanned Impact: The Story of Britain's Post-War Landscape*. Cambridge University Press.

Bravo, O. and Silenzi, J.C. (2002). Strip cropping in the semi-arid region of Argentina: control of wind erosion and soil water accumulation. *Soil Science* 167 (5): 346–352.

Brega Filho, D. and Mancuso, P.C. (2003). *Reúso de Água*, 579. São Paulo, Brazil: Universidade de São Paulo-Facultade de Saúde Pública.

Brodhagen, M., Goldberger, J.R., Hayes, D.G. et al. (2017). Policy considerations for limiting unintended residual plastic in agricultural soils. *Environmental Science Policy* 69: 81–84.

Carlson, C.L. and Adriano, D.C. (1991). Growth and elemental content of two tree species growing on abandoned coal fly ash basins. *Journal of Environmental Quality* 20: 581–587.

Cermák, P. (2008). Forest reclamation of dumpsites of coal combustion by-products (CCB). *Journal of Forest Science* 54 (6): 273–280.

Chaney, R.L., Malik, M., Li, Y.M. et al. (1997). Phytoremediation of soil metals. *Current Opinion in Biotechnology* 8: 279–284.

Chang, Y.C., Nawata, A., Jung, K., and Kikuchi, S. (2011). Isolation and characterization of an arsenate-reducing bacterium and its application for arsenic extraction from contaminated soil. *Journal of Industrial Microbiology and Biotechnology* 39: 37–44. https://doi.org/10.1007/s10295-011-0996-6.

Chauhan, S. and Ganguly, A. (2011). Standardizing rehabilitation protocol using vegetation cover for bauxite waste (red mud) in eastern India. *Ecological Engineering* 37 (3): 504–510.

Chauhan, S. and Silori, C.S. (2010). Rehabilitation of red mud bauxite wasteland in India (Belgaum, Karnataka). *Ecological Restoration* 28 (1): 12–14.

Chen, X., Pei, T., Zhou, Z. et al. (2015). Efficiency differences of roadside greenbelts with three configurations in removing coarse particles (PM10): a street scale investigation in Wuhan, China. *Urban Forestry & Urban Greening* 14 (2): 354–360. https://doi.org/10.1016/j.ufug.2015.02.013.

Cheung, K.C., Wong, J.P.K., Zhang, Z.Q. et al. (2000). Revegetation of lagoon ash using the legume species *Acacia auriculiformis* and *Leucaena leucocephala*. *Environmental Pollution* 109: 75–82.

Clewell, A.F. and Aronson, J. (2006). Motivations for the restoration of ecosystems. *Conservation Biology* 20 (2): 420–428.

Craig, K. (2018). A review of the chemistry, pesticide use, and environmental fate of sulfur dioxide, as used in California. *Reviews of Environmental Contamination and Toxicology* 246: 33–64.

Curtis, P., Slay, C., Harris, N. et al. (2018). Classifying drivers of global forest loss. *Science* 361 (6407): 1108–1111. https://doi.org/10.1126/science.aau3445.

Damalas, C.A. and Eleftherohorinos, I.G. (2011). Pesticide exposure, safety issues, and risk assessment indicators. *International Journal of Environmental Research and Public Health* 8 (5): 1402–1419.

D'Antonio, C.M. and Meyerson, L.A. (2002). Exotic plant species as problems and solutions in ecological restoration: a synthesis. *Restoration Ecology* 10 (4): 703–713.

Das, M., Agarwal, P., Singh, R., and Adholeya, A. (2013). A study of abandoned ash ponds reclaimed through green cover development. *International Journal of Phytoremediation* 15: 320–329.

Das, N. and Chandran, P. (2011). Microbial degradation of petroleum hydrocarbon contaminants: an overview. *Biotechnology Research International* 2011: 941810.

Davies, L., Bates, J.W., Bell, J.N. et al. (2007). Diversity and sensitivity of epiphytes to oxides of nitrogen in London. *Environmental Pollution* 146 (2): 299–310. https://doi.org/10.1016/j.envpol.2006.03.023;Dikshitulu, S., Baltzis, B.C., Lewandowski, G.A., and Pavlou, S. (1993). Competition between two microbial populations in a sequencing fed-batch reactor: theory, experimental verification, and implications for waste treatment applications. *Biotechnology Bioengineering* 42 (5): 643–656.

Dillon, A.J.P., Camassola, M., Henriques, J.A.P. et al. (2008). Generation of recombinants strains to cellulases production by protoplast fusion between *Penicillium echinulatum* and *Trichoderma harzianum*. *Enzyme Microbial Technology* 43 (6): 403–409.

Dris, R., Gasperi, J., Rocher, V. et al. (2015). Microplastic contamination in an urban area: a case study in Greater Paris. *Environmental Chemistry* 12 (5): 592–599. https://doi.org/10.1071/EN14167.

Duncan, B.N., Lamsal, L.N., Thompson, A.M. et al. (2016). A space-based, high-resolution view of notable changes in urban NOx pollution around the world (2005-2014). *Journal of Geophysical Research Atmospheres* 121 (2): 976–996. https://doi.org/10.1002/2015jd024121.

Dushenkov, S., Skarzhinskaya, M., Glimelius, K. et al. (2002). Bioengineering of a phytoremediation plant by means of somatic hybridization. *International Journal of Phytoremediation* 4: 117–126.

Ellstrand, N.C. and Elam, D.R. (1993). Population genetic consequences of small population size: implications for plant conservation. *Annual Review of Ecology and Systematics* 24 (1): 217–242.

Escobedo, F.J., Kroeger, T., and Wagner, J.E. (2011). Urban forests and pollution mitigation: analyzing ecosystem services and disservices. *Environmental Pollution* 159 (8-9): 2078–2087. https://doi.org/10.1016/j.envpol.2011.01.010.

Fantozzi, F., Monaci, F., Blanusa, T., and Bargagli, R. (2015). Spatio-temporal variations of ozone and nitrogen dioxide concentrations under urban trees and in a nearby open area. *Urban Climate* 12: 119–127. https://doi.org/10.1016/j.uclim.2015.02.001.

FAO (2020). *Global Forest Resources Assessment 2020 – ey findings*. Rome: Food and Agriculture Organization of the United Nations https://doi.org/10.4060/ca8753en.

Francesconi, K., Visoottiviseth, P., Sridokchan, W., and Goessler, W. (2002). Arsenic species in an arsenic hyperaccumulating fern, Pityrogramma calomelanos: a potential phytoremediator of arsenic-contaminated soils. *The Science of the Total Environment* 284: 27–35.

Franks, S.J., Avise, J.C., Bradshaw, W.E. et al. (2007). The resurrection initiative: storing ancestral genomes to capture evolution in action. *BioScience* 57 (9): 760–768.

Fuller, S. and Gautam, A. (2016). A procedure for measuring microplastics using pressurized fluid extraction. *Environmental Science & Technology* 50 (11): 5774–5780. https://doi.org/10.1021/acs.est.6b00816.

Garbisu, C., Hernández-Allica, J., Barrutia, O. et al. (2002). Phytoremediation: a technology that uses green plants to remove contaminants from polluted areas. *Reviews on Environmental Health* 17 (3): 173–188.

Gill, A.M. and Moore, P.H.R. (1997). *The Fire Ecology of Australian Plants*. Cambridge University Press.

Gleba, D., Borisjuk, N.V., Borisjuk, L.G. et al. (1999). Use of plant roots for phytoremediation and molecular farming. *Proceedings of the National Academy of Sciences* 96: 5973–5977.

Goettsch, B., Hilton-Taylor, C., Cruz-Pin~on, G. et al. (2015). High proportion of cactus species threatened with extinction. *Nature Plants* 1: 15142. https://doi.org/10.1038/nplants.2015.142.

Golmakani, S., Moghaddam, M.V., and Hosseini, T. (2008). Factors affecting the transfer of radionuclides from the environment to plants. *Radiation Protection Dosimetry* 130 (3): 368–375.

Gupta, D.K., Rai, U.N., Sinha, S. et al. (2004). Role of Rhizobium (CA-1) inoculation in increasing growth and metal accumulation in Cicer arietinum L. growing under fly-ash stress condition. *Bulletin of Environmental Contamination and Toxicology* 73: 424–431.

Gupta, A.K. and Sinha, S. (2008). Decontamination and/or revegetation of fly ash dykes through naturally growing plants. *Journal of Hazardous Materials* 153: 1078–1087.

Gutenmann, W.H., Bache, C.A., Youngs, W.D., and Lisk, D.J. (1976). Selenium in fly ash. *Science* 174: 966–967.

Haferburg, G. and Kothe, E. (2007). Microbes and metals: interactions in the environment. *Journal of Basic Microbiology* 47: 453–467.

Havens, K., Kramer, A.T., and Guerrant, E.O. (2014). Getting plant conservation right (or not): the case of the United States. *International Journal of Plant Sciences* 175: 3–10.

Higgs, E. (2003). *Nature by Design: People, Natural Process, and Ecological Restoration*. MIT Press.

Himadri Ghosh, IndiaSpend.com. (2016, June 4). In just 30 years, India has lost large forests to 23,716 industrial projects. Scroll.In.

Hirabayashi, S. (2021). Technical specifications of urban forests for air purification: a case study in Tokyo, Japan. *Trees, Forests and People* 4: 100078. https://doi.org/10.1016/j.tfp.2021.100078.

Hobbs, R.J. and Harris, J.A. (2001). Restoration ecology: repairing the Earth's ecosystems in the New Millennium. *Restoration Ecology* 9 (2): 239–246.

Hobbs, R.J., Higgs, E., and Hall, C.M. (2018). Defining novel ecosystems. In: *Novel Ecosystems*, 15–31. John Wiley & Sons.

Hobbs, R.J. and Suding, K.N. (2009). *New Models for Ecosystem Dynamics and Restoration*. Oxford University Press.

Howden, R. and Cobbett, C.S. (1992). Cadmium-sensitive mutants of *Arabidopsis thaliana*. *Plant Physiology* 100: 100–107.

Hozhabralsadat, M.S., Heidari, A., Karimian, Z. et al. (2022). Assessment of plant species suitability in green walls based on API, heavy metal accumulation, and particulate matter capture capacity. *Environmental Science and Pollution Research* 29: 68564–66858. https://doi.org/10.1007/s11356-022-20625-z.

Hurley, R., Woodward, J., and Rothwell, J.J. (2018). Microplastic contamination of river beds significantly reduced by catchment-wide flooding. *Nature Geoscience* 11 (4): 251–257. https://doi.org/10.1038/s41561-018-0080-1.

Jacoby, R., Peukert, M., Succurro, A. et al. (2017). The role of soil microorganisms in plant mineral nutrition - current knowledge and future directions. *Frontiers in Plant Science* 8: 1617. 1-19.

Jambhulkar, H.P. and Juwarkar, A.A. (2009). Assessment of bioaccumulation of heavymetals by different plant species grown on flyash dump. *Ecotoxicology and Environmental Safety* 72: 1122–1128.

Jiang, C.Y., Sheng, X.F., Qian, M., and Wang, Q.Y. (2008). Isolation and characterization of aheavy metal-resistant Burkholderia sp. from heavy metal-contaminated paddy field soil and its potential in promoting plant growth and heavy metal accumulation in metal-polluted soil. *Chemosphere* 72: 157–164.

Jones, J.D.G. and Dangl, J.L. (2006). The plant immune system. *Nature* 444 (7117): 323–329.

Jordan, W.R., Gilpin, M.E., and Aber, J.D. (2015). *Restoration Ecology: A Synthetic Approach to Ecological Research*. Cambridge University Press.

Jusaitis, M. and Pillman, A. (1997). Revegetation of waste fly ash lagoons. I. Plant selection and surface amelioration. *Waste Management Research* 15 (3): 307–321.

Juwarkar, A.A. and Jambhulkar, H.P. (2008). Restoration of fly ash dumps through biological interventions. *Environmental Monitoring and Assessment* 139: 355–365.

Kadlec, R.H. and Knight, R.L. (1996). *Treatment Wetlands*. CRC Press.

Kathiresan, K. and Bingham, B.L. (2001). Biology of mangroves and mangrove ecosystems. *Advances in Marine Biology* 40: 81–251.

Kaushik, N., Kumar, K., Kumar, S. et al. (2007). Genetic variability and divergence studies in seed traits and oil content of Jatropha (Jatropha curcas L.) accessions. *Biomass and Bioenergy* 31: 497–502.

Khan, F.I. and Abbasi, S.A. (2000). Attenuation of gaseous pollutants by greenbelts. *Environmental Monitoring and Assessment* 64: 457–475.

Khaniabadi, Y.O., Goudarzi, G., Daryanoosh, S.M. et al. (2017). Exposure to PM(10), NO(2), and O(3) and impacts on human health. *Environmental Science and Pollution Research* 24 (3): 2781–2789. https://doi.org/10.1007/s11356-016-8038-6.

Komara, L.L., Choesin, D.N., and Syamsudin, T.S. (2016). Plant diversity after sixteen years post coal mining in East Kalimantan, Indonesia. *Biodiversitas* 17: 531–538. https://doi.org/10.13057/biodiv/d170220.

Kostić, O., Jariæ, S., Gajić, G. et al. (2018). Pedological properties and ecological implications of substrates derived 3 and 11 years after the revegetation of lignite fly ash disposal sites in Serbia. *Catena* 163: 78–88.

Kostic, O., Mitrovic, M., Knezevic, M. et al. (2012). The potential of four woody species for the revegetation of fly ash deposits from the 'Nikola Tesla-A' thermoelectric plant (Obrenovac, Serbia). *Achieves of Biological Science, Belgrade* 64: 145–158.

Krzaklewski, W., Pietrzykowski, M., and Wos, B. (2012). Survival and growth of alders (*Alnus glutinosa* (L.) Gaertn. and *Alnus incana* (L.) Moench) on fly ash technosols at different substrate improvement. *Ecological Engineering* 49: 35–40.

Kumar, P., Kumar, A., Patil, M. et al. (2020). Plant diversity development under planted woody species on coal mine spoil in a dry tropical environment, India: a case study. *Journal of Ecological Engineering* 21 (2): 228–243. https://doi.org/10.12911/22998993/116330.

Lamb, D., Erskine, P.D., and Parrotta, J.A. (2005). Restoration of degraded tropical forest landscapes. *Science* 310 (5754): 1628–1632.

Lavorel and Garnier (2002). Predicting changes in community composition and ecosystem functioning from plant traits: revisiting the Holy Grail. *Functional Ecology* 16 (5): 545–556.

Lim, J.A., Lee, D.W., and Heu, S. (2014). The interaction of human enteric pathogens with plants. *Journal of Plant Pathology* 30: 109–116.

Liu, H., Whitehouse, C.A., and Li, B. (2018a). Presence and persistence of salmonella in water: the impact on microbial quality of water and food safety. *Frontiers in Public Health* 6: 159.

Liu, J., Cao, Z., Zou, S. et al. (2018b). An investigation of the leaf retention capacity, efficiency and mechanism for atmospheric particulate matter of five greening tree species in Beijing, China. *Science of the Total Environment* 616-617: 417–426. https://doi.org/10.1016/j.scitotenv.2017.10.314.

Lu, P., Hong, Y.F., Hong, Q. et al. (2008). Construction of a stable genetically engineered microorganism for degrading HCH & methyl parathion and its characteristics. *Huan Jing Ke Xue* 29 (7): 1973–1976.

Ma, L.Q., Komar, K.M., Tu, C. et al. (2001). A fern that hyper-accumulates arsenic. *Nature* 409: 579.

Mahon, A.M., O'Connell, B., Healy, M.G. et al. (2017). Microplastics in sewage sludge: effects of treatment. *Environmental Science & Technology* 51 (2): 810–818. https://doi.org/10.1021/acs.est.6b04048.

Marques, A.P.G.C., Rangel, A.O.S.S., and Castro, P.M.L. (2009). Remediation of heavy metal contaminated soils: phytoremediation as a potentially promising clean-up technology. *Critical Reviews in Environmental Science and Technology* 39: 622–654.

McKay et al. (2005). *Local Adaptation Cues in Vegetation*. American Naturalist.

McMillan, N.A., Kunkel, K.E., Hagan, D.L., and Jachowski, D.S. (2019). Plant community responses to bison reintroduction on the Northern Great Plains, United States: a test of the keystone species concept. *Restoration Ecology* 27 (2): 379–388.

McMinn, J.W., Berry, C.R., and Horton, J.H. (1982). Ash basin reclamation with commercial forest species. *Reclamation and Revegetation Research* 1: 359–365.

Meers, E., Van Slycken, S., Adriaensen, K. et al. (2005). The use of the saline grass Puccinellia distans with associated bacteria for remediation of heavy metal contaminated sites. *Plant and Soil* 276 (1–2): 235–242.

Mekuto, L., Ntwampe, S.K.O., and Mudumbi, J.B.N. (2018). Microbial communities associated with the co-metabolism of free cyanide and thiocyanate under alkaline conditions. *3 Biotech* 8 (2): 93.

Miura, S. (2016). The effects of radioactive contamination on the forestry industry and commercial mushroom-log production in Fukushima, Japan. In: *Agricultural Implications of the Fukushima Nuclear Accident Tokyo* (ed. T. Nakanishi and K. Tanoi), 145–160. Springer.

Mohammadi, M.J., Geravandi, S., Vosoughi, M. et al. (2015). An association between air quality and COPD in Ahvaz, Iran. *Jundishapur Journal of Chronic Disease Care* 4 (1): e26621. https://doi.org/10.5812/jjcdc.26621.

Morariu, F., Mâsu, S., Lixandru, B., and Popescu, D. (2013). Restoration of ecosystems destroyed by the fly ash dump using different plant species. *Scientific Papers: Animal Science and Biotechnologies* 46 (2): 180–184.

Mulhern, D.W., Robel, R.J., Furness, J.C., and Hensley, D.L. (1989). Vegetation of waste disposal areas of a coal fired power plant in Kansas. *Journal of Environmental Quality* 18: 285–292.

Naz, R., Khan, M.S., Hafeez, A. et al. (2024). Assessment of phytoremediation potential of native plant species naturally growing in a heavy metal-polluted industrial soils. *Brazilian Journal of Biology* 84: e264473. https://doi.org/10.1590/1519-6984.264473.

Neina, D. (2019). The role of soil pH in plant nutrition and soil remediation. *Applied and Environmental Soil Science* 2019: 1–9.

Netondo, G.W., Onyango, J.C., and Beck, E. (2004). Sorghum and salinity: II. Gas exchange and chlorophyll fluorescence of sorghum under salt stress. *Crop Sciences* 44: 806–811.

Neumann, G., Teras, R., Monson, L. et al. (2004). Simultaneous degradation of atrazine and phenol by Pseudomonas sp. strain ADP: effects of toxicity and adaptation. *Applied Environmental Microbiology* 70 (4): 1907–1912.

Nobel, P.S. (2002). *Remarkable Agaves and Cacti*. Oxford University Press.

Oguis, G.K., Gilding, E.K., Jackson, M.A., and Craik, D.J. (2019). Butterfly pea (Clitoria ternatea), a cyclotide-bearing plant with applications in agriculture and medicine. *Frontiers in Plant Science* 10: 645. 1–23.

Ojha, A., Yaduvanshi, S.K., Pant, S.C. et al. (2013). Evaluation of DNA damage and cytotoxicity induced by three commonly used organophosphate pesticides individually and in mixture, in rat tissues. *Environmental Toxicology* 28 (10): 543–552.

Pal, A., Kulshreshtha, K., Ahmad, K.J., and Yunus, M. (2000). Changes in leaf surface structures of two avenue tree species caused by auto-exhaust pollution. *Journal of Environmental Biology* 21 (1): 15–21.

Palmer, T.M., Stanton, M.L., Young, T.P. et al. (2008). Breakdown of an ant-plant mutualism follows the loss of large herbivores from an African savanna. *Science* 319 (5860): 192–195.

Pandey, V.C., Ancona, V., Roy, M., and Ranđelović, D. (2023b). *Aromatic Plant-Based Phytoremediation*. Elsevier: https://doi.org/10.1016/C2022-0-00341-X.

Pandey, V.C. and Bajpai, O. (2019). Phytoremediation: from theory towards practice. In: *Phytomanagement of Polluted Sites* (ed. V.C. Pandey and K. Bauddh), 1–49. Amsterdam: Elsevier https://doi.org/10.1016/B978-0-12-813912-7.00001-6.

Pandey, V.C., Bajpai, O., and Singh, N. (2016). Plant regeneration potential in fly ash ecosystem. *Urban Forestry & Urban Greening* 15: 40–44.

Pandey, V.C., Gajic, G., Lebrun, M., and Mahajan, P. (2023a). *Designer Cropping Systems for Polluted Land*. Elsevier: https://doi.org/10.1016/C2021-0-02751-6.

Pandey, V.C., Gajic, G., Sharma, P., and Roy, M. (2022). *Adaptive Phytoremediation Practices: Resilience to Climate Change*. Elsevier: https://doi.org/10.1016/C2020-0-00560-8.

Pandey, P.K., Kass, P.H., Soupir, M.L. et al. (2014). Contamination of water resources by pathogenic bacteria. *AMB Express* 4: 51.

Parihar, P., Singh, S., Singh, R. et al. (2015). Effect of salinity stress on plants and its tolerance strategies: a review environmental science and pollution. *Research* 22: 4056–4075.

Phillips, I.R. and Courtney, R. (2022). Long term field trials demonstrate sustainable nutrient supply and uptake in rehabilitated bauxite residue. *Science of the Total Environment* 804: 150134.

Pietrzykowski, M., Krzaklewski, W., and Wo´s, B. (2015). Preliminary assessment of growth and survival of green alder (*Alnus viridis*), a potential biological stabilizer on fly ash disposal sites. *Journal of Forestry Research* 26 (1): 131–136. https://doi.org/10.1007/s11676-015-0016-1.

Pietrzykowski, M., Wo´s, B., Paja˛k, M. et al. (2018a). The impact of alders (*Alnus* spp.) on the physico-chemical properties of technosols on a lignite combustion waste disposal site. *Ecological Engineering* 120: 180–186.

Pietrzykowski, M., Wo´s, B., Paja˛k, M. et al. (2018b). Reclamation of a lignite combustion waste disposal site with alders (*Alnus* sp.): assessment of tree growth and nutrient status within 10 years of the experiment. *Environmental Science and Pollution Research* 25: 17091–17099. https://doi.org/10.1007/s11356-018-1892-7.

Piha, M.I., Vallack, H.W., Michael, N., and Reeler, B.M. (1995). A low input approach to vegetation establishment on mine and coal ash wastes in semi-arid regions. II. Lagooned pulverized fuel ash in Zimbabwe. *Journal of Applied Ecology* 32: 382–390.

Piñeiro, V., Arias, J., Dürr, J. et al. (2020). A scoping review on incentives for adoption of sustainable agricultural practices and their outcomes. *Nature Sustainability* 3 (10): 809–820.

Piñeiro, V., Arias, J., Elverdin, P. et al. (2021). *Achieving Sustainable Agricultural Practices: From Incentives to Adoption and Outcomes*. International Food Policy Research Institute.

Qi, Y., Yang, X., Pelaez, A.M. et al. (2018). Macro- and micro- plastics in soil-plant system: effects of plastic mulch film residues on wheat (Triticum aestivum) growth. *Science of the Total Environment* 645: 1048–1056. https://doi.org/10.1016/j.scitotenv.2018.07.229.

Rahneshan, Z., Nasibi, F., and Moghadam, A.A. (2018). Effects of salinity stress on some growth, physiological, biochemical parameters and nutrients in two pistachio (*Pistacia vera* L.) rootstocks. *Journal of Plant Interactions* 13: 73–82.

Ram, L.C., Jha, S.K., Tripathi, R.C. et al. (2008). Remediation of fly ash basins through plantation. *Remediation* 18: 71–90.

Raskin, I., Nanda-Kumar, P.B.A., Dushenkov, S., and Salt, D.E. (1994). Bioconcentration of heavy metals by plants. *Current Opinion in Biotechnology* 5: 285–290.

Rayu, S., Nielsen, U.N., Nazaries, L., and Singh, B.K. (2017). Isolation and molecular characterization of novel chlorpyrifos and 3, 5, 6-trichloro-2-pyridinol-degrading bacteria from sugarcane farm soils. *Frontiers in Microbiology* 8: 518.

Rensing, C., Kues, U., Stahl, U. et al. (1992). Expression of bacterial mercuric ion reductase in *Saccharomyces cerevisiae*. *Journal of Bacteriology* 174: 1288–1292.

Richards, P. and VanWey, L. (2015). Where deforestation leads to urbanization: how resource extraction is leading to urban growth in the Brazilian Amazon. *Annals of the Association of American Geographers* 105 (4): 806–823.

Rillig, M.C. (2012). Microplastic in terrestrial ecosystems and the soil? *Environmental Science & Technology* 46 (12): 6453–6454. https://doi.org/10.1021/es302011r.

Robinson, N.J., Procter, C.M., Connolly, E.J., and Guerinot, M.L. (1999). A ferric-chelate reductase for iron uptake from soils. *Nature* 397: 694–697.

Roy, A., Bhattacharya, T., and Kumari, M. (2020). Air pollution tolerance, metal accumulation and dust capturing capacity of common tropical trees in commercial and industrial sites. *Science of the Total Environment* 722: 137622. https://doi.org/10.1016/j.scitotenv.2020.137622.

Sakellaris, I.A., Tolis, E.I., Saraga, D.E., and Bartzis, J.G. (2017). VOCs, PAHs and ions measurements in an office environment in the vicinity of a small industry. *Fresenius Environmental Bulletin* 26 (1): 292–300.

Salt, D.E., Smith, R.D., and Raskin, I. (1998). Phytoremediation. *Annual Review of Plant Physiology and Plant Molecular Biology* 49 (1): 643–668.

Samuelsen, A.I., Martin, R.C., Mok, D.W.S., and Mok, M.C. (1998). Expression of the yeast FRE genes in transgenic tobacco. *Plant Physiology* 118: 51–58.

Shabbir, M., Singh, M., Maiti, S. et al. (2018). Removal enactment of organophosphorous pesticide using bacteria isolated from domestic sewage. *Bioresource Technology* 263: 280–288.

Shaw, P.J.A. (1992). A preliminary study of successional changes in vegetation and soil development on unamended fly ash (PFA) in southern England. *Journal of Applied Ecology* 29: 728–736.

Shi, B., Qu, Y., Li, H. et al. (2022). *Pseudomonas simiae* augments the tolerance to alkaline bauxite residue in *Atriplex canescens* by modulating photosynthesis, antioxidant defense enzymes, and compatible osmolytes. *Environmental Science and Pollution Research* 29 (16): 24370–24380.

Siegenthaler, A., Buttler, A., Grosvernier, P. et al. (2014). Discrepancies in growth measurement methods of mosses: an example from two keystone species grown under increased CO_2 and N supply in a restored Peatland. *American Journal of Plant Sciences* 5 (15): 515249.

Simonich, S.L. and Hites, R.A. (1994). Importance of vegetation in removing polycyclic aromatic hydrocarbons from the atmosphere. *Nature* 370: 49–51.

Singh, S.K. (2007). Global Agriculture Information Network (GAIN). Report IN7047, India Biofuels Annual, pp. 5–12.

Singh, M., Singh, V.P., and Prasad, S.M. (2019). Nitrogen alleviates salinity toxicity in Solanum lycopersicum seedlings by regulating ROS homeostasis. *Plant Physiology and Biochemistry* 141: 166–176.

Smičiklas, I. and Šljivić-Ivanović, M. (2016). Radioactive contamination of the soil assessments of pollutants mobility with implication to remediation strategies. In: *Soil Contamination - Current Consequences and Further Solutions* (ed. M.L. Larramendy and S. Soloneski). Intech Open https://doi.org/10.5772/64735.

Smith, S.E. and Read, D. (2008). *Mycorrhizal Symbiosis*, 3e. Academic Press.

Sommer, F., Dietze, V., Baum, A. et al. (2018). Tire abrasion as a major source of microplastics in the environment. *Aerosol and Air Quality Research* 18 (8): 2014–2028. https://doi.org/10.4209/aaqr.2018.03.0099.

de Souza Machado, A.A., Kloas, W., Zarfl, C. et al. (2018). Microplastics as an emerging threat to terrestrial ecosystems. *Global Change Biology* 24 (4): 1405–1416. https://doi.org/10.1111/gcb.14020.

Srivastava, M., Ma, L.Q., Singh, N., and Singh, S. (2005). Antioxidant responses of hyper-accumulator and sensitive fern species to arsenic. *Journal of Experimental Botany* 56: 1335–1342.

Suding, K.N., Higgs, E., Palmer, M. et al. (2015). Committing to ecological restoration. *Science* 348 (6235): 638–640.

Świątek, B., Woś, B., Chodak, M. et al. (2019). Fine root biomass and the associated C and nutrient pool under the alder (*Alnus* spp.) plantings on reclaimed technosols. *Geoderma* 337: 1021–1027.

Tang, T., Hurraß, J., Gminski, R., and Mersch-Sundermann, V. (2012). Fine and ultrafine particles emitted from laser printers as indoor air contaminants in German offices. *Environmental Science and Pollution Research International* 19 (9): 3840–3849. https://doi.org/10.1007/s11356-011-0647-5.

Tilman, D. (1999). The ecological consequences of changes in biodiversity: a search for general principles. *Ecology* 80 (5): 1455–1474.

Tripathi, R.D., Vajpayee, P., Singh, N. et al. (2004). Efficacy of various amendments for amelioration of fly-ash toxicity: growth performance and metal composition of Cassia siamea Lamk. *Chemosphere* 54: 1581–1588.

Tzoulas, K., Korpela, K., Venn, S. et al. (2007). Promoting ecosystem and human health in urban areas using Green Infrastructure: a literature review. *Landscape and Urban Planning* 81: 167–178.

Uzarowicz, L., Kwasowski, W., ´Spiewak, O., and ´Switoniak, M. (2018). Indicators of pedogenesis of technosols developed in an ash settling pond at the Belchatów thermal power station (central Poland). *Soil Science Annual* 69 (1): 49–59. https://doi.org/10.2478/ssa-2018-0006.

Vachova, P., Vach, M., Skalicky, M. et al. (2022). Reclaimed mine sites: forests and plant diversity. *Diversity* 14: 13. https://doi.org/10.3390/d14010013.

Van Rensburg, L., Morgenthal, T.L., Van Hamburg, H., and Michael, M.D. (2003). A comparative analysis of the vegetation and topsoil cover nutrient status between two similarly rehabilitated ash disposal sites. *The Environmentalist* 23: 285–295.

Viippola, V., Whitlow, T.H., Zhao, W. et al. (2018). The effects of trees on air pollutant levels in peri-urban near-road environments. *Urban Forestry & Urban Greening* 30: 62–71. https://doi.org/10.1016/j.ufug.2018.01.014.

Voget, J. and Fischer, B. (2012). Explaining the relationship between the Urban Forestry and Community Sustainability Programs of Tree City USA Municipalities in Indiana. School of Public and Environmental Affairs. Bloomington Urban Forestry Research Group White Paper, pp. 1–23.

Walker, L.R. (1995). Revegetation after catastrophic disturbance: lessons from Mount St. Helens. In: *Long-term Studies of Vertebrate Communities* (ed. M.L. Cody and J.A. Smallwood), 194–219. Academic Press.

Wang, C., Zhou, Z., Liu, H. et al. (2017). Application of acclimated sewage sludge as a bio-augmentation/bio-stimulation strategy for remediating chlorpyrifos contamination in soil with/without cadmium. *Science Total Environment* 579: 657–666.

Ware, J.H., Thibodeau, L.A., Speizer, F.E. et al. (1981). Assessment of the health effects of atmospheric sulfur oxides and particulate matter: evidence from observational studies. *Environmental Health Perspectives* 41: 255–276. https://doi.org/10.1289/ehp.8141255.

Xie, Q.E., Yan, X.L., Liao, X.Y., and Li, X. (2009). The arsenic hyperaccumulator fern *Pteris vittata* L. *Environmental Science & Technology* 43: 8488–8495.

Yan, X., Wang, J., Song, H. et al. (2020). Evaluation of the phytoremediation potential of dominant plant species growing in a chromium salt–producing factory wasteland. *Environmental Science and Pollution Research* 27: 7657–7671. https://doi.org/10.1007/s11356-019-07262-9.

Yli-Pelkonen, V., Viippola, V., Kotze, D., and Setälä, H. (2017). Greenbelts do not reduce NO_2 concentrations in near-road environments. *Urban Climate* 21: 306–317. https://doi.org/10.1016/j.uclim.2017.08.005.

Yuanfan, H., Jin, Z., Qing, H. et al. (2010). Characterization of a fenpropathrin-degrading strain and construction of a genetically engineered microorganism for simultaneous degradation of methyl parathion and fenpropathrin. *Journal of Environmental Management* 91 (11): 2295–2300.

Zevenbergem, C., Bradley, J.P., Shyam, A.K. et al. (2000). Sustainable ash pond development in India–a resource for forestry and agriculture. *Waste Materials in Series* 1: 533–540.

Zhang, G.S. and Liu, Y.F. (2018). The distribution of microplastics in soil aggregate fractions in southwestern China. *Science of the Total Environment* 642: 12–20. https://doi.org/10.1016/j.scitotenv.2018.06.004.

Zhang, G., Ma, L., Beuchat, L.R., and Erickson, M. (2009). Lack of internalization of *Escherichia coli* O157:H7 in lettuce (*Lactuca sativa* L.) after leaf surface and soil inoculation. *Journal of Food Protection* 72: 2028–2037.

6

Plantation Forestry for Ecorestoration

A Working Premises

Manoj Kumar Jhariya[1], *Abhishek Raj*[2], *Arnab Banerjee*[3], *Surendra Singh Bargali*[4], *and Kiran Bargali*[4]

[1] *Department of Farm Forestry, Sant Gahira Guru Vishwavidyalaya, Ambikapur, Chhattisgarh, India*
[2] *Department of Forestry, Pt. Deendayal Upadhyay College of Horticulture and Forestry, Dr. Rajendra Prasad Central Agriculture University, Samastipur, Bihar, India*
[3] *Department of Environmental Science, Sant Gahira Guru Vishwavidyalaya, Ambikapur, Chhattisgarh, India*
[4] *Department of Botany, DSB Campus, Kumaun University, Nainital, Uttarakhand, India*

6.1 Introduction

One-third of the global area is under forest cover which represents more than 2/3rd of the biodiversity at the terrestrial level. Forest is the key economic revenue for 1.6 billion individuals as the majority of humanity is dependent upon forest resources. Therefore, sustainable forest management (SFM) practices have a significant contribution toward poverty alleviation and sustainable development (SD), as well as fulfilling basic human needs (Jhariya et al. 2019a, b). In the present context, the rapid rate of deforestation has caused a significant alteration in the forest habitat and its associated bioresources (Awasthi et al. 2022a, b; Manral et al. 2022; Pandey et al. 2023a). This, in turn, has hindered the smooth functioning of the forest ecosystem and ecological functioning under the anthropogenic pressure of the rising human population (Thakrey et al. 2022; Samal et al. 2022; Bisht et al. 2022, 2023). Such a problem has necessitated for conservation of forest resources through ecorestoration approaches (Raj et al. 2023a, b).

Ecorestoration is an integrated approach that focuses on biodiversity, ecosystem functioning, as well as restoring the quality of habitat (Banerjee et al. 2023a, b). Although various SFM activities have been promoted toward the conservation of tropical forests, all the efforts have gone in vain due to altered land-use practices in the form of conversion of forest land to agricultural land as well as the rapid growth of urban population (OECD 2019). About 50% of the agricultural land in the tropics was procured at the cost of depletion of the forest cover within a span of 20 years (1980–2000) (WRI 2017). Further, various forms of degraded and degenerated forests harbor diver's life forms which get degenerated due to the agricultural conversion of forest lands (FAO 2020b).

Biodiversity and Ecosystem Services on Post-Industrial Land, First Edition.
Edited by Vimal Chandra Pandey.

The major focus of plantation forestry is to manage and sustain the forest cover across various ecosystems to fulfill the requirements of the upcoming human generation. It also aims toward the restoration of various forest areas through proper policy and legal framework under the leadership of national forest management. Plantation forestry also aims toward developing ecosystem resilience as well as maintaining ecosystem services for overall biota and humankind. Further, a well-managed ecosystem and its services help combat natural disasters as well as help adapt to changing climatic conditions followed by human use of natural resources. This, in turn, promotes the sustainable management of land and other valuable resources (Jhariya et al. 2022a, b; Karki et al. 2021, 2022).

Ecosystem services are the key issues for various ecosystems as it is directly and indirectly related to human well-being. In this connection, the forest is valued significantly for providing various forms of resources and maintaining ecosystem balance through various forms of ecosystem services (Yadav et al. 2022). SFM practices in the form of plantation forestry often play a significant role in the regulation of climate, biodiversity, carbon dioxide (CO_2) mitigation, as well as water purification. Such ecosystem services are of utmost importance for the survival and existence of humankind. Ecorestoration is an approach that commits to the regeneration of degraded forests as well as the renewal of ecosystem services. This, in turn, helps in the proper restoration of the land, water, and other ecosystem components and thus matches the whole world's ecological significance, healthy for the upcoming generation (Haj-Amor et al. 2022).

At present time, all over the world, human civilization is suffering from the crisis of global climate change and the greenhouse effect. Combating such mega events, plantation forestry in degraded and degenerated forest areas is becoming of utmost importance across the globe (Samal et al. 2022; Pandey et al. 2023a, b). Under this situation, various countries across the globe have ambitious targets for planting billions or trillions of trees in degenerated habitats. The long-term goal of such activities is to maintain the hydrological balance, protection from soil erosion and degradation, and conservation of bioresources along with poverty alleviation (Khatun et al. 2010; Manral et al. 2023).

Climate change is a big issue for the present-day environment. Both in the developing and developed world, it is creating negative consequences over agriculture, biodiversity, ecosystem, and prolonged ecological consequences. For combating such problems, tree plantation has become the eco-friendliest solution to the problem as well as it helps maintain the climatic balance of the earth's ecosystem (GCEC 2014). Further, one needs to be specific about the plantation scheme for screening specific plants in specific places as well as their survivability rate is also important. It also performs various other vital ecological services in the context of the overall sustainability and well-being of the earth's ecosystem. From a human context, tree plantation often helps in sustainable livelihood generation and socioeconomic upliftment of the people (Roy et al. 2022).

Restoration forestry is a scheme that helps in restoring the degraded forest area, which helps in an overall increase in the forest cover. This process is achieved by two modes, i.e., one through the anthropogenic plantation and the other through the ecological succession of the species. Species selection is an important aspect in terms of maximizing the benefits of plantation forestry. For instance, the monoculture practice of planting exotic and commercial species may harm the indigenous vegetation, which leads to the destruction of the local biodiversity (FAO 2020a, b; Banerjee et al. 2023b). Further, a mixed approach of planting native and non-native species may promote ecorestoration effectively in the concerned site and subsequently generate livelihood for the local community dwellers. It indirectly reduces the pressure on the intact forest resources for the local people in terms of timber and other non-timber forest products.

The present chapter showcases the role, extent, scenario, and prospects of plantation forestry toward the restoration of ecosystems toward climate change adaptation, mitigation, and moving forward for sustainable development.

6.2 Plantation Forestry: Global Area Context

A global forest area reported 4.06 billion ha (31%) of the total geographical land area. However, this reported area is equal to 0.52 hectares/person, which is not distributed equally among people around the globe. Tropical forests recorded the highest cover (45%) followed by boreal, temperate, and subtropical forests in the world. Similarly, five countries, *viz.*, the Russian Federation, China, Brazil, the United States, and Canada recorded 54% of the forest cover, which is more than half of the global forest cover. The scope and area distribution of plantation forestry are also estimated in the world. In brief, plantation forestry comprises one or two species, even-aged with a regular spacing, which is intensively managed for productive purposes. However, "other planted forests" comprised 55% of total planted forests, which are not managed intensively, and are also similar to natural forests. "Other planted forests" are purposely used for soil water protection and ecosystem restoration. Plantation forests cover 131 m ha, which is 45% of the total area of planted forests and 3% of the world's forest area. South America shared the highest area of plantation, which represents 99 and 2% of the total planted-forest area and total forest area. Whereas, Europe shared the lowest area of plantation, which is only 6% and 0.4% of total planted-forest area and total forest area. Similarly, an introduced species is composed of 44% of the total plantation forests in the world (FAO 2020a).

6.3 Expansion of Plantation Forestry to Industrial Forest Plantation

The plantation recorded approximately 5% of the world forest area in the year 2000. This has produced approximately 35% of roundwood on 0.33% of world land areas. However, two-thirds of the roundwood production can be achieved after managing forest plantations properly (Christie 2008). Similarly, the plantation area reached up to 290 m ha with 1.84% of the average annual increment by the year 2015. The average annual increments, *viz.*, 2.51, 2.38, and 2.17, were recorded in North and Central America, South America, and the Asia regions of the world, respectively (FAO 2016). Further, industrial wood production depends on half the plantations in many regions of Asia, Africa, and Central America, whereas the remaining plantation is utilized for fuelwood, desertification, and watershed protection (Siry et al. 2005). Similarly, plantations increased rapidly in most of the countries of Asia and Africa, especially in densely populated regions, which also promotes afforestation activity in degraded land. Plantation has expanded in countries of South America, North America, and Oceania, where population density is low with large pasture land and stable populations (Rudel 2009). South America has expanded its market share in industrial plantations. This country has promoted plantation areas and expected from 10.7 to 26.7 m ha by the year 2050. Industrial plantations used 71% of timber harvesting from extensive plantation areas in South America. However, *Eucalyptus* plantations expanded for the industrial need of pulp mills in this region (Daigneault et al. 2008; Phillips 2013). Further, the southern hemisphere mostly comprises fast-growing industrial plantations which reflected a range of growth varied between 30 and 40 m^3 ha^{-1} per year, whereas 10–15 m^3 ha^{-1} per year in the northern hemisphere, respectively (Siry et al. 2005). A goal for enhancing forest area or tree plantation in different regions of the world is depicted in Table 6.1 (United Nations Department of Economic and Social Affairs 2021).

Table 6.1 Goal for enhancing forest area or tree plantation in different regions of the world.

Serial number	Country	Identified goal for promoting forest area or tree plantation
1	Australia	Forest area expansion by plantation of 20 million trees by the year 2020. Also, the projected plantation of a billion trees in the entire region of the country by the year 2030
2	India	Achieve a plantation goal by adding 20 000 hectares of forests and tree cover annually in the entire region of the country
3	Sri Lanka	Forest area expansion by increasing forest cover to 32% of the entire region of the country
4	Japan	A goal to preserve approximately 25 million hectares of forest areas in the country
5	New Zealand	Target to achieve a plantation goal by adding one billion trees in the country between the years 2018 and 2028
6	Bulgaria	A rigid goal to retain or preserve forested areas in the entire region of the country
7	Madagascar	A goal to preserve or restore approximately four million hectares of forests by upcoming 2030
8	Thailand	A goal to enhance forest cover up to 55% in the country
9	Nigeria	A goal to enhance forest cover from 6% to 25% by the year 2030
10	Suriname	Target to conserve present level of forest cover at 93% in the country
11	Kenya	A goal to enhance forest cover up to 10% in the country
12	Côte d'Ivoire	A goal to enhance forest cover up to 20% in the country by 2030
13	Brazil	Forest area expansion by two million hectares of plantation in the country
14	Jamaica	Target to reduce forest losses in the country
15	Myanmar	A goal to enhance forest cover up to 30% in the country
16	Philippines	A target of restoring 1.50 million hectares of degraded forest area in the country
17	Panama	A target of restoring one million hectares of degraded forest area between 2015 and 2035

Source: Adapted from United Nations Department of Economic and Social Affairs (2021).

6.4 Species Grown in Forest Plantations

A variety of forest species are utilized in plantation forestry, which depends on locality types and market demands. The tree species such as *Acacia*, *Pinus*, *Populus*, and *Eucalyptus* are specially used under the plantation program (FAO 2010). The selected species have quality wood grain and an acceptable growth rate, which makes viable plantations for industrial needs. *Acacia* and *Eucalyptus* plantations expanded continuously to fulfill the pulp and paper demands of the world (FAO 2010). Similarly, species like willows, *Poplars*, and aspens are used for short rotation forestry (SRF) and planted in boreal regions of the world. The social, environmental, and technical constraints do not appear while establishing these species under the plantation program. These species supply timber and wood materials sustainably for industrial needs (Weih 2004). Besides site factors and climate, many tree dimensions and parameters such as size, bark characteristics, form, stand density, etc. affect the technological adoption of species and utility for industrial needs.

6.5 Ecosystem Services Through Plantation Forestry

Plantation forestry delivers uncountable ecosystem services other than roundwood and many industrial products. However, the Millennium Ecosystem Assessment (MEA) has categorized forest-based ecosystem services into four types, *viz.*, provisioning, regulating, supporting, and cultural services. Provisioning services comprise timber, fuelwood, and NTFPs including food, fiber, medicine, and other genetic resources. Air purification, water regulation, watershed management, soil fertility maintenance, erosion control, waste treatment, pest and disease regulation, climate security, and pollination are regulating services. Similarly, primary production, effective nutrient cycling, and soil formation are supporting services, whereas ecotourism development, recreation, and aesthetic, religious, and spiritual values are cultural services (MEA 2005). These services provided by plantation forestry can ensure ecosystem health, people's wealth, and environmental sustainability. The delivery of ecosystem services increases day by day due to the expansion of plantation areas (Innes 2013; Szulecka et al. 2016). Incorporating multiple species in a plantation enhances biodiversity, which intensifies ecosystem services including soil, food, and climate security at the global scale. In ancient times, many industrial markets were incapable of capturing forests based on non-product benefits (Stephens and Grist 2014). Land users and natural resource managers can protect and maintain a variety of ecosystem services by using market-based instruments (Pirard 2012). This approach surely maintains forest-based services, which also offer financial incentives for achieving many conservation goals (Deal et al. 2012). Therefore, the scope and importance of plantation forestry in terms of ecosystem services increase to ensure future sustainability goals (Stephens and Grist 2014). Case studies on ecosystem services provided by plantation forestry in different countries are depicted in Table 6.2 (United Nations Department of Economic and Social Affairs 2021).

Table 6.2 Ecosystem services through plantation forestry in different countries: case studies.

Country	Program	Plantation activities	Ecosystem services
China	"A green-miracle" program of the Saihanba Mechanized Forest Fram	In this program, plantation activities have been ensured by enhancing forest cover from 11.40% to 80%. These activities added the value of forest stock volume up to 10.12 million cubic meters	Plantation activities provide approx. 137 m cubic meters of clean water to the nearby regions of Tianjin and Beijing. These activities further delivered 545 000 tons of pure oxygen along with greater forest carbon storage of 7.69 million tons. Effective forest management promotes ecotourism, which helps in creating jobs to alleviate poverty in this region
Ghana	MTS (Modified Taungya System) program	The country allotted many degraded forest areas for plant production in parallel to forest plantation activities under MTS (Modified Taungya System) program	This activity delivers many ecosystem services including food and nutritional security for forest-dependent peoples in this region
Thailand	Forest restoration program	Approximately 1.60 m ha of five-kilometer broad buffer zones have been created by Royal Forest Department (2019) in between national forest reserves (NFR) and communities for 15 000 community forests in this country	This activity delivers many ecosystem services including food security and the provision of NTFPs (Non-timber forest products) for 18 000 forest-dependent communities residing in or around the forest areas
Australia	A program for Tiwi Islands which supports a community through SFM	The country promotes plantation areas and connects in ecological-based research for sustainable environmental and socioeconomic development	Plantation activities provide shelter, food, medicinal plants, tools, and weapons, and maintain social, cultural, and spiritual needs. Also, provide jobs for future generations under this program in the country
Australia	A program of "Growing a Better Australia: A Billion Trees for Jobs and Growth"	In 2018, the country's NFIP (National Forest Industries Plan) consigned approximately AU$20 million over four years to strengthen the country's plantation resources by adding a billion trees in the upcoming decade	The program ensures multifarious ecological services including the delivery of many forest products for industrial growth and supports scientific research activities for a sustainable environment

New Zealand	"One Billion Trees Program" in the country	The country initiated an eco-friendly program for adding approximately one billion trees by the year 2028	This activity can improve socioeconomic, cultural, and environmental quality in the country
Senegal	A program related to the management of transboundary forests in the country	The country has signed an MoU (memorandum of understanding) with the Gambia for sustainable management of transboundary forests	Delivers uncountable ecosystem services by promoting carbon storage, flux, and stabilization. Managing forests and plantations ensures food and job security for local people.
Côte d'Ivoire	Cocoa forest initiatives based on eco-friendly programs in the country	The country promotes zero-deforestation agriculture practices along with forest plantation activities	Sustainable cocoa production improves people's livelihood and enhances socioeconomic status. Surprisingly, the coffee-cocoa plantation provided approximately ~€6 million to the MoWF (Ministry of Water and Forests) to equip forestry services that help monitoring, management, and protection of forests

Source: Adapted from United Nations Department of Economic and Social Affairs (2021).

6.6 Plantation Forestry Through a Reforestation Approach

Reforestation is an important approach that addresses various important critical issues related to environmental sustainability and the overall well-being of the earth's ecosystem. Considering the event of climate change and global warming, it now has gained worldwide attention as the United Nations (UN) has declared the 2020 decade as a decade for ecosystem restoration (FAO 2020b). The approach of reforestation has gained wide attention through the work of Wangari Maathai in terms of green belt development in Kenya in 1977. This activity ensured him to be a Nobel Laureate in 2004. Further, such activity helped plant more than 45 million tree species which helped in the ecorestoration of the environment and reduced the poverty at the global level. Various countries across the world have diverse approaches for reforestation through plantation schemes and hence increased the vegetal cover in the recent past helping combat desertification (Banerjee et al. 2023b).

Various initiatives from various sectors were undertaken under the banner of plantation forestry to combat climate change as well as target the carbon sequestration process for the well-being of the environment (Bargali et al. 2019). In the corporate sector, some initiatives have been undertaken in the form of plantation forestry to reduce their GHG emissions for the well-being of society and as corporate social responsibilities (CSR) (Prasad et al. 2021a, b). Various international agreements across the globe have committed to the restoration of degraded forest land and natural ecosystems. In this connection, the Bonn Challenge initiated by the German government in collaboration with IUCN has targeted the restoration of the forest area of about 210 million hectare areas across 60 countries. Further, the scheme targeted to achieve the restoration of 350 million hectares of land by 2030 (IUCN 2019). Considering the mitigation of climate change, and working in compatibility with the excerpt of COP26, various nations are also promoting plantation forestry to reduce their GHG emission. However, the issue of reforestation is a challenging one as it depends upon proper species and site selection as various case studies are available about the failures of the project implementation and lack of proper output. One of the major events in such a failure program includes the spreading of exotic species across the world. Due to such ecological invasion, various negative consequences were also observed in various environmental segments in terms of groundwater depletion and the development of exotic species under the banner of ecological invasion. Such developments become costly as eradication of such exotic species becomes highly problematic both from a time and habitat condition perspective (WRI 2017; FAO 2020a).

Combating natural disasters in the era of climatic extremes is a big issue nowadays. In this connection, planting tree species has proved to be successful in some cases across the world. For instance, in China, the grain for the Green program was initiated among the people of the flood-affected areas where people were promoted for plantation in terms of providing economic subsidies and food materials. Such a type of plantation would help combat the hazardous impact of floods in the western part of China. The program also helped in the development of a soil carbon pool with rapid reduction of surface runoff and erosive forces of water and wind (United Nations Department of Economic and Social Affairs 2021).

6.7 Ecological Restoration and Sustainability

Sustainable development is the key to the well-being of human civilization. In this context, the restoration of degraded lands becomes a key initiative in which plantation forestry is a key component. The process of ecorestoration undergoes some sequential steps and processes (Figure 6.1). It involves screening the suitable reasons for the problem, planning, and strategy formulation in harmony with nature, technological integration, and public participation to get the desired outputs (Jhariya et al. 2021a, b).

Restoration has the inherent value of giving economic output, creating employment opportunities as well as combating the changing climate. Further, at present time, due to the intense pressure of agriculture, and the human population, the planet is surviving in the crisis of arable land, and hence major initiatives have been undertaken for the restoration of degraded land (GCEC 2018). As human civilization's existence and survivability depend upon nature, nature restoration is of utmost importance and hence provides economic outputs (FAO 2020b). As per the database, six billion hectares of land area across the globe is severely degraded and more than two billion hectares reflect the need for ecorestoration (OECD 2019). As per the resolution of the Bonn Challenge, the ecorestoration of 160 million hectares of degraded land across 40 countries of the world may contribute 84 billion dollars of economic output on an annual basis (WRI 2017; IUCN 2019). While considering economic output, ecorestoration also provides benefits for all the major ecosystems of the world that

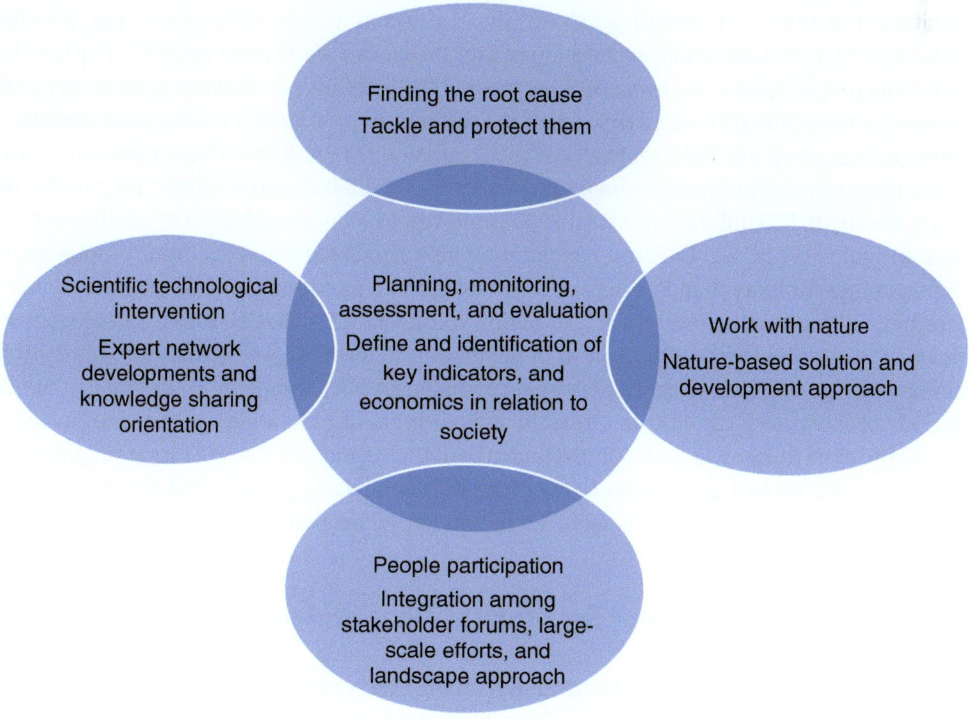

Figure 6.1 Steps toward ecological restoration and sustainable development.

ultimately lead to the well-being of human civilization. According to one research report, the ecorestoration of degraded agricultural land may uplift the economic gain of the farming community up to US$35–40 billion on an annual basis and is capable of feeding more than 175 million people annually for 15 years. This in turn provides additional benefits in terms of coping with climate change as well as the reduction of GHG emissions (GCEC 2014).

6.8 Forest Restoration

From a global perspective, forest cover has its existence over 33% of the geographical area of the earth's surface representing more than 2/3rd of the biodiversity (FAO 2020a). At present times, the rapid degradation of the forest area is associated with heavy loss of species, malfunctioning of ecosystem services, and other ecological principles. Anthropogenic pressure in terms of the human population is gradually increasing day by day and hence restoration of the degraded forest land, as well as forest conservation, becomes the key issue for sustainability. Forest restoration helps restore the biological diversity as well as ecosystem functioning that influences the well-being of human civilization (Meena et al. 2022). Mono-species plantation in comparison to mixed plantation has some inherent problems (Bargali et al. 1992a, b and Bargali et al. 1993a, b). This is because although mono-plantation species are higher in biomass turnover, the diversity of the area gets reduced. This in turn would impact the ecosystem services. The major focus should be given to establishing the balance between ecosystem stability and the biodiversity of a habitat. Therefore, the selection of proper species that are habitat-compatible and overall good for the ecosystem needs to be given top priority (FAO 2020b). In this direction, exploration of the germplasm of the plant species is required to select the proper species for restoration purposes. The interlinkage between biodiversity and ecosystem functioning creates a new era of the restoration process that incorporates the capability among species to fight against ecological invasion (MEA 2005). The restoration of the forest has approximated an economic gain of about 170 billion dollars annually, an increase in forest area up to 350 million hectares till 2030, and would sequester 5 billion of additional CO_2 sink annually (GCEC 2014). Forest restoration is also associated with generating employment opportunities for more than 3 lakh people with an annual investment of US$235 on a per hectare basis along with sequestering 1.38 billion tonnes of CO_2 (IUCN 2019). Climate-smart practices may enhance the effectiveness of plantations in terms of productivity and combating climate change. Further, selective harvesting, reducing species competition, and longer fallow periods of harvesting can also improve the effectiveness of plantation forestry and, in turn, would promote the growth and biomass turnover of the vegetation. These restoration practices can be applied up to a geographical area of 1.9 billion hectares globally and can remove the CO_2 load from the atmosphere of about 1.5 billion tons (The Nature Conservancy 2017).

6.9 Plantation Forestry for Climate Change Mitigation

Plantation forestry maximizes biodiversity; intensifies ecosystem services and helps in GHG reduction for climate security. Climate-resilient plantation management stabilizes and reduces excessive greenhouse gases by capturing atmospheric carbon and storing them into plant parts as biomass through the carbon sequestration process. These practices not only minimize C footprints and climate change but also maximize energy and biomass production along with greater carbon management in the environment. Carbon storage in the standing trees can be used as biomass, which is also a greater alternative to fossil fuels for energy (FAO 2016). Plantation management for biomass production also requires low fossil fuel inputs as compared to other land-use practices (St Clair et al. 2008). Plantation forestry expansion, increasing tree standing volume and growth rates, forest loss reduction, greater usage of long-lived forest products, and using many wood-derived products can replace fossil fuel consumption and reduce climate change problems in the world (McEwan et al. 2020). Therefore, plantation establishments are a viable strategy that ensures greater carbon storage and flux due to the higher carbon sink potential of planted forests (Parks et al. 1997). Policymakers also favor climate resilient plantation practices in degraded and wasteland areas for ecorestoration *via* carbon management. Further, avoiding deforestation activity and scientific management of plantation forestry maximize carbon uptake as biomass and help in climate change mitigation for a sustainable world (Christie 2008; Khatun et al. 2010; Michetti and Rosa 2012).

6.10 Socioeconomic Perspective of Restoration

Under the scenario of changing climate, the development of adaptive capacity and resilience of the ecosystem is a major step toward changing climate. In the restoration process, proper identification of the policy of restoration is very important as it would give different results in different regions. Further, the major objective may be to conserve the existing vegetal patterns and attributes or focus on restoring the processes and services of the ecosystem. Apart from ecological background, forest restoration has socioeconomic implications in terms of upliftment of the rural livelihood and prosperity. In this context, proper management steps of forests need to be explored that have a significant contribution to community upliftment. Therefore, integrating the socioeconomic perspective of forest management and protection is also an important aspect of sustainability.

Considering the tropical forest scenario, it invaluably helps foster biodiversity as well as the diverse form of ecosystem services that lead to overall sustainability in terms of economic output. But, tropical deforestation is a severe issue that leads to the degradation of forest areas as well and it also hinders livelihood and life-supporting services along with loss of biodiversity. Reforestation practices were found to be not fruitful in all the cases as plantation activities have generated some conflicts between the local community stakeholders and the forest department in terms of benefit sharing. Therefore, proper care should be taken to integrate the diverse concept of the restoration process, which will restore the ecosystem on one hand and build capacity, promote the conservation of bioresources, development of social ownership, and overall participatory approach on the other. A case study from south India (Kerala) revealed the importance of agroforestry practices and their potential toward increasing forest cover, and biomass conservation, along with the involvement of local people for sustainable management of forest resources. The results of the work revealed the potential role of agroforestry practice in forest rehabilitation, carbon stock development, and management of the landscape.

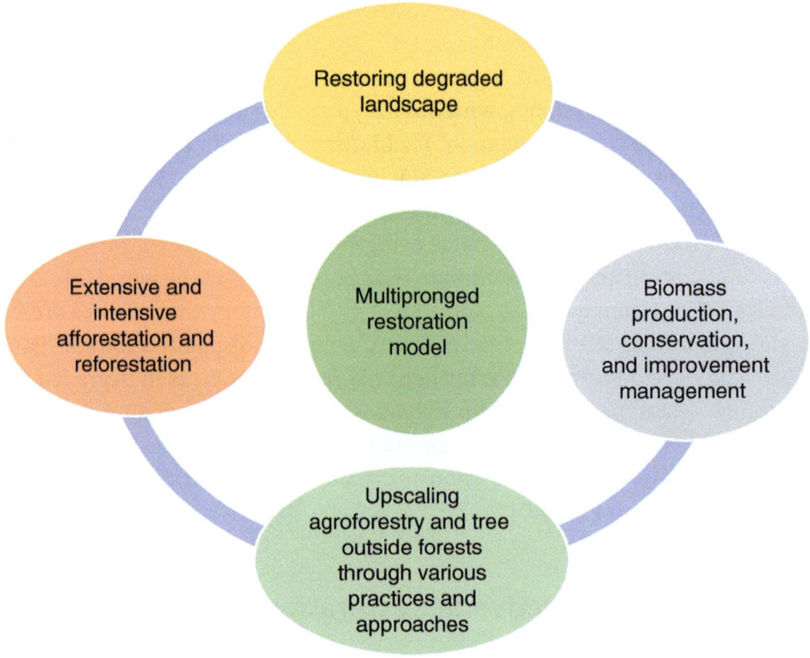

Figure 6.2 Model toward biomass restoration for the landscape.

Agroforestry and plantation forestry have shown significant potential in terms of biomass conversion through effective utilization of the degraded landscape, livestock development, upscaling, and promoting the growth of the species along with an increase in forest areas through extensive and intensive practices (Figure 6.2).

Ecorestoration works toward restoring the degraded landscape to move toward functionality and forest development for increasing forest cover. But there is some inherent barrier in terms of natural calamities, and disease outbreaks, which need to be given due consideration for the survivability of planted species. The restoration process needs to be promoted where the natural successional process is slow and fast recovery of vegetation is hindered. The importance of indigenous species becomes significant in terms of their successful establishment along with the diverse ecosystem services that they provide (Petit and Montagnini 2006).

6.11 Future Perspective

Considering the role of plantation forestry in various spheres of the environment, the potential seems to be increasing day by day, which needs to be explored properly. The multipurpose use of plantation forestry in terms of industrial forest plantation, ecosystem services, ecological restoration of degraded habitats as well as mitigating changing climate has brightened the future of plantation forestry for the well-being of human civilization (Jhariya et al. 2018). However, specific policies based on the specific areas along with proper monitoring are the key to the success of plantation forestry. Research and developmental activities toward the role of plantation forestry in mitigating climate change are the need of the hour as more investment and sustainability approaches can be developed by adopting such strategies (Raj et al. 2020). More research and development focusing on the role of plantation forestry need to be done in the upcoming times.

6.12 Conclusion

Overall on a global basis, plantation forestry plays a significant role in various aspects of the ecology of ecosystems. It has gotten wide attention due to its multipurpose use along with the socioeconomic upliftment of the people in the concerned areas. Considering the benefits of plantation forestry, it has immense potential for the well-being of human civilization as well as protecting environmental hazards along with the negative consequences. Proper policy formulation, strategy designing, and plan-do-check-act cycle need to be implemented at the grassroots level of the plantation activity for its successful output. Another major advantage of plantation forestry is that it is the nature-based solution to big problems such as climate change, global warming, and environmental pollution, which is very much required for achieving environmental sustainability of different ecosystems of the world. However, the potential of plantation forestry needs to be explored through research and developmental activities and promote human civilization to a better sustainable world.

References

Awasthi, P., Bargali, K., Bargali, S.S., and Jhariya, M.K. (2022a). Structure and functioning of *Coriaria nepalensis* wall dominated Shrublands in degraded hills of Kumaun Himalaya. I. Dry matter dynamics. *Land Degradation & Development* 33 (9): 1474–1494. https://doi.org/10.1002/ldr.4235.

Awasthi, P., Bargail, K., Bargali, S.S. et al. (2022b). Nutrient partitioning and dynamics in *Coriaria nepalensis* wall dominated Shrublands of degraded hills of Kumaun Himalaya. *Frontiers in Forests and Global Change* 5: 913127. https://doi.org/10.3389/ffgc.2022.913127.

Banerjee, A., Jhariya, M.K., Bargali, S.S., and Palit, D. (2023a). Ecorestoration for environmental sustainability – an introductory framework. In: *Ecorestoration for Sustainability* (ed. A. Banerjee, M.K. Jhariya, S.S. Bargali, and D. Palit), 1–47. Scrivener-Wiley.

Banerjee, A., Jhariya, M.K., Bargali, S.S., and Palit, D. (2023b). *Ecorestoration for Sustainability*, 1–441. Wiley-Scrivener.

Bargali, S.S., Padalia, K., and Bargali, K. (2019). Effects of tree fostering on soil health and microbial biomass under different land use systems in central Himalaya. *Land Degradation & Development* 30 (16): 1984–1998. https://doi.org/10.1002/ldr.3394.

Bargali, S.S., Singh, R.P., and Joshi, M. (1993a). Changes in soil characteristics in eucalypt plantations replacing natural broad leaved forests. *Journal of Vegetation Science* 4: 25–28.

Bargali, S.S., Singh, S.P., and Singh, R.P. (1992a). Structure and function of an age series of eucalypt plantations in Central Himalaya, I. Dry matter dynamics. *Annals of Botany* 69: 405–411.

Bargali, S.S., Singh, R.P., and Singh, S.P. (1992b). Structure and function of an age series of eucalypt plantations in Central Himalaya, II. Nutrient dynamics. *Annals of Botany* 69: 413–421.

Bargali, S.S., Singh, S.P., and Singh, R.P. (1993b). Pattern of weight loss and nutrient release in decomposing leaf litter in an age series of eucalypt plantations. *Soil Biology and Biochemistry* 25: 1731–1738.

Bisht, S., Bargali, S.S., Bargali, K. et al. (2022). Influence of anthropogenic activities on forest carbon stocks – a case study from Gori Valley, Western Himalaya. *Sustainability* 14: 16918. https://doi.org/10.3390/su142416918.

Bisht, S., Rawat, G.S., Bargali, S.S. et al. (2023). Forest vegetation response to anthropogenic pressures: a case study from Askot Wildlife Sanctuary, Western Himalaya. *Environment, Development and Sustainability* https://doi.org/10.1007/s10668-023-03130-2.

Christie, S. (2008). Energy, chemicals and carbon: future options for the Eucalyptus value chain. *Southern Forests: A Journal of Forest Science* 70: 175–182.

Daigneault, A.J., Sohngen, B., and Sedjo, R. (2008). Exchange rates and the competitiveness of the United States timber sector in a global economy. *Forest Policy and Economics* 10: 108–116. https://doi.org/10.1016/j.forpol.2007.07.001.

Deal, R.L., Cochran, B., and LaRocco, G. (2012). Bundling of ecosystem services to increase forestland value and enhance sustainable forest management. *Forest Policy and Economics* 17: 69–76. https://doi.org/10.1016/j.forpol.2011.12.007.

FAO. (2010). Global Forest Resources Assessment 2010, Main Report. FAO forestry paper no. 163; Food and Agricultural Organisation of the United Nations, Rome, Italy, p. 378. http://www.fao.org/docrep/013/i1757e/i1757e.pdf.

FAO (2016). *Global Forest Resources Assessment 2015. How are the world's forests changing?* 2e, 56. Rome, Italy: Food and Agricultural Organisation of the United Nations http://www.fao.org/3/a-i4793e.pdf.

FAO (2020a). *Global Forest Resources Assessment 2020 – Key Findings*. Rome: FAO https://doi.org/10.4060/ca8753en.

FAO (2020b). *Strategy of the UN Decade on Ecosystem Restoration*. Rome: UN FAO.

GCEC (2014). The New Climate Economy: Better Growth Better Climate. Better Growth, Better Climate: The New Climate Economy Report IMF-WB Annual meetings, 8 October 2014, pp. 1-18.The Global Commission on the Economy and Climate, 2014.

GCEC (2018). Unlocking the Inclusive Growth of the 21st Century: Accelerating Climate Action in Urgent Times. Pp. 207. The Global Commission on the Economy and Climate, 2018.

Haj-Amor, Z., Araya, T., Kim, D.G. et al. (2022). Soil salinity and its associated effects on soil microorganisms, greenhouse gas emissions, crop yield, biodiversity and desertification: a review. *Science of the Total Environment* 843: 156946. https://doi.org/10.1016/j.scitotenv.2022.156946.

Innes, J. (2013). What will we use the forests for in the future. *IUFRO News* 42:1.

IUCN (2019). Second Bonn Challenge Progress Report. Switzerland, 2019, p. 80. https://doi.org/10.2305/IUCN.CH.2019.06.en

Jhariya, M.K., Banerjee, A., and Meena, R.S. (2022a). Importance of natural resources conservation: moving toward the sustainable world. In: *Natural Resources Conservation and Advances for Sustainability* (ed. M.K. Jhariya, A. Banerjee, R.S. Meena, and S.N. Meena), 3–27. Elsevier, https://doi.org/10.1016/B978-0-12-822976-7.00027-2.

Jhariya, M.K., Banerjee, A., Meena, R.S. et al. (2021a). *Sustainable Intensification for Agroecosystem Services and Management*. Springer Nature Singapore https://doi.org/10.1007/978-981-16-3207-5.

Jhariya, M.K., Banerjee, A., Meena, R.S., and Yadav, D.K. (2019a). *Sustainable Agriculture, Forest and Environmental Management*, 606. Springer Nature Singapore https://doi.org/10.1007/978-981-13-6830-1.

Jhariya, M.K., Banerjee, A., Yadav, D.K., and Raj, A. (2018). *Leguminous trees an innovative tool for soil sustainability*. In: *Legumes for Soil |Health and Sustainable Management* (ed. R.S. Meena, A. Das, G.S. Yadav, and R. Lal), 315–345. Springer https://doi.org/10.1007/978-981-13-0253-4_10.

Jhariya, M.K., Meena, R.S., and Banerjee, A. (2021b). *Ecological Intensification of Natural Resources for Sustainable Agriculture*, 655. Springer Nature Singapore. https://doi.org/10.1007/978-981-33-4203-3.

Jhariya, M.K., Meena, R.S., Banerjee, A., and Meena, S.N. (2022b). *Natural Resources Conservation and Advances for Sustainability*. Elsevier, Academic Press https://doi.org/10.1016/C2019-0-03763-6.

Jhariya, M.K., Yadav, D.K., Banerjee, A. et al. (2019b). Sustainable forestry under changing climate. In: *Sustainable Agriculture, Forest and Environmental Management* (ed. M.K. Jhariya, A. Banerjee, R.S. Meena, and D.K. Yadav), 285–326. Singapore: Springer Nature https://doi.org/10.1007/978-981-13-6830-1.

Karki, H., Bargali, K., and Bargali, S.S. (2021). Spatial and temporal trends in soil N-mineralization rates under the agroforestry systems in Bhabhar belt of Kumaun Himalaya, India. *Agroforestry Systems* 95: 1603–1617. https://doi.org/10.1007/s10457-021-00669-9.

Karki, H., Bargali, K., and Bargali, S.S. (2022). Dynamics of fine root and soil nitrogen in *Mangifera indica* based agroforestry systems in Central Himalaya, India. *Land Degradation & Development* 33 (17): 3523–3538. https://doi.org/10.1002/ldr.4406.

Khatun, K., Valdes, P.J., Knorr, W., and Chaturvedi, R.K. (2010). Assessing the mitigation potential of forestry activities in a changing climate: a case study for Karnataka. *Forest Policy and Economics* 12: 277–286. https://doi.org/10.1016/j.forpol.2009.12.001.

Manral, V., Bargali, K., Bargali, S.S. et al. (2022). Relationships between soil and microbial biomass properties and annual flux of nutrients in Central Himalayan forests, India. *Land Degradation & Development* 33: 2014–2025. https://doi.org/10.1002/ldr.4283.

Manral, V., Bargali, K., Bargali, S.S. et al. (2023). Seasonal dynamics of soil microbial biomass C, N and P along an altitudinal gradient in Central Himalaya, India. *Sustainability* 15: 1651. https://doi.org/10.3390/su15021651.

McEwan, A., Marchi, E., Spinelli, R., and Brink, M. (2020). Past, present and future of industrial plantation forestry and implication on future timber harvesting technology. *Journal of Forestry Research* 31 (2): 339–351.

MEA (Millennium Ecosystem Assessment) (2005). *Ecosystems and Human Well-being Synthesis*, 155. Washington, DC: Island Press https://www.millenniumassessment.org/documents/document.356.aspx.pdf.

Meena, R.S., Yadav, A., Kumar, S. et al. (2022). Agriculture ecosystem models for CO_2 sequestration, improving soil physicochemical properties, and restoring degraded land. *Ecological Engineering* 176: 106546. https://doi.org/10.1016/j.ecoleng.2022.106546.

Michetti, M. and Rosa, R. (2012). Afforestation and timber management compliance strategies in climate policy. A computable general equilibrium analysis. *Ecological Economics* 77: 139–148. https://doi.org/10.1016/j.ecolecon.2012.02.020.

OECD (2019). Biodiversity: Finance and the Economic and Business Case for Action. Report prepared for the G7 Environment Ministers' Meeting (2019).

Pandey, R., Bargali, S.S., Bargali, K. et al. (2023a). Fine root dynamics and associated nutrient flux in Sal dominated forest ecosystems of Central Himalaya, India. *Frontiers in Forests and Global Change* https://doi.org/10.3389/ffgc.2022.1064502.

Pandey, R., Bargali, S.S., Bargali, K., and Pandey, V.C. (2023b). Temporal variability in fine root dynamics in relation to tree girth size in sub-tropical *Shorea robusta* forests. *Land Degradation & Development.* 34 (5): 1522–1537. https://doi.org/10.1002/ldr.4550.

Parks, P.J., Hall, D.O., Kristrom, B. et al. (1997). An economic approach to planting trees for carbon storage. In: *Economics of Carbon Sequestration in Forestry*, 1e (ed. R.A. Sedjo, R.N. Sampson, and J. Wisniewski), S9–S21. Boca Raton: CRC Press.

Petit, B. and Montagnini, F. (2006). Growth in pure and mixed plantations of tree species used in reforesting rural areas of the humid region of Costa Rica, Central America. *Forest Ecology and Management* 233 (2–3): 338–343.

Phillips, R.B. (2013). Global dynamics of the pulp and paper industry 2013. *New Zealand Journal of Forestry* 58: 34–42. http://nzjf.org.nz/free_issues/NZJF58_3_2013/647990C9-D846-45bd-8EE1-DA9B5622A907.pdf (accessed 29 August 2018).

Pirard, R. (2012). Market-based instruments for biodiversity and ecosystem services: a lexicon. *Environmental Science and Policy* 19–20: 59–68. https://doi.org/10.1016/j.envsci.2012.02.001.

Prasad, R., Jhariya, M.K., and Banerjee, A. (2021a). *Advances in Sustainable Development and Management of Environmental and Natural Resources: Economic Outlook and Opinions Volume I*, 1–437. CRC Press- a Taylor and Francis Group, Apple Academic Press Inc.

Prasad, R., Jhariya, M.K., and Banerjee, A. (2021b). *Advances in Sustainable Development and Management of Environmental and Natural Resources: Economic Outlook and Opinions Volume II*, 1–428. CRC Press- a Taylor and Francis Group, Apple Academic Press Inc.

Raj, A., Jhariya, M.K., Banerjee, A. et al. (2023a). Forest biodiversity conservation and restoration: policies, plan and approaches. In: *Ecorestoration for Sustainability* (ed. A. Banerjee, M.K. Jhariya, S.S. Bargali, and D. Palit), 317–350. Wiley-Scrivener.

Raj, A., Jhariya, M.K., Banerjee, A. et al. (2023b). Managing forests for offsetting carbon footprints. In: *Ecorestoration for Sustainability* (ed. A. Banerjee, M.K. Jhariya, S.S. Bargali, and D. Palit), 267–294. Wiley-Scrivener.

Raj, A., Jhariya, M.K., Yadav, D.K., and Banerjee, A. (2020). *Climate Change and Agroforestry Systems: Adaptation and Mitigation Strategies*, 383. Apple Academic Press Inc., CRC Press- a Tayler and Francis Group https://doi.org/10.1201/9780429286759.

Roy, O., Meena, R.S., Kumar, S. et al. (2022). Assessment of land use systems for CO_2 sequestration, carbon credit potential and income security in Vindhyan region, India. *Land Degradation & Development* 33 (4): 670–682. https://doi.org/10.1002/ldr.4181.

Rudel, T.K. (2009). Tree farms: driving forces and regional patterns in the global expansion of forest plantations. *Land Use Policy* 26: 545–550. https://doi.org/10.1016/j.landusepol.2008.08.003.

Samal, B., Singh, L., and Jhariya, M.K. (2022). Carbon storage, mitigation and sequestration potential of *Haldina cordifolia* and *Mitragyna parvifolia*in tropical dry deciduous environment of Chhattisgarh, India. *Ecological Engineering* 175: 106490. https://doi.org/10.1016/j.ecoleng.2021.106490.

Siry, J.P., Cubbage, F.W., and Ahmed, M.R. (2005). Sustainable forest management: global trends and opportunities. *Forest Policy and Economics* 7: 551–561. https://doi.org/10.1016/j.forpol.2003.09.003.

St Clair, S., Hillier, J., and Smith, P. (2008). Estimating the pre-harvest greenhouse gas costs of energy crop production. *Biomass Bioenergy* 32: 442–452. https://doi.org/10.1016/j.biombioe.2007.11.001.

Stephens, M.L. and Grist, P. (2014). Market failure for plantations: past experience and emerging trends for delivering wood production and ecosystem services in Australia. *International Forestry Review* 16: 205–215. https://doi.org/10.1505/146554814811724784.

Szulecka, J., Obidzinski, K., and Dermawan, A. (2016). Corporate-society engagement in plantation forestry in Indonesia: evolving approaches and their implications. *Forest Policy and Economics* 62: 19–29. https://doi.org/10.1016/j.forpol.2015.10.016.

Thakrey, M., Singh, L., Jhariya, M.K. et al. (2022). Impact of disturbance on biomass, carbon, nitrogen storage in vegetation and soil properties in tropical dry deciduous forest in Chhattisgarh, India. *Land Degradation & Development* 33: 1810–1820. https://doi.org/10.1002/ldr.4263.

The Nature Conservancy (2017). Lands of Opportunity: Unleashing the Full Potential of Natural Climate Solutions. Printed in the UK on Forest Stewardship Council certified paper, pp. 1–37. https://www.nature.org/content/dam/tnc/nature/en/documents/TNC_NCS_LandsofOpportunity_2017.pdf

United Nations Department of Economic and Social Affairs (2021). Global forest goals report 2021, (Sales No.: E.21.IV.3) New York: United Nations. https://www.un.org/esa/forests/wp-content/uploads/2021/04/Global-Forest-Goals-Report-2021.pdf.

Weih, M. (2004). Intensive short rotation forestry in boreal climates: present and future perspectives. *Canadian Journal of Forest Research* 34: 1369–1378. https://doi.org/10.1139/x04-090.

WRI (2017). How Can Restoring Degraded Landscapes Deliver Financial Returns? https://www.wri.org/insights/how-can-restoring-degraded-landscapes-deliver-financial-returns.

Yadav, V.S., Gupta, S.R., Yadav, S.S. et al. (2022). Carbon sequestration potential and CO_2 fluxes in a tropical forest ecosystem. *Ecological Engineering* 176: 106541. https://doi.org/10.1016/j.ecoleng.2022.106541.

7

Soil Biodiversity and Plant-Microbes Interactions on Post-Industrial Land

Aneta Kowalska[1] *and Anna Grobelak*[2]

[1] *Faculty of Science and Technology, Jan Dlugosz University, Czestochowa, Poland*
[2] *Faculty of Infrastructure and Environment, Czestochowa University of Technology, Czestochowa, Poland*

7.1 Loss of Soil Biodiversity on Post-Industrial Land

The loss of biodiversity and ecosystem services as a result of the industrial and economic activity of the human community, as well as the related deepening impoverishment of soils, are an important global problem. Human activity and continuously increasing demand for a variety of raw materials contribute to the changing natural environment. Mankind contributed to the alternations in local biodiversity in all ecosystems: marine, terrestrial as well as freshwater ecosystems (Valdez et al. 2023). Due to the high demand for natural resources, the problem of biodiversity loss is expected to worsen (Arneth et al. 2020). However, the functionality of the economy would not be possible without raw materials trapped in the lithosphere (Xavier et al. 2021). Therefore, the presence of raw materials often determines the location of many industrial plants (e.g., steelworks or power plants) (Kousehlar and Widom 2020). Hence, the areas of particularly high concentration of mines are the most transformed areas on our planet as a result of human activities. Each mining activity adversely influences the natural environment to a greater or lesser extent. This impact primarily affects the top layer of the earth's crust, but mining inevitably changes the hydrosphere, atmosphere, and biosphere (Warr and Grathoff 2021). Among mining activities, borehole mines have the lowest environmental impact of all mining methods. They do not contribute to land damage or excessive pollution. The most destructive impact on the natural environment is the open-cast mine and the associated removal of the top layer of soil (overburden) under which the mineral is located (Singh and Narzary 2021). However, despite many economic reasons and the possibility of obtaining raw materials necessary for the production of renewable energy (e.g., metals), mining may pose a threat to biodiversity. Land use change, which is particularly important in open cast mining, as well as environmental pollution and climate change, have a negative impact on land cover and habitat degradation leading to biodiversity (Tiamgne et al. 2022). Such influence of mining on biodiversity and ecosystems results from its huge interference with the environment. The indirect effects of mining operations on biodiversity are presented in Figure 7.1.

Biodiversity and Ecosystem Services on Post-Industrial Land, First Edition.
Edited by Vimal Chandra Pandey.

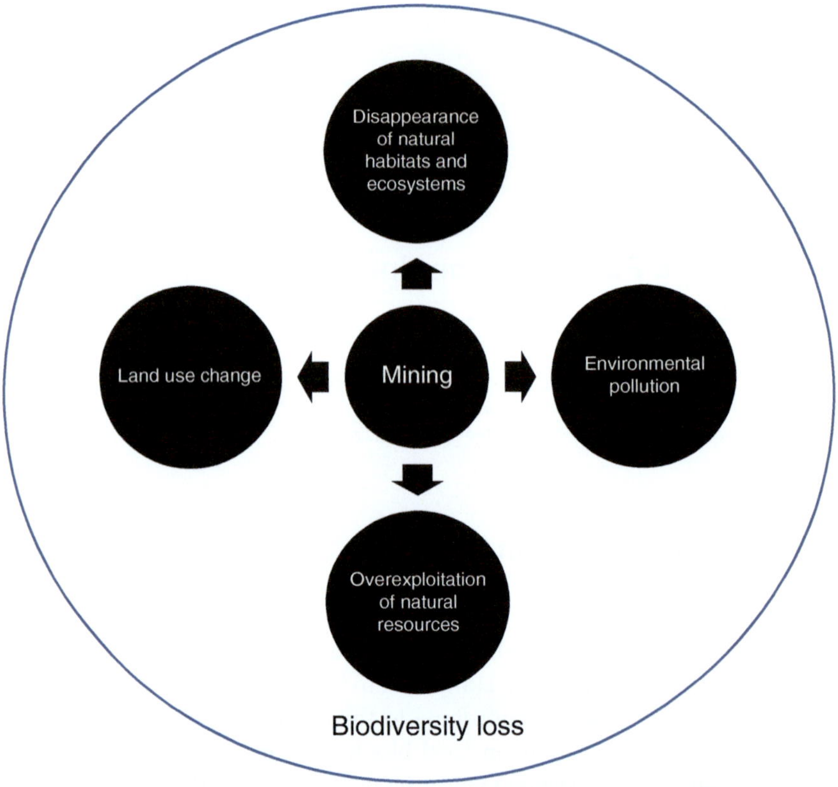

Figure 7.1 The initial influence of mining on biodiversity.

Biodiversity is essential to human well-being as it provides the services that sustain economies and societies, which explains the conservation struggles of individual species. Presently, the loss in biodiversity is observed at higher rates than it was predicted (Valdez et al. 2023). Post-mining areas, especially open-cast mines left without proper reclamation, generate areas of wasteland (Pandey and Mishra 2022). It has been noticed that post-mining areas despite their highly destructive influence on environmental habitats, and initial disturbances in biodiversity may potentially become green spaces promoting biodiversity in some cases (Salisbury et al. 2021).

7.2 Biodiversity as an Opportunity for Sustainable Transformation of Mining Regions

The recultivation of post-industrial areas is necessary to improve the soil quality and restoration of soil cover. It results also from the perspective of geoesthetic and social expectations (Ignatyeva et al. 2020). Post-mining soils usually represent areas with changed hydrological and climatic characteristics. In addition, the ability to penetrate pollutants in the soil profile alters, as well as their penetration into groundwater.

Thus, the spread of pollutants in the ecosystem is much easier compared to a stabilized environment (Van der Heyde et al. 2020).

Soil biodiversity is extremely important as it ensures balance in the soil (Yang et al. 2021). In addition, biodiversity protects the soil against its degradation and increases the regenerative capacity of the soil (Furey and Tilman 2021). The loss of soil biodiversity negatively affects soil quality, reduces its self-renewal capacity, reduces soil production capacity, and accelerates soil depletion processes. In addition, preserving biodiversity entails several positive aspects, including noise reduction, soil temperature reduction, protection of soil against erosion, air purification, improvement of water filtration, improvement of micro and mesoclimate, soil carbon sequestration, improvement of landscape values of post-industrial areas, and related social benefits (Tibbett et al. 2020).

Increasing biodiversity in rehabilitated post-mining areas can significantly accelerate the process of restoring the natural environment (Ahirwal and Pandey 2021). The selection of technology, the vegetation used, and the microbiological richness of the soil are of great importance in the process of reclamation of post-mining areas (Worlanyo and Jiangfeng 2021). Moreover, it was noted that in Europe there are many post-mining environments left to their own devices without implementation of appropriate reclamation to restore the ecosystem (Chugh et al. 2023). Observations conducted in these areas have shown that new species of living organisms can spontaneously gather there, creating a new ecosystem in completely unexpected systems. This leads to the formation of dynamic mosaic systems of ecosystems (Young et al. 2022).

Post-industrial areas often show high ecotoxicity affecting soil depletion. Indigenous species often become extinct due to drastically changing soil chemistry and its toxic effects (Klik et al. 2022). Changing the chemical properties of the soil in the areas of economic zones tends to support the development of new species that adapt to the new environmental conditions (Mukhopadhyay et al. 2021). This may be due to the resistance of living organisms to pollutants present in the soil in toxic concentrations. In particular, post-mining areas such as heaps are conditions for primary succession (Kaneda et al. 2020). This is because they are not covered with a soil layer and are devoid of vegetation cover. All this makes the area almost devoid of nutrients (Sudarmadji and Hartati 2021). It turns out, however, that despite the harsh conditions prevailing in post-industrial areas, fauna and flora may rebuild themselves, which has been observed in a way by the ability of plant species to colonize such areas. The long-term lack of development of post-mining areas promotes the succession of vegetation (Zhang et al. 2023). This makes it possible to spontaneously reconstruct a compact and durable vegetation cover after several years. Environmental conditions prevailing in post-industrial areas make them great objects of research on the course of ecological processes (Shu et al. 2023).

7.3 Treatments Increasing the Biodiversity of Post-Industrial Soil

One of the most common techniques for improving biodiversity in post-industrial areas is planting – both in the form of trees and smaller vegetation (Salisbury et al. 2021). Such a procedure is a good foundation for the development of sustainable habitats and allows for minimizing natural losses in areas highly exploited by humans. The direction of forest reclamation in post-industrial areas is dictated by the desire to create as many habitats as possible, which allows to increase in the number of species in these areas (Broughton et al. 2020). Passive reclamation involving the spontaneous restoration of ecosystems can be an effective way to ensure biodiversity through the natural succession of post-industrial areas (Conradi et al. 2021). In heavily exploited, degraded, or contaminated areas, represented mainly by post-mining and post-industrial areas, the main measure aimed at securing biodiversity is technical reclamation. It assumes a large interference in the environment, which is associated with the need to prepare the soil base (including soil import), planting, and sowing plants (Prach and Tolvanen 2016). Supporting the spontaneous restoration of depleted areas is mainly aimed at accelerating the processes of natural succession and thus accelerating the reconstruction of ecological habitats (Frouz 2021). Such treatments are dictated by the conditions prevailing in post-industrial areas, which include, among others, high pollution hindering the reconstruction of natural ecosystems, or hydrological conditions significantly deteriorated by human activity (Ethis-Eriakha and Akemu 2023).

In the case of rehabilitation of the post-coke area, it is common practice to use an embankment with by-products but also to clean the excavated soil or construct technosol. As a result of the conducted research, it was noticed that such treatments resulted in different patterns of soil fauna colonization. It has been suggested that the biodiversity of soil fauna in such areas is determined by the physico-chemical properties of the soil rather than by colonization limits (Colombini et al. 2022). In turn, the reclamation of post-mining areas significantly facilitates early ecological succession and accelerates the development of microorganisms (Worlanyo and Jiangfeng 2021).

7.4 Post-Industrial Areas as a Refuge of Biodiversity

Although the taxon diversity in post-industrial areas may not be such large as for natural environments, the functional diversity can tip the balance in favor of post-industrial areas (Gallagher et al. 2018). It was noted that post-industrial areas can be a rich habitat for a variety of species. High soil pollution in post-industrial areas, including high concentrations of heavy metals, makes it a good environment for the development of metallophytes, which are excellent at dealing with the presence of heavy metals. High resistance to harsh environmental conditions makes post-industrial areas a habitat for rare species, supporting biodiversity on a global scale (Rahmonov et al. 2021).

Extreme environmental conditions in post-industrial areas lead to microbial selection and increase the population of microorganisms resistant to the prevailing environmental conditions. For example, in the area of post-industrial areas managed with waste from the chemical industry, greater activity of bacteria and fungi capable of functioning in an environment with alkaline pH was observed. This suggests that environmental conditions strongly affect the population of soil microorganisms and their activity. Such a pattern of changes taking place in post-industrial soils may affect the biodiversity of ecological habitats located in post-industrial areas in two ways – by favoring the development of organisms capable of growing in extreme conditions (e.g., high soil salinity, presence of pollutants) (Kwiatkowska et al. 2022). Soil contamination profile determines the microbial population (Okrasińska et al. 2022). Different soil contaminants in the soil may be favorable for specific mycobiotic species, which results from shaping the soil communities by permanent soil contamination (Okrasińska et al. 2022). It has been suggested that the short-term impact of contamination leads to a reduction in the population and diversity of microbes, while permanent contamination leads to the successive selection of specific and highly diverse microbial communities that show remarkable adaptability to environmental contaminants (Bourceret et al. 2016).

It was noticed that the soil population of microorganisms responds differently to the presence of pollutants in the soil. The soil microbial population is closely related to the soil's chemical properties, and according to this, the *Sulfuritalea* and *Pseudomonas* are usually present in soil contaminated with polycyclic aromatic hydrocarbons (PAHs). In turn, *Nitrospira* and *Bryobacter* often occur in areas contaminated with heavy metals (Yang et al. 2022). The bacterial and fungal community in soil contaminated with diazepam and oxazepam appeared to have an accelerated resilience of the fungal community while the bacteria population showed a high evolution. It revealed that *Brevibacillus* strains increased their number in the studied habitat (Crampon et al. 2021).

In addition, soil microorganisms beside the performing basic functions in the soil such as a nutrient cycling, also play a significant role in the renewal of vegetation, which is of key importance for post-industrial areas, as well as in establishing the ecosystem balance (Barapatre et al. 2021).

7.5 Plant-Growth-Promoting Microbes and Their Feasibility for Recultivation of Post-Industrial Lands

The soil is a complex system in which plant roots, enzymes, and rhizosphere microorganisms cooperate enabling and supporting each other's development. Such dependence positively affects the preservation of biodiversity and supports the restoration of ecological habitats in post-industrial areas (Acharya et al. 2023). Thus, the possibilities of plant growth in post-industrial areas may be improved by soil microorganisms that promote plant growth. The presence of such interactions positively supports the development of biodiversity on post-industrial soils, increasing the potential for the restoration of ecological habitats. Arbuscular mycorrhizal fungi (AMF) may facilitate plant production in post-industrial soils, which has been observed by Balacco et al. (2023). Therefore, it has been observed that AMF more preferably colonized plant roots in highly contaminated soil in comparison to the control. It resulted in higher plant production, and higher plant biomass in post-industrial soils contaminated with heavy metals (Balacco et al. 2022).

The interactions between the plant and microorganisms can positively affect both populations in the soil environment. Plants can support the development of microorganisms, as well as the presence of a microbial population in the soil can reduce the toxicity of soil to plants by increasing their growth and development. Therefore, it has been noticed that the fungi population is more susceptible to plant support in comparison to bacteria (Bourceret et al. 2016). Among the direct mechanisms affecting plant development is the modulation of the concentration of phytohormones in plants. It has been noted that some soil microorganisms can reduce the toxicity of pollutants present in the soil, increase plant production through nitrogen fixation, and improve phosphate solubility (Das et al. 2022). Moreover, some endophytic bacteria and rhizobacteria reduce the toxicity of metals conserving the biodiversity of plant species in post-industrial areas.

The co-presence of plants and soil microorganisms in post-industrial areas is an important element in the formation of balanced areas transformed by man. Biodiversity of soil fauna in exploited areas depends strictly on environmental conditions and habitat possibilities for plants. The species composition of both plants and soil microorganisms in post-industrial areas is highly dynamic (Woźniak et al. 2022).

In post-industrial areas, time is an important factor for biodiversity. It was observed that spoil heaps covered with different types of plants (*Calamagrostis epigejos*, *Chamaenerion palustre*, and *Tussilago farfara*) showed different activity and species distribution of soil microorganisms in the top layer. This indicates soil stabilization by developing plants, which determines the composition and functioning of the soil microbial community (Stefanowicz et al. 2015).

Soil microbial activity and microbiome-plant interactions also result from the presence of nutrients in the soil. A positive correlation was observed between the activity of microorganisms and plant growth and the content of organic carbon in the soil (Lasota et al. 2020, Stefanowicz et al. 2020).

In the context of post-industrial areas, plants can also determine the environment of the rhizosphere by colonizing polluted areas. As a result, plants influence the microbial structure (Sasse et al. 2018). In addition, plant-soil microorganism interactions can also be based on the secretion of substances structurally similar to organic pollutants, supporting the processes of decomposition of pollutants present in the soil (Sasse et al. 2018). Collaboration between plants and microorganisms may result from the secretion of organic acids by plant roots, which facilitate the bioavailability of organic pollutants for microorganisms, affecting their proliferation (Gao et al. 2010). A positive effect of native plant growth-promoting bacteria on the development of *Sulla spinosissima* (L.) in soils contaminated after mining was also observed. This was achieved by inoculating the soil with *Rhodococcus qinhshengii* LMR340, which resulted in an increase in plant biomass of more than 40% compared to non-inoculated plants (Oubohssaine et al. 2022). This indicates the extremely important role of biodiversity in contaminated and emerging areas. Interspecies cooperation positively supports the renewal of ecosystems.

7.6 Summary and Conclusions

Post-industrial landscapes are a difficult and demanding area of revitalization. They are also an important element affecting biodiversity and the renewal of natural ecosystems, providing habitats for various species of living organisms. Mining processes, industrial activity, and thus interference with the physicochemical properties of the soil and the local climate, in most cases, lead to a reduction in biodiversity. In turn, it was noticed that heavily polluted and degraded post-industrial areas with demanding physicochemical properties may constitute important foundations for the development of extreme organisms capable of functioning in difficult conditions. It has also been observed that recreating ecosystems can foster the emergence of completely new, unexpected systems of organisms. The biodiversity of soil microorganisms in post-industrial areas is dependent on the presence of toxic substances and nutrients. Moreover, the species composition of vegetation covering the soil as well as the species composition of soil microorganisms (especially rhizosphere microorganisms) determines each other. The presence of certain species of microorganisms can reduce the toxic stress of plants resulting from high soil contamination, which supports the development of vegetation and the establishment of ecosystem balance and also accelerates the reconstruction of the natural environment. Therefore, plant-microbial interactions may be of particular importance for the species composition of post-industrial areas.

Acknowledgments

The research has been funded by Czestochowa University of Technology internal statutory grant.

References

Acharya, S.M., Yee, M.O., Diamond, S. et al. (2023). Fine scale sampling reveals early differentiation of rhizosphere microbiome from bulk soil in young Brachypodium plant roots. *ISME Communications* 3 (1): 54.

Ahirwal, J. and Pandey, V.C. (2021). Restoration of mine degraded land for sustainable environmental development. *Restoration Ecology* 29 (4): e13268.

Arneth, A., Shin, Y.J., Leadley, P. et al. (2020). Post-2020 biodiversity targets need to embrace climate change. *Proceedings of the National Academy of Sciences* 117 (49): 30882–30891.

Balacco, J.R., Vaidya, B.P., Hagmann, D.F. et al. (2023). Mycorrhizal infection can ameliorate abiotic factors in urban soils. *Microbial Ecology* 85: 100–107.

Barapatre, A., Burnase, N., Das, R., and Jaiswal, S. (2021). Harnessing the potential of microbes for rejuvenating soils from mining sites: an initiative for environmental balance and value addition. In: *Innovations in Sustainable Mining: Balancing Environment, Ecology and Economy*, 149–181. Cham: Springer International Publishing.

Bourceret, A., Cébron, A., Tisserant, E. et al. (2016). The bacterial and fungal diversity of an aged PAH- and heavy metal-contaminated soil is affected by plant cover and edaphic parameters. *Microbial Ecology* 71: 711–724. https://doi.org/10.1007/s00248-015-0682-8.

Broughton, R.K., Parry, W., and Maziarz, M. (2020). Wilding of a post-industrial site provides a habitat refuge for an endangered woodland songbird, the British Willow Tit Poecile montanus kleinschmidti. *Bird Study* 67 (3): 269–278.

Chugh, Y.P., Schladweiler, B.K., and Skilbred, C. (2023). Sustainable and responsible mining through sound mine closure. *International Journal of Coal Science & Technology* 10 (1): 14.

Colombini, G., Watteau, F., and Auclerc, A. (2022). Technosol rehabilitation strategies drive soil physico-chemical properties and fauna diversity on a former coking plant area. *Applied Soil Ecology* 177: 104542.

Conradi, T., Henriksen, M.V., and Svenning, J.C. (2021). Global change, novel ecosystems and the ecological restoration of post-industrial areas: the case of a former brown coal mine in Søby, Denmark. *Applied Vegetation Science* 24 (3): e12605.

Crampon, M., Soulier, C., Sidoli, P. et al. (2021). Dynamics of soil microbial communities during diazepam and oxazepam biodegradation in soil flooded by water from a WWTP. *Frontiers in Microbiology* 12: 742000.

Das, P.P., Singh, K.R., Nagpure, G. et al. (2022). Plant-soil-microbes: a tripartite interaction for nutrient acquisition and better plant growth for sustainable agricultural practices. *Environmental Research* 214: 113821.

Ethis-Eriakha, C.B. and Akemu, S.E. (2023). Ecological restoration of various ecosystems: implications for biodiversity conservation and natural resource management. *Ecorestoration for Sustainability* 245–266.

Frouz, J. (2021). Soil recovery and reclamation of mined lands. In: *Soils and Landscape Restoration*, 161–191. Academic Press.

Furey, G.N. and Tilman, D. (2021). Plant biodiversity and the regeneration of soil fertility. *Proceedings of the National Academy of Sciences* 118 (49): e2111321118.

Gallagher, F., Goodey, N.M., Hagmann, D. et al. (2018). Urban re-greening: a case study in multi-trophic biodiversity and ecosystem functioning in a post-industrial landscape. *Diversity* 10 (4): 119.

Gao, Y., Ren, L., Ling, W. et al. (2010). Desorption of phenanthrene and pyrene in soils by root exudates. *Bioresource Technology* 101 (4): 1159–1165.

Ignatyeva, M., Yurak, V., and Pustokhina, N. (2020). Recultivation of post-mining disturbed land: review of content and comparative law and feasibility study. *Resources* 9 (6): 73.

Kaneda, S., Angst, Š., and Frouz, J. (2020). Development of nutrient uptake by understory plant *Arrhenatherum elatius* and microbial biomass during primary succession of forest soils in post-mining land. *Forests* 11 (2): 247.

Klik, B., Holatko, J., Jaskulska, I. et al. (2022). Bentonite as a functional material enhancing phytostabilization of post-industrial contaminated soils with heavy metals. *Materials* 15 (23): 8331.

Kousehlar, M. and Widom, E. (2020). Identifying the sources of air pollution in an urban-industrial setting by lichen biomonitoring – a multi-tracer approach. *Applied Geochemistry* 121: 104695.

Kwiatkowska, E., Joniec, J., and Kwiatkowski, C.A. (2022). Involvement of soil microorganisms in C, N and P transformations and phytotoxicity in soil from post-industrial areas treated with chemical industry waste. *Minerals* 13 (1): 12.

Lasota, J., Błońska, E., Łyszczarz, S., and Tibbett, M. (2020). Forest humus type governs heavy metal accumulation in specific organic matter fractions. *Water, Air, & Soil Pollution* 231: 1–13.

Mukhopadhyay, R., Sarkar, B., Jat, H.S. et al. (2021). Soil salinity under climate change: challenges for sustainable agriculture and food security. *Journal of Environmental Management* 280: 111736.

Okrasińska, A., Decewicz, P., Majchrowska, M. et al. (2022). Marginal lands and fungi–linking the type of soil contamination with fungal community composition. *Environmental Microbiology* 24 (8): 3809–3825.

Oubohssaine, M., Sbabou, L., and Aurag, J. (2022). Native heavy metal-tolerant plant growth promoting rhizobacteria improves *Sulla spinosissima* (L.) growth in post-mining contaminated soils. *Microorganisms* 10 (5): 838.

Pandey, M. and Mishra, A. (2022). Assessment of land reclamation and landscape dynamics using geospatial techniques in open cast coal mines of Korba, Chhattisgarh, India. *Plant Archives* 22 (2): 456–461.

Prach, K. and Tolvanen, A. (2016). How can we restore biodiversity and ecosystem services in mining and industrial sites? *Environmental Science and Pollution Research* 23: 13587–13590.

Rahmonov, O., Cabała, J., and Krzysztofik, R. (2021). Vegetation and environmental changes on contaminated soil formed on waste from an historic Zn-Pb ore-washing plant. *Biology* 10 (12): 1242.

Salisbury, A., Gallagher, F.J., Parag, H.A. et al. (2021). Plant diversity increases in an urban wildland after four decades of unaided vegetation development in a post-industrial site. *Urban Ecosystems* 24: 95–111.

Sasse, J., Martinoia, E., and Northen, T. (2018). Feed your friends: do plant exudates shape the root microbiome? *Trends in Plant Science* 23 (1): 25–41.

Shu, C., Ruda, M., Dzhumelia, E. et al. (2023). Restoring soil cover and plant communities with arbuscular mycorrhizal fungi as an essential component of DSS for environmental safety management in post-industrial landscapes. *Agronomy* 13 (5): 1346.

Singh, K.N. and Narzary, D. (2021). Geochemical characterization of mine overburden strata for strategic overburden-spoil management in an opencast coal mine. *Environmental Challenges* 3: 100060.

Stefanowicz, A.M., Kapusta, P., Błońska, A. et al. (2015). Effects of *Calamagrostis epigejos*, *Chamaenerion palustre* and *Tussilago farfara* on nutrient availability and microbial activity in the surface layer of spoil heaps after hard coal mining. *Ecological Engineering* 83: 328–337.

Stefanowicz, A.M., Kapusta, P., Zubek, S. et al. (2020). Soil organic matter prevails over heavy metal pollution and vegetation as a factor shaping soil microbial communities at historical Zn–Pb mining sites. *Chemosphere* 240: 124922.

Sudarmadji, T. and Hartati, W. (2021). Post mining evaluation of forest land rehabilitation and potential ecosystem recovery. *Joint Symposium on Tropical Studies (JSTS-19) 4-6 September, 2019, Tanjung Redeb, Berau Regency, East Kalimantan Province, Indonesia*, (pp. 388–396). Atlantis Press.

Tiamgne, X.T., Kalaba, F.K., and Nyirenda, V.R. (2022). Mining and socio-ecological systems: a systematic review of Sub-Saharan Africa. *Resources Policy* 78: 102947.

Tibbett, M., Fraser, T.D., and Duddigan, S. (2020). Identifying potential threats to soil biodiversity. *PeerJ* 8: e9271.

Valdez, J.W., Callaghan, C.T., Junker, J. et al. (2023). The undetectability of global biodiversity trends using local species richness. *Ecography* 2023 (3): e06604.

Van Der Heyde, M., Bunce, M., Dixon, K. et al. (2020). Changes in soil microbial communities in post mine ecological restoration: implications for monitoring using high throughput DNA sequencing. *Science of the Total Environment* 749: 142262.

Warr, L.N. and Grathoff, G.H. (2021). Geoscientific applications of particle detection and imaging techniques with special focus on monitoring clay mineral reactions. In: *Handbook of Particle Detection and Imaging*, 945–962. Cham: Springer International Publishing.

Worlanyo, A.S. and Jiangfeng, L. (2021). Evaluating the environmental and economic impact of mining for post-mined land restoration and land-use: a review. *Journal of Environmental Management* 279: 111623.

Woźniak, G., Malicka, M., Kasztowski, J. et al. (2022). How important are the relations between vegetation diversity and bacterial functional diversity for the functioning of novel ecosystems? *Sustainability* 15 (1): 678.

Xavier, L.H., Giese, E.C., Ribeiro-Duthie, A.C., and Lins, F.A.F. (2021). Sustainability and the circular economy: a theoretical approach focused on e-waste urban mining. *Resources Policy* 74: 101467.

Yang, Z.N., Liu, Z.S., Wang, K.H. et al. (2022). Soil microbiomes divergently respond to heavy metals and polycyclic aromatic hydrocarbons in contaminated industrial sites. *Environmental Science and Ecotechnology* 10: 100169.

Yang, G., Roy, J., Veresoglou, S.D., and Rillig, M.C. (2021). Soil biodiversity enhances the persistence of legumes under climate change. *New Phytologist* 229 (5): 2945–2956.

Young, R.E., Gann, G.D., Walder, B. et al. (2022). International principles and standards for the ecological restoration and recovery of mine sites. *Restoration Ecology* 30: e13771.

Zhang, Q., Ma, J., Gonzalez-Ollauri, A. et al. (2023). Soil microbes-mediated enzymes promoted the secondary succession in post-mining plantations on the Loess Plateau, China. *Soil Ecology Letters* 5 (1): 79–93.

8

Afforestation of Former Asbestos Mines in Quebec, Canada

An Efficient Nature-Based Climate Solution that also Leads to Gains in Biodiversity

Nicolas Bélanger[1], Laurence Grimond[1], Rim Khlifa[1], Simon Bilodeau-Gauthier[2], and David Rivest[3]

[1] *Laboratoire sur la science des données, Université du Québec (TÉLUQ), Montréal, Québec, Canada*
[2] *Direction de la recherche forestière, Ministère des Ressources naturelles et des Forêts, Québec, Québec, Canada*
[3] *Département des sciences naturelles, Université du Québec en Outaouais, Gatineau, Québec, Canada*

8.1 Introduction

Earth's surface air temperature has increased by about 1 °C since the late 19th century. It will continue to increase by at least 2 °C by 2100 due to increased radiative forcing, which is caused by increasing atmospheric concentrations of greenhouse gases (GHGs), such as CO_2, CH_4, and N_2O (IPCC 2021). Climate models further predict that the effects of climatic phenomena are exacerbated by increases in the frequency, intensity, and duration of storms, floods, and droughts. A warmer and more chaotic climate will disrupt the major ecological balances of the planet. In particular, climate change will be accompanied by more frequent forest fires and periods of water stress, together with the continued acidification of ocean surface waters and melting of the ice caps, thereby endangering many animal and plant species (Allen et al. 2015; Barbero et al. 2015; Liang et al. 2022; Stewart-Sinclair et al. 2020). When combined with habitat destruction and overexploitation of resources, global loss of biodiversity is now occurring at an unprecedented rate (Caro et al. 2022). Natural disasters are becoming increasingly common and are attributable, in many cases, to human activity (IPCC 2021). Model projections appear to have underestimated the frequency and intensity of some of these disasters. In some regions of the world, we can further expect the loss of emerged land (due to the rise in sea level) and a substantial drop in agricultural production, causing food crises, human migrations, and conflicts (McDowell 2013; Anderson et al. 2020). Climate change will most likely exert more direct and negative effects on infrastructure and elevate risks of transmitting animal diseases that are dangerous to humans (Rocklöv and Dubrow 2020).

The Paris Agreement, which was adopted in 2015 and follows the United Nations Framework Convention on Climate Change (UNFCCC), aims to reduce GHG emissions that would limit global warming to 1.5 °C, thereby mitigating their impacts on the

Biodiversity and Ecosystem Services on Post-Industrial Land, First Edition.
Edited by Vimal Chandra Pandey.

environment and human wellbeing. Yet, given climate inertia, the long lifetime of GHGs in the atmosphere, and the failures of many countries to limit and lower GHG emissions, it has already been established that climate change will exceed the temperature threshold set by the Paris Agreement (IPCC 2021). For example, under the RCP 4.5 scenario, which is an intermediate climate-warming trajectory, maximum CO_2 emissions should be reached by 2040 and subsequent declines should be observed. This is the most probable scenario, considering the decline in fossil fuel reserves in the absence of very restrictive climate policies. By 2100, CO_2 emissions would be half of those of 2050, whereas CH_4 emissions would increase until 2050 and then decrease by about a quarter by 2100. The temperature increase could be as much as 2.6 °C by 2100, sea level could increase by as much as 0.63 m, and many animal and plant species are not likely to adapt to these changes (IPCC 2021).

Forest conservation, ecological restoration of disturbed areas, and agroecology (e.g., agroforestry) are some of the nature-based solutions to reduce atmospheric GHGs and mitigate climate change (Griscom et al. 2017, 2019; Bastin et al. 2019; Lewis et al. 2019; Seddon 2022). Forest conservation practices promote greater carbon sequestration than other land uses such as agriculture, livestock farming, and urban development. Reforestation after forest harvesting, avoidance of forest conversion to other uses, and land use and development that more closely emulates the structure, functioning, and disturbances of natural environments, are all strategies that can maximize carbon stores, thereby helping mitigate climate change. Afforestation of land that has long been devoid of trees, together with healthier soil management through the recovery of residual materials with fertilizing properties, the use of biochar and the avoidance of nitrogen-based (industrial, synthetic) fertilizers, are other ways to absorb more atmospheric CO_2 and reduce emissions of CH_4 and N_2O. The latter GHGs, respectively, have about 28 and 273 times greater global warming potential than CO_2 over 100 years (Byrne and Goldblatt 2014; Forster et al. 2021). According to Griscom et al. (2017), an equivalent CO_2 flux of several billion tons to the atmosphere could be avoided annually by 2030, if such nature-based solutions were to be implemented. This would represent more than one-third of the reductions that would be required by 2030 for warming not to exceed 2 °C. This avoidance could be conducted in a cost-effective way while assuring food security, timber needs, and biodiversity conservation.

Disturbed sites, such as highly mineralized urban sites, former mines and sand/gravel pits, and abandoned farmlands, abound around the world. These sites are often unvalued and, thus, neglected, given that they provide few or no ecosystem services (Evans et al. 2013). Sometimes, they are associated with environmental complications (Wright et al. 2018). Ecological restoration technologies could be used in an attempt to reintegrate these sites into the natural landscape and to restore, at least in part, the ecological services that are lost during commercial operations, e.g., habitats for wildlife, and a multitude of other social, cultural, environmental and economic benefits. To restore an ecosystem, the re-establishment of trees as dominant species is often the best option to restore the dominant flows of energy and matter (Singh et al. 2022) and the three-dimensional structure that is necessary to house other organisms (e.g., birds; Nero 2021). In many cases, these disturbed sites exhibit physical, chemical and biological limitations to the recovery of trees. Sources of tree propagules may be limited on such sites; it is frequently necessary to plant tree seedlings as a first step to re-establishment (Singh et al. 2022). Finding native tree species and genotypes that are resistant to establishment constraints that are imposed by these disturbed sites is needed to assure successful afforestation. Such challenging conditions include soils with poor aeration and organic matter content, the presence of toxicants (e.g., heavy metals and hydrocarbons) and high salinity (Pietrzykowski 2019). Plant stress that is imposed by soils in these environments can be reduced by using symbiotic microorganisms and organic amendments, but optimal mixtures of materials and application rates remain to be determined (Antonelli et al. 2018; Ohsowski et al. 2018).

To combat climate change, the Canadian Government has opted to fund nature-based solutions. The Natural Climate Solutions Fund (4 billion dollars CAD over 10 years) includes the 2 Billion Trees Commitment, the Nature Smart Climate Solutions Fund and the Agricultural Climate Solutions program. The research case study that is presented here has triggered a large initiative, which is funded by the 2 Billion Trees Commitment, for the afforestation of former asbestos mines in southern Quebec, starting in 2024. We thus provide a short historical perspective of asbestos mining in the country as well as review the main research findings over the past six years as a means to portray the optimal restoration technologies of asbestos mines, including technologies that lead to successes in regard to plant and animal biodiversity.

8.2 Historical Background of Asbestos Mining in Canada

Asbestos mining in Canada began in the late 19th century, but it did not become a significant industry until the early 20th century. The first asbestos mine in Canada, the Jeffrey Mine, opened in Asbestos (now Val-des-Sources), Quebec, in the late 1870s. Asbestos production in Canada grew steadily over the ensuing years and reached its peak in the mid-20th century, when the nation became one of the world's leading producers of the mineral (Savard 2017). The Province of Quebec was the primary hub for asbestos mining in the country, but limited asbestos mining activities took place in other jurisdictions, i.e., Newfoundland and Labrador, British Columbia and Yukon Territory (Kuyek 2003). Concurrently, asbestos mining expanded around the globe to meet industrial demands. Russia, South Africa, Australia, China, the United States, Zimbabwe, Kazakhstan, and Brazil were amongst the countries most actively involved in the asbestos industry.

Asbestos was initially considered a "magic mineral" and was thus given the name "white gold." It was rather inexpensive to mine and manufacture, and quite abundant in various locations around the world, making it appealing for a range of industries and uses (Frank 2006; IARC Working Group on the Evaluation of Carcinogenic Risks to Humans 2012). Key reasons for its widespread use for many decades have included: (i) heat resistance, withstanding extremely high temperatures without melting or burning. This property has made it valuable for use in heat-resistant materials and applications, such as insulation for furnaces, boilers, and pipes; (ii) fire resistance, making it an excellent material for fireproofing and flame-retardant applications. It was used in fireproof clothing, building materials, and automotive components; (iii) chemical resistance made it appropriate for various chemical applications; (iv) effective insulation, thus being widely used in building insulation, electrical insulation, and automotive insulation to prevent the transfer of heat or electrical current; (v) durable and long-lasting, making it suitable for roofing materials, pipes and flooring; (vi) sound absorption properties, which made it useful for acoustic insulation in buildings, theatres and sound studios; and (vii) strong and flexible, making it suitable for reinforcing materials like cement, plastic and rubber.

Despite its useful properties, health risks were associated with asbestos exposure, including the development of serious respiratory diseases (asbestosis) and cancers (mesotheliomas, lung cancer, ovarian and laryngeal cancers). These risks began to be documented through scientific research (Brofy et al. 2007; Goldberg and Luce 2009). A growing public awareness of health risks of asbestos exposure, together with a movement advocating for the ban of asbestos products, subsequently led to changes in attitudes towards asbestos use (Savard 2017). The asbestos industry began facing many lawsuits and legal challenges from individuals who had been exposed to asbestos and who developed related health problems. Subsequent legal actions placed severe financial pressure on asbestos companies. In turn, the industry markedly slowed in the late 20th century and early 21st century. Eventually, the economic viability of asbestos mining became increasingly unsustainable as the global demand for asbestos products continued to decline. Heightened public perception of the long-term health consequences of asbestos exposure outweighed its benefits, which increased pressure on governments to take action. This first resulted were stricter regulations on asbestos mining and production, and then an eventual ban on asbestos in many countries, including Canada (Government of Canada 2018; Lin et al. 2019). In practice, the implementation of regulations and bans was made possible because of the international support of the scientific community, activists, and asbestos victims (Ruff 2017). The last two asbestos mines that operated in Canada (including the Jeffrey Mine) ceased operations in 2011, resulting in adverse socioeconomic impacts, including unemployment and mental health problems in asbestos mining communities (Allen et al. 2018).

Open-pit mining operations of asbestos have drastically transformed the landscape of southern Quebec (van Horssen 2016). The legacy of this industry is now reduced to lands that are characterized by immense open pits, waste rock (overburden), and tailings piles, which in turn are linked to further health and environmental problems (Figure 8.1). This is especially true of the surroundings of the towns of Thetford Mines, Black Lake, and Val-des-Sources in southern Quebec, with abandoned mining sites covering a total area of 2308 hectares, which is a bit larger than the total surface area of the urban area of the towns themselves (BAPE 2020). Over time, these sites have accumulated about 800 million tons of material. Research has recently shown that massive erosion of asbestos mine wastes is incurring high sediment accumulation rates and alkalinization (magnesium), together with chromium and nickel enrichment, in a chain of lakes near the former mining operations (Jacques and Pienitz 2022). This pollution has been carried as far as 25 km downstream. Encroachment of wastes by surrounding natural forests is very limited, given the "hostile" nature of these wastes and the erosion of the steep pile banks. Because of the human health-risk challenges that are associated with the mineral itself, ecological restoration of these sites also requires special handling precautions, which delays much of the work.

Since 1995, costs of restoration in Quebec are to be assumed by the owners under the Mining Act (RSQ, c. M-13.1). Unfortunately, most asbestos mines had already ceased operations when these requirements became effective, which means that closed mines were not

(a)

(b)

(c)

Figure 8.1 Waste rock and tailings piles in close proximity to the town of Black Lake (a), a tailings pile ending on a small lake (b), and a tailings pile (aerial view) showing substantial erosion of its steep banks into a stream (c). *Source:* Photos by Nicolas Bélanger.

covered by these new obligations (BAPE 2020; Lévesque et al. 2020). In many cases, owners went bankrupt and abandoned their sites, thereby placing the responsibility on the government to restore these sites (Simard 2019). At other mines, some operations are ongoing (e.g., granite quarrying) and, as such, waste piles are undergoing restoration (by the owners) through the establishment of vegetation in various ways. Revegetation of such wastes can mitigate some environmental and socioeconomic impacts (e.g., dust control), and it can also bring back some ecosystem services, such as (i) restoring soil quality and nutrient cycling by re-establishing soil – plant – microbe interrelationships; (ii) contributing to carbon (C) sequestration and acting as a solution for mitigating climate change; and (iii) providing habitats for wildlife (Harris 2003; Ciccarese et al. 2012; Martín-Moreno et al. 2016).

8.3 Typical Ecological Restoration

Landscapes of asbestos communities in Quebec are not very different from those of other former mining communities. Indeed, most extractive industries face many mine closure challenges (Bainton and Holcombe 2018). This situation arises mainly because most mine wastes have physical and chemical properties that would restrict plant establishment and growth (Shrestha and Lal 2006). For instance, waste material is generally coarse, contains no or low concentrations of organic matter, has no microbial communities capable of activating nutrient cycling, and has high levels of toxicants, such as heavy metals. This description is no different for asbestos mine wastes. Yet, asbestos wastes are distinctive, given that magnesium largely dominates the chemical signature of the mineral group, i.e., various forms of chrysotile $[(Mg)_3Si_2O_5(OH)_4]$. The presence of these minerals can potentially create plant nutrient imbalances (i.e., calcium and potassium) and, thus, it must be considered when preparing a re-vegetation plan.

Several soil enrichment and soil reconstruction techniques (e.g., inorganic fertilizers, stockpiled topsoil, organic amendments, such as compost and manure) have been tested as a means of maximizing plant establishment and growth on mined lands, for which success varies (Sopper 1993; Brown et al. 2007; Shrestha et al. 2009; Carson et al. 2014). Whole soil reconstruction is likely the optimum approach for maximizing plant development and carbon sequestration potential of mine wastes under most situations. Residual wastes (e.g., biosolids) often have high carbon content and are available in large quantities at low cost, if any. In fact, many governments are aiming to recycle these materials in large amounts as an amendment for agricultural soils and, more recently, as material for soil reconstruction and ecological restoration of mine (and other industrial) sites. Biosolids, e.g., promote soil ecosystem recovery and plant productivity on mine sites by providing a substrate with an adequate pH, while increasing organic carbon, microbial activity and diversity, nutrient and moisture availability, and offering a relatively porous and stable medium for root development (Gardner et al. 2010; Cele and Maboeta 2016; Antonelli et al. 2018; Carabassa et al. 2020). A single application can lead to positive effects on both plant growth and ecosystem carbon sequestration.

Using residual wastes efforts have been made to reconstruct soils over asbestos mining wastes (either overburden or tailings) in Quebec. The idea has been to mix at least two residual wastes in appropriate proportions with the goal of creating a substrate that can successfully support plant growth. So far, the main goal has been to use these mixtures to grow grasses and forbs quickly and efficiently. As an example of these initiatives, the work reported herein is being conducted on a decommissioned asbestos mine at Black Lake (46°00′48″N, 71°22′06″W). Conditions are similar amongst sites, but approaches can differ depending upon the environmental consultants who are carrying out the restoration work. Yet, restoration techniques are relatively similar within the region and the description provided here likely only deviates slightly from what is being performed at other neighboring mines. The mine of interest operated between 1955 and 2011 for the extraction of chrysotile. A few limited areas are currently mined for granite, whereas the rest of the mine has been undergoing reclamation since closure. The mine is located within the sugar maple (*Acer saccharum*) – yellow birch (*Betula alleghaniensis*) bioclimatic domain (Gouvernement du Québec 2019), which is characterized by a humid continental climate. Geologically, it falls within the Thetford Mines Ophiolite Complex of the Lower Ordovician (479 Ma), with igneous rocks as diverse as granite, granodiorite, monzogranite and peridotite, and metamorphic serpentinite (De Souza et al. 2012).

Typically, soil reconstruction is performed using a mixture of municipal biosolids and deinking sludge that are derived from neighboring municipalities and deinked pulp mills, respectively. These materials are carefully mixed offsite with a tractor loader at a ratio of 1:3 (Figure 8.2a). The mixture is loaded and transported by dumper truck to the site (Figure 8.2b), where it is spread as evenly as possible with an excavator that is equipped with a root rake, at an approximate rate of 1200 Mg ha^{-1} (300 Mg ha^{-1} for biosolids; 900 Mg ha^{-1} for sludge) (Figure 8.2c).

The application rate is the maximum allowed by government authorities based mainly upon total heavy metals and nitrogen loading. At this rate, the target is equivalent to the mine wastes being covered to a depth of ~20 cm with the mixture (Figure 8.3a). De-inking sludge, which is enriched in calcium, is used to mitigate the antagonistic effects of magnesium over calcium uptake by plant roots. Once the reconstructed soil is in place, common wild oat (*Avena sativa*) is seeded as a cover crop at a rate of 80 kg ha^{-1}. A mixture of 55% Timothy grass (*Phleum pratense*), 30% red clover (*Trifolium pratense*) and 15% alsike clover (*Trifolium hybridum*) seeds is also applied at a rate of 20 kg ha^{-1} using a broadcast spreader mounted on a caterpillar all-terrain vehicle (Figure 8.3b). Oats act as nurse plants in this protocol. Soil reconstruction occurs during the entire snow-free season (May–October), whereas seeding is performed either in the spring or autumn.

If seeded in spring, grasses and clovers will create a dense cover that reaches about 60 cm in height by autumn, whereas this cover increases in density and reaches maturity with grasses over 90 cm in height by the end of the second growing season (Figure 8.4). If seeded in autumn, a sparse cover is established a few weeks later. As snow arrives in mid-November, the main growth spurt is triggered the next spring and maximum biomass is attained in the autumn of that year or the next, depending upon the growing season. In all cases, however, plant growth is fast and effective, and has given no indication of foliar toxicities (e.g., chromium or nickel) or nutrient imbalances (e.g., calcium) over the years (N. Bélanger, *personal observations*).

Figure 8.2 Biosolids and sludge are mixed with a tractor loader (a), then loaded and transported to the site with a dumper truck (b) before being applied carefully with a root rake that is attached to an excavator (c). *Source:* Photos by Olivier Nadeau (Viridis Environnement) and Nicolas Bélanger.

(a)

(b)

Figure 8.3 Freshly applied mixture on the plateau of a tailings pile (a), and application of the seed mixture shortly thereafter with a broadcast spreader mounted on an all-terrain vehicle (b). *Source:* Photos by Olivier Nadeau (Viridis Environnement) and Nicolas Bélanger.

(a)

(b)

Figure 8.4 Grasses and forbs in the spring (mid-May) after they were seeded the previous autumn on the plateau of a waste rock pile (a), and grasses and forbs during their third growing season in mid-June (b). *Source:* Photos by Nicolas Bélanger.

8.4 Afforestation

There are many challenges that are associated with growing trees on old mine sites (Macdonald et al. 2015). This is largely due to constraints that are imposed on plants by the wastes, as previously discussed. Even in cases where soils are reconstructed, there can be problems regarding insufficient rooting depth of these soils, their very well-drained nature (thus limiting water availability), their low fertility, their content of toxicants (e.g., biosolids contain low amounts of heavy metals, such as cadmium and lead), and their poor stability, causing them to lose volume over time through wind and water erosion. These issues can be more severe for trees, which tend to be more resource-demanding than grasses and forbs, given their more extensive growth. In this respect, very few attempts have been made at afforestation on former asbestos mine wastes until recently.

In 2017, our team initiated an experiment, which consisted of four waste rock plateaus that were reclaimed between 2012 and 2015 by seeding grasses and clover onto soils reconstructed with a mixture of municipal and deinking sludge (as described in Section 8.3). Tree species that were tested included green alder (*Alnus viridis* subsp. *crispa*), tamarack or eastern larch (*Larix laricina*), white spruce (*Picea glauca*), three cultivars of miyabeana or Japanese willow (*Salix miyabeana*; SX61, SX64 and SX67) and three hybrid poplar clones (*Populus maximowiczii* × *Populus balsamifera*; M × B-915302, M × B-915303 and M × B-915311). These shrubs and trees were planted on three microsites in early June. The first microsites were prepared by excavating 50 × 50 cm mounds with a height of about 30 cm using a backhoe loader. The second microsites were prepared by applying glyphosate (regulation number 28486, 4.67 L ha^{-1} diluted in 100 L ha^{-1}) to a 30 cm radius around the center of the planting microsite less than 24 hours before planting. The third microsites consisted of planting cuttings or seedlings without site preparation. In this case, shrubs or trees were planted directly into the grasses and clovers, which was considered the control. Full details of the experimental design, field measurements, statistical analyses and results are provided in Grimond (2022).

Survival rates of all shrub and tree species are presented in Table 8.1, including hybrid poplar clones as a group (first growing season) and individually (fourth growing season). Results highlight generally low survival rates, but mostly very low survival for shrubs. Green alder and miyabeana willow exhibited very low survival rates under the control and herbicide treatments after one growing season (0–24%) and on all microsites after four growing seasons (0–9%). These survival rates were statistically lower than all of the tree species that were tested. After four years, regrowth on some of the willow stools was observed, despite the original willow shoots being dead (Figure 8.5a). Substantial browsing was observed on those shoots, likely from deer and hare, given that the presence of feces was evidence of their activity in the study plots. Miyabeana willow has been reported as being targeted by herbivores (Larsson 1998; Tanaka and Nakamura 2015). Yet, grasses likely provided a suitable environment for these herbivores and as such, their increased presence is bound to be associated with greater browsing on the planted shrubs and trees (Parsons et al. 2007). Like many other restoration cases, there are compromises that need to be accepted in regard to bringing about diversity in both plants and wildlife in the process. Green alder is quite sensitive to herbivory, below-ground competition and low soil moisture (Matthews 1992; Asmara et al. 2022). When acting individually or in combination, these factors likely led to almost complete total mortality of the species in the study

Table 8.1 Survival rates (%) of all shrubs and tree species after one and four growing seasons on the three planting microsites.

	First growing season			Fourth growing season		
Species	**Control**	**Herbicide**	**Mounds**	**Control**	**Herbicide**	**Mounds**
Green alder	0	10	77	0	0	4
Miyabeana willow	24	6	69	1	5	9
Tamarack	40	64	95	12	29	51
White spruce	25	77	96	13	12	52
Hybrid poplars	87	60	99	22	0	50
M × B-915302	—	—	—	10	0	54
M × B-915303	—	—	—	13	0	33
M × B-915311	—	—	—	13	0	63

Values are calculated based on the ratio of the number of trees alive to the number of trees being monitored. In the case of hybrid poplar clones, results are shown for all clones combined (i.e., Hybrid poplars, *P. maximowiczii* × *P. balsamifera*) and individual clones (i.e., M × B-915302, -915303, -915311).
Source: Adapted from Grimond (2022).

(a) (b)

Figure 8.5 Regrowth of a miyabeana willow stool (a) and a tamarack seedling growing on a mound (b) that had succumbed to drought on the plateau of a waste rock pile. *Source:* Photos by Laurence Grimond.

plots. Tamarack, white spruce and hybrid poplars exhibited statistically similar survival rates; the trees differed markedly from the shrubs, i.e., green alder and miyabeana willow. No significant differences in survival rates were apparent among the three hybrid poplar clones.

Survival rates were significantly greater on mounds than the remaining treatments for all shrub and tree species after one growing season (69–99%) and for tamarack, white spruce and hybrid poplar clones after four growing seasons (50–53%). In a forestry context, mounding has been shown to be effective in enhancing seedling survival and growth by suppressing competing plants and permitting good root development, particularly for hybrid poplars (Bilodeau-Gauthier et al. 2011; Mc Carthy et al. 2017; Thiffault et al. 2020). Yet, mounds have also been shown to promote low soil moisture (Löf and Birkedal 2009). Although mounds in the study plots have likely reduced competition for light and increased soil volume for root development (and thus, nutrient uptake), mounds were difficult to construct because of the very rocky wastes that are available for construction. As such, mounds remained small (height, 30 cm; width, 60 cm) and were mostly composed of waste. In this respect, water stress likely remained an issue and explains, at least in part, the overall decrease in survival between the first- and fourth-growing seasons (Table 8.1; Figure 8.5b). Since no significant effect was detected between control and herbicide treatments, the application of one dose of glyphosate was apparently ineffective in limiting competition from grasses. This is especially true for miyabeana willow and poplar clones for which the herbicide treatment resulted in complete mortality after four years (Table 8.1). Perhaps root systems of grasses survived and consequently, competition for trees and shrubs returned just as vigorous shortly thereafter. Thus, a more sustained control of competing vegetation, using chemical or mechanical methods, during the first few years after planting and over a larger surface around the trees and shrubs would likely have resulted in greater success with regard to survival (DesRochers and Sigouin 2015).

According to Grimond (2022), hybrid poplar clones were, on average, 100 cm tall when planted, and those that survived reached an overall mean height of 312 cm after four years of growth (Figure 8.6). This corresponds to a growth rate of 53 $\text{cm}\,\text{yr}^{-1}$. Some individuals attained stem heights greater than 500 cm, which corresponds to a growth rate of 100 $\text{cm}\,\text{yr}^{-1}$. This growth rate is comparable or superior to growth rates that have been reported in other mine reclamation studies (Casselman et al. 2006; Larchevêque et al. 2015), in forest environments that are characterized by acidic/poor soils (Bilodeau-Gauthier et al. 2011), and in plantations for biomass production growing at a similar latitude (Larocque 1999). They are slightly lower than hybrid poplars that were grown in a forest setting with support of liming and fertilization (Bilodeau-Gauthier et al. 2022) or in agricultural soils of greater fertility (Labrecque and Teodorescu 2005). The clones that were tested in this experiment all had the same parents (i.e., *P. maximowiczii* × *P. balsamifera*) and in turn, they did not exhibit significant differences in terms of survival and growth. There was no effect of microsite or clone on stem height, but all clones exhibited greater diameters on mounds than in the control (Figure 8.6). The effect of weeding on hybrid poplar diameter was demonstrated in other cultural contexts (Pinno and Bélanger 2009). Tamarack and white spruce growth was not monitored systematically in this experiment, but those that survived were seemingly in good health

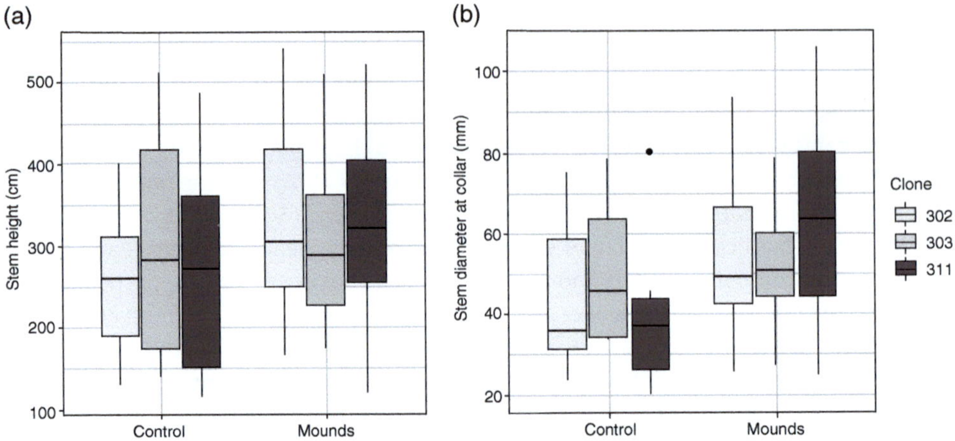

Figure 8.6 Box-and-whisker plots of stem height (a) and stem diameter at the collar (b) of three hybrid poplar clones (M × B-915*302*,-915*303*,-915*311*) growing on the two planting microsites (no hybrid poplar survival on phytocide microsites; not shown). The gray-scale boxes represent the values between the 25th and the 75th percentiles, the horizontal line indicates the median (50th percentile), and the vertical whiskers represent the 10th and 90th percentiles. Outliers (outside the percentile range) are indicated by black dots. *Source:* Grimond (2022)/with permission of Université de Montréal.

and some specimens showed considerable growth over the four years of monitoring (Figure 8.7).

Despite encouraging results with hybrid poplars, tamarack and white spruce, the outcome was considered very unsatisfactory as a whole, with survival rates ranging from 0 to 30%. It could be surmised that areas previously reclaimed with grasses and forbs are not suitable for planting trees, given that these plants, notably the grasses, strongly compete for light, water and nutrients. Moreover, their roots occupy much of the soil space to begin with, thereby inhibiting root growth and development of newly planted trees (Messier et al. 2009). Indeed, technosols that are colonized by grasses and forbs represent abiotic (low soil moisture, limited rooting space) and biotic (competition, herbivory) stressors that do not favor tree survival. A new restoration approach was subsequently proposed, i.e., one in which trees are planted on mining wastes that have not been previously reclaimed, as a means of yielding greater tree survival and growth rates. This new approach to afforestation of former asbestos mines was experimentally tested (full details are provided in Grimond et al. (2023), including results after three years of growth). Implementation of the improved protocol will begin in 2024 through a large initiative that is funded by the Two Billion Trees Commitment of the Canadian Government. The technology is described here along with the six-year results that have recently affirmed interest in this nature-based solution.

With this new technology, afforestation is possible on both the plateaus of the waste rock piles and on tailings of former asbestos mines. Specifically, the same base mixture of municipal and deinking sludge that is described in Section 8.3 is either created and applied at the same rate (i.e., 300 and 900 Mg ha^{-1}, respectively), or the base mixture is remixed with Class B metal-contaminated soils (Beaulieu 2021) with a tractor loader at a 1:1 volume

Figure 8.7 Example of a tamarack plot exhibiting good growth after four years. *Source:* Photo by: Laurence Grimond.

ratio. Because the base mixture is applied to the fields at the same rate when constructing the biosolids/sludge/class B soil technosol, it results in a thickness of about 40 cm compared to 20 cm for biosolids/sludge technosol. Once evenly spread with a bulldozer, windrows with a height of 50 cm (B technosol) and 1 m (BSC technosol) are created with a small excavator, resulting in 75 cm-wide rows with 1 m spacing between rows (Figure 8.8).

Five species were tested thus far, i.e., jack pine (*Pinus banksiana*, JP), red pine (*Pinus resinosa*, RP), white spruce (WS), tamarack (TAM), and three clones of hybrid poplar (*P. maximowiczii* × *P. balsamifera*, M × B-102216; [*P. deltoides* × *P. nigra*] × *P. maximowiczii*, DN × M-915508; and *P. maximowiczii* × *P. nigra*, M × N-103124). Tamarack and poplar clones are typically bare-root seedlings, whereas spruce and pine seedlings are containerized (310 cm^3), except for jack pine seedlings, which are in 110 cm^3 containers. Height of these seedlings is typically between 100 and 150 cm for poplar clones, 30–40 cm for spruce, 20–30 for tamarack and red pine, and 10–15 cm for jack pine. Seedlings are planted on the windrow tops at a minimum spacing of 1 m, ideally in the spring (late May to early June), although autumn could also be considered; the latter remains to be tested. Plant competition is expected to be minimal within the first year of planting, but it could become important during the second growing season. As such, weed control is recommended (see further discussion on this topic later in this section).

As a whole, the research with trees that are planted on windrows has provided some valuable insights regarding the best practices for promoting the establishment and growth of tree plantations in former asbestos mines. Three-year results showed much higher survival for most species compared to those planted in grasses, suggesting a promising

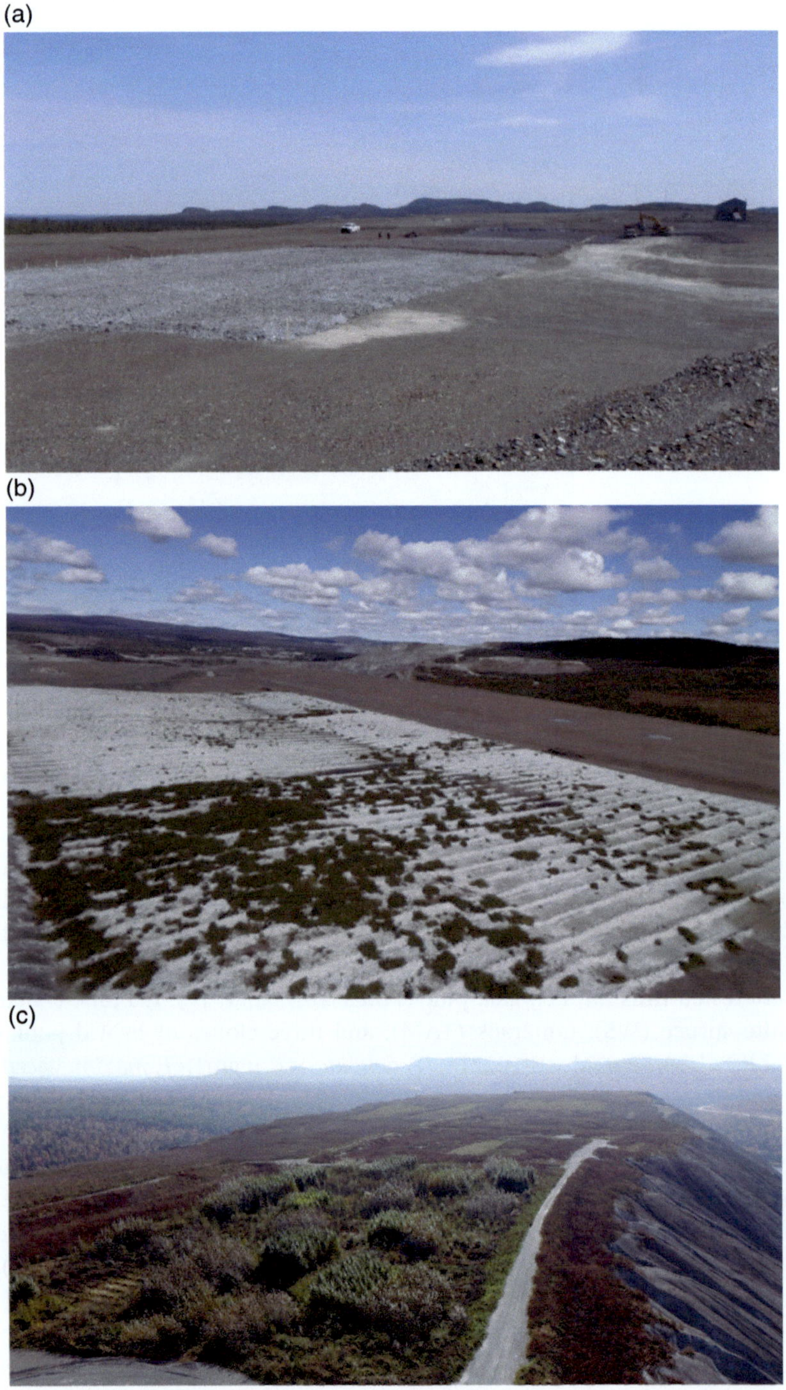

Figure 8.8 Homogeneous application of the base mixture on a tailings plateau (a), followed by the creation of windrows with an excavator (b). Note the absence of vegetation on the plateau in (a) because of a hostile growth environment. Plants growing on the fresh windrows (about 20 days) in (b) originate from seeds in human waste (e.g., tomato, cucumber, and squash plants). In (c), the same experimental plantation after six years of growth (bottom half) is shown. Restoration with grasses and herbs around the plantation and steep banks of the pile to the right can be observed. The darkening of the vegetation of the plantation is due to leaf senescence (picture taken on October 4). *Source:* Photos by Simon Bilodeau-Gauthier and Nicolas Bélanger.

Table 8.2 Mean survival rates (%) of tree species growing after three growing seasons.

Species	Third growing season
Jackpine	14.4 (5.88)
Red pine	28.8 (4.17)
Tamarack	31.4 (8.76)
White spruce	46.5 (4.51)
M×N-103124	93.5 (1.07)
M×B-102216	90.2 (4.66)
DN×M-915508	91.6 (3.92)

Values are calculated based upon the ratio of the number of trees alive divided by the number of trees being monitored. In the case of hybrid poplar clones, results are shown for individual clones (i.e., M×N-103124, M×B102216, and DN×M-915508). Means include both mixtures/technosols, whereas values in parentheses are standard deviations.
Source: Adapted from Grimond (2022).

technology (Table 8.2), whereas a recent survey in mid-June of 2023 (i.e., sixth year of growth) not only confirms the growth trend but also validates the technology as a reliable one, at least in the mid-term. For one, hybrid poplars are performing well on both waste rock and tailings piles, although their height and diameter growth is slightly greater on waste rock piles (Figure 8.9). Plateaus of waste rock piles under study are at an elevation lower than the plateaus of tailings piles. The former are not only characterized by less windy conditions, but the technosols also have higher moisture. Very little (if any) and no new mortality in hybrid poplars is observed in the sixth year of growth. The largest trees reach over 7 m in height, regardless of clone, and diameters at breast height are consistently greater than 7 cm (Figure 8.10a). Conversely, mortality of tamarack, white spruce and jack pine is very high on the plateaus of waste rock piles (some study plots exhibited 100% mortality), whereas tamarack performs very well in the plateaus of the tailings piles, followed by white spruce and then jack pine (Figure 8.9). Tamarack and white spruce survival and growth on the tailings are spatially and temporally consistent, and trees are vigorous. The best-performing larches are over 3.5 m in height (Figure 8.10b) and generally compete well for light even in the presence of native poplars and willow shrubs, which are aggressively encroaching upon the site. Differences between the technosols are small. Furthermore, the addition of Class B contaminated soils to a mixture of municipal biosolids and deinking sludge can be advantageous with regard to tree survival and growth, likely because this technosol has greater thickness/volume, which holds more water and nutrients and offers more space for root systems to develop and expand (Grimond et al. 2023).

White spruce grows more slowly than tamarack, but it appears healthy and performs very well under high shade/competition. The case of white spruce is particularly interesting, not so much due to its growth performance, but rather because the restoration work involving white spruce seems to be leading towards a stand that resembles one at the onset of secondary succession in the boreal forest. Late successional boreal forests are typically

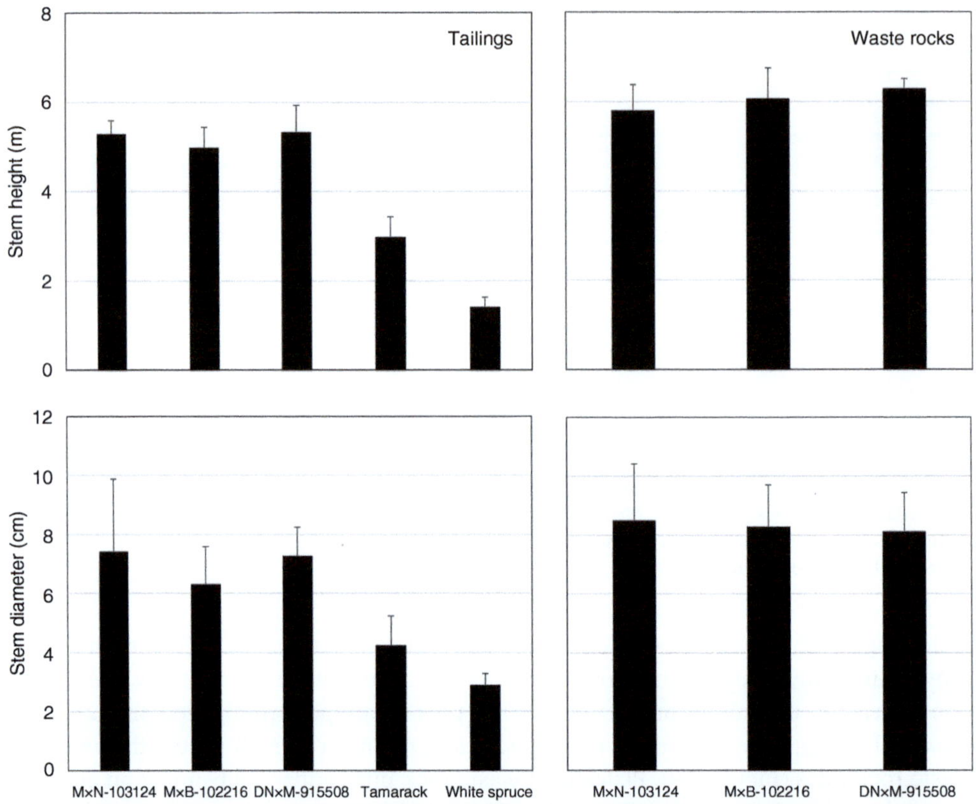

Figure 8.9 Mean stem height and diameter of three hybrid poplar clones (M × N-103124, M × B102216, and DN × M-915508), tamarack and white spruce in their sixth growing season (mid-June). Diameters are at breast height for poplars and at the collar for tamarack and white spruce. Means include both mixtures/technosols, whereas the error bars are standard deviations. Neither conifer data are shown for the waste rocks due to their high mortality and small sample size, nor are they shown for pine species for the tailings for the same reason.

dominated by coniferous trees such as spruce, pine, and fir, but pioneer species such as trembling aspen (*Populus tremuloides*) or white birch (*Betula papyrifera*), which are faster-growing and shade-intolerant, generally dominate after a disturbance such as fire or logging (Bergeron and Fenton 2012). This is due to a significant increase in available light as a result of the loss of canopy cover. Sunlight stimulates the growth of herbaceous plants, shrubs, and tree seedlings (e.g., aspen and birch) that prosper under high light levels in open conditions. In time, tree species composition may shift towards more shade-tolerant species like spruce and fir. This gradual transition represents the ongoing ecological succession process in the boreal forest. A more diverse understorey community of plants, including ferns, mosses, and various herbaceous species, also becomes established during the process. At the mine sites, when white spruce is planted on technosols, its growth is slow and produces little shade, which allows native poplar trees and willow shrubs to encroach the plots aggressively (Grimond et al. 2023). It should be noted, however, that pioneer species at the mine sites volunteer naturally after the later succession species, as opposed to

Figure 8.10 Example of hybrid poplar (a) and tamarack (b) with substantial growth in their sixth year of growth (mid-June). The gray stake is 1.0 m and the trees reach respectively over 7 m and 3.5 m in height. *Source:* Photos by Nicolas Bélanger.

natural regeneration in the boreal forest. Given that these pioneer tree species grow rapidly, they quickly dominate the site (i.e., within four or five years), thereby creating shade for spruce trees as is the case in a second succession boreal stand (Figure 8.11). At these sites, most tree growth can be quickly attributed to the performance of trembling aspen and balsam poplar (*P. balsamifera*), but a closer study of the understorey reveals a high density of healthy white spruce. One could assume that spruce will continue to develop under high shade and eventually become dominant in the stands after the poplars die off (Bergeron and Fenton 2012).

Survival rates are generally lowest in red and jack pines and the trees remain small under strong competition from native poplars (Figure 8.12). At first glance, one could compare this situation to white spruce, but the status and denouement of pine species in these restoration areas are not as positive. Pine species are not considered late successional species; rather, they are pioneer species that establish and thrive in open areas after a disturbance. They generally grow fast under low competition and thus high availability of resources (Vasiliauskas and Chen 2002). The rapid ability of many pines (including red and jack pines) to colonize open areas following a wildfire through seed masting from their serotinous cones, confirms their pioneer status in many forest ecosystems. Red and jack pines in the restored asbestos mine areas will likely be outcompeted by the native poplar trees that are effectively encroaching the sites. Given that pine seedling stocks are amongst the smallest in height (10–15 cm and 20–30 cm for jack and red pine, respectively) makes them more susceptible to competition. Many pine trees are showing signs of poor foliar health. Pine

Figure 8.11 From left to right, increasing competition intensity for white spruce in their sixth year of growth (mid-June). Competing trees are mostly native poplars. On the right, spruce trees are the understorey trees, with native poplars that have encroached upon the site being the dominant trees. The gray stake is 1.0 m and the best-performing trees reach close to 2 m in height. *Source:* Photos by Nicolas Bélanger.

Figure 8.12 Example of a red pine seedling under high competition and exhibiting signs of foliar nutrient deficiencies/toxicities (to be determined) in their sixth year of growth (mid-June). The gray stake is 1.0 m and the trees seldom reach more than 50 cm in height. *Source:* Photo by Nicolas Bélanger.

species generally prefer ammonium (NH_4^+) to nitrate (NO_3^-) as a nitrogen source (Lavoie et al. 1992; Malagoli et al. 2000). This preference is due to the way these trees have adapted to their natural environments, which are often characterized by acidic and nutrient-poor soils that are common in many coniferous forest ecosystems where NH_4^+ is the main nitrogen source. Biosolids are generally dominated by NH_4-nitrogen (Schroder et al. 2008).

However, the technosols that are constructed from these residues at asbestos mine sites are NO_3-dominated (Grimond et al. 2023), suggesting that nitrogen deficiencies could be an issue for pines. Foliar analyses are required to identify if poor foliar health is due to nutrient deficiencies, nutrient imbalances or toxicities.

Windrow technosols are composed of a mixture of municipal biosolids, deinking sludge, and contaminated soils representing a reliable reclamation technology for plateaus of waste rock and tailings piles in asbestos mines. Tree selection is crucial, however. Hybrid poplars can be successfully planted under both situations and offer great potential for afforestation. With regard to conifers, namely tamarack and white spruce, they offer substantial potential for ecological restoration as well, but sites and species should be selected carefully. Most importantly, wet areas should be avoided if planning to plant white spruce and pine species should be managed with greater care (see below). Mixing of slightly contaminated soils, if available, or other similar mineral material, also should be considered to provide more depth to the technosol, thereby permitting root development and easier access to water and nutrients. Cultural practices should include thorough weed control as a means of optimizing seedling survival and growth. Both chemical and mechanical means of weed control could be considered depending on available local resources. Inoculation of forest soils with mycorrhizas was shown to be beneficial to jack pine and white spruce, for example (Onwuchekwa et al. 2014; Nadeau et al. 2018). Mycorrhizal inoculation of the technosols prior to planting could also benefit the conifers. Some trees are inoculated in the tree nursery prior to being planted at the mine sites. However, this has not been tested directly in the context of afforestation of former asbestos mines.

The technology for afforestation that is presented here shows promise, but there are two ways that tree growth could be compromised in the mid-term. First, water and nutrient demand will increase as trees enter later stages of development, which in turn will increase growth. Considering that technosols have a limited available volume for roots, more important limitations for these resources could be observed very soon. For example, canopy closure in hybrid poplar plantations has already occurred after 5–6 years, which suppresses most of the competing understorey vegetation (Figures 8.15 and 8.18), but this is when intraspecific competition would begin. Guittonny-Larchevêque et al. (2016) showed that for similar mine settings in north-western Quebec, roots of trees that were planted in technosols would not develop in the underlying substrates. This depth limitation remains to be tested in the context of afforestation of former asbestos mines in southern Quebec, but if that is the case, one can assume that trees will compete with one another for soil space with time. Thus, growth could be slowed considerably. Concomitant thinning could occur due to intraspecific competition, together with interspecific competition with native poplars and willows, for example.

Second, as trees mature, gain height and become fewer in number through natural thinning, they could become more vulnerable to wind-throw. The plateaus often exhibit very high winds because of their exposure. The permanent vertical displacement of stems by strong winds is referred to as "lodging," which is quite common in agriculture (Kong et al. 2022). It can lead to yield loss in crops by changing leaf orientation to sunlight, thereby lowering light assimilation rates (Setter et al. 1997). In the first two years of growth, we observed lodging of hybrid poplars (15–30°) due to high winds on the tailings piles. This effect was more pronounced on the biosolids/sludge technosol, which is shallower and offers less anchorage than technosols with Class B contaminated soil. Shallow rooting systems are not uncommon in forests (Strong and La Roi 1983) and lodging in the plantations disappeared over time (at five to six years, poplar stems are fully vertical). However, there is always a possibility that wind-throw risks would be high for trees that put on substantial height growth, but less radial growth under high intraspecific competition (Pinno and Bélanger 2009). The forester's rule-of-thumb is that a diameter at breast height (DBH in cm, at 1.3 m height) over height (in m) ratio (d/h) of 0.9 or above typically represents low susceptibility to wind-throw for hybrid poplars. Below 0.9, trees become particularly vulnerable to wind damage. Of course, some clones with particular genotypes are more prone to wind-throw than others (Klocko et al. 2013). A typical tree in its sixth year of growth on the plateaus of tailings piles at asbestos mine sites is about 7.25 cm in DBH and 5.25 m in height, equalling a d/h ratio of almost 1.4. Thus far, there has been no indication of wind-throw problems in those plantations, but monitoring of trees will continue to identify any issue with wind damage and assess the long-term success of this afforestation technology, both with hybrid poplars and conifers.

8.5 Carbon Sequestration

There is growing popularity in the use of biosolids and sludge to restore highly disturbed areas (Carabassa et al. 2020). The reasons are threefold. First, until recently, landfilling has been the most viable option for disposing of these wastes. With increasing waste generation, and environmental and health hazards that are associated with landfilling of these wastes, this disposal option is being revisited and new management approaches are now being deployed. Notably, biosolids are used as a soil amendment in agriculture (Collivignarelli et al. 2020), and more recently, they have been used in ecological restoration. Second, waste products that are used to construct technosols for restoration likely represent a sound climate solution because the carbon that they store is probably in a steady state with environmental conditions and can thus be considered a stable carbon sink. Moreover, this practice of moving the wastes away from landfills can potentially decrease GHG emissions. Third, the recycling of these wastes to build new soils can promote plant growth and, thus, carbon sequestration in disturbed areas.

In the context of asbestos mines in southern Quebec, Khlifa et al. (2023) observed very steady organic-carbon contents of technosols six years after they were constructed and seeded with grasses and herbs. Soil organic carbon (SOC) was estimated at 130 Mg C ha^{-1}. This is a significant pool of carbon when compared to natural environments. According to Tremblay and Ouimet (2000), the SOC content of boreal, mixed and deciduous forest soils in Quebec is 44 Mg C ha^{-1}, on average. Out of a sample of 284, less than 3% of the forests that were sampled in this study exhibited SOC contents above 100 Mg C ha^{-1} and less than 2% had SOC contents similar to the technosols that were sampled by Khlifa et al. (2023). Based upon these data, it can be concluded that technosols at the asbestos mine sites hold as much as two to threefold more SOC than do most of the adjoining deciduous forests.

Khlifa et al. (2023) measured relatively high summer CO_2 emission rates from the technosols shortly after their construction, but emissions returned within about 18 months to values similar to forest soils in southern Quebec, i.e., about 300–600 mg CO_2 m^{-2} h^{-1}, depending upon soil temperature (Royer-Tardif et al. 2019; Bélanger et al. 2021). The balance between carbon inflows and outflows determines SOC content. Yet, the initially large CO_2 emissions of the technosols (as high as 3000 mg CO_2 m^{-2} h^{-1}) were too small and too short-lived to have any substantial effect on their SOC content over six years (Khlifa et al. 2023).

Although organic-C in these soils appears to be in a steady state, a promising outcome would be for technosols to start accumulating carbon over time. This remains to be verified, but based on observations of afforestation of sites where technosols were built as windrows, the main premise is that the planted trees and those encroaching the sites will add more carbon through below-ground (roots) and above-ground (leaves, twigs, branches) litters than would grasses and herbs (Nave et al. 2013; Barcena et al. 2014). There is evidence that hybrid poplar leaf litter is accumulating in large amounts, mostly between the windrows, and this process will likely lead to new carbon in the soil in the form of a forest floor (Figure 8.13). Yet, as has been observed in many other studies and given the harsh growing conditions that prevail on mine sites, it may take as many as 10–20 years before a net carbon sink is observed in technosols (Trlica 2010; Antonelli et al. 2018; Carabassa et al. 2020). This is about the same time needed for agricultural soils when they are planted with forest trees (Laganière et al. 2010).

(a) (b)

Figure 8.13 Hybrid poplar leaves accumulating between windrows. The gray surface in (a) is the windrow on which leaves do not tend to accumulate because of their inclination. The stem in (b) is from a hybrid poplar clone with a potential for a high diameter growth, sometimes reaching 11–12 cm at breast height within 5–6 years. *Source:* Photo credits: Nicolas Bélanger.

Despite the uncertainties of whether technosols will act as a sink for carbon in the near future or not, data on CH_4 emissions in these soils provide further encouragement that they have the potential for mitigating climate change (Khlifa et al. 2023). Standard disposal of biosolids and sludge in landfills is linked to significant emissions of CH_4 released by anaerobic digestion (Lou and Nair 2009). Yet, data indicate that technosols that are built from biosolids and sludge at asbestos mines in southern Quebec act as small CH_4 sinks, and not as sources. This observation also appears to hold relatively well even in the wettest waste rock piles, where anaerobic digestion could occur. These data suggest that the use of biosolids and sludge for reclamation of asbestos mines in southern Quebec could potentially lower radiative forcing by CH_4, a GHG that is 25–32 times more powerful than CO_2 and which has been creeping upward since the 1980s; a recent (steep) increase has been observed (Nisbit et al. 2019). However, landfilling of biosolids can also produce N_2O by the co-oxidation of ammonia by methanogenic soil bacteria, although to a lesser extent (Wang et al. 2017). The benefits of a decrease in CH_4 emissions associated with the use of biosolids for restoration purposes (relative to landfilling) could thus be overshadowed if technosols favor N_2O emissions. Fresh (wet) biosolids can be large sources of N_2O (Roman-Perez and Hernandez-Ramirez 2022) and very preliminary data suggest that fresh technosols at asbestos mine sites emit large amounts of N_2O, but the duration of these large emissions is unknown (N. Bélanger, unpublished). It is not uncommon that soils acting as CH_4 sinks can be sources of N_2O (Wang et al. 2005). Further monitoring of soil conditions and gas efflux is thus required to determine if carbon accumulates in the new soils and to assess GHG dynamics as a means to fully evaluate their potential as a climate solution.

Section 8.4 demonstrated that windrow technosols on asbestos mine sites, whether they are built on tailings or waste rocks, represent a reliable reclamation technology. This is the case when tree species are carefully selected. Hybrid poplars offer the highest potential compared to the conifers, although some conifers also have potential for afforestation of asbestos mines (e.g., tamarack and white spruce) if control of plant competition is performed. Yet, at this stage of stand development, it is difficult to assess the amount of carbon that could be stored in the mid-term by the trees. Not only are the trees too young to make reliable projections on growth, but the few allometric relationships that are available in the literature also limit calculations of carbon stocks. The only available allometric equations that allow relatively robust estimates of tree carbon stocks to be made are for hybrid poplars, especially clone DN × M-915508. Truax et al. (2018) measured height and DBH for this clone at various planting densities at three agricultural sites in southern Quebec with relatively contrasting yields. These data are available at eight and fourteen years, whereas planting stocks are identical to the ones that are used for asbestos mine restoration. Our projections suggest that a DBH of 14–19 cm could be achieved in 14 years, with 16–17 cm being our best estimate (Figure 8.14). This considers that high-density planting at 2500 stems per hectare for restoration purposes (and not forestry) would lead to some intraspecific competition and, thus, it is expected that growth rates would decline relative to more conventional stem densities in hybrid poplar plantations (Truax et al. 2018). These diameter estimates were then used as input into allometric equations that were developed by Fortier et al. (2017) for the same clone and sites. This yielded respective above-ground biomass and coarse root biomass estimates of about 80 kg/tree and 12 kg/tree over 14 years, which

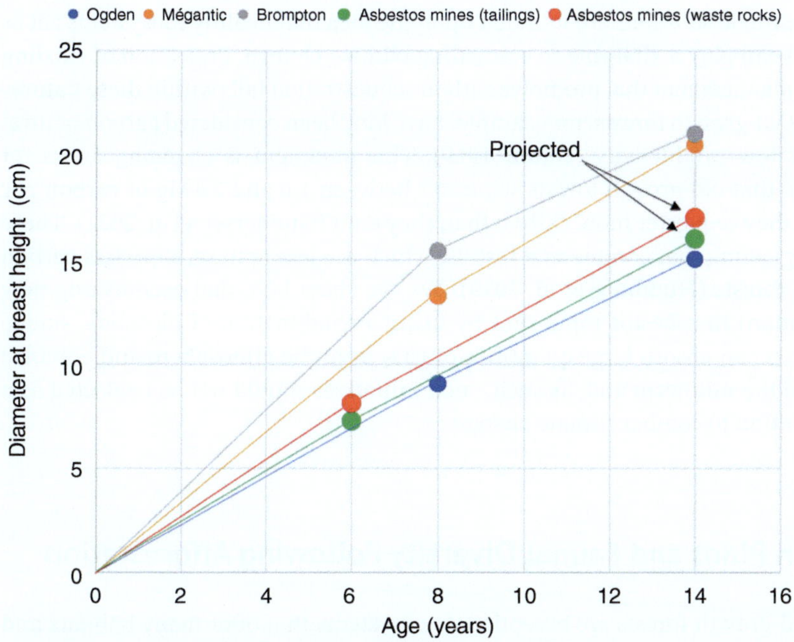

Figure 8.14 Projections of diameter at breast height (DBH) of hybrid poplar clone DN × M-915508 based on measurements after three years of growth and in its sixth year of growth at asbestos mine sites. These data were constrained to diameter measurements (8 and 14 years) of that same clone growing at similar rates on agricultural sites (Ogden, Mégantic and Brompton) in southern Quebec. *Source:* Adapted from Truax et al. (2018).

can then be transformed to a hectare basis, i.e., 230 Mg of biomass per hectare. A rule-of-thumb is that carbon is half of tree biomass (Lamlom and Savidge 2003), meaning that a DN × M-915508 plantation with a density of 2500 stems per hectare should accumulate an amount of carbon similar to that stored in the technosols over fourteen years, i.e., 115 vs. 130 Mg ha^{-1}.

As discussed in Section 8.4, it is difficult to forecast the growth trajectory of these plantations given uncertainties that are related to intraspecific competition and possible wind damage/wind-throw. Yet, it is safe to assume that the technosols and hybrid poplars will collectively sequester the equivalent of almost 250 Mg ha^{-1} of carbon after 15 years. This is impressive considering the hostile growing conditions that prevailed prior to restoration (Figure 8.8a). It is even more impressive if one considers that maple forests in Canada, which are typical of the study area, store on average 253 Mg ha^{-1} of carbon, including soil carbon (Shaw et al. 2014). In just 15 years, a hybrid poplar forest that was planted on technosols created on asbestos mine plateaus of tailings and waste rocks could store a quantity of carbon equivalent to that of neighboring natural forests. Additional carbon sequestration is expected thereafter because these plantations are not established for commercial interests (i.e., harvesting), but for environmental benefits. These are however shorter-lived than maple trees and in this respect, a replacement strategy should be considered in 30–50 years when poplars begin to die off and start to release CO_2 to the atmosphere during their decomposition.

Nature-based climate solutions are founded upon the idea that healthy ecosystems act as carbon sinks and can play a vital role in combating climate change. Protection of existing forests and their management that promotes carbon sequestration falls within these nature-based solutions. Old-growth forests, for example, have long been considered carbon neutral because of their low productivity (Odum 1969). This paradigm is changing as recent research suggests that old-growth forests sequester between 1.6 and 2.4 Mg of carbon per hectare and that they sequester more carbon than they emit (Gundersen et al. 2021). There is more work suggesting that commercial forests, which are less mature, store less carbon than old-growth forests (Hudiburg et al. 2019). Yet, we show here that establishing new forests (afforestation) in asbestos mine sites by using a combination of biosolids, sludge and hybrid poplars can absorb large quantities of CO_2 from the atmosphere and enhance carbon storage in the mid-term and, as such, such initiatives should not be neglected as a nature-based solution to combat climate change.

8.6 Gains in Plant and Faunal Diversity Following Afforestation

On one hand, old-growth forests are exceptional ecosystems that offer many habitats and shelter much biodiversity (Paillet et al. 2010). On the other hand, tree plantations are generally perceived by the public as systems that do not provide suitable habitats for encouraging plant and animal biodiversity (Horák et al. 2019; Lévesque et al. 2020). Concerns about maintaining biodiversity are considerable because the effects of intensive tree planting (in a silvicultural context) on biodiversity remain poorly documented. This unfavorable perception of plantations is nevertheless called into question. Although establishment of plantations to replace natural forests often causes a loss of biodiversity, the scientific literature suggests that this option is far from disastrous and more favorable when establishing mixed species plantations (Wang et al. 2022). In the context of disturbed sites, notably that of mines, the level of restoration effectiveness is measured by starting from references to natural systems, including biodiversity, but these references should also be later adapted based upon the restoration challenges that lie ahead (Marshall et al. 2022).

Because further restoration work on asbestos mines in Quebec is very likely to be funded through climate change mitigation programs, restoration efforts can aim for carbon sequestration, but such initiatives should not be performed to the detriment of biodiversity and other environmental benefits (Matos et al. 2022). Thankfully, afforestation initiatives at asbestos mines in Quebec suggest this is also leading to biodiversity gains, i.e., plants and wildlife. For one, there is evidence that native poplars and willows are aggressively encroaching onto the sites once the technosols have been set in place on the plateaus of waste rocks and tailings (Figure 8.15). Such encroachment of native trees and shrubs does not occur if plateaus are left unmanaged (Figure 8.8a). The technosols thus play a major role in facilitating native plants to colonize the sites. It is believed that the proximity of the piles to the natural forests (sometimes within 200 m) is another factor allowing this dynamic encroachment. Restoration success has also been driven by the combination of soil amendments and proximity to natural forests in other geographic contexts (e.g., König et al. 2022).

Figure 8.15 A trembling aspen bluff in control plot (no trees planted) with trees over 4 m on the plateau of an asbestos tailings pile six years after technosols were put in place (a). The main native trees and shrubs encroaching the plateaus are trembling aspen (b), balsam poplar (c) and Bebb's or beaked willow (*Salix bebbiana*) (d). *Source:* Photos by Nicolas Bélanger.

Encroachment of native trees and shrubs provides additional litter to the soil and adds diversity/complexity to the plantation. This can create competition for light, nutrients and water, but it could also develop complementarity between plant species. Further monitoring is required to gain a full portray of interactions between planted trees and those naturally establishing on the sites, particularly the effects of poplars on the development of (planted) conifers. As a whole, however, encroachment of poplars and willows should be viewed as a positive by-product of the restoration process in terms of socio-economic and environmental benefits. This response will lead to more diverse and complex systems (Macdonald et al. 2015), while also eliminating the need to plant mixed tree species plantations or to plant trees altogether in areas where encroachment is very dynamic (Grimond et al. 2023).

Many other plants with important ecological roles are also observed in the restoration (Figure 8.16). For one, common milkweed (*Asclepias syriaca*) is establishing well in many restored areas. It is a favorite of declining monarch butterfly (*Lepidoptera nymphalidae*) populations east of the Rocky Mountains, given that it provides a host and nectar for the species during their northeasterly migration (Pocius et al. 2017). The monarch caterpillar feeds exclusively on milkweed leaves and the restoration of monarch habitat with milkweed species is thus essential to increase the population (Oberhauser et al. 2016). Red raspberry (*Rubus idaeus*), eastern bluestar (*Amsonia tabernaemontana*) and comfrey (*Symphytum officinale*) are perennial herbaceous plants that also grow very well in the restored areas. Raspberry provides food (berries) for terrestrial and avian fauna (Titterington et al. 1979; Crawford 1982; Steventon and Major 1982), whereas flowers of all of these perennials are rich in nectar and pollen, making them attractive to pollinators, notably bees (Gagnon et al. 1991; Gegear and Laverty 1995; Neal 1998; Moran 2023). Staghorn sumac (*Rhus typhina*) was also found on the sites, although in lesser abundance. It is a native North American shrub or small tree that has been used for restoration purposes, given that it offers shelter and food for many bird species (Galbraith-Kent and Handel 2007). Staghorn sumac is well adapted to disturbed environments and can thrive in a variety of soil types and conditions. Most of these plants are often the first to colonize abandoned fields and disturbed areas. They have relatively deep and fibrous root systems that stabilize soil, thereby providing soil structure and erosion control. As such, these plants could help stabilize the windrows and maintain their form. They could also improve soil quality by adding organic matter and nutrients from their litters.

Common mullein (*Verbascum thapsus*) is a biennial herbaceous that is a plant native to Eurasia. It was introduced into North America during early European colonization for medicinal purposes, mostly for respiratory ailments (Turker and Gurel 2005). Mullein is an efficient seed producer and can be considered invasive in some regions, particularly in areas with disturbed soils (Wilbur and Hufbauer 2012). As such, it should be managed carefully and not allowed to become a dominant species that potentially displaces native vegetation. The fact that vegetative reproduction does not occur in common mullein can help in controlling it (Gross 1980). Thus far, it has not been the most aggressive colonizer of the plateaus of waste rocks and tailings piles (Figure 8.17a). It does have a deep tap root, which can help anchor unstable soils (Turker and Gurel 2005). Common reed (*Phragmites australis* subsp. *australis*) is significantly more vigorous than the American subspecies (*P. australis* subsp. *americanus*). Thus, the former is considered invasive in Canada (Belzile

(a) (b)

(c) (d)

Figure 8.16 Some other plant species establishing on the plateaus of asbestos waste rock and tailings piles after technosols were put in place include but are not limited to common milkweed (a), red raspberry (b), eastern bluestar (c) and staghorn sumac (d). *Source:* Photos by Nicolas Bélanger.

(a) (b)

Figure 8.17 Some plant species establishing on the plateaus of asbestos waste rock and tailings piles after technosols were put in place are considered invasive species. These include, but are not limited to common mullein (a) and common reed (b). *Source:* Photo credits: Nicolas Bélanger.

et al. 2010; Catling and Mitrow 2011). Due to large allocation of above-ground biomass, *Phragmites* blocks light for other plants and as such, they are strong competitors and decrease plant diversity rapidly when invading wet environments. Thus far, common reed has not been an aggressive intruder into the restored sites, but its presence was observed in some of the wettest areas (Figure 8.17b). As trees mature and create shade with their large canopies, these species will likely disappear. In their sixth year of growth, hybrid poplars are already efficiently suppressing grasses and forbs to the point that the ground surface is almost free of vegetation (Figure 8.18). More monitoring is required to assess the extent of development of these invasive species.

During the sixth year of growth, an attempt was made to monitor wildlife in the plantations by assessing the bird species that were present. None of the authors are wildlife specialists, but birds are probably the easiest to monitor due to the available artificial intelligence applications that recognize species based on vocalizations. The application Merlin Bird ID (Cornell Lab) was used for such an undertaking, once in mid-June and once in early September. In total, eight bird species were identified, some in both plateau types and others in just one plateau type. These are displayed in Table 8.3, along with some details of their habitat as well as on what plateau(s) they were observed. All of these birds are passeriforms. They are common species of southern Quebec and none are considered endangered (Les Oiseaux du Québec 2023). Some species are mostly associated with forest and wooded areas, e.g., cedar waxwing, whereas others also use wooded and shrub areas, e.g., American robin and song sparrow, but none of these species are limited to such

Figure 8.18 In their sixth year of growth, the fast-growing hybrid poplars are creating shade with their closed canopies, thus leaving very little room for understory vegetation. Note that this picture was taken in mid-June on the plateau of tailings pile at 11 a.m., i.e., under full sunlight. *Source:* Photo by Nicolas Bélanger.

Table 8.3 Bird species detected using the Merlin Bird ID application in the plantations of the plateaus of waste rock and tailings piles.

Species	Main habitats	Waste rocks	Tailings
American robin (*Turdus migratorius*)	Forests, open fields, and urban settings	×	×
Red-winged blackbird (*Agelaius phoeniceus*)	Open fields and wetlands	×	
Song sparrow (*Melospiza melodia*)	Forests and open fields	×	×
Yellow warbler (*Setophaga petechia*)	Wooded areas, shrublands, wetlands, and open fields	×	×
Cedar waxwing (*Bombycilla cedrorum*)	Forests, wooded areas, and open fields		×
Common yellowthroat (*Geothlypis trichas*)	Shrublands, wetlands, and open fields	×	×
Indigo bunting (*Passerina cyanea*)	Forests and open fields	×	
American crow (*Corvus brachyrhynchos*)	Forests, open fields, and urban settings	×	

These data are no indication of the frequency of use of the plantations due to the limited sampling.

habitats and often use open fields such as grasslands and farmlands. A few surveys were casually conducted a few hundred meters away from plantations in areas (i) that were not restored thus far or (ii) which have been restored with grasses and forbs. Out of the eight species identified in plantations, only American crow vocalizations were detected in areas that have not been restored, whereas yellow warbler, cedar waxwing and indigo bunting were not detected in areas that had been restored with grasses and forbs. The sampling neither allows us to differentiate statistically the bird species that are using the areas restored with grasses and forbs or which are planted with trees, nor their frequency of use. A more thorough sampling is required to do so, which should encompass vocalizations in the center of the plantations, at their edges and away from the edges at various distances. At this point, these are anecdotal data and additional sampling is planned to follow the use of these different habitats by birds. Surveys of feces will also reveal the extent of use of the plantations by deer, hare and other terrestrial mammals (thus far, only observed in areas with grasses and forbs). However, it can be said with confidence that restoration with grasses and forbs as well as trees leads to more bird activity on the plateaus and is another indication of restoration success, regardless of what technology is being used.

8.7 Conclusion

Attempts at restoring abandoned asbestos-mining sites with trees confirm that such an endeavor may be successful when undertaken with proper techniques and adequate species. The tree plantations show they could provide multiple benefits compared with the heavily disturbed state of the sites prior to restoration, including but not limited to carbon sequestration (in soils and both planted and naturally encroaching vegetation) and newly created habitats for wild fauna and flora. They thus show to be viable climate-smart and nature-based approaches to address the current challenges of climate change and biodiversity loss that humanity is facing worldwide. In regard to carbon sequestration, plantations also show more promise than sites restored with grasses and forbs, whereas it is too early to say in regard to biodiversity.

There are nonetheless certain parameters that must be well controlled for plantation success to be met, because a former mining site is, at its core, a very hostile environment for plants. In light of the results presented here, a few steps are essential for a successful ecological restoration.

1) Create an appropriately reconstructed soil, or technosol, in which the trees shall be planted. This is an absolute necessity because the underlying background material (e.g., waste rocks or tailings) is highly compacted, offering no substrate for planting, and is entirely devoid of resources, and also displaying some heavy metals. The technosols should provide the organic matter and nutrients necessary for tree establishment and development, as well as sufficient rooting depth to prevent eventual water stress or wind-throw. The materials used for technosols at Asbestos sites comprise municipal biosolids, deinking sludge, and lightly contaminated soil.
2) Plant tree seedlings in the newly created and bare technosols, rather than on already-vegetated areas among grasses and forbs. The latter are strongly competitive for water,

nutrients, light and soil space, and as such, high mortality was observed in the first year for most tree and shrub species when planted in that context. In addition, technosols of a few years old settle down somewhat, thus losing volume and offering a shallower soil for new seedlings – and that shallow space is also filled with grass and forb roots. It is therefore recommended to plant the trees soon after technosol creation, when the available rooting volume is at its largest and competitors are absent.

3) Choose the species to be planted wisely. The ideal candidate is resistant to harsh environmental conditions, notably drought and strong winds, and grows quickly, both under- and aboveground. Hybrid poplars perform the best among seven other shrub or tree species tested. They show high survival and fast growth during the first six years, rooting strongly into the technosols and already closing their canopy.

Future research directions for asbestos mine sites restored with trees include monitoring of: (i) stand development, i.e., stand structure and diversity, tree growth, root development in the technosols and the material underneath it, carbon pools, foliar nutrition (and levels of metal contaminants), and susceptibility to wind-throw, (ii) carbon reserves in technosols, and (iii) herbaceous plants, insects and wildlife, with emphasis for the latter on birds, small rodents and larger mammals such as deer.

Acknowledgments

The work was supported by a grant to N. Bélanger and D. Rivest (2019-GS-262, 733) through Fonds de recherche du Québec – Nature et technologies (FRQNT), Fonds de recherche du Québec – Société et culture (FRQSC), and Fonds vert of the Quebec Government in collaboration with Viridis Environnement and Granilake. Many students and research professionals were involved with field and laboratory work. We also thank the staff at Viridis Environnement and Granilake, namely Olivier Nadeau, for providing site access and images of field operations as well as helping with field logistics. Thanks also to William F.J. Parsons for the English revision of the book chapter.

References

Allen, C.D., Breshears, D.D., and McDowell, N.G. (2015). On underestimation of global vulnerability to tree mortality and forest die-off from hotter drought in the Anthropocene. *Ecosphere* 6: 1–55.

Allen, L.P., Baez, J., Stern, M.E.C. et al. (2018). Trends and the economic effect of asbestos bans and decline in asbestos consumption and production worldwide. *International Journal of Environmental Research and Public Health* 15: 531.

Anderson, R., Bayer, P., and Edwards, D. (2020). Climate change and the need for agricultural adaptation. *Current Opinions in Plant Biology* 56: 197–202.

Antonelli, P.M., Fraser, L.H., Gardner, W.C. et al. (2018). Long term carbon sequestration potential of biosolids-amended copper and molybdenum mine tailings following mine site reclamation. *Ecological Engineering* 117: 38–49.

Asmara, D.H., Allaire, S., van Noordwijk, M., and Khasa, D.P. (2022). Tree establishment on post-mining waste soils: species, density and mixture effects. *Canadian Journal of Forest Research* 52: 79–89.

Bainton, N. and Holcome, S. (2018). A critical review of the social aspects of mine closure. *Resource Policy* 59: 468–478.

BAPE. (2020). L'état des lieux et la gestion de l'amiante et des résidus miniers amiantés. Bureau d'audiences publiques sur l'environnement, Rapport d'enquête et d'audience publique. https://voute.bape.gouv.qc.ca/dl?id=00000156531 (accessed August 2023).

Barbero, R., Abatzoglou, J.T., Larkin, N.K. et al. (2015). Climate change presents increased potential for very large fires in the contiguous United States. *International Journal of Wildland Fire* 24: 892–899.

Barcena, T.G., Kiaer, L.P., Vesterdal, L. et al. (2014). Soil carbon stock change following afforestation in Northern Europe: a meta-analysis. *Global Change Biology* 20: 2393–2405.

Bastin, J.-F., Finegold, Y., Garcia, C. et al. (2019). The global tree restoration potential. *Science* 365: 76–79.

Beaulieu, M. (2021). Guide d'intervention - Protection des sols et réhabilitation des terrains contaminés. Québec. https://www.environnement.gouv.qc.ca/sol/terrains/guide-intervention/guide-intervention-protection-rehab.pdf (accessed August 2023).

Bélanger, N., Collin, A., Khlifa, R., and Lebel-Desrosiers, S. (2021). Balsam fir and American beech influence soil respiration rates in opposite directions in a sugar maple forest near its northern range limit. *Frontiers Forests and Global Change* 4: 664584.

Belzile, F., Labbé, J., LeBlanc, M.-C., and Lavoie, C. (2010). Seeds contribute strongly to the spread of the invasive genotype of the common reed (*Phragmites australis*). *Biological Invasions* 12: 2243–2250.

Bergeron, Y. and Fenton, N.J. (2012). Boreal forests of eastern Canada revisited: old growth, nonfire disturbances, forest succession, and biodiversity. *Botany* 90: 509–523.

Bilodeau-Gauthier, S., Paré, D., Messier, C., and Bélanger, N. (2011). Juvenile growth of hybrid poplars on acidic boreal soil determined by environmental effects of soil preparation, vegetation control, and fertilization. *Forest Ecology and Management* 261: 620–629.

Bilodeau-Gauthier, S., Palma Ponce, G., Miquel, J.-C. et al. (2022). Growth and foliar nutrition of a hybrid poplar clone following the application of a mixture of paper mill biosolids and lime mud. *Canadian Journal of Forest Research* 52: 117–128.

Brofy, J.T., Keith, M.M., and Shieman, J. (2007). Canada's asbestos legacy at home and abroad. *International Journal of Occupation and Environmental Health* 13: 235–242.

Brown, S., DeVolder, P., Compton, H., and Henry, C. (2007). Effect of amendment C:N ratio on plant richness, cover and metal content for acidic Pb and Zn mine tailings in Leadville, Colorado. *Environmental Pollution* 149: 165–172.

Byrne, B. and Goldblatt, C. (2014). Radiative forcing at high concentrations of well-mixed greenhouse gases. *Geophysical Research Letters* 41: 152–160.

Carabassa, V., Domene, X., Díaz, E., and Alcañiz, J.M. (2020). Mid-term effects on ecosystem services of quarry restoration with Technosols under Mediterranean conditions: 10-year impacts on soil organic carbon and vegetation development. *Restoration Ecology* 28: 960–970.

Caro, T., Rowe, Z., Berger, J. et al. (2022). An inconvenient misconception: climate change is not the principal driver of biodiversity loss. *Conservation Letters* 15: e12868.

Carson, A.W., Rutherford, P.M., and Burton, P.J. (2014). Desulphurized tailings serve as a useful soil supplement for mine reclamation. *Canadian Journal of Soil Science* 94: 529–541.

Casselman, C.N., Fox, T.R., Burger, J.A. et al. (2006). Effects of silvicultural treatments on survival and growth of trees planted on reclaimed mine lands in the Appalachians. *Forest Ecology and Management* 223: 403–414.

Catling, P.M. and Mitrow, G.L. (2011). The recent spread and potential distribution of *Phragmites australis* subsp. australis in Canada. *The Canadian Naturalist* 125: 95–104.

Cele, E.N. and Maboeta, M. (2016). A greenhouse trial to investigate the ameliorative properties of biosolids and plants on physicochemical conditions of iron ore tailings: implications for an iron ore mine site remediation. *Journal of Environmental Management* 165: 167–174.

Ciccarese, L., Mattsson, A., and Pettenella, D. (2012). Ecosystem services from forest restoration: thinking ahead. *New Forests* 43: 543–560.

Collivignarelli, M.C., Abbà, A., and Benigna, I. (2020). The reuse of biosolids on agricultural land: critical issues and perspective. *Water Environment Research* 92: 11–25.

Crawford, H.S. (1982). Food selection and digestibility by tame white-tailed deer in central Maine. *The Journal of Wildlife Management* 46: 974–982.

De Souza, S., Tremblay, A., Ruffet, G., and Pinet, N. (2012). Ophiolote obduction in the Quebec Appalachians, Canada – $^{40}Ar/^{39}Ar$ age constraints and evidence for syntectonic erosion and sedimentation. *Canadian Journal of Earth Sciences* 9: 91–110.

DesRochers, A. and Sigouin, M.-E. (2015). Effect of soil mounding and mechanical weed control on hybrid poplar early growth and vole damage. *Écoscience* 21: 278–285.

Evans, D.M., Zipper, C.E., Burger, J.A. et al. (2013). Reforestation practice for enhancement of ecosystem services on a compacted surface mine: path toward ecosystem recovery. *Ecological Engineering* 51: 16–23.

Forster, P., Storelvmo, T., Armour, K. et al. (2021). The Earth's energy budget, climate feedbacks, and climate sensitivity (Chapter 7). In: *Climate Change 2021: The Physical Science Basis. Contribution of Working Group I to the Sixth Assessment Report of the Intergovernmental Panel on Climate Change* (ed. V. Masson-Delmotte et al.). Cambridge University Press https://www.ipcc.ch/report/ar6/wg1/chapter/chapter-7/.

Fortier, B., Truax, B., Gagnon, D., and Lambert, F. (2017). Allometric equations for estimating compartment biomass and stem volume in mature hybrid poplars: general or site-specific? *Forests* 8: 309.

Frank, A.L. (2006). The history of the extraction and uses of asbestos. In: *Asbestos: Risk Assessment, Epidemiology, and Health Effects* (ed. R.F. Dodson and S.P. Hammar), 1–7. CRC Press.

Gagnon, M., Gingras, J., and De Oliveira, D. (1991). Honey bee (Hymenoptera: Apidae) foraging behavior and raspberry pollination. *Journal of Economic Entomology* 84: 457–460.

Galbraith-Kent, S.L. and Handel, S.H. (2007). Lessons from urban lakeshore restoration project in New York City. *Ecological Restoration* 25: 123–128.

Gardner, W.C., Broersma, K., Naeth, A. et al. (2010). Influence of biosolids and fertilizer amendments on physical, chemical and microbiological properties of copper mine tailings. *Canadian Journal of Soil Science* 90: 571–583.

Gegear, R.J. and Laverty, T.M. (1995). Effect of flower complexity on relearning flower-handling skills in bumble bees. *Canadian Journal of Zoology* 73: 2052–2058.

Goldberg, M. and Luce, D. (2009). The health impact of nonoccupational exposure to asbestos: what do we know? *European Journal of Cancer Prevention* 18: 489–503.

Gouvernement du Québec. (2019). Forêt ouverte. https://www.foretouverte.gouv.qc.ca (accessed August 2023).

Government of Canada. (2018). Prohibition of asbestos and products containing asbestos regulations (SOR/2018-196). https://laws-lois.justice.gc.ca/eng/regulations/SOR-2018-196/page-1.html (accessed August 2023).

Grimond, L. (2022). Boisement d'une ancienne mine de chrysotile du sud du Québec: essais de divers types de technosols et d'essences ligneuses. M.Sc. thesis. Département de géographie, Université de Montréal.

Grimond, L., Rivest, D., Bilodeau-Gauthier, S. et al. (2023). Novel soil reconstruction leads to successful afforestation of a former asbestos mine in southern Quebec, Canada. *New Forests* 45: 477–503. https://doi.org/10.1007/s11056-023-09989-3.

Griscom, B.W., Adams, J., Ellis, P.W., and Fargione, J. (2017). Natural climate solutions. *Proceedings of the National Academy of Science* 114: 11645–11650.

Griscom, B.W., Lomax, G., Kroeger, T. et al. (2019). We need both natural and energy solutions to stabilize our climate. *Global Change Biology* 25: 1889–1890.

Gross, K.L. (1980). Colonization by *Verbascum thapsus* (mullein) of an old-field in Michigan: experiments on the effects of vegetation. *Journal of Ecology* 68: 919–927.

Guittonny-Larchevêque, M., Bussière, B., and Pednault, C. (2016). Tree-substrate water relations and root development in tree plantations used for mine tailings reclamation. *Journal of Environmental Quality* 45: 1036–1045.

Gundersen, P., Thybring, E.E., Nord-Larsen, T. et al. (2021). Old-growth forest carbon sinks overestimated. *Nature* 591: E21–E23.

Harris, J.A. (2003). Measurements of the soil microbial community for estimating the success of restoration. *European Journal of Soil Science* 54: 801–808.

Horák, J., Brestovanská, T., Mladenovic, S. et al. (2019). Green desert?: Biodiversity patterns in forest plantations. *Forest Ecology and Management* 433: 343–348.

van Horssen, J. (2016). *A Town Called Asbestos. Environmental Contamination, Health, and Resilience in a Resource Community*. UBC Press.

Hudiburg, T.W., Law, B.E., Moomaw, W.R. et al. (2019). Meeting GHG reduction targets requires accounting for all forest sector emissions. *Environmental Research Letters* 14: 095005.

IARC Working Group on the Evaluation of Carcinogenic Risks to Humans (2012). Asbestos (chrysotile, amosite, crocidolite, tremolite, actinolite and anthophyllite). In: *Arsenic, Metals, Fibres and Dusts*, vol. 100C, 219–309. International Agency for Research on Cancer Monographs on the Evaluation of Carcinogenic Risks to Humans.

IPCC (2021). Climate change 2021: the physical science basis. Contribution of working group I to the sixth assessment report of the intergovernmental panel on climate change, climate change 2021: the physical science basis. In: *Contribution of Working Group I to the Sixth Assessment Report of the Intergovernmental Panel on Climate Change* (ed. V. Masson-Delmotte et al.). Cambridge University Press https://www.ipcc.ch/report/ar6/wg1/.

Jacques, O. and Pienitz, R. (2022). Asbestos mining waste impacts on the sedimentological evolution of the Bécancour chain of lakes, southern Quebec (Canada). *Science of the Total Environment* 807: 151079.

Khlifa, R., Rivest, D., Grimond, L., and Bélanger, N. (2023). Stability of carbon pools and fluxes of a technosol along a 7-year reclamation chronosequence at an asbestos mine in Canada. *Ecological Engineering* 186: 106839.

Klocko, A., Stanton, B., van Oosten, C., and Strauss, S.H. (2013). Green revolution plantations: could short trees be a big thing? *ISB New Report* May 2013:7–9.

Kong, X., Peng, P., Li, L. et al. (2022). Wind lodging-associated yield loss is reduced by wheat genetic diversity. *European Journal of Agronomy* 133: 126441.

König, L.A., Medina-Vega, J.A., Longo, R.M. et al. (2022). Restoration success in former Amazonian mines is driven by soil amendment and forest proximity. *Philosophical Transactions of the Royal Society B* 378: 20210086.

Kuyek, J. (2003). Asbestos mining in Canada. A brief Presented to the International Ban Asbestos Conference. MiningWatch Canada. https://miningwatch.ca/sites/default/files/asbestos_mining_in_canada_0.pdf (accessed August 2003).

Labrecque, M. and Teodorescu, T.I. (2005). Field performance and biomass production of 12 willow and poplar clones in short-rotation coppice in southern Quebec (Canada). *Biomass Bioenergy* 29: 1–9.

Laganière, J., Angers, D.A., and Paré, D. (2010). Carbon accumulation in agricultural soils after afforestation: a meta-analysis. *Global Change Biology* 16: 439–453.

Lamlom, S.H. and Savidge, R.A. (2003). A reassessment of carbon content in wood: variation within and between 41 North American species. *Biomass Bioenergy* 25: 381–388.

Larchevêque, M., Desrochers, A., Bussière, B., and Cimon, D. (2015). Planting trees in soils above non-acid-generating wastes of a boreal gold mine. *Ecoscience* 21: 217–231.

Larocque, G.R. (1999). Performance and morphological response of the hybrid poplar DN-74 (*Populus deltoides* x *nigra*) under different spacings on a 4-year rotation. *Annals of Forest Science* 56: 275–287.

Larsson, S. (1998). Genetic improvement of willow for short-rotation coppice. *Biomass Bioenergy* 15: 23–26.

Lavoie, N., Vézina, L.-P., and Margolis, H.A. (1992). Absorption and assimilation of nitrate and ammonium ions in jack pine seedlings. *Tree Physiology* 11: 171–183.

Lévesque, A., Bélanger, N., Poder, T.G. et al. (2020). From white to green gold: digging into public expectations and preferences for ecological restoration of asbestos mines in southeastern Quebec, Canada. *The Extractive Industries and Society* 7: 1411–1423.

Lewis, S.L., Wheeler, C.E., Mitchard, E.T.A., and Koch, A. (2019). Restoring natural forests is the best way to remove atmospheric carbon. *Nature* 568: 25–28.

Liang, Y., Bi, H., Huang, H. et al. (2022). Contribution of warm and moist atmospheric flow to a record minimum July sea ice extent of the Arctic in 2020. *Cryosphere* 16: 1107–1123.

Lin, R.-T., Chien, L.-C., Jimba, M. et al. (2019). Implementation of national policies for a total asbestos ban: a global comparison. *Lancet Planetary Health* 3: 341–348.

Löf, M. and Birkedal, M. (2009). Direct seeding of *Quercus robur* L. for reforestation: the influence of mechanical site preparation and sowing date on early growth of seedlings. *Forest Ecology and Management* 258: 704–711.

Lou, X.F. and Nair, J. (2009). The impact of landfilling and composting on greenhouse gas emissions – a review. *Bioresource Technology* 16: 3792–3798.

Macdonald, S.E., Landhäusser, S.M., Skousen, J. et al. (2015). Forest restoration following surface mining disturbance: challenges and solutions. *New Forests* 46: 703–732.

Malagoli, M., Dal Canal, A., Quaggiotti, S. et al. (2000). Differences in nitrate and ammonium uptake between Scots pine and European larch. *Plant and Soil* 221: 1–3.

Marshall, A.R., Waite, C.E., Pfeifer, M. et al. (2022). Fifteen essential science advances needed for effective restoration of the world's forest landscapes. *Philosophical Transactions of the Royal Society B* 378: 20210065.

Martín-Moreno, C., Martín Duque, J.F., Nicolau Ibarra, J.M. et al. (2016). Effects of topography and surface soil cover on erosion for mining reclamation: the experimental spoil heap at El Machorrro mine (central Spain). *Land Degradation and Development* 27: 145–159.

Matos, F.A.R., Edwards, D.P., Magnago, L.F.S. et al. (2022). Invasive alien acacias rapidly stock carbon, but threaten biodiversity recovery in young second-growth forests. *Philosophical Transactions of the Royal Society B* 378: 20210072.

Matthews, R.F. (1992). *Alnus viridis* subsp. *crispa*. Fire Effects Information System. Forest Service, United States Department of Agriculture. https://www.fs.usda.gov/database/feis/plants/shrub/alnvirc/all.html (accessed August 2023).

McCarthy, R., Rytter, L., and Hjelm, K. (2017). Effects of soil preparation methods and plant types on the establishment of poplars on forest land. *Annals of Forest Science* 74: 47.

McDowell, C. (2013). Climate-change adaptation and mitigation: implications for land acquisition and population relocation. *Development Policy Review* 31: 677–695.

Messier, C., Coll, L., Poitras-Larivière, A. et al. (2009). Resource and non-resource root competition effects of grasses on early- versus late-successional trees. *Journal of Ecology* 97: 548–554.

Moran, K. (2023). The Bellarmine Bee Bed: organizing a native plant garden using feedback from the local community. Undergraduate thesis. Department of Environmental Science, Bellarmine University. https://scholarworks.bellarmine.edu/ugrad_theses/108

Nadeau, M.B., Laur, J., and Khasa, D.P. (2018). Mycorrhizae and rhizobacteria on Precambrian rocky gold mine tailings: I. Mine-adapted symbionts promote white spruce health and growth. *Frontiers in Plant Science* 9: 1267.

Nave, L.E., Swanston, C.W., Mishra, U., and Nadelhoffer, K.J. (2013). Afforestation effects on soil carbon storage in the United States: a synthesis. *Soil Science Society of America Journal* 77: 1035–1047.

Neal, P.R. (1998). Pollinator restoration. *Trends in Ecology and Evolution* 13: 132–133.

Nero, B.F. (2021). Structure, composition and diversity of restored forest ecosystems on mine-spoils in South-Western Ghana. *PLoS ONE* 16: e0252371.

Nisbit, E.G., Manning, M.R., Dlugokencky, E.J. et al. (2019). Very strong atmospheric methane growth in the 4 years 2014-2017: implications for the Paris agreement. *Global Biogeochemical Cycles* 33: 318–342.

Oberhauser, K.S., Wiederholt, R., Diffendorfer, J.E. et al. (2016). A trans-national monarch butterfly population model and implications for regional conservation priorities. *Ecological Entomology* 42: 51–60.

Odum, E.P. (1969). The strategy of ecosystem development. *Science* 164: 262–270.

Ohsowski, B.M., Dunfield, K., Klironomos, J.N., and Hart, M.M. (2018). Plant response to biochar, compost, and mycorrhizal fungal amendments in post-mine sandpits. *Restoration Ecology* 26: 63–72.

Onwuchekwa, N.E., Zwiazek, J.J., Quoreshi, A., and Khasa, D.P. (2014). Growth of mycorrhizal jack pine (*Pinus banksiana*) and white spruce (*Picea glauca*) seedlings planted in oil sands reclaimed areas. *Mycorrhiza* 24: 431–441.

Paillet, Y., Bergès, L., Hjältén, J. et al. (2010). Biodiversity differences between managed and unmanaged forests: meta-analysis of species richness in Europe. *Conservation Biology* 24: 101–112.

Parsons, M.H., Lamont, B.B., Koch, J.M., and Dods, K. (2007). Disentangling competition, herbivory, and seasonal effects on young plants in newly restored communities. *Restoration Ecology* 15: 250–262.

Pietrzykowski, M. (2019). Tree species selection and reaction to mine soil reconstructed at reforested post-mine sites: Central and eastern European experiences. *Ecological Engineering* 142: 100012.

Pinno, B.D. and Bélanger, N. (2009). Competition control in juvenile hybrid poplar plantations across a range of site productivities in central Saskatchewan, Canada. *New Forests* 37: 213–225.

Pocius, V.M., Debinski, D.M., Pleasants, J.M. et al. (2017). Milkweed matters: Monarch butterfly (*Lepidoptera nymphalidae*) survival and development on nine midwestern milkweed species. *Environmental Entomology* 46: 1098–1105.

Regroupement Québec Oiseaux. (2023). Liste des oiseaux du Québec. https://s1.yapla.com/media/CPYCn9TslMHKJSLwzd8PIVrDo/chorq-liste-oiseaux-quebec-2023-12.pdf (accessed August 2023).

Rocklöv, J. and Dubrow, R. (2020). Climate change: an enduring challenge for vector-borne disease prevention and control. *Nature Immunology* 21: 479–483.

Roman-Perez, C.C. and Hernandez-Ramirez, G. (2022). Nitrous oxide production and nitrogen transformations in a soil amended with biosolids. *Canadian Journal of Soil Science* 102: 505–518.

Royer-Tardif, S., Whalen, J., and Rivest, D. (2019). Can alkaline residuals from pulp and paper industry neutralize acidity in forest soils without increasing greenhouse gas emissions? *Science of the Total Environment* 663: 537–547.

Ruff, K. (2017). How Canada changed from exporting asbestos to banning asbestos: the challenges that had to be overcome. *International Journal of Environmental Research and Public Health* 14: 1135.

Savard, S. (2017). Débats politiques et prise de parole citoyenne: la création de la Société nationale de l'amiante. *Journal of the Canadian Historical Association* 28: 157–191.

Schroder, J.L., Zhang, H., Zhou, D., and Basta, N. (2008). The effect of long-term annual application of biosolids on soil properties, phosphorus, and metals. *Soil Science Society of America Journal* 72: 73–82.

Seddon, N. (2022). Harnessing the potential of nature-based solutions for mitigating and adapting to climate change. *Science* 376: 1410–1416.

Setter, T.L., Laureles, E.V., and Mazaredo, A.M. (1997). Lodging reduces yield of rice by self-shading and reductions in canopy photosynthesis. *Field Crops Research* 49: 95–106.

Shaw, C.H., Hilger, A.B., Metsaranta, J. et al. (2014). Evaluation of simulated estimates of forest ecosystem carbon stocks using ground plot data from Canada's National Forest Inventory. *Ecological Modelling* 272: 323–347.

Shrestha, R.K. and Lal, R. (2006). Ecosystem carbon budgeting and soil carbon sequestration in reclaimed mine soil. *Environment International* 32: 781–796.

Shrestha, R.K., Lal, R., and Jacinthe, P.-A. (2009). Enhancing carbon and nitrogen sequestration in reclaimed soils through organic amendments and chiseling. *Soil Science Society of America Journal* 73: 1004–1011.

Simard, M. (2019). The mining industry in Quebec: status, trends and stakes. *Canadian Studies* 85: 193–217.

Singh, A.N., Raghubanshi, A.S., and Singh, J.S. (2022). Plantations as a tool for mine spoil restoration. *Current Science* 82: 1436–1441.

Sopper, W.E. (1993). *Municipal Sludge Use in Land Reclamation*. Lewis Publishers, 176 pages.

Steventon, J.D. and Major, J.T. (1982). Marten use of habitat in a commercially clear-cut forest. *Journal of Wildlife Management* 46: 175–182.

Stewart-Sinclair, P.J., Last, K.S., Payne, B.L., and Wilding, T.A. (2020). A global assessment of the vulnerability of shellfish aquaculture to climate change and ocean acidification. *Ecology and Evolution* 10: 3518–3534.

Strong, W.L. and La Roi, G.H. (1983). Rooting depths and successional development of selected boreal forest communities. *Canadian Journal of Forest Research* 13: 577–588.

Tanaka, M. and Nakamura, M. (2015). Spatially distinct responses within willow to bark stripping by deer: effects on insect herbivory. *Naturwissenschaften* 102: 46.

Thiffault, N., Elferjani, R., Hébert, F. et al. (2020). Intensive mechanical site preparation to establish short rotation hybrid poplar plantations – a case-study in Québec, Canada. *Forests* 11: 785.

Titterington, R.W., Crawford, H.S., and Burgason, B.N. (1979). Songbird responses to commercial clear-cutting in Maine spruce-fir forests. *Journal of Wildlife Management* 43: 602–609.

Tremblay, S., and Ouimet, R. (2000). Modèle de prédiction du stock de carbone organique emmagasiné dans la couverture morte des sols forestiers minéraux du Québec. Direction de la recherche forestière, Ministère des ressources naturelles, Gouvernement du Québec. Note de recherche forestière no. 104.

Trlica, A. (2010). Mitigation of climate change through land reclamation with biosolids: carbon storage in reclaimed mine soils, life cycle analysis of biosolids reclamation, and ecosystem services with reforestation. Master's thesis. School of Forest Resources, University of Washington.

Truax, B., Fortier, B., Gagnon, D., and Lambert, F. (2018). Planting density and site effects on stem dimensions, stand productivity, biomass partitioning, carbon stocks and soil nutrient supply in hybrid poplar plantations. *Forests* 9: 293.

Turker, A.U. and Gurel, E. (2005). Common mullein (*Verbascum thapsus* L.): recent advances in research. *Phytotherapy Research* 19: 733–739.

Vasiliauskas, S. and Chen, H.Y.H. (2002). How long do trees take to reach breast height after fire in northeastern Ontario? *Canadian Journal of Forest Research* 32: 1889–1892.

Wang, Y., Xue, M., Zheng, X. et al. (2005). Effects of environmental factors on N_2O emission from and CH_4 uptake by the typical grasslands in the Inner Mongolia. *Chemosphere* 58: 205–215.

Wang, X., Jia, M., Lin, X. et al. (2017). A comparison of CH_4, N_2O and CO_2 emissions from three different cover types in a municipal solid waste landfill. *Journal of the Air and Waste Management Association* 67: 507–515.

Wang, C., Zhang, W., Li, X., and Wu, J. (2022). A global meta-analysis of the impacts of tree plantations on biodiversity. *Global Ecology and Biogeography* 31: 576–587.

Wilbur, H.D. and Hufbauer, R.A. (2012). Timing control efforts to limit seed set of common mullein (*Verbascum thapsus*). *Invasive Plant Science and Management* 5: 390–394.

Wright, I.A., Paciuszkiewicz, K., and Belmer, N. (2018). Increased water pollution after closure of Australia's longest operating underground coal mine: a 13-month study of mine drainage, water chemistry and river ecology. *Water Air and Soil Pollution* 229: 55.

9

Bauxite Mine Restoration and Management

Shailesh Kumar Yadav[1], *Arnab Banerjee*[1], *Manoj Kumar Jhariya*[2], *Abhishek Raj*[3], *Surendra Singh Bargali*[4], *and Vimal Chandra Pandey*[5]

[1] *Department of Environmental Science, Sant Gahira Guru Vishwavidyalaya, Ambikapur, Chhattisgarh, India*
[2] *Department of Farm Forestry, Sant Gahira Guru Vishwavidyalaya, Ambikapur, Chhattisgarh, India*
[3] *Department of Forestry, Pt. Deendayal Upadhyay College of Horticulture and Forestry, Dr. Rajendra Prasad Central Agriculture University, Samastipur, Bihar, India*
[4] *Department of Botany, DSB Campus, Kumaun University, Nainital, Uttarakhand, India*
[5] *CSIR-National Botanical Research Institute, Lucknow, Uttar Pradesh, India*

9.1 Introduction

Mining is a process that helps extract the required minerals below the earth's surface. Minerals are the key factors in the prosperity of various economic activities, and well-being of the human civilization. Bauxite is a mineral from which aluminum is extracted. The process of bauxite mining involves surface mining of the crust as bauxite is generally found near the surface region. However, it may vary on a case-to-case basis (Sidiki 2022) (Figure 9.1). The open-cast mining process involves the removal of the topsoil layer and the excavation of the mineral layer. This therefore requires clear felling over the land surface and hence rapid rate of deforestation takes place over the area (Kumar et al. 2017). The process of bauxite mining involves sequential steps, which include exploration, development, active mining, disposal of waste rock material, ore extraction, beneficiation, disposal of tailings, reclamation of the site, and closure of the mine. The present chapter deals with the various approaches to the management of bauxite mining along with their sustainable management.

9.2 Impact of Bauxite Mining

The overall impact of bauxite mining is negative when we consider it in terms of its environmental perspective (Figure 9.2). Such mining activity alters the environmental scenario of an area through altered land use. The impact may vary depending upon the type of mineral, the methodology used for excavation of the bauxite as well as the surface area involved in the mining process. It also has severe consequences for the societal structure of

Biodiversity and Ecosystem Services on Post-Industrial Land, First Edition.
Edited by Vimal Chandra Pandey.

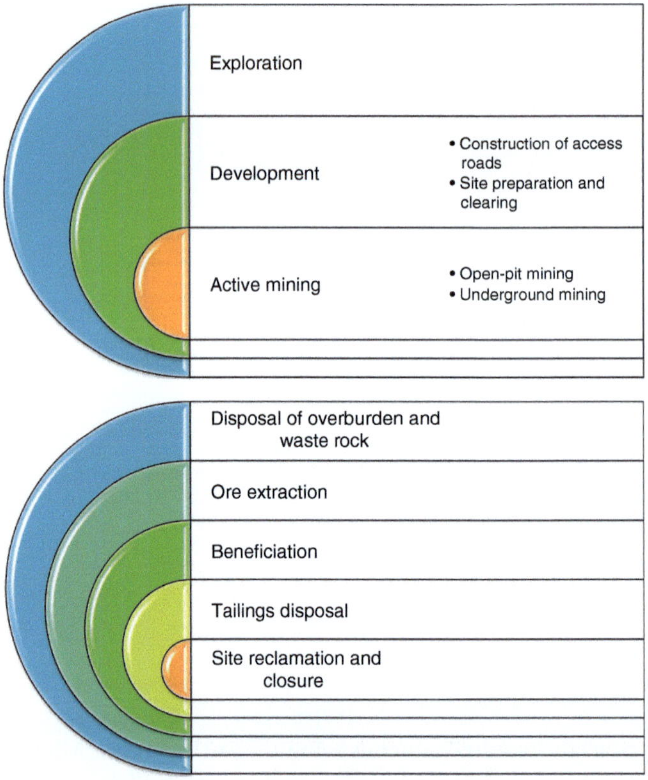

Figure 9.1 Phases of bauxite mining.

the area. Surface mining of bauxite causes alteration of the chemical, biological, and physical nature of the land surface as well as has a negative impact on the biota of the area. The main reason behind this is the loss of functionality of the soil ecosystem causing a socioeconomic impact on the local stakeholders of the concerned area (Riscky et al. 2020).

9.2.1 Social Impact

The impact of bauxite mining on society is immense as it both has short-term and long-term impacts. On a short-term basis, bauxite mining over an area causes the generation of huge amounts of dust to which the local people get exposed. Further, it also gives rise to various forms of vector-borne diseases. On a long-term basis, there is rapid contamination of water and food resulting in respiratory lung alignments. Such events are also associated with various forms of heavy metal toxicity through various forms of renal and cardiovascular disorders among children and adult individuals. Further, the problem aggravates in terms of neurological disorders leading to the formation of cancer and subsequent death. Noise produced during bauxite mining has also severe negative consequences in terms of loss of hearing and various other forms of noise-related disorders. Cumulatively, all these cause mental ill health among the local individual (Mara 2022).

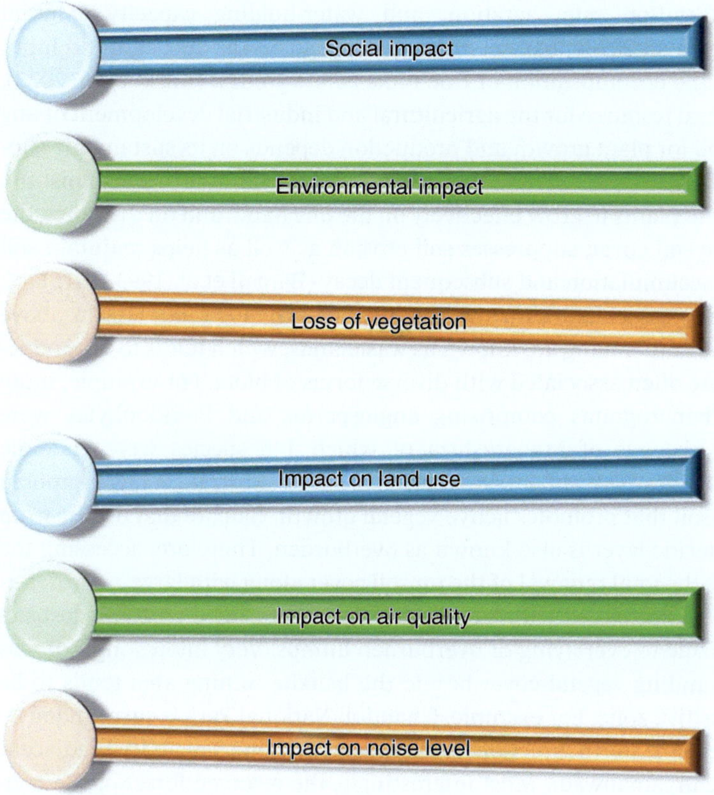

Figure 9.2 Impact of bauxite mining.

9.2.2 Environment Impact

The consequences of bauxite mining were evident in the form of soil, air, and water pollution. The dust emitted during bauxite mining reduces visibility and it also contaminates the food chain. Further, it also causes various forms of cardiorespiratory diseases on a chronic basis. In aquatic systems, bauxite and heavy metals reach in sediments of rivers and gradually enter the food chain. Plant species tend to uptake heavy metals from the soil which leads to food chain contamination causing safety issues. Habitat destruction is also a significant environmental impact of bauxite mining that causes severe loss of biota of the tropical forest ecosystem (Lee et al. 2017).

9.2.3 Loss of Vegetation, Forest Fragmentation, and Biodiversity Loss

Population pressure, agricultural expansion/intensification, development of infrastructure, and industrialization are the major causes of land degradation (Davidar et al. 2010; Awasthi et al. 2022a, b) and adversely affect the biodiversity and physicochemical properties of soil to a great extent (Gosain et al. 2015; Bisht et al. 2022, 2023). Biodiversity improves the soil structure, infiltration rate, aeration, and water-holding capacity (Manral et al. 2022, 2023; Pandey et al. 2023a, b). Soil nutrients depend on the aboveground input received from litter and the decomposition of fine roots below ground (Karki et al. 2022). Soil is an important natural resource for the agricultural and industrial development of any nation. The quality of soil for plant growth and production depends on its sustainable supply of plant nutrients (Bargali et al. 2018, 2019). Soil gives support in terms of moisture, nutrient, and anchorage to plants to grow effectively on the one hand, and on the other, the plant provides protective soil cover, suppresses soil erosion as well as helps maintain soil nutrients through litter accumulation and subsequent decay (Bargali et al. 1993, 2015).

Bauxite ore is generally associated with the top level of plateaus (700–1100 m above MSL). The outcrops of bauxite mining are known as wastelands, which refers to unproductive lands. These areas are often associated with diverse forms of biota. For example, more than 300 species of Phanerograms comprising angiosperms and Pteridophytes were recorded in the rocky plateaus of Maharashtra of which 146 species were endemic (Watve 2013). Further, studies have shown weathering of rocky substratum has promoted the gradual build-up of soil that promotes active vegetal growth. Bauxite that occurs from 5 to 8 m in the upper lateritic layer is also known as overburden. Therefore, accessing the bauxite mineral requires the total removal of the topsoil cover along with large-scale deforestation. Another problem associated with bauxite mining is associated with the destruction of the vegetal layer due to overlaying of overburden dumps. Very interestingly, it was observed that the surrounding vegetal cover beside the bauxite mining area tends to be diverse and is an ecosensitive zone. For example, Chandoli National Park is surrounded by the Udgiri bauxite mine in the north direction, Radhanagari Wildlife Sanctuary is adjacent to the bauxite mine at Durgamanwadi. Most interestingly, the reserved forest of the concerned area lies within a 10 km buffer zone. The area is characterized by dense coverage of vegetation followed by the formation of a narrow corridor between the two national parks. As per the recommendations given by the Western Ghat Ecology Expert panel, most of the mine area falls under an ecosensitive zone and hence recommendations for phasing out mining activity in the area were already recommended in 2016 (Lee et al. 2017).

9.2.4 Impact on Land Use

The term land use refers to the current use of land. It was observed that due to bauxite mining, there are severe alterations in the land use of the mining site. First of all, the entire vegetal cover gets reduced under surface mining of bauxite. Mostly forest areas, fertile agricultural land gets jeopardized by the activity of bauxite mining (Nagendran et al. 2021).

9.2.5 Impact on Air Quality

Various operational processes lead to the generation of fugitive dust that increases the load of suspended particulate matter (SPM) to a considerable level. Various processes associated with mining activity from mineral excavation to transportation at required sites are the major source of fugitive emission of air pollution. Mining activity is also associated with a secondary source of pollution through the emission of gaseous and particulate pollutants due to the burning of fossil fuel in vehicles associated with mining. Primary gaseous pollutants and various forms of particulate matter are emitted in huge amounts through bauxite-mining activity. Another negative consequence includes the generation of windblown dust due to heavy traffic in the mined area with highly mechanized and heavy vehicles. Such types of emissions not only impair the air quality as well as impair human health through the mechanism of heavy metal toxicity (Rosli et al. 2021).

9.2.6 Impact on Noise Level

Surface mining of bauxite is often associated with the generation of diverse forms of noise and high vibrations that may lead to hearing impairment in the people engaged in such work. Activities such as blasting, drilling, deployment of heavy machines, and other associated mining operations often generate a significant amount of noise. Hearing impairment of mine workers may be temporary and permanent. However, such heavy noises are often associated with major health issues like heart disease, high blood pressure, and the development of ulcers. Various works have reported that such high-frequency noise often reduces working efficiency, irritation, abnormal behavior, fatigue, and socially irresponsible behavior that lead to increased incidence of accidents.

9.3 Approaches Toward Management of the Impact of Bauxite Mining

9.3.1 Rehabilitation

Rehabilitation is a process that works toward bringing back the function and fertility of the ecosystem with some modifications in terms of altered land use and change in species composition (Raj et al. 2023a,c). The nature of the ecosystem becomes simpler form but becomes more productive. For example, grazing land replaces the woodland area. However, the situation is reversed in terms of less productive such as hybrid systems, for example, the plantation of Eucalyptus in place of an understory stratum of weed grass (Figure 9.3).

Various approaches in terms of ecorestoration, revegetation, and reclamation were used to restore the quality of degraded land (Yadav et al. 2022). According to Practice of Ecological Restoration in Australia (Standards Reference Group SERA 2016), the most used term is restoration and rehabilitation. Rehabilitation is the key process after mining through the plantation of screened species.

Figure 9.3 Process of rehabilitation.

9.3.2 Restoration

The difference between restoration and rehabilitation becomes a crucial factor regarding the framing of policy, legal framework, and various government mechanisms. Restoration is a process that aims to reframe the functionality and structural configuration of an ecosystem from a disturbed state. Further, the process undergoes a succession phase for the recolonization of plant species that leads to the development of identicalness with the original ecosystem (Figure 9.4).

9.3.3 Ecorestoration

The process of ecorestoration involves a sequential development process that initiates from land degradation to the post-mining climax phase (Figure 9.5). First, the forest ecosystem is denuded for mining activity and excavation of minerals followed by plantation activity after post-mining. Then natural succession process leads to the development of a climax community. In this process, the degraded ecosystem may reach to near natural condition of the old ecosystem (Riscky et al. 2020; Raj et al. 2023b).

9.4 Restoration Targets and Objectives

The restoration process should involve specific aims and objectives which include going back to the pre-mining stage through a proper restoration and reclamation process. This has been reported in Australia through the establishment of native vegetation that would

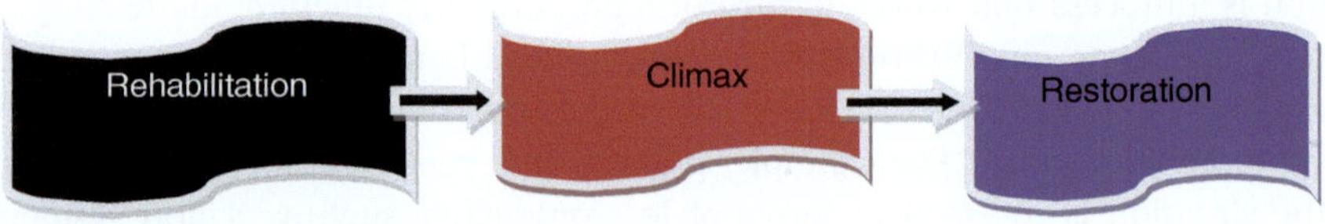

Figure 9.4 Process of restoration.

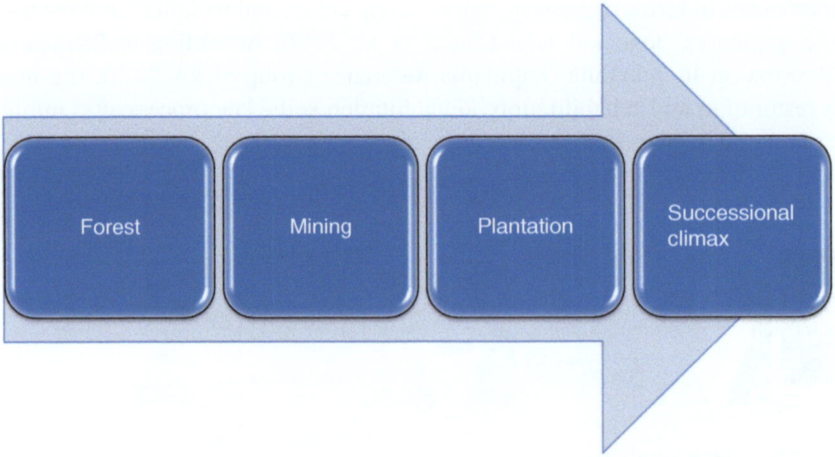

Figure 9.5 Process of ecorestoration.

lead to self-sustainability. Another major objective includes rehabilitation for improving the pre-mining condition and, third, converting the degraded land to some sort of beneficial land use.

9.5 Stages of Restoration Planning and Implementation

Proper planning and its subsequent implementation are the keys to the success of the restoration process (Figure 9.6). First, there should be some specific objectives and goals on which the rehabilitation process would be aimed. After that, with the specification of goals and objectives, rehabilitation planning was done for its successful execution. Various approaches and techniques were implemented to fulfill the completion criteria. Processes such as designing for floral and faunal recolonization, landscape architecture, and soil development followed by screening of suitable species for plantation are usually done

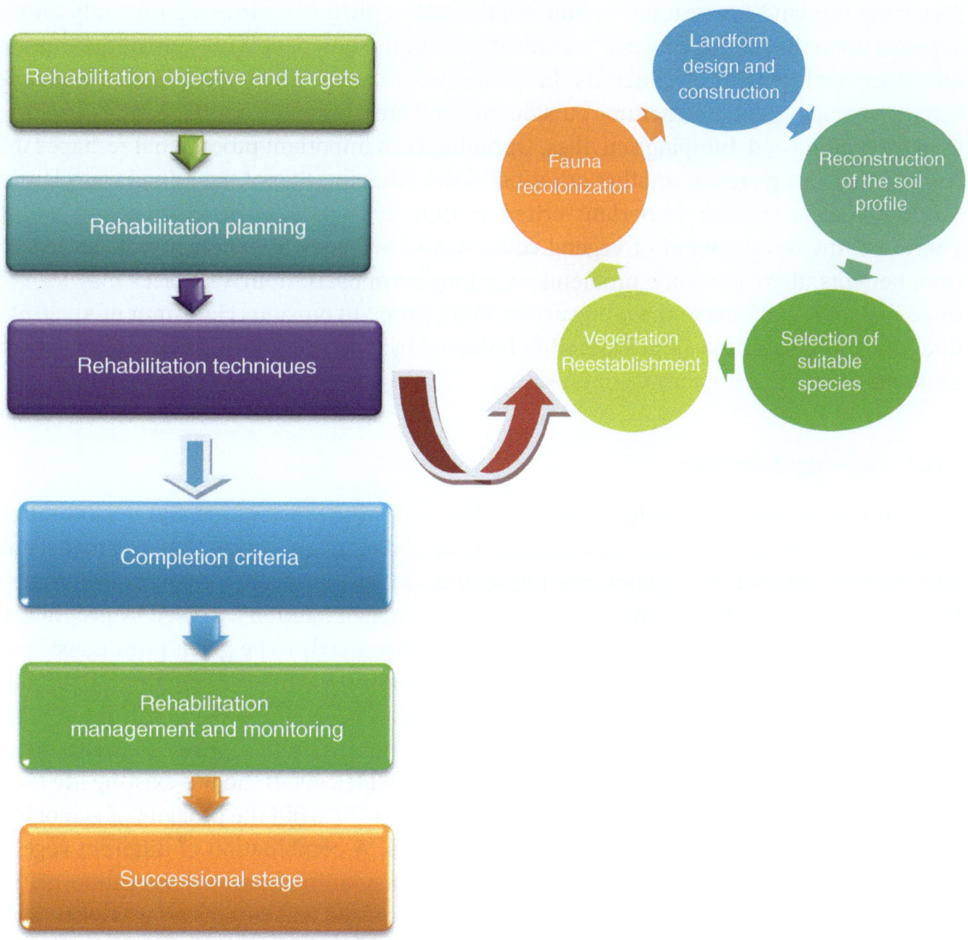

Figure 9.6 Stages of rehabilitation.

under the process of rehabilitation techniques. After the processes were implemented at the ground level, proper management and monitoring were implemented to increase the effectiveness of the planning process. At last, the target ecosystem undergoes various successional process that leads to recolonization of flora and fauna at the degraded site.

9.6 Restoration Implementation

This is a major step forward for bringing back the originality and functionality of the ecosystem from a degraded state. It involves diverse processes such as the development of land architecture, establishing optimum growth medium for plants, various forms of reclamation of soil habitat, screening of suitable species, and evaluation of recolonization rate of flora and fauna along with their subsequent management (Jhariya et al. 2022a, b).

9.6.1 Land Preparation

Preparing the land for restoration and rehabilitation purposes varies significantly and it depends upon the mining process. Technically, dragline spoil piles do not provide an option for selective placement of materials. In comparison, truck/shovel operations provide the benefit of placing stable structures outside the land area. Software applications can also be used for hauling and dumping activities. Dumping is an important process that reshapes the land. Surface roughness is another criterion in the rehabilitation of the mined wastelands.

Roughness is another important criterion that helps to hold both seed and water. Therefore, the development of vegetal cover would be better under such surfaces. Apart from benefits, there are some problems on a long-term basis. Rough surfaces may lead to slope instability and make the land surface more prone to erosion (Haj-Amor et al. 2022). Roughness increases the slope instability followed by a higher flow of concentrate, which ultimately leads to a higher rate of erosion.

9.6.2 Species Selection

Screening of species is dependent upon the objectives of the restoration and rehabilitation process, the probability of success followed by desired land use. It was observed that a specific form of vegetation is sometimes required to achieve a specific function in terms of biogeochemical cycling of nutrients, carbon, and nitrogen fixation followed by groundwater hydrology. The substratum or growth medium also needs to be given prior consideration due to altered land use and practices during mining activity. Screening of suitable species that are adaptive to the existing ecological condition and will also fulfill the rehabilitation aims and objectives. While going through the restoration process, species documentation and inventorization are required to have an idea about the pre-existing life form of vegetation that existed before the mining of the area. To avoid the problem of ecological invasion, the non-native should be generally avoided. A combination of different vegetal layers would be good for indigenous ecosystems. Depending upon the condition of the growth medium, rare or threatened species may be used for vegetation establishment. Cover crop plantation along with indigenous vegetation can be used to protect the soil layer from erosion activities at the initial phase till the indigenous vegetation becomes capable enough to protect the soil. Under this process, a suitable amount of fertilizer needs to be applied to promote cover crops without hampering the growth of indigenous vegetation. Oats were used as cover crops along with fertilizer application over native seeds in areas of Western Australia (Petersen and Brooks 1996; Riscky et al. 2020).

9.6.3 Establishment of a Plant Growth Medium

Establishing a plant growth medium is a challenging task as it leads to the development of a suitable habitat for the growth of vegetation. The area should have some inherent characteristics in terms of optimum infiltration capacity and available water capacity, optimum level of soil aeration, proper rooting depth for plants, efficient cycling of nutrients, deferred from soil problems with good microbial load that would support plant growth.

For preparation of a suitable growth medium, topsoil may be placed over excavated overburden or mill tailings. To achieve the target of the optimum root zone, soil material and overburden need to be separated and treated. If there is a deficit of suitable topsoil, the subsoil material can be used after significant reclamation by chemical, physical, and biological modes. Topsoil is a key factor for planning and management of the ecorestoration process (Banerjee et al. 2023). Decision on the conservation of topsoil needs to be taken based on type, nature, and distribution of soil, and overburden dump before the initiation of mining. The concept of conservation of the topsoil emerges in the case when the amelioration of the overburden dump is not possible or the cost of amelioration is not affordable. Another problem is the presence of weed seeds in the topsoil, which may hamper the development of beneficial or indigenous vegetation restoration. However, if the cost associated with negative consequences due to the spreading of weed species was found to be greater than the amelioration of the subsoil, then such topsoil needs to be removed. Another alternative for the problem is to go for the application of strips of overburden along with topsoil on an alternate basis. Indigenous vegetation in the form of shrubs and trees can be directly applied as seed sowing in the overburden. In this condition, the topsoil acts as a bed for grass seeds to undergo growth over an overburdened area in association with indigenous vegetation that would reduce the competition between grass species and indigenous species. It was observed that topsoil is advantageous in comparison to overburdened material as it provides a congenial environment for plant growth as well as least treatment is required for topsoil to be used as substratum for plant growth.

In preparing for soil substratum for optimum growth of plants, the entire soil profile is recommended only in cases when all the horizons are satisfactory or made satisfactory through chemical processes. However, mixing the subsoil content is usually not recommended due to abnormalities in the chemical structure, and physical configuration of soil particles. One strategy can be implemented by the application of microbial and nutrient-rich A-horizon with the overburden dump for optimum growth of plants. It is usually observed that topsoil acts as a seed bank and the application of such a thin layer of topsoil during the post-mining phase may increase the development of vegetal cover. Under the rehabilitation process, the topsoil should be maintained as a seed bank in order to promote successful plant establishment. Regarding maintenance of the soil seed bank, collection of topsoil should be done in time when there is a higher availability of seed. Due consideration should be given to the burning vegetation before pre-mining activity as it may influence the germination potential of seeds. Further, it is better to re-spread the topsoil in a specific area for its effective conservation.

Research reports reflected significant results in the form of the fact that there is significant deterioration of the seed bank during the rehabilitation process after bauxite mining in the western part of Australia from the topsoil layer. But, most interestingly, the research revealed that some of the lost seeds on indigenous vegetation were restored in the rehabilitation areas Topsoil deterioration takes place in the initial year of bauxite mining.

9.6.4 Physical Amelioration

Physical amelioration of soil includes reduction of soil compaction through heavy machinery and improvement of the water infiltration rate along with prevention of soil erosion. Ripping of the area through engineered waterways may be done on a graded way or over a contour area. Ripping activity is dependent upon several factors of rehabilitation such as type, nature, depth of overburden, and topsoil along with types of instruments used in the rehabilitation process. Mixing rock with topsoil during ripping activity is important for reducing erosion. Another important aspect is the avoidance of intermixing saline waste material with the topsoil.

Reduction in erosion can be achieved through intermixing of rock material with surface soil (Howard and Lowe 2014), which would enhance the salt infiltration up to a considerable depth (Jennings et al. 1993). However, the major issue is achieving the proper combination of soil mix and rock material (Howard and Lowe 2014).

9.6.5 Chemical Amelioration

While considering the chemical amelioration process, the problems related to soil pH, loss of fertility, alkalinity, and acidity need to be considered. While working on the rehabilitation process, the recommended pH value for agricultural areas and crops may not be applicable. The situation is that the indigenous vegetation in the rehabilitated area may survive under extremes of pH. However, a baseline database on soil properties along with ameliorative activities is recommended for highly acidic pH. Sodic soil comprising greater than 6% of cation exchange capacity is another hurdle that needs to be overcome in the restoration process. In the case of topsoil and overburdened soil, clay dispersion makes them highly erosive along with a lesser water infiltration rate and a high level of tunnel erosion.

Solutions to these problems include the application of gypsum in the topsoil and giving time for mixing the gypsum with topsoil. Saline waste materials are the biggest problem for the mining area. It is better not to have saline waste over the rehabilitated landforms. The main aspect of these includes managing the seepage of saline materials from the base of waste dumps. Various technologies such as screening and use of salt-tolerant vegetation, increased mixing of rocky waste material, and encapsulation may be used to arrest the salinity problem over an area. Further, the regulation of groundwater movement is a key factor that needs to be monitored precisely. Under chemical amelioration, bringing back the fertility of soil is another biggest challenge. Usually, the topsoil of a certain depth is placed over a waste heap. This creates a completely different environment than the soil properties of the target ecosystem. Therefore, fertilization may be done in order to avoid nutrient deficiency for the growing plants. The major problem is that a high amount of nutrient load is present below and aboveground biomass and a very small amount is available for the plant uptake in the form of organic matter (Westman 1978). Therefore, the sole application of chemical fertilizer may not be the solution for this. There should be a proper nutrient budget as well as a load of nutrients in a required amount that needs to be added for proper plant growth and nutrient cycling. However, nutrients are added in the initial phases to re-establish the level of nitrogen for promoting the development of grasses in order to prevent erosion. Up to 300 $kg\,ha^{-1}$ application of nitrogen and phosphorus-rich fertilizers is recommended for application. However, the application rate may vary from site to site basis. Sequential steps include ripping, fertilizing, and seeding in the rehabilitation program. During the succession and development phase of the ecosystem, application of chemical fertilizers may sometimes become essential. Indigenous vegetation and eguminous species capable of fixing atmospheric nitrogen can be used as seed mix. Often the application of organic fertilizers is recommended in place of inorganic fertilizers.

9.6.6 Biological Amelioration

The most significant method of biological amelioration is the establishment of vegetation. Compatibility of the vegetation with rehabilitation objectives as well as land use is an important aspect of biological amelioration. The establishment of plant species can be done through various plant propagules present in the topsoil followed by seed sowing, and harvesting plants in rehabilitated areas. Further transplantation of nursery-raised seedlings or from natural areas along with soil substratum helps in the invasion of vectors from the surrounding areas.

Supplementation of seed material can be done through seeds collected from the surrounding area or near the mining area. Seed sowing is a cost-effective method for species establishment. Seeding in comparison to seed sowing is much better in terms of similarity to the original vegetation. Seeding also reduces labor costs as well as the heterogeneous nature of vegetation stand develops.

9.6.6.1 Seed Supply

Seeds are either purchased or collected and hence quality control is a big issue. The initiation of collection of native seeds should be started one to two years prior to use so that source and amount can be determined. The local-based collection is usually recommended as the seeds would be compatible with the local environmental setup as well as in terms of genetic makeup. Proper care should be taken about the seeds; they should be cleaned and stored in a safe place to avoid pest populations.

9.6.6.2 Seed Treatment

Seed treatment is an important process that needs to be done to increase seed germination efficiency. The treatment process includes treatment with heat, exposure to smoked water, etc. (Floradata 2001). For areas with precipitation uncertainty, limited seeds are to be treated to maintain the viability of seeds for a longer time. Microbial inoculation and lime application may also aid the process.

9.6.6.3 Ecosystem Succession

For establishing a sustainable ecosystem, natural ecological succession is one of the key processes that need to be given key consideration. Pioneer species need to be incorporated into the seed mix. However, species that occur in the later phase of development if found fruitful can also be employed for seeding purposes. The value of relative abundance would show a fluctuating nature. Among early colonizing species, the higher seed rate may outfit the species under indigenous diversity.

9.6.6.4 Seeding Rate

Field trials are essential for determining the seeding rates. 1–3 $kg\,ha^{-1}$ is the recommended seeding rate for indigenous vegetation in bauxite-mined and other sand mining areas. The survival rates for coarse and fine seeds are 5–10% and 1–5%, respectively, through direct seeding methods. Considering more than 70% survivability rate, the sowing rates for coarse and fine seed species are 2–4 $kg\,ha^{-1}$ and 0.1–1.0 $kg\,ha^{-1}$, respectively. Various factors need to be considered while going for seeding activity in the rehabilitated area. For surface stabilization, the seeding rate for grass species is 10–20 $kg\,ha^{-1}$.

9.6.6.5 Seed Spreading

Various forms of seed-spreading methods are available for seeding the degraded land or over the overburden soil. The rate of spread of species should be homogenous and fulfill the desired target of rehabilitation. The species seeding should be implemented before the onset of the monsoon or after the monsoon break. The indigenous species do require optimum conditions for seed germination and hence need to be seeded at a specific time of the year. Species having problems with the release of seeds can be reintroduced in the rehabilitation area to prevent soil erosion through their sustained growth. Plantation of nursery-raised seedlings is another method through which rehabilitation programs can be made successful. However, it is only applicable when the seeding method does not give appropriate and sustainable results. However, the availability of good-quality seedlings is important in this connection. The process has some inherent benefits in terms of proper use of seeds, inoculation through mycorrhiza, and other modes under the revegetation program. The major hindrance includes higher costs in terms of nursery operation and maintenance.

Ecological invasion in the rehabilitated area is another issue for bauxite-mined wastelands. For example, it was reported that there is a gradual development and spread of orchid species in the rehabilitation area of bauxite mining (Grant and Koch 2006). Such an invasion process can be promoted by improving the interconnectivity of the bauxite-mined areas with their surrounding vegetation. In this connection, rehabilitation of some species becomes problematic and hence tissue culture method can be adopted for their recolonization over the concerned area. But future exploration needs to be done for this particular aspect.

From a plant science point of view, it was observed that plants undergo symbiotic relationships with microorganisms, fungi, actinomycetes, and other tiny forms of organisms for the benefit of the soil ecosystem. Ecto- and endo-mycorrhiza are also found in the soil rhizospheric zone encompassing the roots of the plants. At the initial phase of the restoration and rehabilitation process, researchers have revealed that the indigenous vegetation often forms a strong association with vesicular-arbuscular mycorrhiza (VAM) fungi. Further, these VAM fungi act as phosphate-solubilizing organisms that aid plant growth under degraded habitats as most of them are phosphorus-deficient. It was also observed that the beneficial role of VAM fungi seems to be missing in areas under surface mining of bauxite. Researchers have further reported that the early rehabilitation process involves the colonization of limited ectomycorrhizal fungi. Also, in the successional phase of restoration, lack of specific mycorrhiza often hampers the recolonization process. Therefore, to recolonize mycorrhiza, topsoil should be conserved and returned to the original site immediately after post-mining. Production of large stockpiles should be kept at a minimum and revegetation should be promoted as early as possible.

9.6.7 Faunal Recolonization

Recolonization of the faunal species in the degraded land is a key issue in the restoration and rehabilitation process. It was observed that animals tend to recolonize if the nature of the vegetal cover of restored areas reflects similarity with the vegetation of the surrounding areas. Researchers have revealed that under restored habitat, the fulfillment of habitat conditions remains unfruitful for a long period. However, various methods are available for faunal recolonization in the restored habitat. Technologies such as the transplantation of grasses in degraded habitats, and reutilization of vegetal cover through re-spreading and chipping provide suitable habitat conditions for diverse faunal groups. The development of nest boxes promotes the growth of mammalian and avian species. Logwood materials for the growth of ground-dwelling faunal forms, building up of surface boulders for suitable habitat for reptile species, development of perches to be used by raptors and other avifauna, creating habitat for various invertebrate and reptilian species through the establishment of old dead trees, etc. However, in such restoration approaches, proper care must be taken to fulfill the objective of restoration and rehabilitation process followed by proper land uses.

9.6.8 Monitoring

Monitoring is an important process to evaluate the success of the restoration process and its effective implementation. Various methodologies are available for effective monitoring of the process of restoration and rehabilitation of the mine area (Fletcher and Erskine 2013). For restoration and rehabilitation projects measures such as diversity of flora, vegetal cover, and diversity of arthropod population in the soil were used in various rehabilitation projects (Ruiz-Jaen and Aide 2005). Methods involve transect, point count, and plot study to evaluate the vegetal cover, species counting, and densities. Landscape function analysis (LFA) is the most widely adopted methodology for the measurement of ecosystem functions and services. Further trajectory analysis was carried out for the evaluation of the soil ecosystem (Tongway et al. 2003; Tongway and Hindley 2004; Erskine et al. 2013). According to the recommendation of the Society for Ecological Restoration (SER 2004), nine different attributes need to be considered while assessing the success of the ecorestoration process. This includes assessment of similarity in diversity and structure of degraded ecosystem to that with reference sites, monitoring of indigenous species, evaluating the presence of functional groups for long-term stability of the ecosystem, assessing the capacity of the physical environment for sustainability of the population, normal functioning of the ecosystem, integration with the natural habitat, evacuation of the threats toward the restoring ecosystem, developing ecosystem resiliency toward natural disasters and development of self-sustaining system.

9.7 Sustainable Bauxite Mining

A clean and healthy environment is important for the survival and growth of all living beings. Our present environment and ecology are threatened by anthropogenic activities including mining, quarrying, and other industrial activities. The growing concern about environmental degradation has prompted the world to think on a new dimension and address the environment-related issues more scientifically. Sustainable development of any region mainly depends on three important aspects: social progress, economic growth, and protection of the environment. Increased pressure on land to satisfy the greed of human beings accelerated the interventions in the form of quarrying, mining, and unscientific conversion of land use, etc., which in turn leads to significant land modifications, thereby influencing the biophysical system and adversely affecting the environmental stability and ecological security (Kamble and Bhosale 2019).

In most of the literature, the emphasis was given to sustainable ways of mining. However, the concern is that "the socio-economic upliftment that should logically follow the implementation of ensuring that mining operation is conducted responsibly is not visible in many mining areas and communities where the mining companies operate. The study of Ololade and Annegarn (2016) found that the environmental and social costs associated with mining were high, while economic benefits to surrounding communities were low." Ololade and Annegarn (2016) also documented that "the perceptions of community and corporations were found to contrast sharply: the reality experienced by community members fell well short of the optimistic scenarios presented in the corporate social responsibility reports of the mining companies, which has implications for the mining industry in the area" (Ololade and Annegarn 2016).

According to Pretty and Odeku (2017), "good environmental governance is an important pillar in the global partnership for sustainable development." Environmental governance may be described "as how people exercise authority over nature (Hassan 2007)." It is pertinent to mention that "natural resource management was internationalized by the 1972 Stockholm Declaration 2004, echoing the earlier internationalization of human rights by the 1948 Universal Declaration of Human Rights (UDHR) UN General Assembly in Paris 10 December 1948" (Hassan 2007). First principle of Stockholm declaration encompasses the consideration of healthy environment as human right "man has the fundamental right to freedom, equality and permits a life of dignity and well-being, and he bears a solemn responsibility to protect and improve the environment for present and future generations" (Pretty and Odeku 2017).

9.8 Rehabilitation and Restoration in Surguja – A Case Study from Chhattisgarh, India

Mainpat area situated in Surguja division of Chhattisgarh is the northern hilly region of the state. The area is highly rich in bauxite ore and hence there is rapid excavation of bauxite from the Mainpat area. A huge area has already been denuded and deforested due to bauxite mining. Further, suitable plantation activities besides the bauxite mining area have been shown in Figures 9.7–9.10 in which the major focus was given on successful adoption of

Figure 9.7 Bamboo plantation site.

Figure 9.8 Burnt plantation.

Figure 9.9 River site plantation.

Figure 9.10 Fuel wood plantation.

rehabilitation and restoration approaches followed by plantation of screened species for livelihood generation for the local people.

Table 9.1 represents the plantation details of the restored habitat of the bauxite mine area of Mainpat, Surguja. Successful plantation of four different species was recorded and average girth class indicates most of the species were under mature condition and can be considered Tree population except Bamboo. Apart from bamboo, most of the plantation was found to be six to seven years old. The survival percentage of most of the species ranged between 70% and 80%. The height of the plantation ranged between 350 and 2000 cm.

Table 9.2 represents the phytosociological attributes of various species. Higher density and abundance values were recorded by *Acacia auriculiformis* and the lowest were recorded for bamboo plantation.

9.9 Conclusion

Surface mining of bauxite has some inherent drawbacks in terms of having environmental impact, social impact, altered land use, loss of soil fertility, and soil erosion. Now combating such diverse problems raised due to bauxite mining is a big challenge and hence it requires a holistic approach of reclamation, restoration, rehabilitation, and ecorestoration. In terms of effective management of bauxite mining, proper care should be taken both in pre-mining, during mining, and in the post-mining phase. The major important issue includes soil amelioration as well as maintaining the top soil layer to be applied during the post-mining phase. Mineral excavation should simultaneously be accompanied by rehabilitation and revegetation activity. The ecorestoration approach includes the selection of suitable species and the preparation of soil substratum for promoting plant growth. Effective monitoring in terms of landscape analysis followed by analyzing the diverse attributes for assessing the success of plantation for ecorestoration is required. A case study of the Mainpat area reflecting successful ecorestoration has also been shown in this chapter. Above all, sustainable management of the impacts of bauxite mining demands a holistic approach to restoration, rehabilitation, and ecorestoration.

9.10 Future Directives

Mining is such an issue that various policies and debates came to the forefront of being devoid of land for mining purposes. Mining in areas under legal protection, and less protected ecosystems poses a significant threat of environmental degradation for human civilization. The major problem is that most of the mining areas or activities across the world take place in areas with more conservation value along watershed areas. Most of the developed and developing nations lack appropriate measures to combat the impact of mining along with the ecorestoration of the mined wastelands. Also, there are lacunae of strict implementation of law for the protection of mined wastelands, corruption in terms of funding, and lack of awareness of civil society. Proper policy framework along with strict

Table 9.1 Plantation details in restored habitat of a Bauxite mine area.

Serial number	Division	Compartment number	Area (ha)	Latitude and longitude	Species name	Year of plantation	Number of individuals	Average girth (cm)	Average height (cm)	Survival percentage
1	Sarguja	191, 395	6.5	22.817 83.261	*Acacia auriculiformis*	2008–2009	7000	45–50	1900–2000	75–80
2	Sarguja	11/1	20	22.777 83.165	*Acacia auriculiformis*	2015–2016	8000	50–60	1000	80
3	Sarguja	P 2297	25	22.828 83.167	*Eucalyptus globulus*	2015–2016	27 000	30–40	800	80
4	Sarguja	P 2296	40	22.833 83.158	*Bamboo* spp.	2020–2021	16 000	9	350	80–85

Table 9.2 Phytosociology of plantation in bauxite mine area of Mainpat, Surguja.

Serial number	Species name	Year of plantation	Frequency	Density (ha^{-1})
1	*Acacia auriculiformis*	2008–2009	100	8
2	*Acacia auriculiformis*	2015–2016	100	8.9
3	*Eucalyptus globulus*	2015–2016	100	8
4	*Bamboo* sp.	2020–2021	100	3.3

legal enforcement is required for effective management of the negative impact of mining. Surface mining-oriented policy, law, and act need to be framed without any loopholes followed by their strict implementation is the need of the hour. More research and development toward screening of suitable species based upon habitat quality, environmental conditions, and feasibility of application needs to be done in the upcoming times for sustainable management of the impacts of bauxite mining.

References

Awasthi, P., Bargali, K., Bargali, S.S., and Jhariya, M.K. (2022a). Structure and functioning of *Coriarianepalensis* wall dominated Shrublands in degraded hills of Kumaun Himalaya. I. Dry matter dynamics. *Land Degradation & Development* 33 (9): 1474–1494. https://doi.org/10.1002/ldr.4235.

Awasthi, P., Bargail, K., Bargali, S.S. et al. (2022b). Nutrient partitioning and dynamics in *Coriarianepalensis* wall dominated Shrublands of degraded hills of Kumaun Himalaya. *Frontiers in Forests and Global Change* 5: 913127. https://doi.org/10.3389/ffgc.2022.913127.

Banerjee, A., Jhariya, M.K., Bargali, S.S., and Palit, D. (2023). Ecorestoration for environmental sustainability– an introductory framework. In: *Ecorestoration for Sustainability* (ed. A. Banerjee, M.K. Jhariya, S.S. Bargali, and D. Palit), 1–47. Scrivener-Wiley.

Bargali, K., Manral, V., Padalia, K. et al. (2018). Effect of vegetation type and season on microbial biomass carbon in Central Himalayan forest soils, India. *Catena* 171 (12): 125–135. https://doi.org/10.1016/j.catena.1018.07.001.

Bargali, S.S., Padalia, K., and Bargali, K. (2019). Effects of tree fostering on soil health and microbial biomass under different land use systems in central Himalaya. *Land Degradation & Development* 30 (16): 1984–1998. https://doi.org/10.1002/ldr.3394.

Bargali, S.S., Shukla, K., Singh, L. et al. (2015). Leaf litter decomposition and nutrient dynamics in four tree species of dry deciduous forest. *Tropical Ecology* 56 (2): 57–66.

Bargali, S.S., Singh, S.P., and Singh, R.P. (1993). Pattern of weight loss and nutrient release in decomposing leaf litter in an age series of eucalypt plantations. *Soil Biology and Biochemistry* 25: 1731–1738.

Bisht, S., Bargali, S.S., Bargali, K. et al. (2022). Influence of anthropogenic activities on forest carbon stocks – acase study from Gori Valley, Western Himalaya. *Sustainability* 14: 16918. https://doi.org/10.3390/su142416918.

Bisht, S., Rawat, G.S., Bargali, S.S. et al. (2023). Forest vegetation response to anthropogenic pressures: a case study from Askot Wildlife Sanctuary, Western Himalaya. *Environment, Development and Sustainability*. https://doi.org/10.1007/s10668-023-03130-2.

Davidar, P., Sahoo, S., Mammen, P.C. et al. (2010). Assessing the extent and causes of forest degradation in India: where do we stand? *Biological Conservation* 143: 2937–2944.

Erskine, P., Fletcher, A., and Seaborn, B. (2013). Opportunities and constraints of functional assessment of mined land rehabilitation. In: *Mine Closure 2013* (ed. M. Tibbett, A. Fourie, and C. Digby). Perth: Australian Centre for Geomechanics.

Fletcher, A. and Erskine, P. (2013). Rehabilitation closure criteria assessment using high resolution photogrammetrically derived surface models. In: *UAV-g2013, Rostock* (ed. G. Grenzdörffer and R. Bill). Germany: International Society for Photogrammetry and Remote Sensing.

Gosain, B.G., Negi, G.C.S., Dhyani, P.P. et al. (2015). Ecosystem services of forests: carbon Stock in vegetation and soil components in a watershed of Kumaun Himalaya, India. *International Journal of Ecology and Environmental Science* 41 (3–4): 177–188.

Grant, C.D. and Koch, J.M. (2006). Ecological aspects of soil seed-banks in relation to bauxite mining. II. Twelve year old rehabilitation mines. *Australian Journal of Ecology* 22 (2): 177–184.

Haj-Amor, Z., Araya, T., Kim, D.G. et al. (2022). Soil salinity and its associated effects on soil microorganisms, greenhouse gas emissions, crop yield, biodiversity and desertification: a review. *Science of the Total Environment* 843: 156946. https://doi.org/10.1016/j.scitotenv.2022.156946.

Howard, E.J. and Lowe, S.M. (2014). Innovative rehabilitation of marine dredge spoil. In: *Mine Closure 2014* (ed. A.B. Fourie and M. Tibbett). Perth: Australian Centre for Geomechanics.

Jennings, B., Barrett-Lennard, E.G., Hillman, B.J. and Emrose, M. (1993). *Mine Waste Management in Arid Areas*, Report no. 110, Minerals and Energy Research Institute of Western Australia.

Jhariya, M.K., Banerjee, A., and Meena, R.S. (2022a). Importance of natural resources conservation: moving toward the sustainable world. *Natural Resources Conservation and Advances for Sustainability*. https://doi.org/10.1016/B978-0-12-822976-7.00027-2 3–27.

Jhariya, M.K., Meena, R.S., Banerjee, A., and Meena, S.N. (2022b). *Natural Resources Conservation and Advances for Sustainability*. Elsevier, Academic Press. https://doi.org/10.1016/C2019-0-03763-6.

Kamble, H.P. and Bhosale, S.M. (2019). Environmental impact of Bauxite mining: areview. *International Journal of Research in Applied Science and Engineering Technology* 7 (1): 86–90.

Karki, H., Bargali, K., and Bargali, S.S. (2022). Dynamics of fine root and soil nitrogen in *Mangiferaindica* based agroforestry systems in Central Himalaya, India. *Land Degradation & Development* 33 (17): 3523–3538. https://doi.org/10.1002/ldr.4406.

Kumar, A., Jhariya, M.K., Yadav, D.K., and Banerjee, A. (2017). Vegetation dynamics in Bishrampur collieries of Northern Chhattisgarh, India: eco-restoration and management perspectives. *Environmental Monitoring and Assessment* 189 (8): 371. https://doi.org/10.1007/s10661-017-6086-0.

Lee, K.Y., Ho, L.Y., Tan, K.H. et al. (2017). Environmental and occupational health impact of bauxite mining in Malaysia: review. *IIUM Medical Journal Malaysia* 16 (2): 137–150.

Manral, V., Bargali, K., Bargali, S.S. et al. (2022). Relationships between soil and microbial biomass properties and annual flux of nutrients in Central Himalayan forests, India. *Land Degradation & Development* 33: 2014–2025. https://doi.org/10.1002/ldr.4283.

Manral, V., Bargali, K., Bargali, S.S. et al. (2023). Seasonal dynamics of soil microbial biomass C, N and P along an altitudinal gradient in Central Himalaya, India. *Sustainability* 15: 1651. https://doi.org/10.3390/su15021651.

Mara, L. (2022). Quality of social impact assessment in bauxite mining sector. Masterthesis. Aalborg University Environmental Management and Sustainability Science.1–51. http://www.plan.aau.dk/.

Mortlock, W. and Lloyd, M. (2001). *A Guide to Collection, Storage and Propagation of Australian Native Plant Seed.* Australian Centre for Mining Environmental Research Limited (ACMER).

Nagendran, P., Rafikul, I., and Muhammad, F.A. (2021). Environmental impact assessment for Malaysian bauxite mining industry. *International Journal of the Analytic Hierarchy Process* 13 (1): 2–26. https://doi.org/10.13033/ijahp.v13i1.851.

Ololade, O.O. and Annegarn, H.J. (2016). Mining conflicts and corporate social responsibility: titanium mining in Kwale, Kenya. *The Extractive Industries and Society* 3 (6): 485–493.

Pandey, R., Bargali, S.S., Bargali, K. et al. (2023a). Fine root dynamics and associated nutrient flux in Sal dominated forest ecosystems of Central Himalaya, India. *Frontiers in Forests and Global Change*. https://doi.org/10.3389/ffgc.2022.1064502.

Pandey, R., Bargali, S.S., Bargali, K., and Pandey, V.C. (2023b). Temporal variability in fine root dynamics in relation to tree girth size in sub-tropical *Shorearobusta* forests. *Land Degradation & Development* 34 (5): 1522–1537. https://doi.org/10.1002/ldr.4550.

Petersen, A.E. and Brooks, D.R. (1996). Environmental management practices at RGC's Eneabba operation in the dry heath sand-plains of Western Australia. In: *Environmental Management in the Australian Minerals and Energy Industries: Principles and Practices* (ed. D.R. Mulligan), 571–582. Sydney: UNSW Press.

Pretty, M.M. and Odeku, K.O. (2017). Harmful mining activities, environmental impacts and effects in the mining communities in South Africa: a critical perspective. *Environmental Economics* 8 (4): 14–24.

Raj, A., Jhariya, M.K., Banerjee, A. et al. (2023a). Forest biodiversity conservation and restoration: policies, plan and approaches. In: *Ecorestoration for Sustainability* (ed. A. Banerjee, M.K. Jhariya, S.S. Bargali, and D. Palit), 317–350. Wiley-Scrivener.

Raj, A., Jhariya, M.K., Banerjee, A. et al. (2023b). *Land and Environmental Management Through Forestry*, 1–372. Wiley-Scrivener.

Raj, A., Jhariya, M.K., Banerjee, A. et al. (2023c). Managing forests for offsetting carbon footprints. In: *Ecorestoration for Sustainability* (ed. A. Banerjee, M.K. Jhariya, S.S. Bargali, and D. Palit), 267–294. Wiley-Scrivener.

Ricksy, P., Maman, T., Takumi, S., and Keitaro, T. (2020). Post bauxite mining land soil characteristics and its effects on the growth of *Falcatariamoluccana* (Miq.) Barneby & J. W. Grimes and *Albiziasaman*(Jacq.) Merr. *Applied and Environmental Soil Science* 2020: 6764380. 8 pages. https://doi.org/10.1155/2020/6764380.

Rosli, N., Rahman, R.A., Razelan, I.S.M. et al. (2021). Impact of bauxite mining on quality of life: an analysis of road users. *Earth and Environmental Science* 682 (2021): 012040. https://doi.org/10.1088/1755-1315/682/1/012040.

Ruiz-Jaen, M.C. and Aide, T.M. (2005). Restoration success: how is it being measured? *Restoration Ecology* 13 (3): 569–577.

SER (Society for Ecological Restoration) (2004). *The SER International Primer on Ecological Restoration*. SER International Science & Policy Working Group.

SERA (Society for Ecological Restoration Australasia) (2016). *National Standards for the Practice of Ecological Restoration in Australia*. SERA. http://www.seraustralasia.com.

Sidiki, S. (2022). An analysis of civil society organizations advocating for adequatebauxite mining in Boke Region (Western Guinea). *International Journal of Environmental Protection and Policy* 10 (3): 39–47. https://doi.org/10.11648/j.ijepp.20221003.11.

Tongway, D.J. and Hindley, N.L. (2004). *Landscape Function Analysis: Procedures for Monitoring and Assessing Landscapes*. Canberra: CSIRO Sustainable Ecosystems.

Tongway, D., Hindley, N., and Seaborn, B. (2003). *Indicators of Ecosystem Rehabilitation Success: Stage Two— Verification of EFA Indicators, Final Report*. Kenmore, Queensland: Australian Centre for Mining Environmental Research.

Watve, A. (2013). Status review of rocky plateaus in the northern Western Ghats and Konkan region of Maharashtra, India with recommendations for conservation and management. *Journal of Threatened Taxa* 5 (5): 3935–3962. https://doi.org/10.11609/JoTT.o3372.3935-62.

Westman, W.E. (1978). Inputs and cycling of mineral nutrients in a coastal subtropical eucalypt forest. *Journal of Ecology* 66: 513–531.

Yadav, S.K., Banerjee, A., Jhariya, M.K. et al. (2022). Ecorestoration of bauxite mined wasteland. In: *Natural Resources Conservation and Advances for Sustainability*, 173–193. Elsevier. https://doi.org/10.1016/B978-0-12-822976-7.00028-4.

10

Role of the Local Government in Re-Use of Post-Industrial Sites in Poland

Katarzyna Sadowy

SGH Warsaw School of Economics, Warsaw, Poland

10.1 Introduction

Re-use processes of the formerly built-up land constitute a crucial challenge in the world, and certainly in Europe. As stated by the European Environment Agency (EEA) in 2023 "about 80% of Europe's land surface has been shaped by human activities: covered with buildings, roads, and industrial infrastructure, or used for agriculture. The way we use land constitutes one of the main drivers of environmental degradation and climate change." (EEA 2023). For decades, urbanization, industrialization, and urban sprawl were based on the constantly increasing land take. Sites that lost their function, underperformed, or were located in the degrading areas of the cities, were left behind, while other greenfield areas were sealed and built up. Apart from the growing sense of environmental responsibility, European countries simply are reaching the end of the line of such practices. It is therefore the relevant question of who and how can introduce new, less wasteful practices, who can and should decide the future of the degraded areas, and in what way the necessary costs should be met, especially in cases of historical contamination of the ground. Roles and motivations of the private and public sectors differ, as well as the role of the national and local levels of public administration.

Growing or decreasing need for new investments depends on several factors, demography included. In 1960, European population (counting the 27 current Member States) was 354.5 million and grew for decades, reaching 446.8 million on 1 January 2022 (European Commission 2023, p. 9). During the period of the COVID-19 pandemic, there was the first decline in the population since 1960, but natural population change (difference between live births and deaths) has been negative since 2012. European society is aging, with an increasing share of people over 65 (Eurostat 2023a). Demographic trends differ significantly between the Member States. For example, the loss of population in absolute terms due to the COVID-19 pandemic was highest in Italy (611 000) and Poland (304 000), but in both cases, it meant, respectively, 1% and 0.8% of the population. In relative terms, the loss was greatest in Croatia, (−4.8%, −196 000) and Greece (−2.4%, −259 000). All these countries,

Biodiversity and Ecosystem Services on Post-Industrial Land, First Edition.
Edited by Vimal Chandra Pandey.

together with Bulgaria, Latvia, Lithuania, Hungary, and Romania are projected to experience further population decline in the next years (European Commission 2023, p. 9).

At the same time, the European Union (EU) has experienced another trend, which today is assessed as devastating for the environment and increasing the risks involved in the climate crisis. It seals the soil and increases land take when urban and other artificial land developments are introduced on land that was previously agricultural, forest, or had other semi-natural characteristics. In 2018, in the EU and the UK, 97.903 km^2 of soil was sealed, mainly in cities and towns (42% and 30% sealed, respectively). Europe also lost cropland (14% of the sealed soil), grassland ecosystems (3%), and forests (2%) (EEA 2023). Since 2000, the main drivers of land take were industrial and commercial land use, together with an extension of residential areas and construction sites. In the period 2012–2018, land take in EU Member States (EU28) was on average 539 km^2 per year. It decreased, in comparison to the period 2000–2006, when it was on average over 1000 km^2 per year (EEA 2019), still there is a large gap between the current situation and the EU's goal of the "no net land take" in 2050, included in the 7th Environment Action Programme and the Land Degradation Neutrality target of the United Nations Development Goals (SDG 15.3). The net land-take concept is based on the assessment of the land-take together with land return to non-artificial land categories (re-cultivation). While some land was re-cultivated in the EU-28 in the period 2000–2018, 11 times more land was taken. A very small share of the previously urbanized and soiled land was restored as open and environmentally active land, but the re-cultivation of land increased from 2012 to 2018, mostly in Luxembourg, the Netherlands, the UK, and Belgium (EEA 2019).

All these phenomena, together with the improved understanding of risks posed by urban heat islands, floods, and other climate-related impacts on human health, make the re-use of the formerly urbanized and soiled land a priority. Post-industrial sites constitute a stock, which is at the same time attractive and challenging. Attraction lies first with the scale. Europe, which was heavily industrialized in the 19th and 20th centuries, ceased to be so in the early 21st century. Cities that relied on manufacturing for economic development and which supported their role as the capital cities had to reinvent themselves (Sadowy 2023). Deindustrialization left in its wake several sites, from very large to small ones, quite often located within the range of the city center or at least with a convenient transportation system close by. On the other hand, post-industrial sites require significant investment for adaptive re-use practices (remediation of the contaminated soil, management of the waste and rubbles, etc.), which can be too expensive for such sites to compete with the greenfield sites. Another challenge for the companies willing to invest in post-industrial sites might be the lack of information about the exact scale of the future costs, due to the lack of also expensive investigations and specialized research regarding several sites. Therefore, market-driven processes have failed in the long run to effectively include several post-industrial sites in land-use processes. The situation is even more difficult in the case of re-cultivation of the land, which might be a desirable solution in shrinking or aging cities, at risk from the climate crisis. The creation of green, recreational areas in the former post-industrial site requires funds but is not profitable. The role of the public sector is therefore strategic. The chapter discusses the tools available for various levels of public administration in Poland, with special attention paid to the local government.

Costs, possible new uses, and incentives for the re-cultivation or re-use depend significantly on the characteristics of the sites: their location, state, scale, ownership, and neighborhood. Post-industrial sites differ in their characteristics even more than the original industrial sites of which they are remnants. Like the industrial sites before them, they are of varied scale: from small warehouses or workshops tucked in the corner of the inner yards in the tenement houses to mid-size production sites encompassing the buildings that served as places of production, offices, and storages; to the large scale entities, which dominated the urban landscape over several urban blocks, with large chimneys towering over factories and housing dedicated for the workers. Depending on the type of production, they were more or less harmful to the environment, emitting various amounts of pollution to air, water, and soil. Today, they vary in the amount of toxic or harmful compounds still present on the site. Depending on the use and period of construction, post-industrial buildings are of a varied architectural character. On all these differences, after almost 30 years of deindustrialization, there is another historical layer of differences: the level of neglect, ownership, and accountability for their current state. Some industrial sites remained vacant for longer periods, while others were adapted, partially or in total for new uses. Some abandoned buildings became the prey of thieves, who removed valuable elements like steel construction or installations. Some sites remained the property of the same entity, but in many cases, the change of ownership took place. Currently, some industrial sites have been completely rid of former buildings and transformed by the construction of new ones; other sites were adapted for new uses with the best possible restoration of the existing building stock, and others remain dilapidating, waiting for new phases in their history. Several studies recommend a rational approach following the character of the post-industrial site: to select to most appropriate direction of land re-use and to minimize remediation costs by adjusting the direction of land use to the types of pollution caused by previous industrial activity (DeSousa 2002; Linkov et al. 2006; Skalny and Białecka 2015).

Thus, the following elements are relevant for the presented analysis: character, scale, and location of the post-industrial site; ownership and accountability for the site; environmental problems resulting from former, industrial pollution; and legal regulations placing the local government within the system of the remediation, recultivation or revitalization processes. The findings show the situation in Poland, a country in which post-industrial sites constitute a complex problem. Industrialization was the main source of economic development from the late 19th century until the 1980s. Since the 1990s, Poland underwent a profound transition, encompassing institutional, political, and economic systems, from a centrally planned economy to a free market, modelled on Western Europe and the United States; from strict political control of the party dependent on the Soviet Union (USSR) to democracy. This transition was accompanied also by the cultural evolution of the perception of post-industrial heritage.

One of the crucial results of the transition, initiated in 1989, was the reconstruction of the local government and the principle of subsidiary, giving the municipalities the power to deal with all local problems, which can be governed on their level (Izdebski 2018; Izdebski and Kulesza 2004; Jaworska-Dębska et al. 2019). New responsibilities and new challenges accompanied institutional change. Post-industrial sites have been located mostly in cities, so governance of their transformation has depended to a large extent on the urban local government. The role of the municipalities, the limitations, and the potential for improvement remain relevant for the future of post-industrial areas.

10.2 Post-Industrial Sites in Poland

10.2.1 Waves of Industrialization and Economic Development

Poland, like several European countries, experienced urbanization as a phenomenon triggered by industrialization. There were three waves of industrial development in Poland. First, belated in comparison to such countries as England or France, took place in the late 19th century, when the country was still divided between three occupants: Prussia, Austria, and Russia (since the late 18th century). Occupied areas remained on the margins of the occupying empires, but some industrial spillovers occurred. In the part of the country occupied by Russia, the change in the economic policy became a strong trigger for industrialization. It encompassed the removal of the customs taxes and investments in railways, connecting Warsaw (Polish capital city, introduced in the Russian-occupied area) with large Eastern markets. It was the main source of industrial activities in Warsaw before the First World War (WWI 1914–1918). In 1918, Poland regained independence after over a century of occupation and a new government focused on stimulating the second wave of industrialization. Various methods were in place, including governmental orders for goods or invitations for Western engineers. Several factories were modernized or constructed in the years 1919–1939 when the Second World War (WWII) started.

After WWII (1939–1945), Poland, a formally independent country, remained in the Eastern Block, a group of countries under the strong influence of the USSR, together with Bulgaria, Czechoslovakia, Eastern Germany, Hungary, Romania, and Yugoslavia. A centrally planned economy was introduced under the control of the leading political party (PZPR). Societal, political, and economic goals were, to some extent, dictated by the USSR. Industrialization and the leading role of the working class (at least declaratively) constituted the core of politics in Poland and the Eastern Block. This triggered the third wave of industrialization, accompanied by urbanization and the supporting role of coal mining in the Silesia region. In Warsaw and several other large cities, the administrative borders were significantly enlarged to provide areas for new industrial investments and large-scale, prefab housing. Pre-war industrial sites, looted and/or destroyed by German and Russian armies were to some extent adapted for production purposes, but post-war production required much larger sites that surrounded the pre-war industrial districts.

Until 1989, all large industrial sites were state-owned. The transition in Poland started in 1989, with the end of the socialist regime. Its first period encompassed the years 1989–2004, starting with the reconstruction of the democratic institutions (including local government) and privatization of the manufacturing companies. The end of this period can be identified as 2004, the year when Poland became a Member State of the EU. Since then, several policies started to be created in accordance with the EU standards while local government has become one of the beneficiaries of EU funds. The exchange of good practices and international cooperation became much easier. However, the country benefitted from and struggled with the heritage of the first transition period. It was characterized by rapid privatization (Lewandowski and Szomburg 1991; Leksykon prywatyzacji 1998); Western and global companies entering the local markets and competing on the ground which was still new for local entrepreneurs. There was a new legal framework, on national, regional, and local levels. A large number of new small and medium enterprises have been created with a high dynamic of successes and failures. Former, state-owned companies were often dissolved. Ownership of several industrial (and post-industrial) sites changed several times over, making the accountability for the contamination difficult to prove. Abandoned sites could also be used as illegal waste dumps, adding new difficulties in finding entities responsible and holding them accountable.

Areas, including post-industrial sites, were also important parts of municipal and public policies. A new legal framework for revitalization was finally created in 2015, after many years of debates. However, earlier programs, similar to those created within the rule of the new Act of Revitalization, have focused mostly on socio-economic aspects, which in Poland created a legal basis for formal revitalization programs and financing. Environmental degradation and involved challenges have remained mostly under the control of agencies responsible for environmental issues. It has been part of the long-criticized "silos structure" of public institutions, responsible separately for the environment, economic development, societal problems, and heritage within specialized entities and rarely cooperating. Post-industrial sites and their revitalization created one of the bases for such cooperation which is developing, although still slowly.

10.2.2 Definitions of the Post-industrial Areas

The separation of economic and environmental responsibilities of public agencies is reflected also in legal definitions and their use. In Polish law and practice, there are three processes pertinent to the analyzed problem: remediation, recultivation, and revitalization. As pointed out, e.g., by Wałkowski (2013) and Kawęcka (2021), they are, to some extent, mistakenly used as synonyms. Remediation concerns technological and biological processes aiming at the removal of pollution existing in the ground and soil. Recultivation is usually applied to polluted agricultural and forest areas, as the process which allows the reintroduction of such type of production (food, wood, etc.) on the land (GUS, nd. Słownik pojęć). It may encompass the construction of infrastructure (e.g., roads), so it differs from the European approach discussed in the Introduction, where recultivation is part of the no net land take as the "reverse land-take process," which occurs when artificial areas are returned to non-artificial land categories through recultivation and renaturalization (Eurostat 2023b, p. 272). Revitalization encompasses all activities necessary for new development of the under-used or degraded area, and in practice, it focuses mostly on the socio-economic aspects. In the Revitalization Act there is a definition of the "degraded area," where at least one of the following problems is present (Rev. Act 2015, Art. 9):

- **economic:** in particular low level of entrepreneurship, poor condition of local businesses;
- **environmental:** in particular low quality standards of the environment, the presence of waste posing a threat to the life and health of people, or the state of the environment;
- **spatial and functional:** in particular insufficient technical and social infrastructure or its poor condition; lack of access to basic services or their low quality, failure to adapt

urban character to changing functions; infrastructure not adapted to the needs of people with special needs; scarcity or poor quality of public space;
- **technical:** in particular degradation of the technical condition of the public facilities and housing, low energy efficiency.

In Polish law, there is no precise definition of a post-industrial site. The problem is addressed most directly in the Revitalization Act (Art. 10, ust. 3 and Art. 9 ust.1), which states that such areas may be included in the revitalization programs, together with areas of former ports, mines, military bases and sites formerly used by the railway companies. This approach focuses on the socio-economic characteristics of such sites, which may be heavily contaminated or completely free of any remaining pollution, simply unused or underused. The definition of contamination of dangerous substances is precisely presented in the Act of Environmental Protection (2001) and is defined by the harm they may pose to human health and/or life (not to other forms of life). It is broadly consistent with another definition, included in the law about work safety regulations (Kawęcka 2021).

In 2004, the government published the Programme for the Post-Industrial Sites (2004), in which the authors encompassed the following definitions (pp. 3–4):

- "post-industrial area" (*teren poprzemysłowy*), the area of previous industrial economic activities, which is no longer present on the site. The area was defined as unused or underused.
- "degraded area" (*teren zdegradowany*), the area that is either polluted or "unfavorably" changed from the original form,
- "contaminated area" (*teren zanieczyszczony*), the area on which the presence of specific chemical substances are higher than those allowed in a specific act.

Polish Standard (PN-G-7800:2002) defines the recultivation of the post-industrial areas as all activities and works leading the post-mining areas and waste-covered land to a state enabling their rational use for economic purposes, industrial or other. According to Statistics Poland (GUS), devastated areas consist of land "that has completely lost its use value as a result of human activity or other factors" (GUS, Devastated areas). It is therefore mostly economic perspective. Socio-economic goals generally dominate over strictly environmental perspectives. Environmental risks are discussed mostly in the context of human health and life, and improvement of the contaminated or degraded areas is either a path to further economic development or a tool for the betterment of living conditions.

10.2.3 Stock of the Post-Industrial Areas and Sources of Information

It is very difficult to assess the stock of the post-industrial sites which need remediation. The main problem constitutes the overlap between several types of sites which are included in various statistics. There are accounts of the areas that need remediation and other, of the areas which need revitalization. Both types encompass post-industrial sites, but in both cases, together with other types of contaminated or abandoned sites. Both types of statistics, of the abandoned sites and the contaminated areas, require constant monitoring and updating, as their size and number may decrease or increase depending on the public policies, and actions of the private sector (individuals and companies).

According to the official statistics in 2004, the stock of devasted and degraded areas in Poland encompassed over 70 000 ha (BDL, nd.), that is, approximately 0.22% of the total area of the country, 31 393 136 ha. The problem decreased over time, and in 2020, the area

Table 10.1 Devasted and degraded areas in Poland in the years 2004–2014.

	2004	2005	2006	2007	2008	2009	2010
Devastated and degraded areas in Poland (ha)	70 683	64 978	65 143	64 373	63 735	62 077	61 161
of which							
devastated	62 053	59 346	59 272	58 369	57 854	55 848	54 520

	2011	2012	2013	2014	2015	2016	2017	2018	2019	2020
Devastated and degraded areas in Poland (ha)	64 011	64 343	61 958	62 774	63 374	64 651	62 038	61 863	62 089	62 482
of which										
devastated	56 868	57 040	55 443	56 130	56 531	57 854	53 743	54 701	54 964	54 721

Source: Adapted from BDL.

decreased to less than 62.5 thousand ha. Devastated land constituted the majority of the problematic areas and decreased similarly. However, the trend has not been consistent during the whole period (Table 10.1).

Bank of the Local Data (*Bank Danych Lokalnych*, BDL) is the most extensive database providing data aggregated for the country and voivodeships. Several more detailed analyses and studies attempted to assess the extent of existing post-industrial areas and the problems posed by them. After the deindustrialization in the 1990s, some studies were undertaken for the largest industrialized cities in Poland. According to the authors, in Łódź, the former center of the textile industry in Poland, there were approximately 200 ha of postindustrial land, owned by large private companies (Liszewski and Wolaniuk 1998). In Warsaw, which during the centralized economy period had several large factories (e.g., steel works, prefabricated building industry, machinery, electronics, and spirit and food industry), it was assessed that approximately 500 ha of post-industrial sites were underused or unused and it constituted over 20% of the industrialized areas from 1985 (Misztal 1996). Another study was prepared in the years 2007–2008 for further development of revitalization practices. According to the report in Poland over 115 000 ha of the problematic areas in need of revitalization were identified (Jarczewski 2009, p. 291; Jarczewski 2010, p. 31). In Polish cities in 2009, there were over 115 000 ha of degraded or devasted areas, of which post-industrial sites constituted approximately 19% (Jarczewski 2009, p. 291).

Domański showed in 2000 that in almost 60% of the 463 investigated cities in Poland, some unused post-industrial sites were located. The largest share was located in the largest cities, with over 100 000 inhabitants – in almost 80% of such cities, post-industrial sites were found. The presence of the post-industrial sites decreased proportionally with the size of the cities. In cities with less than 100 000 but at least 50 000 inhabitants, the situation was similar to the group of larger cities (over 70% of cities with unused post-industrial sites), while only 1/3 of the cities with less than 2000 inhabitants declared such a problem (Table 10.2).

During the first phase of transition, the largest post-industrial stock was located in Silesia, wherein the functional area of the city of Katowice alone was over 1000 ha of

Table 10.2 Share of the cities in which post-industrial, unused sites were found in 2000 according to the number of inhabitants.

Size of the city (number of inhabitants)	Share of the cities in which post-industrial, unused sites were found (%)
100 000 and more	79.4
50 000–99 999	73.5
20 000–49 999	65.5
10 000–19 999	69
5000–9999	44
2000–4999	49.5
Less than 2000	33.3

Source: Adapted from Domański (2000).

postindustrial land (Wrona et al. 1996; Domański 2000). In the Silesia region, there is a large share of the former mines, apart from the former manufacturing sites. In the 20th century, Silesia was a hub of coal mining and heavy industry, especially in the 1970s and 1980s. In Silesian voivodeship in 2009, over 11 000 ha of degraded areas were identified, constituting the largest concentration of such areas in the country. More than half, namely 6.600 ha were located in 14 cities of the Silesian Metropolitan Union (Górnośląski Związek Metropolitalny Silesia) (Krupanek 2009). In Katowice (the largest city of the region), degraded and post-industrial sites constituted approximately 11% of the urban area, while in other cities, the share differed from 6% to 11% (in Chorzów, Świętochłowice, and Bytom), and from 3% to 6% in other cities (Zagórska 2014). According to the report authored by experts from the World Bank, in 2022, in Poland, there were approximately 800 000 ha of post-industrial sites in need of remediation, recultivation, and/or revitalization (World Bank 2022).

In early years 2000 public administration started the process of creating the databases on voivodeship and national levels. Silesian region was at the forefront of this initiative. With the support of EU funds, they launched the project to create an open internet platform with data regarding post-industrial and degraded areas (OPI – TPP, nd.). The main goals were the following (Zagórska 2014):

- To intensify and speed up the process of the transformation of the degraded areas;
- To coordinate all actions for the sustainable development;
- To improve the management of the environment in the Silesian voivodeship;
- To ensure free-of-charge access to information for the citizens.

The main stakeholders in creating the database were the office of the Marshall and Central Mining Institute (Główny Instytut Górnictwa). The voivodeship database was used for the Voivodeship Programme for the Transformation of the Post-Industrial and Degraded Areas (WPPTP). In 2008, it was accepted by the regional assembly. The program consisted of tools enabling both initial and full evaluation of the areas and their classification. The further goal was the optimization of the remediation and re-cultivation processes and channeling of financial resources to the areas that are potentially the greatest threat to the environment and people's health (Zagórska 2014). A comprehensive database on the environment in the Silesian Voivodeship, with particular emphasis on post-industrial and degraded areas provided an integrated tool for obtaining, processing, and sharing environmental data. It was built on the basis of the Geographic Information System (GIS) in compliance with the National Terrain Information System (OPI-TPP, nd.). Today, it constitutes (Brożek 2016):

- Regional database with a complex information about post-industrial sites;
- The system allowing coordination of data gathering and decreases the overlaps in the work of various public entities;
- Integrated tool to gather and disseminate information.

The system allows the creation of a report on a specific subject and access to both current and historical data. Various groups of users have different access to the system. Citizens and investors (non-registered users) may find data and use basic analytical tools; municipalities have access to the possibility to edit data; a representative of the marshal's office coordinates activities on the platform, may add new functionalities and layers and edit data as an administrator; main administrator of the system cannot edit data, is responsible for the operation of the website and access levels of other users. The main challenges perceived in 2016 at the voivodeship level were the accuracy and updating of the data, as well as the involvement and active attitude of the local government (Brożek 2016). During the following years, the system has been upgraded to ORSIP 2.0, a new version of the Open Regional Spatial Information System of the Silesian Voivodeship, which enables access to spatial data in the form of interactive maps and related services. For this purpose, a new digital environment has been implemented to present harmonized sets of spatial data in map applications dedicated to residents, entrepreneurs, and public administration. The user can find, e.g., data regarding geodesy and cartography, nature and environmental protection, spatial development, transport, tourism, culture, and health. As stated on the website "The ORSIP 2.0," project aims to quickly browse and filter information resources, improve the management of the Silesian Voivodeship at all levels of public administration, through efficient data flow between administrative units, institutions, and business entities (ORSIP 2.0).

On the national level since 2016, there has been a platform providing data on historical pollution of the ground and imminent threats of environmental damage and damage to the environment. Through the platform, environmental protection authorities and environmental protection inspection authorities have direct access to all data contained in the register of direct threats of environmental damage and environmental damage (according to the Act of 13 April 2007 on preventing in the environment and their repair). They also have access to all data contained in the register of historical pollution of the earth's surface (under the Act of 27 April 2001 Environmental Protection Law). The authorities have to apply to be registered to have access to the data. Gathering the data is the responsibility of the local government, on the poviat level.

10.3 Role of the Voivodeships and Municipalities in the Management of Post-Industrial Areas

10.3.1 Tools and Methods of Local Management and Governance

The basis for the local government in Poland is the European Self-Government Charter (1985), the Constitution of the Republic of Poland, and the relevant Acts on local government. As stated in the Charter, "local authorities, acting within the limits of the law, are to be able to regulate and manage public affairs under their responsibility in the interests of the local population. Consequently, the Charter considers that public responsibilities should be exercised preferably by the authorities closest to the citizens, the higher level being considered only when the co-ordination or discharge of duties is impossible or less efficient at the level immediately below." In Poland, in accordance with the provisions of the Constitution, a three-tier structure of self-government was introduced. It consists of a commune/municipality (*gmina*), a poviat (*powiat*) and a voivodeship (*województwo*). The scope of responsibility of local government is broad. For the lowest level of local governance, spatial planning, management of the real estate and environmental protection are among the most crucial.

Spatial planning reflects the institutional structure of the local government. Local legal regulations, documents, and plans must be in accordance with the regulations and plans of the higher level of governance. Thus, local strategies and local plans cannot be contrary to the plans of the voivodeship, while voivodeship spatial plans must reflect the regulations and goals included in the national plan for spatial development. Voivodeship is a singular element of the administrative structure because of its two-fold nature: it is the highest level of the local government structure (represented by the voivodeship marshal) but also a local representation of the state (governor or *wojewoda* acting for the government). Taking all this into account, the actual decisions regarding the re-use, land take, and all other pertinent decisions, are taken on the basic local level, by the representatives of the municipality or commune.

Regarding the remediation and re-use of post-industrial sites, municipalities can steer the direction of spatial and economic development through various methods, the main being the spatial planning regulations. Under the regulations ruling for the majority of the post-1989 period, there were two types of documents that created the legal framework on the local level. One was the "Studium" which encompassed the whole city and consisted of the analysis of the current state and the directions for future development. It provided general direction, consistent with the document on strategic development. It was not generally binding law, but it was binding for all administrative decisions, especially such as the strongest tools for spatial management: local spatial development plans. The plans constituted the legal basis of spatial order. After the change in legal regulations Spatial Act 2003 with amendements municipalities are required to create not later than in 2026 new master plans, encompassing whole administrative area of each municipality and constituting the generally binding local law. The plans are to be intertwined with strategic documents for local development.

Municipalities have therefore two important tools to shape the future environmental character of post-industrial sites: ownership and local law, expressed in the local spatial development plans. Through the plans, municipalities can control: the function or functions to be located on the sites; the share of the green area in a total area of the site; architectural character of new buildings and/or of the adaptive re-use, necessary measures to be taken for the sake of environment. In the case of municipally owned sites, local administration can act in a twofold way. First, it is equally possible to guarantee the future character of the site using spatial planning processes; second, municipalities may invest directly in the land they own. This approach is especially important in the case of post-industrial sites which are to be recultivated and, after remediation if necessary, adapted for recreational and environmentally rich areas. Additionally, during the last few years, Polish municipalities broadened and improved their participative processes. In that manner, they can bring the decisions about post-industrial sites to a truly local level, involving local communities in the decision process. Most often it is fulfilled by the process of consultation, which may become quite complex, encompassing a series of meetings, walks, and workshops. Another tool available for the citizens is the civic budget. Share of the public yearly spending of the municipality is decided through them. Inhabitants may propose projects, which (if supported by the necessary number of signatures) are subsequently voted on by the community. Those who receive the highest number of votes are funded within the limits of the civic budget funds.

Municipalities can also stimulate the development processes, e.g. through short-term and long-term strategic and operational plans; financial instruments (taxes, rents, and other financial incentives), and infrastructural investments. In the 1990s, the overall economic approach was based on the profound trust in the self-regulating mechanisms of the market. Since then, the role of the support of the public sector for the development of specific economic activities has been better understood and more often promoted by researchers and practitioners (Karpiński 2020).

Municipalities may undertake direct interventions, solely using public funds (own and EU funds) or in partnership with the private entity. They can also execute other forms of public interventions to use post-industrial areas for public purposes, using, in this respect, e.g., pre-emption rights to buy the strategic property, or to be involved in the mixed-capital companies, concerted action zones, and land intervention zones. They may, e.g., purchase the property to block the undesirable development or to prepare it for sale, using their funds for demolition, remediation, and other methods. Regional investment may also be important regarding infrastructural investments, which may induce the private sector to invest in post-industrial sites.

10.3.2 Problems and Barriers Encountered by the Municipalities in Governance

Decentralization and democratic processes initiated in 1989 encountered from the start the barrier, which persists even today in Poland: the discrepancy between the range of the local government responsibilities and the funds, which in several cases are not sufficient to fulfill them. With the membership in the EU, local governments, both urban and rural, have gained both larger financial capacity and expertise, thanks to the numerous projects. Several of them were located in the post-industrial areas. However, financial limitations, as well as other barriers are still found on the local level. In 2005, Ostręga and Uberman (2005) identified the following bottlenecks:

- dispersion and lack of consistency between the legal acts regulating reclamation processes and development;
- limitation of the recultivating processes mostly to agricultural and forest uses;
- lack of taking into account some significant factors characterizing post-industrial areas (cultural, environmental, social, and spatial factors), too much focus on few chosen elements;
- problems in establishing actual ownership of the site;
- a too-small share of the areas included in the local spatial development plans;
- financial problems, including debts of owners (users) of land and facilities requiring revitalization;
- lack of sufficient incentives to interest potential investors in the post-industrial areas;
- insufficient practices of public-private partnership.

To some extent, legal problems have been remedied by the novelization of several acts and the new Revitalization Act (2015). Local government also gained experience and expertise in dealing with problematic areas, through direct investments, spatial planning, and public-private partnerships. However, the last possibility is still much limited, especially for the sites that are too small for such a venture. Experimental, bottom-up forms of cooperation between the municipalities and local community or entrepreneurship are yet to be developed. Local administration often fears the risks involved. In 2000, Domański presented the mutual perception between municipalities and private companies as opponents that must be overwhelmed, instead of partners with whom grander things can be achieved. In such a case, a weak opponent is a better choice, while in cooperation, a strong partner fits the bill much better (Domański 2000). In the 2020s, the situation is still present and mutual mistrust is difficult to be taken out of the equation. Several representatives of the local government also fear any suspicions of favors for the private sector or the corruption accusations, therefore finding it safer to avoid cooperation altogether or to propose conditions that the private sector finds difficult to accept.

Remediation, recultivation, or revitalization are complex processes and they need time and resources to gain the experience. However, as presented above, several municipalities, both large and small, have been actively involved in various ventures of post-industrial re-use. A growing base of good practices is also helpful for other communes. Such examples must be present in their own country, first, for the linguistic reasons and lack of barriers to learning about them in detail, and, second, because the legal, societal, and institutional context is very similar, making it easier to transfer the knowledge from one municipality to another. Still, it is necessary to remember that the public sector is usually less informed about market trends and has often little experience in market-oriented ventures. This is exactly why public-private partnership, often used in the post-industrial sites in other European countries, is so important for effective development.

Spatial planning is a very time-consuming and rather expensive process, which also limits the possibilities of the local government to effectively steer the re-use of the post-industrial sites in the desired directions. Approximately, 1/3 (namely 31.7%) of the area was encompassed in the local spatial development plans in 2022 (BDL, nd). On one hand, the obligations to have the master plans ready in 2026 may improve the situation, on the other hand, short time allowed to the municipalities for the preparations of the plans may result in some hasty and underinvestigated solutions regarding the most complex areas.

10.3.3 Environment, Health, and Economy: Goals of the Re-Use of Post-Industrial Sites

The varied characteristics of post-industrial sites offer different opportunities for their re-use. Large sites present better economic opportunities for office, commerce, and residential uses, especially if they are not contaminated and are located within the city center. Large sites on the outskirts of the city, especially well connected to the railways or highways, are attractive for the logistic hubs. Contaminated sites located in the less expensive neighborhoods provide an opportunity for new green areas. These new uses introduce either elements of higher quality of life (greenery, recreation, services) or are areas of new forms of pollution (light and noise in the logistic hubs). Main adaptation practices of post-industrial sites in Poland encompass: new forms of production, office districts, commerce, services, and residential and recreational areas (green areas and sports facilities). In Silesia, the largest number of new jobs on the postindustrial sites was created in commerce, industrial parks, and a smaller share in the cultural institutions (Jarczewski and Huculak 2010).

Involvement of the public sector is necessary in the investments focused on the improvement of the environment, which cannot be profitable in financial terms. According to Domański's study (2000), based on the analysis of 101 Polish cities, the first period of transition encompassed the type of re-use of the post-industrial sites, which might be called the low-hanging fruit. The most direct re-use of the site consists of a new form of manufacturing, and 59.3% of the re-use carried out solely by the private sector consisted of such adaptations (share of the re-used area). Large-scale commercial investments were also popular (shopping malls, supermarkets) and 32% of areas of private re-use investments had such a character. 8.7% of the re-used sites were adapted for services. Proportions were very different in the case of the municipalities' financial and/or ownership involvement. They also encompassed manufacturing, commerce, and services (respectively 29.4%, 24.4%, and 34% of the re-used areas), and also other functions, such as recreational and residential (12.2%), at the time absent from the private ventures.

In time, residential investments became more popular for the private sector, especially in the attractive location within the city, improved by the municipal investments. Praga district in Warsaw is a good example of such practices. Construction of the new metro line and launching new cultural institutions and start-up hubs created attractive neighborhoods for residential and commercial investments (Sadowy 2023, pp. 129–141). The private sector profited from the opportunity, while municipally owned sites remained underused or abandoned (Sadowy and Lisiecki 2019). Table 10.3 presents examples of various post-industrial sites and their situation.

The public sector is often connected to the public utility forms of re-use: cultural institutions and recreational areas. Cultural institutions are created on the sites, where buildings

Table 10.3 Examples of the private and public investments in the post-industrial sites.

Former (industrial) use	Current use	Location	Private or public investment
Koneser, spirits factory	Multifunctional up-ended commercial center "Koneser" (hotel, conference center, restaurants stores)	Warsaw, Praga district (close to the city center)	Private
Heat and power plant "Powiśle"	Multifunctional up-ended commercial center "Powiśle" (hotel, conference center, restaurants stores)	Warsaw, city center	Private
Chemical and beauty products "Pollena-Uroda"	Residential, "Bohema"	Warsaw, Praga district (close to the city center)	Private
Służewiec Przemysłowy: industrial district. Various, including machines, electronic devices, prefab elements for housing.	Office district (with housing in the outskirts of the post-industrial site)	Warsaw, Służewiec Przemysłowy (mid-way between the city center and the suburbia)	Private
Brewery	Mixed-use (including housing) "Browary Warszawskie	Warsaw, Wola district (close to the city center)	Private
Textile factory	Multifunctional commercial center "Manufaktura", Museum of modern art	Łódź (close to the city center)	Private (commerce, hotel, services) and public (museum)
Textile Scheibler`s factory	Residential, "Scheibler`s Lofts"	Łódź, not far from the city center	
Textile factory (owned by Girard)	Residential, "de Girard`s Lofts"	Żyrardów, town center (satellite town for Warsaw with good railway connection)	
Chemical products, Solvay (Sodium plant and quarry)	Shopping mall, religious center, green area.	Cracow, outskirts	Private (commerce), public
Heat and power plant	Baltic Philharmonic	Gdańsk, close to the city center	Public
Electric plant	Mazovian Center for Contemporary Art	Radom, city center	
Waste dump from the prefabricated construction process	Park	Czechowice-Dziedzice, close to the residential area	Public
Mine	Park Amelung	Chorzów	Public
Mine	Mixed-use, recreational, cafes, cultural institutions	Chorzów	Public (remediation and most of the investments), private (services in the renovated buildings)

Source: Adapted from various sources.

have high architectural and heritage values and this function offers additional opportunity for visitors to recognize and enjoy them. Green recreational areas constitute a chance to both improve the quality of life in often less attractive and hitherto neglected areas and to remediate the areas that were heavily contaminated by previous functions.

At first, private constituted the strongest group of actors in the case of post-industrial sites. It was welcomed by the public sector, as municipalities focused mostly on the need to improve the socio-economic situation and had very limited funds to invest in the problematic, post-industrial areas. In time, goals and focus became more complex. As the climate crisis increased, local communities' pressure for the improvement of the local, especially for the green recreational areas in the vicinity of, increased parallelly. However, there are examples of public remediation and recultivation of post-industrial areas, which have now been existing for some time. In Czechowice-Dziedzice (Silesia region, approximately 35 000 inhabitants), a large park was created in 2010 in the degraded area. It was used as a facility for the prefab construction site in the 1980s and remained practically a form of waste dump ever since (the technology of the prefab construction was often wasteful in the 1980s with a large amount of debris left). Municipality used their own and European funds for the park of 3.6 ha. In Chorzów (also Silesia region, approximately 100 000 inhabitants), the park was created in the area where mining left behind water reservoirs. They were originally the result of the mining activity in the 1920s, later serving as the source of water for the local mines (KWK Barbara). The park was opened in 2007. In 2009, municipality renovated another post-industrial location with the former headquarters of the mine management, the old assembly hall, the casino, the foremen's house, the mine fire station, and the liquid air warehouse. Today there are a variety of publicly and privately managed uses, mostly cultural institutions, restaurants, and cafes. Park (called Eko-Park) was also created in the formerly industrial district Ursus in Warsaw (with the large tractor factory, closed during the transition period). It covers 5.5 ha and local administration presents it as a "combination of modern architecture, natural green areas, and innovative ecological solutions" (Kamińska 2017). Large part of the area has been recultivated in a way as close to natural as possible, aiming at the improvement of local biodiversity. It encompasses an element very popular in green areas in Poland in the last decade, flower meadows, which do not require constant mowing. Several elements of paths, benches, etc. are produced with the use of reclaimed materials.

One of the most significant and largest re-used post-industrial sites in the area of the former Solvay Sodium Plant in Cracow. The area served in the 19th century as the quarry and in the early 20th century an industrial plant was opened. In the 1930s the settling tanks were constructed. After WWII the site became state-owned and further developed. The waste from the production and the sodium sediments created a characteristic landscape, later monikered as "the White Seas." The production process was assessed as highly harmful to the environment and in 1989 Prime Minister made a decision to close the whole entity, encompassing the main production site (30 ha), quarry (60 ha), settling tanks for alkaline sodic waste (White Seas, 80 ha) and 60 ha of other functions. In the following years, some of the buildings were dismantled, and settling tanks preliminary recultivated.

The first investment was typical for the 1990s and consisted of the large commercial complex "Zakopianka" (named after the nearby highway to the City of Zakopane). In 2011 another large complex was constructed, a religious center commemorating the pope Jan Paweł II and providing space for several events organized by the Catholic church. In 2012 local spatial development plan was introduced. According to its provisions, the White Seas were to constitute a public space, an open green area with elements of a park, as well as sports and recreation infrastructure. In its vicinity, the so-called Small White Seas are located. It is a plot of land with wild green vegetation, of approximately 18 ha. The planning documents specify that this area is to be designated for isolating and arranging greenery.

The latest phase of the Solvay re-use, introducing the provisions of the local plan, differed from the previous two. The commercial center and religious center consisted mostly of new buildings and sealed areas (parking, square). The third phase consists mostly of the remediation and recultivation of the space which became semi-natural and partly recreational areas for inhabitants of Cracow, a city with one of the highest levels of air pollution in the country. On the site of the White Seas, there were five tons of limestone slimes, lime granulate, fragments of flint, slags, and leach. The slopes were reshaped and reinforced to stabilize them, covered with soil, and the grass was sawn. Natural succession of plants followed. In 2022, after 30 years of the last period of production on the site, the study showed that the level of potentially toxic substances is low enough and does not pose a threat to fauna, flora, and humans (Cała and Ostręga 2021). Some waste in the soil restrained the growth of the plants but to various extents, e.g., loam seems to provide stable conditions for the germination and growth of root parts of the plants (Świdwa-Urbańska and Kalisztan 2019). The remediated and recultivated area covers nearly 35 ha and is located in the vicinity of the Wilga River valley.

Re-use of this area reflects a broader trend observed in the adaptive processes of the post-industrial sites. At first, the goal was purely economic and commercial, with large investments of companies operating in the fast-developing market of shopping malls. In Cracow, as in other post-industrial areas and cities, an important element was also the creation of new jobs (Jarczewski and Huculak 2010). In the case of the former Solvay site, the new commercial complex employed approximately 2000 people (Okrutniak 2010). The second investment was carried out by another strong actor on the scene of spatial development in Poland, Catholic church. One venture was market-driven, while the other had a strong top-down character of the stakeholder very difficult to negotiate with or to contest. Current endeavors are much more society-oriented and society-driven, with strong environmental and ecological factors. It concerns also a very difficult part of the area in question, left for the local government to deal with. Since 2015 there have been plans to create the golf club in the area. A company won the tender and was obliged by the contract to build a golf course and a sports and recreation complex for the local community. The company paid the rent, but due to the failure to comply with the provisions of the contract (the works schedule), the contract was terminated in 2020. The company did not withdraw from the land after the court and bailiff proceedings and the takeover of this area by the municipality took place at the end of 2021. In the mid-1990s, when first plans were made for the area due to the lack of understanding to what extent the salinity of the land may determine certain possibilities of development, it was not taken into account (Gliniak and Sobczyk 2014).

This area can be thus the example of a specific delay rent. Abandoned and ignored for a prolonged period, it experienced a natural succession of local species. Pioneer and

Table 10.4 Preferences of the respondents regarding the White Seas Park in Cracow.

Should the following functions/uses be present in the planned park?	Number of confirmation answers
Place of relax	909
Place of high environmental values	852
Place of peace and quiet	686
Place for sport	486
Place for children	303
Place for cultural events	203
Other	123

Source: Adapted from Kraków konsultacje (2021).

ruderal vegetation created a specific landscape, which can be perceived as valuable in itself, protected and further developed due to the careful planning and introduction of other natural elements (Jagiełło and Wowrzeczka 2017; Okrutniak 2010). Due to the lack of private investments, more sustainable forms of re-cultivation can take place, instead of the golf course, hardly a democratic and environment-friendly solution. In White Seas, despite the scarcity of nutrients (nitrogen and potassium), vegetation appeared quite soon. The pioneers were algae creating the coating, enabling mushroom succession, followed by moss and vascular plants. In a few years, self-sown seedlings of the surrounding tree species have entered: birch, aspen, willow, and poplar. The current biodiversity of flora and fauna is astonishing, counting dozens of species (Okrutniak 2010). The municipality also decided to involve the local community on the scale, which was inconceivable in the 1990s. A series of information meetings, workshops with experts, telephone duty hours, and research walks with residents took place to establish the most valued aspects of the area and the goals of its future development. Citizens indicated that the White Seas should be a place of peace and quiet, relax in greenery, and a space of great natural value. Over 1200 respondents answered the online survey, most of them between 26 and 46 years old. Approximately, 250 answered based on their knowledge of the site, while others were acquainted with the material prepared by the municipality. The need to relax rather than to find new attractions and entertainment was prominent, the majority of the respondents sought relaxation and appreciated the value of the natural environment (Table 10.4).

450 respondents declared they would visit the park a few times a week, mostly walking and observing nature. These two options were much more popular than sports or educational activities. Park of the White Seas also gained support through the civic budget (Cracow Civic Budget 2022). The climate crisis and the experience of the COVID-19 pandemic probably influenced the perspective of the respondents. The future of such a trend in post-industrial areas seems to be important. However, it does not exclude other approaches and needs for re-introduction of the unused areas in cities.

10.3.4 New Centralized Policies

The question of the postindustrial sites has not been addressed in Poland in a systematic, global, and long-term oriented manner, being rather a series of experiments on the local level. As discussed above, valuable lessons were learned through the process, and several successful investments have been fulfilled, both by the private and public sectors, with limited but important examples of inter-sectoral cooperation. A transition period was characterized by the strong decentralization and growing role of the local government. On the other hand, on the national level, several centralized actions have been undertaken. Some of them are completed through the specialist parliament acts (*lex specialis*, called *spec-ustawa* meaning "special act"). Special acts override the existing legal and institutional structure, allowing, e.g., large infrastructural investments of roads and highways, which do not have to be carried out according to the complex systems of multi-level spatial planning. In 2023, a special act was proposed regarding some large-scale post-industrial sites.

Several arguments for such actions were presented. One of them consists of the difficulties the local government encounters in dealing with the historical contamination of the site. The control of the highest control organ in the country (NIK) showed that apart from the lack of funds, municipalities also experience problems with the identification of the legal owner, i.e., the person legally obliged to remove the harmful substance and/or waste from the property. It also results from the financial shortcomings of communes, preventing them from the broader investigation regarding accountability and from removing this waste by the so-called substitute execution. Along with assigning specific tasks to mayors of cities, they were not provided with an adequate share of public income (NIK, 2019).

On the other side, the Ministry pursuing the centralized approach emphasizes that the issue of large-scale brownfields is further complicated by issues outside the scope of environmental legislation, such as the presence of degraded infrastructure and buildings, bankruptcy of companies, and the question of determining their legal successors, problems with determining land ownership, trade-in polluted real estate, issues of admissibility of state aid. Therefore, regulations on waste management or protection of the earth's surface turn out to be ineffective in terms of solving the problem of individual areas of this type. National Fund for Environmental Protection and Water Management (NFOŚIGW) has implemented financial support programs (Protection of the Earth's Surface and Removal of Abandoned Waste) but did not guarantee full financing of the costs of removing abandoned hazardous waste, and from 2020 until mid-2023, the Fund did not announce calls for applications for co-financing such removal of abandoned waste, waiting for the recommendations of the Minister of Climate and Environment regarding the conditions for further financing of projects related to abandoned waste (NIK, 2019).

The purpose of the planned regulation was (Ministry of Climate and Environment 2022):

- To recognize, prepare, and assess the scale of problems related to reducing the impact of large-scale degraded areas on the environment and human health;
- To remove or decrease the negative impact on the environment and human health of large-scale degraded areas, including the elimination of waste dumping as a threat to the environment and human health; to clean the contaminated areas for subsequent development;
- To restore large-area degraded areas for re-use or preparation for new functions, if possible.

As the government informed, the main goal of the new act which is to be presented to the parliament will encompass large-scale post-industrial sites. The act has a character of *lex specialis* because it names specifically five large post-industrial sites which are to be immediately encompassed by the remediation, re-cultivation, and/or revitalization processes.

The obligation to gather and publish detailed information on them will remain on the national level, in the agency responsible for the environmental issues, including the identification of the areas in need of remediation which had not yet been identified. The administration accountable for the improvement of the state of the sites and environment will be on the local level. Regulations encompass the process of the renumeration of the undertaken remediation processes. It is yet unclear if the sums paid by the entity responsible for the contamination will be paid to the public sector, will be transferred to the national environmental fund, or the municipalities responsible for the process of recultivation. From an even broader perspective, representatives of the World Bank assess that the re-use of the post-industrial sites will create a crucial element of the green transition of the Polish economy. They declare the will to cooperate both on the national and local levels.

10.4 Conclusions

Management of the post-industrial areas is a complex problem, encompassing economic, social, environmental, cultural, and health aspects. In Poland, it is additionally complicated by the overlap in time between the institutional transition (from a centrally planned, politically driven economy to a democratic and market-driven economy) and deindustrialization processes. Post-industrial sites constitute a threat to the environment and human health, while remediation, recultivation, and/or revitalization can be a very expensive process. Meanwhile, several sites offer very attractive economic opportunities for re-development due to their location, size, and, in some cases, historical or heritage value, which might be used in the marketing of the place (Sklokina et al. 2021).

Various actors and stakeholders have been active in Poland in the re-use of the post-industrial sites since 1989, from the private and public sectors. The role of the citizens is also increasing, placing the inhabitants and local communities at the heart of the decision-making processes. Local government plays an important role in this complex cooperation. Municipalities hold several tools to direct the re-use of the sites toward desired goals, either economic, environmental, or society-oriented. Existing barriers (financial and institutional) and criticism of renunciation or lack of expertise of the local government are present in the public debate and some research studies. However, after 30 years of gaining experience and almost 20 years of the membership in EU, several cases show the ability to build capacity on the most local level, as well as the ability to follow current and future needs, especially regarding the societal challenges and climate crisis. Municipalities can be limited by political calendars and act in the short-term perspective of elections, but they may also be more involved in long-term planning than profit-oriented supra-local companies. They may abstain from cooperation with the private sector, for fear of corruption suspicions, or lack of the necessary knowledge, but public-private partnerships remain one of the most efficient tools for responsible re-use of the post-industrial areas. Local government is also a crucial stakeholder in non-profit investments with strong positive externalities, such as the recultivation of the post-industrial areas and the creation of environmentally and socially valuable sites of biodiversity and recreation.

Acknowledgement

The study was part of the project funded by Dean of Collegium of Management and Finance of SGH Warsaw School of Economics (KZiF/S23:1.18).

References

Act of Environmental Protection (2001). Ustawa z dnia 27 kwietnia 2001 r. Prawo ochrony środowiska, Dz. U. 2001 Nr 62 poz. 627.

BDL (n.d.). Bank Danych Lokalnych, Bank of the local data for Poland. https://bdl.stat.gov.pl/bdlarch/start and https://bankdanych.io/? (accessed 23 May 2023).

Brożek, P. (2016). *Presentation of the OPI-TPP, for Urząd Marszałkowski Województwa Śląskiego*. Katowice: Wydział Ochrony Środowiska.

Cała, A. and Ostręga, M. (2021). *Sposób zagospodarowania terenu znajdującego się w sąsiedztwie doliny rzeki Wilgi i w rejonie ulic Podmokłeji Herberta pn. „Białe Morza" -konsultacje społeczne*. Material for the public consultation process.

Cracow civic budget (2022). https://zzm.krakow.pl/aktualnosci/1260-budzet-obywatelski-2022.html (accessed 25 May 2023).

DeSousa, C.A. (2002). Brownfield redevelopment in Toronto: an examination of past trends and future prospects. *Land Use Policy* 19: 297–309.

Domański B., 2000, Restrukturyzacja terenów poprzemysłowych w miastach (Transformation of derelict industrial land in towns and cities) (w:) Z. Ziobrowski, D. Ptaszycka-Jackowska, A. Rębowska, A. Geissler (red.) Rewitalizacja, rehabilitacja i restrukturyzacja - odnowa miast. Kraków: Instytut Gospodarki Przestrzennej i Komunalnej Oddział w Krakowie, s. 107–142.

European Commission (2023). The impact of demographic change – in a changing environment. Final Commission Staff Working Document, Brussels, 17.1.2023 SWD.

European Environment Agency (2019). Land take assessment. https://www.eea.europa.eu/data-and-maps/indicators/land-take-3/assessment (accessed 25 May 2023).

European Environment Agency (2023). Land use. https://www.eea.europa.eu/en/topics/in-depth/land-use (accessed 21 May 2023).

The Council of Europe (1985). European Self-Government Charter. https://www.coe.int/en/web/conventions/full-list?module=treaty-detail&treatynum=122 (accessed 25 May 2023).

Eurostat (2023a). Demography 2023 edition. https://ec.europa.eu/eurostat/web/interactive-publications/demography-2023) (accessed 25 May 2023).

Eurostat (2023b). Sustainable development in the European Union. Monitoring report on progress towards the SDGs in an EU context. 2023 edition.

Gliniak, M. and Sobczyk, W. (2014). Koncepcja zagospodarowania terenu poprzemysłowego "Solvay". *Edukacja-Technika-Informatyka* 1 (5): 354–359.

GUS (n.d.). Słownik pojęć. https://stat.gov.pl/metainformacje/slownik-pojec/pojecia-stosowane-w-statystyce-publicznej/423,pojecie.html (accessed 21 May 2023).

Izdebski, H. (2018). Prawo do samorządu-prawo obywateli i prawo mieszkańców. *Samorząd Terytorialny* 10: 5–16.

Izdebski, H., & Kulesza, M. (2004). Administracja publiczna. *Zagadnienia ogólne,* Warsaw: Wydział Prawa i Administracji Uniwersytetu Warszawskiego,.

Jagiełło, M. and Wowrzeczka, B. (2017). Dawny kamieniołom Libana w Krakowie w kontekście planowanej rewaloryzacji. *Teka Komisji Architektury, Urbanistyki i Studiów Krajobrazowych* 13 (4): 30–38.

Jarczewski, W. (2009). *Przestrzenne aspekty rewitalizacji śródmieścia, blokowiska, tereny poprzemysłowe, pokolejowe i powojskowe*. (in:) Jarczewski, W., ed. *Rewitalizacja Miast Polskich*. T. 4. Cracow: Instytut Rozwoju Miast.

Jarczewski, W. (2010). *Degradacja techniczna i społeczna*. (in:) Ziobrowski, Z., ed. *Rewitalizacja miast polskich*. T. 9.Cracow: Instytut Rozwoju Miast.

Jarczewski, W. and Huculak, M. (2010). Wpływ rewitalizacji terenów poprzemysłowych na rynek pracy w woj. śląskim. *Problemy Rozwoju Miast* 4: 34–42.

Jaworska-Dębska, B., Olejniczak-Szałowska, E., and Budzisz, R. (2019). *Decentralizacja i centralizacja administracji publicznej: współczesny wymiar w teorii i praktyce*. Wolters Kluwer Polska.

Kamińska, J. (2017). Parki na terenach poprzemysłowych. *Wspólnota. Pismo Samorządu Terytorialnego*, https://wspolnota.org.pl/news/parki-na-terenach-poprzemyslowych (accessed 25 May 2023).

Karpiński, A. (2020). Atrakcyjność dziedzin przemysłu w procesie rozwoju kraju i Mazowsza dawniej i dziś. *MAZOWSZE Studia Regionalne* 32: 11–34.

Kawęcka, A. (2021). Prawno-środowiskowa problematyka inwestycji typu brownfield: nieruchomości poprzemysłowe, na których wystąpiło zanieczyszczenie powierzchni ziemi. *Folia Iuridica Universitatis Wratislaviensis* 10 (2): 106–126.

Kraków konsultacje (2021). Official information form the Cracow municipality about the public consultation for White Seas site. https://www.krakow.pl/aktualnosci/262230,26, komunikat,prezydent_zdecydowal___beda_konsultacje_w_sprawie_bialych_morz. html?_ga=2.95167444.970548485.1658941166-1246597756.1610542306 (accessed 25 May 2023).

Krupanek, J. (2009). *Programowanie przekształceń terenów poprzemysłowych i zdegradowanych w skali regionalnej. Przykład województwa śląskiego*. In: *Innowacyjne rozwiązania rewitalizacji terenów zdegradowanych* (ed. J. Skowronek). Katowice: Centrum Badań i Dozoru Górnictwa Podziemnego Sp. z o.o. w Lędzinach, Instytut Ekologii Terenów Uprzemysłowionych w Katowicach.

Leksykon prywatyzacji (1998). Warsaw: PWE.

Lewandowski, J. and Szomburg, J. (1991). Własność jako próg reformy gospodarczej. *Ekonomiczna alternatywa. Przegląd Polityczny* 14: 25.

Linkov, I., Satterstrom, F.K., Kiker, G. et al. (2006). From comparative risk assessment to multi-criteria decision analysis and adaptive management: recent developments and applications. *Environmental International* 32: 1072–1093.

Liszewski, S. and Wolaniuk, A. (1998). Centra miast polskich w okresie transformacji (przykład Łodzi). *Biuletyn KPZK PAN* 182: 129–151.

Ministry of Climate and Environment (2022). Projekt ustawy o wielkoobszarowych terenach zdegradowanych. https://www.gov.pl/web/premier/projekt-ustawy-o-wielkoobszarowych-terenach-zdegradowanych3 (accessed 25 February 2024).

Misztal, S. (1996). Development of the industrial areas in Warsaw and trends in their restructuring. In: *Restructuring of Derelict Industrial Areas*. Berlin: European Academy of the Urban Environment.

NIK (2019). Informacja o wynikach kontroli. Działania organów administracji publicznej w zakresie usuwania historycznych zanieczyszczeń powierzchni ziemi.

Okrutniak, M. (2010). Rekultywacja Krakowskich Zakładów sodowych Solvay–sukces czy porażka? *Wszechświat* 111 (10–12): 271–276.

OPI-TPP (n.d.). Database: national terrain information system. https://transformacja.slaskie.pl/content/platforma-informacji--tereny-poprzemyslowe-i-zdegradowane-opi--tpp (accessed 25 May 2023).

ORSIP 2.0 (n.d.). Geoportal system. https://geoportal.orsip.pl/zalozenia-i-cele-orsip-20 (accessed 25 May 2023).

Ostręga, A. and Uberman, R. (2005). Formalnoprawne problemy rewitalizacji terenów poprzemysłowych, w tym pogórniczych. *Górnictwo i Geoinżynieria* 29 (4): 115–127.

Programme for the Post-Industrial Sites (2004). Ministersstwo Środowiska. Program Rządowy dla terenów poprzemysłowych. Przyjęty przez Radę Ministrów 27 kwietnia 2004.

Rev. Act (2015). Ustawa z dnia 9 października 2015 r. o rewitalizacji, Dz.U. 2015 poz. 1777.

Sadowy, K. (2023). *Evolutionary Urban Development. Lessons form Central and Eastern Europe*. London: Routledge.

Sadowy, K. and Lisiecki, A. (2019). Post-industrial, post-socialist or new productive city? Case study of the spatial and functional change of the chosen Warsaw industrial sites after 1989. *City, Territory and Architecture* 6: 1–14.

Skalny, A. and Białecka, B. (2015). Kierunki przekształceń terenów poprzemysłowych-analiza przypadków. *Zeszyty Naukowe. Organizacja i Zarządzanie/Politechnika Śląska* 82: 227–239.

Sklokina, I., Brodowicz, D.P., and Sadowy, K. (2021). *(Post)przemysłowe dzielnice Warszawy i Lwowa: znaczenie pamięci i dziedzictwa dla biznesu i przemysłu kreatywnego*. In: *Kultury historyczne Polski i Ukrainy* (ed. T. Stryjek and V. Sklokin). Warszawa: Wydawnictwo Naukowe SCHOLAR.

Spatial Act (2003). Ustawa z dnia 27 marca 2003 r. o planowaniu i zagospodarowaniu przestrzennym, Dz.U. 2003 nr 80 poz. 717.

Świdwa-Urbańska, J. and Kalisztan, P. (2019). Ecotoxicological risk of alkaline sodic waste on plant development. *Geomatics and Environmental Engineering* 13 (4): 109–118.

Wałkowski, D. (2013). Od rekultywacji do remediacji. https://codozasady.pl/p/od-rekultywacji-do-remediacji (accessed 25 May 2023).

World Bank (2022). Press information Sprawiedliwa transformacja w stronę czystej energii szansą na przyciągnięcie inwestycji i stworzenie nowych miejsc pracy w Polsce about the report Wsparcie polskich regionów węglowych w transformacji. https://www.worldbank.org/pl/news/press-release/2022/12/15/just-transition-away-from-coal-to-attract-investment-create-new-jobs-in-poland (accessed 30 May 2023).

Wrona, A., Gołubowicz, J., and Broda, J. (1996). *Główne problemy zagospodarowania terenów zniszczonych dla potrzeb budownictwa w województwie katowickim*. In: *Gospodarka terenami zniszczonymi działalnością człowieka*. Zabrze: Materiały Konferencji, Instytut Podstaw Inżynierii Środowiska PAN.

Zagórska, E. (2014). Działania podejmowane w województwie śląskim w dziedzinie rewitalizacji terenów poprzemysłowych. *Studia Ekonomiczne* 177: 67–76.

11

Restoration of Ecosystem Services of Endangered Wetlands in Post Oil and Gas Exploration Era in the Niger Delta, Nigeria

Aroloye O. Numbere, Victoria C. Obinna, Eberechukwu M. Maduike, and Austin E. Abah

Department of Animal and Environmental Biology, University of Port Harcourt, Nigeria

11.1 Introduction

Ecosystem services are those direct and indirect benefits derived from the ecosystem. The benefits derived from the environment cannot be fully quantified in monetary terms (Fu et al. 2011). However, researchers have tried to quantify and evaluate the cost of ecosystem services in financial terms (Joshi and Negi 2011). The direct benefits of the ecosystem include air purification and climate stabilization, food production, water resources, soil stabilization/formation and crop pollination, and water and soil quality protection. On the other hand, the indirect value of the ecosystem includes the production of medicine, pharmaceuticals, etc.

A wetland is defined as "areas of marsh, fen peatland or water whether natural or artificial, permanent or temporary with water that is static or flowing, fresh, brackish or salt including areas of marine water the depth of which at low tide does not exceed six metres" under the auspices of UNESCO (see Table 11.1). Wetlands are found at the intersection of the land and sea, most especially where the mangrove forest is situated along the coastal areas (Chen et al. 2023). Wetlands serve as a buffer zone for materials from the sea or land. Wetlands have an ample supply of biodiversity because of their fertile and nutrient-rich nature that facilitates the growth and proliferation of micro- to macro-organisms (Mętrak et al. 2023). Wetlands have also been called the "kidneys of the earth" (Mitsch and Gosselink 1986) because of their ability to purify the waters that pass through them.

Wetland is also described as permanently or seasonally flooded with freshwater comprising a mosaic of vegetation and open, shallow water areas. Wetlands occupy the transitional zone between permanently wet and deep-water bodies and generally dry regions. They show large diversity but are made up of the presence of water, soils that differ in composition from those that make up the surrounding catchment, and a plant community dominated by hydrophytes (water-loving plants). Wetlands occupy approximately 6% of the earth's surface and are essential systems in the tropics (Table 11.2). Three major categories of tropical freshwater wetlands are recognized as follows: (i) floating vegetation communities (sudd), e.g., Lake Naivasha Upper Nile, Sudan, (ii) permanent and seasonal herbaceous

Biodiversity and Ecosystem Services on Post-Industrial Land, First Edition.
Edited by Vimal Chandra Pandey.

Table 11.1 Classification of freshwater wetlands according to the RAMSAR convention.

Serial number	Category	Types	Sub-category
1	Riverine	Perennial	Permanent rivers and streams, including waterfalls, inland deltas
		Temporary	Seasonal and irregular rivers and streams. Riverine floodplains, including river flats, flooded river basins, seasonally flooded grassland
2	Lacustrine	Permanent	Permanent freshwater lakes (>9 ha), including shores subject to seasonal or irregular inundation. Permanent freshwater ponds
		Seasonal	Seasonal freshwater lakes (>9 ha), including floodplain lakes
3	Palustrine	Emergent	Permanent freshwater marshes and swamps in inorganic soils, with emergent vegetation whose bases lie below the water table for at least most of the growing season.
			Permanent peat-forming freshwater swamps, including tropical upland valley swamps dominated by plants such as *Papyrus* or *Typha*
			Peatlands include acidophilous, embrogenous (vegetation lying above groundwater level, relying on rainwater), or soligenous (receiving water from slope run-off) mines covered by moss, herbs, or dwarf shrub vegetation and fens of all types. Seasonal freshwater marshes on inorganic soil, including sloughs, potholes, seasonally flooded meadows, sedge marshes, and dambos. Alpine and polar wetlands, including seasonally flooded meadows moistened by temporary waters from snow mat vegetation. Volcanic fumes roles continually moistened by condensing water vapor.
4	Forested		Shrub swamps, including shrub-dominated freshwater marsh, shrub-carr and thickets on inorganic soils. Freshwater forest, wooded swamps on inorganic soils. Forested peatlands, including peat swamp forest

Source: Adapted from Bullock (1993).

Table 11.2 The extent of wetlands in the tropical parts of the continents.

Continent	Wetland area (km^2)
South America	1 200 000
Asia	350 000
Africa	340 000
Australia	2000

Source: Bullock (1993)/with permission of Cambridge University Press.

swamps, are dominated by rooted plants such as *Typha domingensis* or *Phragmites kurka* (e.g., South American floodplains), (iii) swamp forests, have scattered trees. Many trees have pneumatophores, adventitious (aerial), or stilt roots. The swamp forest is the dominant type of ecosystem in the Niger Delta area of Niger.

Because of the sensitive nature of the wetlands, they are vulnerable to contamination; that is why in some quarters, they are called waste lands because of the tendency for their conversion to the less productive system because of their ability to act as environmental filters to waste products originating from land or sea. The soft terrain of wetlands makes them prone to contamination. Wetlands are the breeding ground for many aquatic, terrestrial, and arboreal organisms (Saha et al. 2023). Fishes breed and spawn in wetlands, and birds feed and breed on the area's rich supply of insects and other arthropods. Birds that embark on Trans-Atlantic flights stop to rest and feed before continuing their journey. However, anthropogenic activities have devasted most of the rich and productive wetland areas of the Niger Delta. Oil and gas exploration activities are accompanied by many other associated actions that have decimated the wetlands, such as deforestation, oil spill, land fragmentation, soil contamination, pollution, etc.

11.2 Characterization of Wetlands in the Niger Delta

The Niger Delta is a sedimentary basin that originated in the Eocene period 40–50 million years ago (Chukwuma-Orji et al. 2021). It extends from the southern part of Nigeria toward the equator at the outer edge of the continental shelf in the Gulf of Guinea. It is bound to the west by the Ethiope/Benin River, to the northeast by the Sombreiro River, and to the East by the Imo River. These natural boundaries are also surrounded by five ecological zones, comprising the following in a seaward direction: (i) moist lowland forests, (ii) freshwater swamp forests, (iii) mangrove forests, (iv) mangrove swamps, and (v) saltmarsh and tidal flats along the estuaries and shorelines. The total land area covered by the Niger Delta is approximately 29 000 km^2 or about 3.2% of the total land area of Nigeria. There is a network of other rivers within the delta, which are partly or wholly independent of the Niger and drain part of the flood discharges but are, nevertheless, considered part of the delta system from which the area finally enters the Atlantic through 21 estuaries (Table 11.3) and seven rivers.

The Niger Delta region experiences semi-diurnal tides with two high wagers (HW) and two low waters (LW) in a 25-hour cycle. Approximately, 10 000 km^2 of the Niger Delta is subject to tidal force (Table 11.4).

11.2.1 Soil and Land Characteristics

There are broad relief zones in the Niger Delta as follows:

a) **The beach/barrier ridge islands:** are along the shorelines of the Gulf of Guinea. The barrier ridge is a continuous beach line along the coast with white beach sands, barrier islands, and sand ridges some 4 m above the sea.
b) **The tidal flood plains:** lie behind the beach sand ridges fronting the Gulf and extend for some 38–43 km along the Benin and Sombreiro estuaries, respectively.

Table 11.3 Major estuaries and rivers in the Niger Delta, Nigeria.

Estuaries		Rivers
Benin	Andoni	Ossioma
Escravos	Imo	Ethiope
Forcados	Nun	Jamieson
Ramos	Brass	Ase
Dodo	St. Nicholas	Orashi
Pennington	St. Barbara	Sombreiro
Digatoro	St. Bartholomew	New Calabar
Bengatoro	Sombreiro	Imo
Kulama	New Calabar	
Fish town	Bonny	

Table 11.4 Impact of oil and gas exploration on organisms and environment in the Niger Delta, Nigeria.

Serial number	Exploratory action	Activities	Aspect of environment	Type of impact	Organisms affected
1	Exploration		Marine, wetland, terrestrial	Destruction of vegetation, farmland, and human settlement	Mangroves, fish, wildlife
		Seismic/ geophysical investigation	Wetland, terrestrial	Noise, seismic shooting (dynamite)	Fishes, animal
		Drilling enhanced recovery well	Marine, wetland, terrestrial	Toxic waste accumulation, oil pollution	Mangrove, fish, animal
2	Production/ processing	Gas flaring	Marine, wetland, terrestrial, atmosphere	Air pollution, gas flaring, heat production, acid rain	Mangrove, agriculture
3	Tanker loading	Offshore	Marine, wetland	Water pollution from the ballast, destruction of the seabed by dredging	Mangrove, fish, animal
		Onshore pipeline	Wetland, terrestrial	Land pollution and degradation	
4	Storage depots		Terrestrial	Land pollution	Agriculture, animal
5	Transportation	Offshore: marine craft, ship	Marine, wetland, terrestrial	Effluent discharge	Fish, mangrove, and animal
		Onshore: trailer			
6	Refinery	Refining petrochemicals	Marine, wetland, terrestrial	Air and water pollution	Mangrove, fish, animal, and human

Source: Snowden and Ekweozor (1987).

c) **The riverine flood plains:** extend from the Niger-Anambra confluence to the Nembe-Port Harcourt axis. The riverine floodplain is further divided into (i) upper riverine deltaic plain from the Niger-Anambra Confluence to the Bomadi-Yenagoa axis, a comparatively well-drained narrow flood plain; and (ii) lower riverine deltaic plain, which is poorly drained transitional zone between the riverine floodplain and the mangrove swamps to the south.
d) **The higher fringe plains:** are generally well-drained and lie at the foot of the sandstone hills and ridges. This may be subdivided into (i) Warri Plain to the northwest covering the upper Ase and Ethiope River basins, and (ii) Sombreiro Plain between the Orashi and Sombreiro Rivers to the northeast.

11.2.2 Environmental Sensitivity of the Niger Delta

The Niger Delta is considered environmentally sensitive (Jonah and Francis 2023) because it has abundant biodiversity from micro- to micro-organisms, including invertebrate and vertebrate species. It has numerous life forms, some of which have become endangered or threatened. There are over 4600 plant species, 274 species of mammals, 839 species of birds, 56 species of forest snakes, and 19 species of amphibians in Nigeria. The Niger Delta act as a buffer. The deltaic region acts as a trap and filter for water, which aids the accumulation and formation of fertile land. The Niger Delta is also known to be sensitive because it is endangered due to man's degradational activities, such as oil and gas exploration, leading to oil spills and deforestation. It is in a state of a precipice that may result in the extinction of some species. The Niger Delta is a sink for groundwater because of its high-water table that results from constant rainfall and closeness to water bodies (Jekayinfa et al. 2023). The land is a sink for surface and atmospheric water, which helps replenish the groundwater. Therefore, any entry of pollutants will contaminate the drinking water source.

11.3 Impact of Oil and Gas Exploration on Wetlands of the Niger Delta

11.3.1 Physiological Effect on Organisms

In an aquatic environment, exploratory activities can lead to pipeline blowouts because of equipment failure, engineering failure, natural causes, third-party sabotage, and accidents. The crude oil spillage that results can cause fish death when crude oil clogs the gills of fish, thereby cutting off their oxygen supply. Similarly, pollutants can cause reproductive failures and changes in the gestation period and breeding cycle. Eggs, fry, and adults of many aquatic organisms are killed when the pH of water drops to about 5.0. The level of marine acidification can also disrupt the food chain by killing aquatic plants, insects, and invertebrates used for food by fish. Aquatic plants such as macrophytes (water lily, water lettuce, and hyacinth) are destroyed as food and shelter for fish. Floating oil on the sea is usually transported to the mangrove swamps leading to the oiling of roots, defoliation of trees, or outright killing. Contamination of the wetland can impact humans through the soil-organism-trophic pathways (Figure 11.1).

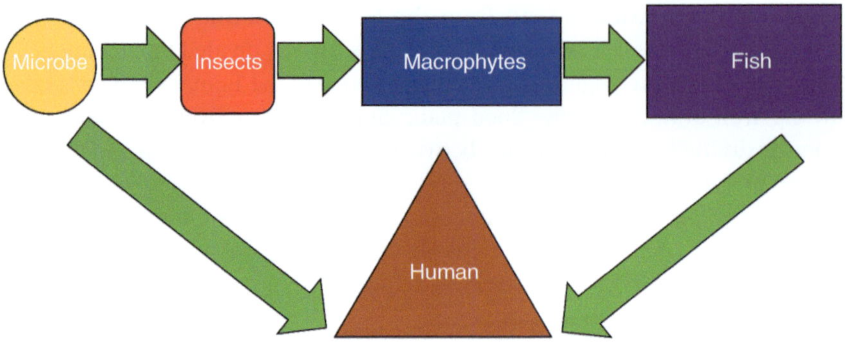

Figure 11.1 Trophic pathway showing the transmission of pollutants up the food chain in wetlands of the Niger Delta, Nigeria.

Wetlands are among the most critical ecosystems on the planet, and they are home to a diverse range of creatures. These environments are vital to biodiversity conservation and provide numerous benefits to humans. Oil and gas exploration and the hazards of spillage and gas flaring have caused several physiological impacts on the animals in the wetland (Figure 11.2). Several of these oil components interact synergistically to affect the animal; this action can sometimes weaken the immune system, making the animal more susceptible to parasites and other disease organisms (Nyeche et al. 2015). The toxic impact of oil and gas exploration on the physiology of animals cannot be reliably assessed due to the intricate nature of crude oil pollution, as harmful compounds rarely occur in isolation.

Figure 11.2 Impact of oil and gas exploration on the wetland at Bodo, Niger Delta, Nigeria. Dead crabs and other organisms were found on the floor. Oil shin is visible on the water, and black crude oil particles are on the soil.

Major contaminants from this exploration usually act as Endocrine Disruptors (EDs). EDs are synthetic or natural chemical compounds of exogenous sources that can cause significant health problems in animal growth and reproduction when released into the aquatic environment by human activity. EDs are associated with interference in the sexual differentiation of marine animals, infertility, and reduction of sperm production. In amphibians, the main effects caused by EDs include changes in hormonal activity during embryonic development, leading to changes in anatomy and behavior. This change in hormonal activities also results in a reduction in reproductive success, as well as gonadal abnormalities, hermaphroditism, and other alterations in the reproductive system (da Silva et al. 2018).

Oil reduces the insulating properties of fur-bearing mammals, such as sea otters, and the water repellency of avian feathers, exposing these creatures to a harsh environment (Ndimele et al. 2018). Birds and mammals will die of hypothermia if they cannot repel water and protect themselves from cold water. Oil can also entrap young sea turtles, who mistake it for food. Dolphins and whales can inhale the oil, which can harm their lungs, immune systems, and reproduction. Many birds and animals absorb oil while cleaning themselves, which can be fatal. Fish, shellfish, and corals may not be directly exposed, but they may come in contact with oil if it is mixed into the water column; shellfish may also be exposed in the intertidal zone. Exposure to oil can lead to the following impairment in adult fish – reduced growth, enlarged livers, alterations in heart and respiration rates, fin erosion, and reproductive dysfunction. Fish eggs and larvae are particularly vulnerable to fatal and sublethal consequences. Even if no deadly effects are identified, oil can render fish and shellfish unfit for human consumption (Kardena and Helmy 2023).

Crude oil spills harm fish hatcheries in coastal waters and pollute valuable fish (Osuagwu and Olaifa 2018).

In an experimental study, crude oil-contaminated water impaired the reproductive hormonal function of rabbits by decreasing the peripheral FSH and LH levels (Nyeche et al. 2015). Brain damage, such as cerebral cortex malfunctions, was also detected in pregnant rats' fetuses exposed to bonny light crude oil (Fischer et al. 2005). Male rats exposed to Nigerian Qua-Iboe Brent crude oil had lower packed cell volume, higher total leukocyte count, and lower absolute lymphocyte count. The exposure generated enhanced extravascular hemolysis in the spleen and excessive deposition of the yellow-brown pigment hemosiderin in the organ. This hemolysis was most likely caused by erythrocyte sequestration in the splenic sinusoids (Igwebuike et al. 2007). Exposure of female rats to crude oil by oral gavage resulted in an irregular estrous cycle, prolonged gestation length, reduced litter size, and increased stillbirths. Animals with chronic exposure at higher doses could not conceive, but the control had standard conception and gestation (Nwaigwe et al. 2011).

11.3.2 Parasitological Impact of Oil Exploration

Parasites are found at the bottom and top of the food chain because they feed at both ends. Parasites are organisms that benefit from their host without the host deriving any benefit. Parasites infest plants, the primary producers, and animals and humans at the top of the food chain (Sarkar et al. 2023). The biological significance of parasites shows that they negatively impact organisms by causing sickness and death, with a ripple effect on the ecosystem (Guo et al. 2023). Whereas some parasites kill their hosts, other successful parasites do not kill their hosts but rather exploit them so that they will still be alive.

Nevertheless, some parasites live symbiotically with other organisms and help maintain energy flow through the food chain. However, this process is truncated by oiling activities that destroy the health of the wetland through deforestation, land fragmentation, hydrocarbon pollution, etc. Thus, eliminating the lower-level organisms in the food chain creates a domino effect on other higher organisms up the food chain, destabilizing the ecosystem. For instance, decomposers that help to recycle nutrient materials by breaking down organic matter are killed by hydrocarbon pollutants. The degradation of mangrove wetlands eliminates soil microbes that inhabit the soil, leading to the soil's barrenness. The death of plants also impacts the parasite population that lives on them.

Interest in the ecology of parasites concerning environmental monitoring has indicated that these organisms are sensitive toward the quality of the macroenvironment. Parasites have been used as indicators of environmental health, bringing to bear the environmental implications of parasites, leading to the establishment of environmental parasitology. This could also be used in a medical context related to the contamination and occurrence of infective parasitic stages in the environment. In dealing with environmental parasitology, parasites as accumulation indicators for selected pollutants, parasites as effect indicators, and the role of parasites interacting with established bioindicators are considered.

It is safe to say that the aquatic environment represents the final repository for many human-generated pollutants associated with anthropogenic activities. Anthropogenic activities such as oil spills lead to elevated concentrations of metals and various organic compounds in the environment and, more so, in the sea, which increases contamination levels in marine ecosystems. Studies have shown significant effects and interactions between parasite levels and the presence and concentration of various pollutants and environmental stressors (Vidal-Martinez et al. 2010); some of the studies that unraveled the impacts of anthropogenic changes on parasites suggest that oil explorations and its consequences in an environment may increase the prevalence of some species of parasites while reducing others. Thus, parasites play the following roles in the wetland environment.

11.3.2.1 As Accumulation Indicators

The biological availability of certain substances commonly accepted to be harmful to the environment can be established by certain species of parasites. Certain groups of endoparasites are excellent accumulators of toxic metals (Sures 2004, 2003; Nachev and Sures 2016) and selected organic pollutants (Le et al. 2014). Therefore, the four major endo helminth taxa (Acanthocephala, Cestoda, Digenea, and Nematoda) have been considered and suggested as sentinels for metal pollution (Sures et al. 2017) because of oil exploration activities. While other parasites outside these groups may fail to thrive in the presence of metal pollution, they will have a high prevalence.

11.3.2.2 Parasites as Effect Indicators

When organisms are exposed to pollutants, physiological responses are activated, resulting from the uptake and accumulation of toxic substances. In ecotoxicological research, physiological, behavioral, or molecular changes are determined as a response to adverse environmental changes, often due to the presence and effects of pollutants (Amiard-Triquet et al. 2015) or habitat disturbance. Parasites as effect indicators will mainly focus on the direct impact of contaminants on the viability and longevity of free-swimming stages such as cercariae or on changes in population and community structure which oil pollution can alter. Physiological responses of organisms to pollutants are a consequence of the uptake and accumulation of toxic substances. The variety of responses ranges from increased stress and protective molecules to a complete breakdown of physiological homeostasis and death of the exposed organism. Bioindicators are species that reflect environmental impact because they respond to habitat alterations with changes in physiology or chemical composition.

Effect bioindicators are organisms that are used to detect environmental impacts through their changes in physiology, chemical composition, behavior, or number (Nachev and Sures 2016).

These bioindicators measure exposure to pollutants at different levels of the organization, from the subcellular to the ecosystem, and are helpful for effect indication.

Parasites such as helminths, crustaceans, and protozoans (mainly ciliates and myxozoans) of both terrestrial and aquatic environments are used as bioindicators of environmental pollution. Vidal-Martinez et al. (2010) established an association between environmental variables and parasites. Earlier works have attributed a weak association between ecological impact and parasite abundance to the wide range of stressor intensities and the inherent variability of parasite responses (Lafferty 1997). Studies have shown significant effects and interactions between parasite levels and the presence and concentration of various pollutants and environmental stressors (Vidal-Martinez et al. 2010).

11.3.3 Entomological Impact of Oil and Gas Exploration

Insects (Arthropoda) make up the largest population of organisms in the world (McGavin and Davranoglou 2023). The mangrove forest and wetland in the Niger Delta have numerous species of arboreal, terrestrial, and aquatic insects that live or hover around it Numbere and Moudingo 2023. Beetles and butterflies are dominant insect species found in a mangrove forest in the Niger Delta. Insects pollinate the flowers and feed on the sap from the mangrove leaves. For instance, black ants are also found in large populations and play a symbiotic role with the mangrove tree, feeding on the tree sap while protecting against external intruders. The ants also help to break down the leaves, which facilitates decomposition. Soil insects help to break down and leave litter which is further decomposed by bacteria and fungi. Oil spill kills large populations of wetland insects by contaminating the environment and blocking their breathing apparatus. Insects, thus, get asphyxiated and die because of poor oxygen circulation due to oil pollution. Economically valuable insects such as bees that build their hives on the branches of mangrove trees are destroyed during deforestation to create right-of-way passage for crude oil pipelines.

Hemipteran species in the families Veliidae, Mesoveliidae, and Hermatobatidae are surface inhabitants but located in estuaries (Andersen 1999; Damgaard and Zettel 2014). Changes in salinity and temperature influence larvae of insects in brackish water.

Examples of these species include larval Diptera (mosquito), Coleoptera, and Hemiptera. Emergent plants (e.g., Spartina species) support certain grasshoppers and plant hoppers. Intertidal regions are also harsh, and insect life is restricted to a few species living in crevices, such as Diptera, Coleoptera, and Hemiptera. The intertidal area is where the wetland system is found in the Niger Delta, and it is a collection of an oil spill carried by the tidal current. Thus, larval stages of insects suffer significantly because the pollutants block their breathing apparatus and kill them (Little 2018). This led to the wiping out of these organisms in the food chain and may have a ripple effect on the trophic dynamics of the ecosystem. Species of insects that are found in the aquatic (freshwater) and wetland environments are:

1) Ephemeroptera
2) Odonata, e.g., dragonflies, have internal gills for breathing in water.
3) Plecoptera has gills.
4) Trichoptera
5) Diptera
6) Hemiptera, e.g., water strider and water boatman (have its legs modified for swimming)
7) Coleoptera
8) Neuroptera
9) In Lepidoptera, the adult is terrestrial, and the larvae are aquatic.

The aquatic insects take in oxygen with an open or closed (no spiracles) tracheal system and gills (i.e., tracheal cuticular lamellar extensions of the body). Therefore, oil particles enter insects' breathing apparatus during oil spills to suffocate them to death. The heavy metal concentration or pollutant level in the soil affects the kind of insect present. An increase in heavy metals in water can lead to a decrease in diversities and loss of insect nymph (e.g., stoneflies) due to reduced oxygen. Exploratory activity can impact the wetland and aquatic temperature, affecting insect development and life cycle time. This is why insects are used for environmental monitoring in degraded streams. Ground-dwelling insects include the following: Hymenoptera (ants), Coleoptera, Diptera, and non-insect Hexapoda, e.g., Dermaptera (earwigs), ground beetles, herbivorous, larvae with legs, etc. These insects have well-developed legs (mole cricket), cicadas, and beetles for digging. They are also flightless and reduced in size in adults (e.g., termites). During an oil spill or soil disturbance, these insects are killed, which impacts the trophic dynamic.

11.3.4 Environmental View of Oil and Gas Exploration

Oil and gas exploration impacts all aspects of the environment, such as air, land, and water (Numbere 2018). Seismic operations lead to land degradation, such as deforestation, fragmentation, and hydrocarbon pollution. During oil well search, dynamites are exploded to create seismic waves to detect oil wells. The exploding dynamite destroys the forest and disintegrates the land. Baroid and other drilling chemicals are flushed into the soil and water, contaminating them. The chemicals percolate through the ground to contaminate groundwater aquifer. Similarly, during erosion, chemicals from oil fields are washed into rivers to contaminate aquatic organisms. Deforestation of mangroves makes the ground bare leading to erosion and soil loss.

The stress response of mangroves, according to Gundlach et al. (1985), is as follows:

1) Tree mortality
2) Defoliation of canopy
3) Root mortality
4) Bark fissuring
5) Seedling mortality
6) Epithelial scurring
7) Lenticel expansion
8) Adventitious pneumatophore
9) Leaf deformities/chlorosis
10) Propagule stunting/bending
11) Leaf stunting
12) Reduction in leaf number
13) The number of lenticels reduces
14) Epiphytic mortality
15) Tree snail density
16) Tree crab density
17) Changes infaunal community

11.4 Sedimentary Environment of the Niger Delta

Since the Niger Delta is a sedimentary basin, it has one of Africa's most extensive sedimentary environments (Xiao and Wang 2023), leading to the geological formation of a vast wetland area. Thus, to understand the wetland ecosystem of the Niger Delta, it is vital to know what the sedimentary environment means. A sedimentary is the sum of the physical, chemical, and biological conditions under which sediment accumulates. The sedimentary environment leads to the formation of rocks through the weathering process. Animals and plants are major sedimentary rock contributors, making them rich in crude oil. The plants in the forest and swamp decay in situ to form peat, which produces lignite from where coal is formed. In addition, molluscs, corals, and other wetland substances use seawater to build their shells and skeletons, which are eventually incorporated into the accumulated sediments. Thus, sediments resulting from the differentiation of raw materials can be categorized into three (Figure 11.3):

1) **Detrital sediments:** are formed by accumulating transported particles of old and new minerals produced during weathering.
2) **Chemical, organic sediments:** are formed from materials manufactured by plants and animals. These two divisions together make up the bulk of sedimentary activities.

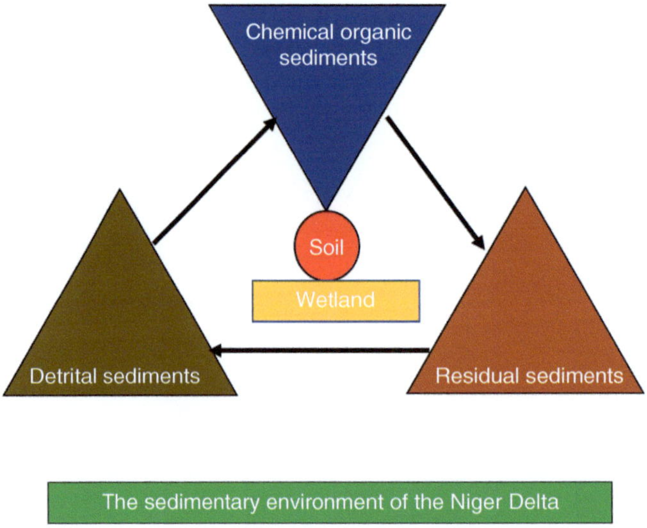

Figure 11.3 The sedimentary environment of the Niger Delta, Nigeria shows the formation of the wetland soil.

3) **Residual sediments:** represent materials left behind at the weathering site because they are challenging to transport following their stratification by plant cover and resistance to transportation.

There are four criteria for classifying sedimentary rocks:

1) Classification based on physical and chemical processes of origin
2) Classification based on provenance
3) Classification based on grain size
4) Classification of chemical composition (organic rocks)

Types of sedimentary environment

1) Marine environment
2) Mixed (transitional) environment
3) Continental (non-marine) environment

11.4.1 Swamp Environment

The swamps and marshes have similar characteristics to the wetland environment. Swamps are bodies of shallow standing water with variations in salinity depending on the swamp's location. Salinity varies from brackish to freshwater and ranges from hypersaline to hyposaline. Generally, the paludal environment is a low wet ground usually occupied by plant life. In the tropics, like the Niger Delta, plants like mangroves are likely to be found. In temperate regions, it is occupied by grass. Deposition in the paludal environment is generally dependent on biological activities. Among the sediments found are silt and mud. Dissolved salts and gas also occur.

Table 11.5 Types of environments and the sub-divisions in the Niger Delta Region, Nigeria.

Serial number	Types of environments	Sub-division	Basis of classification	Species
1	Marine	Littoral zone	Depth of water	Tilapia
		Neritic zone		
		Bathyal zone		
		Abyssal zone		
2	Terrestrial	Deltaic zone	Sediments	Crabs
		Lagoonal zone		
		Littoral zone		
3	Continental	Terrestrial zone: desert and glacial environment	Sand and ice	Desert rats
		Aqueous zone: estuarine, deltaic, lacustrine, and swamp/paludal environments		Invertebrates e.g., barnacles

Oil and gas exploration impacts different aspects of the environment and different organisms in the wetland area. Crude oil affects the physiology of animals in water and land. It also affects the arthropods at the bottom of the food chain. Exploratory activities impact parasitic organisms in the wetland environment, which may lead to their decline.

11.4.2 Shoreline Types in the Niger Delta

Mangrove forests mainly inhabit the shoreline of the Niger Delta and are sensitive to oil spills. According to Gundlach et al. (1985), the area's shoreline is categorized as shown in Table 11.6. The shoreline is categorized based on sediment type, geomorphology, and biota.

11.5 Causes of Wetland Degradation

A wetland is a sensitive area because of its soft bottom and numerous biodiversity that inhabits it. Some anthropogenic activities that degrade the wetland ecosystem in the Niger Delta include but are not limited to the following:

1) **Petroleum industry activities:** crude oil exportation is Nigeria's primary foreign exchange earner (Okotie 2018). Although there are numerous natural resources in Nigeria, crude oil stands out because it is the centripetal force on which the nation's economy revolves. Because of the rich supply of crude oil underneath the earth's crust, many oil-producing companies, such as Shell, NNPC, Chevron, Mobil, and Agip, are doing business there. Some companies are privately owned, while others, such as the NNPC, have the government as a major stakeholder. These companies have thousands of oil fields and tens of thousands of oil wells scattered across the region in onshore and

Table 11.6 Shoreline types and their vulnerability to pollution in the Niger Delta, Nigeria.

Serial number	Shoreline type	Biomass type	Vulnerability to oil spill	Example of vegetation
1	Steep rocky cliff	Low	Low	Water plants
2	Eroding wave-cut platforms	Low	Low	Water plants
3	Flat fine-grained sand beaches	None	Low	Water plants
4	Steeper medium-to-coarse-grained sand beaches	None	Medium	Water plants
5	Exposed tidal flats	Low	High	Mangroves
6	Mixed sand and shell beaches	Low	High	Mangroves
7	Exposed tidal flats	High	High	Mangroves
8	Erosional peat scarps	High	High	Mangroves
9	Sheltered tidal flats	High	High	Mangroves
10	Sheltered salt marshes and mangroves	High	High	Mangroves

offshore environments with over 50 billion barrels of crude oil. The Niger Delta is rich in natural gas and has over 500 trillion cubic meters of natural gas. The liquefied natural gas (LNG) project was established in Bonny in early 2000 to utilize the gas wasted through flaring. The Niger Delta area is made up of three petroleum refineries and three petrochemical plants located at Eleme local government area (old and new Port Harcourt Refineries) and the Warri Refinery located in Ekpan, all in the Niger Delta area.

2) **Non-petroleum-based industry:** the Niger Delta has many non-petroleum-based industries that manufacture various products utilized by the people of the area (Heum et al. 2003). Examples include bottling companies, cement factories, sawmills, steel companies, confectionary or food industries, etc. Most of these industries are situated close to the river, i.e., in coastal areas, to quickly transport raw materials and finished products via the river. The problem is that most of these industries release their waste into the sea resulting in contamination. Some companies also use water from the sea to cool their turbines and later remove the hot water into the river.

For instance, gas turbine plants in the Niger Delta operate near the river to easily transmit diesel fuel to run their industrial machines. Some spillages contaminate the wetlands and the adjoining rivers during the transfer of petroleum products from boats to fuel depots. Similarly, there are numerous sawmill industries in the coastal areas of the Niger Delta area. They were established near the coast for easy transportation of logs of wood needed for producing plywood and other wood products used in the furniture industry. The main problem of the sawmill industry is the massive production of sawdust that is often not recycled but dumped in the river to create a blockage of the water body. These sawdust heaps have obliterated many wetland areas. Continuous deposition saw dust in creeks has converted the river system into land because of years of piling these waste products. On the other hand, the confectionary industry deposits sewage directly into the river, causing algae growth, which increases to cover the river surface through eutrophication. The sewage causes nutrient enrichments that allow the development of macrophytes that increase to cover the wetland and river, leading to an increase in biological oxygen demand (BOD) and a decline in dissolved oxygen (DO).

Aquaculture is another widespread practice in riverine communities that damages many wetlands (Gopal 2013) because of the release of liquid waste from the fish farm and plastic and other solid waste from the ponds. These waste materials block the drainages and convert the wetland area into a polluted environment.

Steel companies produce steel products in a blast furnace using coke, which is derived from the earth's crust. Steel-making operations produce tons of hazardous waste and sludge, often released into the river, increasing heavy metal concentration in the adjoining wetland soils and rivers. Furthermore, effluents from the steel/metal plant contain suspended solids, ammonia, cyanide, phenols, and heavy metals.

Paint production is also another industrial activity that negatively impacts the wetland and coastal areas of the Niger Delta. Paint industries produce waste containing pigment, metals, resins, solvents, and additives in wastewater and sludge. Sometimes, a lot of water is used for flushing the paint mixture and deposited in the river to get the right paint color. The breweries also contribute to the degradation of the wetlands by adding organic waste rich in solvents and color into the environment. The breweries use large quantities of water with the needed alcoholic mixture to produce various brands of intoxicating drinks. The large amounts of water produced contain organic products like sugar, yeast, beer, and malt residues. Waste from alcoholic beverages includes amyl, butyl, propyl alcohol, carbohydrates, yeast, aldehydes, and esters. Hazardous chemicals in the effluents are caustic soda, hypochlorites, and peroxides. Brewery waste is oxygen-demanding and increases the BOD and COD of the receiving water bodies.

Cement factories are also found in the Niger Delta and are established near the river, where they produce liquid waste that hardens to coat benthic areas, mudflats, and mangrove forests. Other industries include glass factories, whose waste produces radioactive components detrimental to the environment and humans.

The food-producing industry in the region produces oil palm and vegetation oil used in cooking. The food industry mainly produces organic biodegradable waste that causes nutrient enrichment (eutrophication). Waste produced from food processing factories includes organic waste containing aldehyde and ketone, fatty acids, sodium bicarbonate, and ammonium bicarbonate. The fruit juice and carbonate/mineral waters are produced by bottling companies situated in the Niger Delta region. The waste products include caustic soda from bottling plants and nutrients.

3) **The upstream urban and industrial development project:** the urbanization of rural areas has also contributed to wetland loss in the Niger Delta because of the establishment of some infrastructural development projects within the wetland, such as residential apartments, construction of bridges, dams, schools, hotels, churches, playgrounds, seaside resorts, industrial complexes, etc. It is common to find that many companies prefer establishing their operational bases near the coast for easy transportation of raw materials and finished goods. Oil servicing firms use the coastal areas as tank farms and booth camps for seismic house parties involved in oil and gas exploration. Thus, hazardous waste is discharged during industrial activities, contaminating

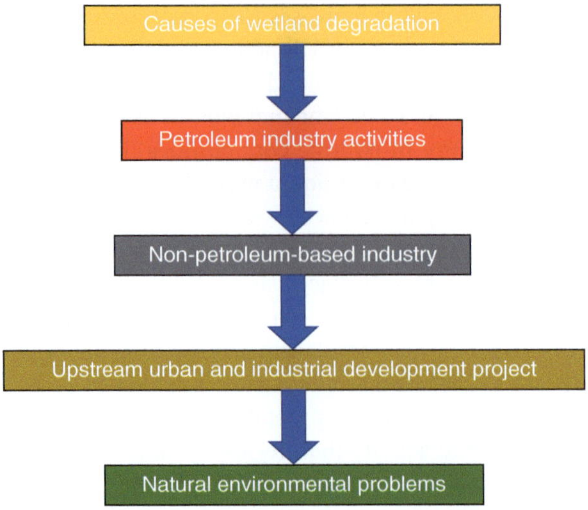

Figure 11.4 The causes of wetland degradation in the Niger Delta region, Nigeria.

wetlands and degrading land. For instance, the steel industry is set up in coastal areas because of the use of water from the river to cool smelted steel. Steel mills are the sources of heavy metals in waste flows. The steel sewage contains hydrochloric acid or sulfuric acid concentration and a high concentration of ferric hydroxide that could affect benthic organisms. Effluents from steel smelters could also contain oil waste. Sawmills and metal plating plants mentioned earlier are also industries that channel the garbage into the aquatic environment. Products from sawmills degrade slowly and impact benthic organisms. The metal plating discharges pollutants such as cyanide and heavy metals (e.g., copper, chromium, and nickel), silver, and gold into the wetland environment.

4) **Natural environmental problems:** the Niger Delta region is spared from hazardous natural disasters such as earthquakes, volcanoes, tsunamis, and hurricanes, which cause significant damage to coastal areas around the world. The causes of wetland degradation are shown in Figure 11.4.

11.6 Ecosystem Services in a Restored Wetland

The restoration of previously polluted wetlands takes a long time. Wetlands gradually spring back to life by providing ecosystem services (Mo et al. 2023) that benefit the environment and humans. The environment benefits when the soil starts to harbor microbes that hasten decomposition, increasing the successional process. Increased soil nutrients mean the growth of plants will commence unhindered. Similarly, naturally or artificially planted mangrove seedlings will start to grow and populate the area quickly. The growth of the seedlings attracts another biodiversity that increases the species diversity in the area. The wetland soil is sensitive to species proliferation because of the rich sediment constantly flushed by tidal currents. Thus, ground dwellers like insects and other invertebrates (crabs, periwinkle, and barnacles) and fingerling of fishes and periwinkle will begin to colonize the area. The growing seedlings

Figure 11.5 A remediated site that was re-polluted in Bodo, Niger Delta, Nigeria. This area will produce no ecosystem services because of the polluted state of the environment.

will once again help purify the environment by absorbing carbon dioxide and releasing oxygen, which helps in air purification. Mangrove trees are major carbon sequestration agencies in wetlands areas globally. The cleansing of the atmospheric air and the cleaning of the soil by the pneumatophores of any pollutants provide a much healthier enjoinment for the human population that depends on seafood for survival. The cleaning of the environment ensures the re-entry and resurgence of microbenthic infauna. The macro- and micro-benthic fauna mainly live in the bottom areas of the newly sprouting mangrove forest. During the earlier oil spill, there was a decline in species diversity, but after the remediation, these species came back to repopulate the area. And may multiply speedily because of the clean environment created compared to the previous polluted state. However, if the cleaned area becomes polluted again, there will be no opportunity to return to ecosystem services (Figure 11.5).

11.7 Sustainable Management of the Wetland in the Niger Delta

Wetland in the Niger Delta is one of the most productive aspects of the coastal/estuarine ecosystem. Therefore, wetlands should be safeguarded and sustainably managed to ensure indigenous species are not threatened or become extinct. To ensure that the wetlands are safe from adverse anthropogenic activities, indigenous people should take deliberate steps to protect the area. Strict protection is thus required because of the sensitivity of the wetland to pollution and contamination and the numerous species inherent in the area. Oiling activities from petroleum and non-petroleum-based industries are one of the biggest threats to the survival of the wetlands in the Niger Delta. The Niger Delta is the country's oil capital and has thousands of oil wells scattered across the zone, making oil spillage a regular occurrence. Daily reports of an oil spill in different locations across the Niger Delta mostly enter the mangrove forest and embed in wetland areas. Thus, to curb ongoing oiling activities in the wetland area, alternative sources of energy (solar and biomass energy) should be relied upon to reduce the quest for oil in the country. For instance, solar energy is gradually being adopted in many places because of the fuel cost to power industrial machines and vehicles. Solar energy is a sustainable energy source because it does not negatively impact the environment compared to crude oil. Strictly regulating industrial activities would significantly reduce the damage caused by oiling activities to the ecosystem.

Furthermore, the government should set up a wetland monitoring agency (WMA) to monitor wetland areas with the power to sanction erring industries that pollute the wetlands. Members of the communities can physically protect the wetlands against oil theft leading to soil spills by creating a buffer zone to prevent entry by vandals. The buffer zones will have a boom that absorbs the oil. Citizens should stop waste disposal in the coastal environment and mangrove forests because it degrades the wetlands. Heavy metal-laden waste contaminates the mangrove forest. Agencies can sustainably manage the wetland area by mapping the entire wetland zone using satellite data and ground truthing. Identifying the wetland areas will ensure the constant monitoring of their health. In addition, stream flow dynamic and tidal current data should be collected to study oil spill movement into the mangrove forest and to provide ways of curbing wetland contamination. The agencies can monitor mapped wetlands well to ensure that any spill can be quickly identified, tracked, and cleaned before it causes damage to the ecosystem. The artisanal refinery is built by some locals who steal crude oil from broken pipes and convert it to oil products using elementary methods that often pollute the environment. The artisanal refinery is also another major factor of wetland degradation in the Niger Delta (Figure 11.6). It needs to be urgently stopped to reduce the destruction of the wetland zone.

Humans are the primary cause of wetland loss due to their encroachment into the mangrove forest to dump waste, build houses, and cut mangrove trees to produce firewood. Wetland guards must be formed to help watch the wetland areas against detrimental human activities to ensure the sustainable development of the wetland area. The wetland guards should be given the legal powers to apprehend defaulters and bring them to book in the law court for possible sanctions or imprisonment. While protecting already standing mangrove forests, those that have been remediated can be repopulated with seedlings from the nursery so that they will recover quickly.

11.8 Policy Development of Wetlands in the Niger Delta

The government should establish strict laws to protect wetland areas. The law will involve the prevention of any human activities in the regions that have threatened or endangered species. In addition, they should prevent resource exploitation in areas that have been overexploited to reduce wetland loss. Nigeria has many moribund environmental protection

(a) (b) (c) (d)

Figure 11.6 Impact of artisanal refinery on wetland and mangrove forest in Bodo, Niger Delta, Nigeria showing (a) polluted soil with withering weeds, (b) pipeline running across a wetland, (c) an area with compete elimination of mangroves, and (d) dark crude oil spots on soil. Spills from broken pipes have wiped out a large area of mangrove forest.

laws, but the problem is more enforcement and implementation. Here those laws should be revived and given legal muscle to be prosecutorial so that the citizens will respect them. There are laws banning oil spills and gas flaring. Still, the government cannot sanction offenders because, most times, they are major stakeholders in the oil industries or have a financial interest that runs counter to the "polluter pays principle." There should be a rejuvenation of laws protecting the sea and the mangrove vegetation in the Niger Delta. Where there are no laws, the government can formulate new laws and give legal backing to protect the environment. Specifically, laws protecting wetland areas may not exist or are found under other environmental regulations, such as the laws for the protection of the sea. We thus suggest that wetland protection laws be promulgated so that they will become operational to help restore the quickly eroding wetland areas in the Niger Delta. There should be massive public enlightenment of this law in schools and electronic media so that the people will be aware and desist from infringing on the law.

11.9 Conclusion and Recommendations

Wetland serves as the resource base of the estuarine environment because of the numerous organisms that migrate there to take refuge and to breed and spawn. The presence of these organisms thus makes them sensitive and vulnerable to pollution. Wetland connects the coastal and terrestrial environments, making them significant in organisms' survival from both land and sea. Therefore, deliberate efforts should be made to stop or reduce pollutants from entering the wetland. The authorities should ban the disposal of hazardous waste in the mangrove forest.

Furthermore, the government should control the over-exploitation of mangrove forests for firewood by locals by establishing monitoring teams and wetland guards. Uncontrolled harvesting of resources and hunting has decimated the mangrove population, making them vulnerable to stochastic events that may wipe them out. Selective exploitation should ensure a win-win situation whereby the locals benefit from their resources and yet help protect them from destruction.

References

Amiard-Triquet, C., Amiard, J.C., and Mouneyrac, C. (ed.) (2015). *Aquatic Ecotoxicology: Advancing Tools for Dealing with Emerging Risks*. Academic Press.

Andersen, N.M. (1999). The evolution of marine insects: phylogenetic, ecological and geographical aspects of species diversity in marine water striders. *Ecography* 22 (1): 98–111.

Bullock, A. (1993). Perspectives on the hydrology and water resource management of natural freshwater wetlands and lakes in the humid tropics. In: *Hydrology and Water Management in the Humid Tropics: Hydrological Research Issues and Strategies for Water Management* (ed. M. Bonnell, M.M. Hufschmidt, and J.S. Gladwell). Paris: UNESCO.

Chen, H., Li, D., Chen, Y., and Zhao, Z. (2023). Spatial–temporal evolution monitoring and ecological risk assessment of Coastal Wetlands on Hainan Island, China. *Remote Sensing* 15 (4): 1035.

Chukwuma-Orji, J.N., Okosun, E.A. and Onoruoiza, A.L., 2021. Palynostratigraphy and paleobathymetric studies of XAD-1 well Niger Delta Basin, Nigeria. *Journal of Mining and Geology*, 57, pp. 193–202.

Damgaard, J. and Zettel, H. (2014). The water bugs (Hemiptera--Heteroptera: Gerromorpha & Nepomorpha) of New Caledonia: diversity, ecology and biogeographical significance. *Zoologia Neocaledonica* 8: 219–238.

Fischer, V.A., Akpuaka, F.C., and Ngokere, A.A. (2005). Effects of bonny light crude oil on the histology of cerebral cortex in fetuses of Wistar Rat. *Journal of Experimental and Clinical Anatomy* 4 (1): 9–11.

Fu, B.J., Su, C.H., Wei, Y.P. et al. (2011). Double counting in ecosystem services valuation: causes and countermeasures. *Ecological Research* 26: 1–14.

Gopal, B. (2013). Future of wetlands in tropical and subtropical Asia, especially in the face of climate change. *Aquatic Sciences* 75: 39–61.

Gundlach, E.R., Kana, T.W. and Boehm, P.D. (1985). Modeling spilled oil partitioning in nearshore and surf zone areas. *International Oil Spill Conference* (Vol. 1985, No. 1, pp. 379-383). American Petroleum Institute, Conference Feb. 1, 1985, Allen press.

Guo, Q., Potter, K.M., Ren, H., and Zhang, P. (2023). Impacts of exotic pests on forest ecosystems: an update. *Forests* 14 (3): 605.

Heum, P., Quale, C., Karlsen, J.E. et al. 2003. Enhancement of Local Content in the Upstream Oil and Gas Industry in Nigeria: A Comprehensive and Viable Policy Approach. SNF Report No. 25/3.

Igwebuike, U.M., Obidike, R.I., Shoyinka, S.V. et al. (2007). Effects of Nigerian Qua Iboe Brent crude oil on rat spleen and haematological parameters. *Veterinary Arhives* 77 (3): 247–256.

Jekayinfa, S.M., Oladunjoye, M.A., and Doro, K.O. (2023). A review of the occurrence, distribution, and impact of bitumen seeps on soil and groundwater in parts of southwestern Nigeria. *Environmental Monitoring and Assessment* 195 (2): 351.

Jonah, I.B. and Francis, I.O. (2023). Geospatial analysis of oil spill in the selected parts of the Niger Delta. *Journal of Environmental and Geographical Studies* 2 (1): 25–53.

Joshi, G. and Negi, G.C. (2011). Quantification and valuation of forest ecosystem services in the western Himalayan region of India. *International Journal of Biodiversity Science, Ecosystem Services & Management* 7 (1): 2–11.

Kardena, E. and Helmy, Q. (2023). Biosurfactants in oil spill cleanup. In: *Marine Surfactants* (ed. S.-K. Kim and K.-H. Shin), 327–374. CRC Press.

Lafferty, K.D. (1997). Environmental parasitology: what can parasites tell us about human impacts on the environment? *Parasitology Today* 13 (7): 251–255.

Le, T.Y., Rijsdijk, L., Sures, B., and Hendriks, A.J. (2014). Accumulation of persistent organic pollutants in parasites. *Chemosphere* 108: 145–151.

Little, D.I. (2018). Mangrove restoration and mitigation after oil spills and development projects in East Africa and the Middle East. In: *Threats to Mangrove Forests: Hazards, Vulnerability, and Management* (ed. C. Makowski and C.W. Finkl), 637–698. Springer.

McGavin, G.C. and Davranoglou, L.R. (2023). *Essential Entomology*. Oxford University Press.

Mętrak, M., Chachulski, Ł., Pawlikowski, P. et al. (2023). Potential role of high-altitude wetlands in preservation of plant biodiversity under changing climatic conditions in the arid Eastern Pamir. *Catena* 220: 106704.

Mitsch, W.I. and Gosselink, I.G. (1986). *Wetlands*. New York: Van Nostrand Reinhold.

Mo, L., Chen, J., and Xie, Y. (2023). Wetlands conservation in Beijing of China: present status and development strategies. *Watershed Ecology and the Environment.* 5: 73–79.

Nachev, M. and Sures, B. (2016). Environmental parasitology: parasites as accumulation bioindicators in the marine environment. *Journal of Sea Research* 113: 45–50.

Ndimele, P.E., Saba, A.O., Ojo, D.O. et al. (2018). Remediation of crude oil spillage. In: *The Political Ecology of Oil and Gas Activities in the Nigerian Aquatic Ecosystem* (ed. P.E. Ndimele), 369–384. Academic Press.

Numbere, A.O. (2018). The impact of oil and gas exploration: invasive nypa palm species and urbanization on mangroves in the Niger River Delta, Nigeria. In: *Threats to Mangrove Forests: Hazards, Vulnerability, and Management* (ed. C. Makowski and C.W. Finkl), 247–266. Springer.

Numbere, A.O. and Moudingo, J.H.E. (2023). Scenarios of nypa fruticans invasion: impacts and management strategies in West and Central Africa. *Journal of Coastal Research* 39 (1): 114–128.

Nwaigwe, A.N., Anya, K.O., Nwaigwe, C.O., and Nwaigwe, C.U. (2011). The effect of Nigerian Qua-iboe Brent crude oil on the reproductive performance of female Wistar albino rats. *Nigerian Veterinary Journal* 32 (3): 187–191.

Nyeche, V.N., Ajuogu, P.K., and Ndor, L. (2015). Toxicological influence of crude oil contaminated water on some reproductive hormones of adult rabbits does in Niger Delta, Nigeria. *IOSR Journal of Agriculture and Veterinary Science* 8 (3): 86–89.

Okotie, S. (2018). The Nigerian economy before the discovery of crude oil. In: *The Political Ecology of Oil and Gas Activities in the Nigerian Aquatic Ecosystem* (ed. P.E. Ndimele), 71–81. Academic Press.

Osuagwu, E.S. and Olaifa, E. (2018). Effects of oil spills on fish production in the Niger Delta. *PloS One* 13 (10): e0205114.

Saha, A.K., Kashaigili, J., Mashingia, F. et al. (2023). Determination of environmental flows in data-poor estuaries—Wami River Estuary in Saadani National Park, Tanzania. *Hydrology* 10 (2): 33.

Sarkar, S., Diab, H., and Thompson, J. (2023). Microplastic pollution: chemical characterization and impact on wildlife. *International Journal of Environmental Research and Public Health* 20 (3): 1745.

da Silva, A.P.A., de Oliveira, C.D.L., Quirino, A.M.S. et al. (2018). Endocrine disruptors in aquatic environment: effects and consequences on the biodiversity of fish and amphibian species. *Aquatic Science and Technology* 6 (1): 35–51.

Snowden, R.J. and Ekweozor, I.K.E. (1987). The impact of a minor oil spillage in the estuarine Niger Delta. *Marine Pollution Bulletin 18* (11): 595–599.

Sures, B. (2003). Accumulation of heavy metals by intestinal helminths in fish: an overview and perspective. *Parasitology* 126 (7): S53–S60.

Sures, B. (2004). Environmental parasitology: relevancy of parasites in monitoring environmental pollution. *Trends in Parasitology* 20 (4): 170–177.

Sures, B., Nachev, M., Selbach, C., and Marcogliese, D.J. (2017). Parasite responses to pollution: what we know and where we go in 'Environmental Parasitology'. *Parasites & Vectors* 10: 1–19.

Vidal-Martinez, V.M., Pech, D., Sures, B. et al. (2010). Can parasites really reveal environmental impact? *Trends in Parasitology* 26 (1): 44–51.

Xiao, Y. and Wang, H. (2023). A new insight on sedimentary facies of the Miocene to Pliocene Agbada Formation, offshore Niger Delta Basin, Nigeria. *Journal of African Earth Sciences* 200: 104859.

12

Carbon Sequestration in Revegetated Coal Mine Soil

A Chronosequence Study in the Gevra Opencast Project, Chhattisgarh, India

Preeti Singh[1]*, Amlan Kumar Ghosh*[2]*, Ebhin Masto*[3]*, Santosh Kumar*[1]*, and Chandini Pradhan*[2]

[1] *ICAR-Indian Agricultural Research Institute, Jharkhand, India*
[2] *Department of Soil Science and Agricultural Chemistry, Institute of Agricultural Sciences, BHU, Varanasi, India*
[3] *Environment, Emission and CRM Section, CSIR - Central Institute of Mining and Fuel Research (Digwadih Campus), PO; FRI, Dhanbad, Jharkhand, India*

12.1 Introduction

In the fiscal year 2017–2018, India's coal production reached 606.89 million tons, with over 92% stemming from open strip mining processes (Singh et al. 2022b). Such mining necessitates the complete removal of vegetation and topsoil to access coal seams, causing widespread soil degradation, reduction in microbial populations, and a substantial decrease in soil organic carbon (SOC). Notwithstanding these adversities, post-mining land reclamation, when adeptly managed, can lead to appreciable SOC sequestration over time (Singh et al. 2015).

Mine soil restoration can be approached from two dimensions: physical, which involves financial implications and aims at achieving a harmonious landform; and biological, which focuses on sustainable vegetation cover that integrates seamlessly with the existing landscape (Singh et al. 2022a). Biological restoration, particularly through revegetation, mitigates erosion risks, offers protection against soil degradation, and plays a pivotal role in SOC restoration. Thereby, it sets the stage for carbon sequestration.

Spontaneous vegetation succession can occur on unmanaged mine soil substrates (Prach and Hobbs 2008), but continuous exploitation needs often prioritize afforestation-based reclamation over natural succession (Macdonald et al. 2015; Sroka et al. 2018). Adverse substrate properties – like elevated temperatures, moisture deficits, low SOC levels, and nutrient-poor conditions – commonly restrict vegetation growth, thereby impeding ecosystem restoration through natural succession (Maiti 2012). Given these constraints, tree species resilient to harsh soil conditions are favored for mine soil substrate revegetation (Horodecki and Jagodziński 2017). Presently, indigenous trees such as *Dalbergia sissoo* Roxb., *Gmelina arborea* Roxb., *Albizia* spp., and *Azadirachta indica, Azadirachta juss* are prioritized due to their ecosystem enhancement abilities, particularly their contributions to carbon and

Biodiversity and Ecosystem Services on Post-Industrial Land, First Edition.
Edited by Vimal Chandra Pandey.

nitrogen storage, nutrient cycling, and soil fertility elevation. Revegetation aids in refining the properties of the mine soil substrate, fostering the colonization of native species.

This study delves into the varying carbon pools and carbon sequestration under tree species, namely *A. indica*, *D. sissoo*, and *G. arborea*, within three chronosequences. We also elucidate the transformations in humic acid's (HA's) molecular properties across these chronosequences. The research's main objective was to uncover the evolving characteristics of SOC post-vegetation establishment on revegetated mine soils and to discern the vegetation's influence on the SOC pool and molecular attributes. The hypothesis posits that the ecosystem carbon pool and SOC reservoir improve with reclamation age and that distinct tree species, owing to differences in litter quality, differentially influence carbon sequestration. The study's findings are anticipated to provide valuable insights into carbon dynamics in post-mining ecosystems and the pivotal role of afforestation in SOC sequestration.

12.2 Analysis and Classification of Organic Residue/Plant Materials: Impacts on Carbon Sequestration in Soil

The amount and quality of carbon stored in soil are influenced by the carbon input from plant materials and its inherent quality, given constant environmental conditions. In this study, leaf litter was examined for its Carbon (C), Nitrogen (N), Lignin, and Polyphenol content, with subsequent calculation of the C/N, Lignin/C, and Lignin/N ratios.

According to a classification system outlined by Palm et al. (2001), organic residues can be divided into four groups:

- Class 1: High N and High Lignin, High N and High Polyphenol
- Class 2: High N and Low Lignin, Low N and High Polyphenol
- Class 3: Low N and High Lignin, High N and Low Polyphenol
- Class 4: Low N and Low Lignin, Low N and Low Polyphenol

Based on these categories, residues were sorted into four classes, with three tree species – *A. indica*, *G. arborea*, and *D. sissoo* – selected for variation in their Lignin content and Lignin: N, as well as Polyphenol: N ratios (Table 12.1).

The Nitrogen content observed ranged from 1.56% in *A. indica* to 1.84% in *G. arborea*. *D. sissoo*, possessing a low carbon content of 42.98%, showed the lowest quantities of Lignin

Table 12.1 Chemical and biochemical characteristics of the leaf litter and their ratios.

Tree species	Carbon (%)	Nitrogen (%)	Polyphenol (%)	Lignin (%)	C/N	Polyphenol/N	Lignin/N
Azadirachta indica	44.87^{f}	1.56^{ab}	13.47^{d}	2.20^{d}	28.84^{g}	8.63^{a}	1.41^{c}
Dalbergia sissoo	42.98^{d}	1.72^{bc}	11.62^{e}	1.38^{c}	24.96^{e}	6.75^{b}	0.80^{b}
Gmelina arborea	46.12^{h}	1.84^{bc}	20.43^{b}	2.58^{e}	25.08^{f}	11.10^{c}	1.40^{c}

Means with the same letter within a column are not significantly different at ($p < 0.05$). C = Carbon, N = Nitrogen.

(1.38%), Polyphenols (11.93%), a C/N ratio of 24.96, and a Lignin/Nitrogen ratio of 0.80. Conversely, *G. arborea* exhibited the highest values of Lignin (2.58%), Polyphenols (20.45%), Carbon (46.12%), and Nitrogen (1.84%) but had a lower C/N ratio (25.007) and Lignin/Nitrogen ratio (1.40).

Limited Nitrogen availability in plant residues impedes microbial decomposition (Waksman and Tenney 1928). Lignin is notably resistant to both chemical and biological degradation compared to other organic materials originating from plants (Singh et al. 2022c). Litter with abundant Nitrogen and a low C/N ratio accelerates decomposition (Parton et al. 2007) while also enhancing the carbon utilization efficiency of decomposers, leading to increased microbial biomass (Cotrufo et al. 2013). Studies have shown that litters with low Lignin concentrations facilitate SOC accumulation owing to their rapid decomposition rates (Zhou et al. 2019).

12.3 Litterfall Dynamics, Seasonal Variations, and Implications for Soil Organic Carbon Sequestration

In a forest ecosystem, most organic matter and nutrients are incorporated in the soil primarily via litterfall, which is essential for sustaining forest productivity (Tesfay 2020). Distinct variations in litterfall were evident among various plant species across different months. The trend showing maximum litterfall from December to April and a dip in June might be attributed to plant adaptations during drier seasons. The pattern of litterfall shows marked seasonal influences. Devi and Yadava (2010) found that cooler, drier periods tend to instigate earlier senescence in plant parts, particularly leaves. Moreover, dry spells increase litter shedding. Similar observations of increased litterfall during the colder and warmer months in North Indian mixed deciduous forests have been documented by researchers such as Dutta and Agrawal (2001).

For all studied species, the peak litterfall was between December and April, with a notable decline during May and June. *G. arborea* registered the highest litterfall at 4635.0 $kg\,ha^{-1}$, trailed by *A. indica* at 2690 $kg\,ha^{-1}$, and the least was seen in *D. sissoo* at 2363.2 $kg\,ha^{-1}$. The primary contributor to organic carbon in the soil is the above-ground plant biomass (Carrera et al. 2009). Thus, it is generally believed that plant biomass significantly influences SOC storage.

Despite expectations, no significant impact of annual litterfall on SOC concentrations was found in the present investigation, the outcomes align with Lajtha et al. 2014. In some evergreen forests, a rapid SOC accumulation was noted over recent years with stable litterfall rates and increased species diversity (Xiang et al. 2022). Hence, only an increase in litter quantity does not assure increased SOC storage (Zhou et al. 2019).

12.4 Impact of Plant Litter Quality on Decomposition Rate and Soil Health in Restored Coal Mine Areas of Chhattisgarh, India

Differences in the decomposition rate of various plant litters can be linked to the diverse substrate qualities of each plant species, coupled with soil-related attributes, especially since climate conditions remained consistent across all plantation sites. To foster enhanced microbial activity on mine spoil soil, it is beneficial to opt for plant species recognized for higher litter deposition and faster decomposition rates, thereby hastening the restoration process. The age of revegetation can significantly impact soil quality via litter breakdown and root-related activities. Our data suggests that the distinct decomposition rates and carbon storage in the restored coal mine soil of Gevra, Chhattisgarh, India stem from litter quality, given the consistency in soil attributes and climate across the plantations.

Numerous studies, including Krishna and Mohan (2017), reveal that plants with higher N content exhibit faster decomposition. This trend was evident in our study, particularly with *G. arborea* and *D. sissoo*, both recognized for their high N content. Conversely, *A. indica*, known for its lower N concentration, showcased slower decomposition rates.

G. arborea's litter witnessed the highest mass loss of 47.65%, retaining 58.65% of its original mass, closely followed by *D. sissoo* with a 51.17% loss, over a year. *A. indica* registered the least loss at 51.37%. Predictions for 95% mass loss spanned from 1310 days (*G. arborea*) to 1498 days (*A. indica*). Observations like C concentrations in residual litter showed negative correlations with remaining weight, whereas N concentrations depicted positive correlations, aligning with results by Singh et al. (1999) and Dutta and Agrawal (2001).

Factors like C/N, lignin/C, polyphenol/N, and lignin/N ratios are indicative of stable humus formation in soils, as supported by various studies (e.g., Singh et al. 2020; Zhang et al. 2008). Notably, litter quality plays a pivotal role in decomposition processes, with our findings underscoring that the lignin/N and C/N ratios bear the most substantial relationship with decomposition rates. This observation aligns with insights from other researchers like Smolander et al. (2008) and Walela et al. (2014), justifying the interplay between these ratios and decomposition rate constants.

12.5 Role of Plant-Microbe Interactions in Soil Carbon Sequestration: Insights from Microbial Biomass Carbon Dynamics

Interactions between plants and microbes, as well as the dynamics of the C and N cycles, are pivotal in SOC sequestration (Singh et al. 2020). Even though microbial biomass carbon (MBC) constitutes a minor segment of SOC, it governs the SOC distribution among inputs, lesser organic carbon fractions, and greater organo-mineral organic carbon. Ruess and McNaughton (1987) highlighted a direct correlation between microbial biomass C and total soil organic C. After 25 years of restoration from the initial dump, a notable difference in soil MBC was observed across all species. Values ranged from 45 to 2974 $\mathrm{ug\,C\,g^{-1}}$ soil, consistently increasing with reclamation duration (Table 12.2). On average, *D. sissoo* showed the highest MBC, followed by *A. indica* and *G. arborea* in surface soil post-25-year reclamation. Other studies, like Singh et al. (2015), found similar observations. Notably, the MBC content in 8-year-old soil was significantly less than the 25-year-old restored mine soil; average values peaked at 1656 $\mathrm{ug\,C\,g^{-1}}$ in soil restored for 25 years. *D. sissoo*, in particular, had superior MBC content, likely owing to its low litter C/N, lignin/N, and Polyphenol/N ratios, enhancing litter decomposition rates and carbon efficiency in microbes and facilitating SOC movement to deeper soil layers.

Table 12.2 Influence of tree species and years of reclamation on MBC (ug C g^{-1} of soil) at two depths.

Depth (cm)	*Azadirachta indica* 8 yrs	14 yrs	25 yrs	Mean T×D	*Dalbergia sissoo* 8 yrs	14 yrs	25 yrs	Mean T×D	*Gmelina arborea* 8 yrs	14 yrs	25 yrs	Mean T×D	Recent dump
0–15	$679^{h¶}$	977^{g}	1449^{d}	$1035^{C£}$	2194^{b}	228^{j}	1338^{e}	1254^{B}	370^{i}	392^{i}	1161^{f}	641^{D}	58^{f}
15–30	347^{i}	995^{g}	1818^{c}	1053^{C}	89^{k}	1480^{d}	2974^{a}	1515^{A}	249^{j}	223^{j}	1199^{f}	557^{E}	45^{g}
Mean T×Y	513^{F}	986^{D}	1633^{B}		1141^{C}	855^{E}	2156^{A}		310^{G}	308^{G}	1180^{C}		

	Azadirachta indica	*Dalbergia sissoo*	*Gmelina arborea*
Mean tree	1044^{B}	1384^{A}	599^{C}
	8 yrs	14 yrs	25 yrs
Mean year	654^{C}	716^{B}	1656^{A}
	0–15 cm	15–30 cm	
Mean depth	977^{B}	1042^{A}	

Means followed by the same letter(s) are not significantly different ($p < 0.05$).
£Capital letters have been used to differentiate the means. T = Tree species (represents the significant difference among the tree species). D = Depth (represents the significant difference at different depths).

In this research, soil MBC content showed an increasing trend from the recent dump to 25 years of restoration. This progression underscores the gradual rejuvenation of soil microbial communities over the reclamation phase. Correlational analyses showcased a significantly strong positive relationship between MBC and total organic carbon (TOC) ($r = 0.46$, $p < 0.01$), resonating with findings from Adeli et al. 2013 and Ingram et al. 2005. This showed that organic matter as the primary nutrient source for soil microbes increases with an increase in SOC content.

12.6 Different Pools of Soil Organic Carbon in Restored Mine Soil

SOC pools are typically categorized into very labile, labile, non-labile, and resistant carbon pools based on their decomposition rate and turnover duration. The distribution of organic carbon, including the very labile and labile pools, peaked for *G. arborea* in eight years of reclamation. In contrast, resistant carbon was most prominent in 25 years of reclamation for *D. sissoo*. Throughout the restoration phase, the very labile carbon made up only 4.10–32.14% of the SOC. The average labile carbon showed a decrease from 11.10% in 8 years to 8.20% in 25 years. Notably, the mean values for very labile and labile pools fluctuated with soil depth, while non-labile and resistant pools showed an increasing trend.

In terms of the non-labile carbon pool, there was a notable decrease for *A. indica* and *G. arborea*, but an increase for *D. sissoo* as the years of reclamation increased. The resistant carbon pool was found highest under *D. sissoo*, followed by *G. arborea* and *A. indica*. The resistant pool, among the four carbon pools, showed a steady increase with increased reclamation years. Chaudhuri et al. (2015) also found a consistent increase in the resistant SOC pool over time, suggesting more substantial humic-like substances in older mine soils. This work confirms that SOC pools play a pivotal role in overall carbon storage in restored ecosystems.

Active carbon, a primary SOC component, originates from plant and root remnants (Liu et al. 2017). These carbons undergo rapid decomposition by soil microbes, facilitating SOC mineralization (Jha et al. 2012). Contrarily, resistant SOC pools, less susceptible to microbial degradation, persist for over 25 years, as validated by multiple studies (Zhang and Zhou 2018). As a consequence, resistant SOC accumulates in the soil as the years of biological reclamation progress (von Lützow et al. 2007). Biological reclamation of mine soil not only improves the active pool of carbon but also increases the resistant pool of SOC due to physical protection provided by soil aggregates (An et al. 2010; Hyvönen et al. 2007; Pandey et al. 2014). Consequently, the SOC content increases, indicating transitions between distinct carbon pools. As the years of restoration progress, the vegetative surface cover expands, enhancing polycondensation (Alexandrova 1980). Correlation studies have unveiled a strong negative link between non-labile C and the leaf Lignin/N ratio. Given its elevated Lignin/N ratio, *A. indica* decomposes more slowly, resulting in lower resistant carbon. On the other hand, *D. sissoo*, with its minimal Lignin/N and polyphenol/N ratios, showed the highest resistant carbon.

12.7 Total Organic Carbon, C Stock, and C Sequestration in Reclaimed Mine Soil

The TOC content in the soil showed an increasing trend along the re-vegetation age continuum. A twofold increase in the soil's TOC content was observed across a re-vegetation age gradient spanning from 0 to 25 years (Table 12.3). Notable variations in TOC were evident in the surface soil between 8 and 25 years post-reclamation. As the depth transitioned from 0–15 cm to 30–45 cm, the SOC content registered a decrease of 76%, 79%, and 59% under the *A. indica*,*D. sissoo*, and *G. arborea* trees, respectively, within the 25-year reclamation. The TOC levels were predominantly elevated in the surface layers compared to the sub-surface layers, likely attributed to the deposition of leaf litter and its subsequent transformation into humus in the topsoil.

The average total C stock was determined to be 70.69, 57.97, and 38.64 $Mg\,C\,ha^{-1}$ for *A. indica*, *D. sissoo*, and *G. arborea*, respectively. The C stock in soil was notably higher

Table 12.3 TOC (%) of soil with the reclamation age under three tree species.

Depth (cm)		0–15	15–30	30–45	45–60	$Mean_{T \times Y}$
Azadirachta indica	8 yrs	2.64^{gh}	2.69^{gh}	1.16^{lm}	1.71^{k}	2.05^{E}
	14 yrs	3.23^{ef}	2.43^{hij}	1.49^{kl}	1.42^{kl}	2.14^{E}
	25 yrs	5.10^{b}	4.41^{c}	4.37^{c}	1.16	4.77^{A}
	$Mean_{T \times D}$	$3.66^{A£}$	3.18^{B}	2.34^{E}	2.78^{C}	
Dalbergia sissoo	8 yrs	2.20^{ij}	2.19^{ij}	2.26^{ij}	2.08^{j}	2.18^{E}
	14 yrs	4.50^{c}	2.97^{fg}	2.53^{hi}	1.45^{kl}	2.86^{D}
	25 yrs	3.67^{d}	2.19^{ij}	4.61^{c}	4.37^{c}	3.71^{C}
	$Mean_{T \times D}$	3.46^{A}	2.45^{DE}	3.13^{B}	2.63^{CD}	
Gmelina arborea	8 yrs	0.81^{n}	0.29^{p}	0.23^{p}	0.21^{p}	0.38^{G}
	14 yrs	1.50^{kl}	0.91^{mn}	0.75^{n}	0.64^{no}	0.95^{F}
	25 yrs	5.65^{a}	4.41^{c}	3.34^{e}	3.89^{d}	4.32^{B}
	$Mean_{T \times D}$	2.65^{CD}	1.87^{F}	1.44^{G}	1.58^{G}	
Recent dump		0.34^{h}	0.24^{h}	0.36^{h}	0.16^{h}	0.27^{C}

	Azadirachta indica	*Dalbergia sissoo*	*Gmelina arborea*	
Mean tree	2.98^{A}	2.92^{A}	1.88^{B}	
	8 yrs	14 yrs	25 yrs	0 yrs
Mean year	1.53^{C}	1.98^{B}	3.27^{A}	0.27^{C}
	0–15 cm	15–30 cm	30–45 cm	45–60 cm
Mean depth	2.96^{A}	2.27^{B}	2.11^{C}	2.11^{C}

Means followed by the same letter(s) are not significantly different ($p < 0.05$).
£Capital letters have been used to differentiate the means. T = Tree species (represents the significant difference among the tree species). D = Depth (represents the significant difference at different depths).

Table 12.4 C stock (Mg C ha^{-1}) of soil with the reclamation age under three tree species

Depth (cm)		0–15	15–30	30–45	45–60	$Mean_{T*Y}$
Azadirachta	8 yrs	59.00^{fghi}	63.41^{fg}	23.90^{qr}	41.71^{mno}	47.01^{E}
indica	14 yrs	57.71^{fghij}	53.28^{hijk}	47.77^{jkilmn}	38.11^{nop}	49.22^{E}
	25 yrs	106.98^{c}	96.16^{d}	118.24^{b}	141.96^{a}	115.83^{A}
	$Mean_{T \times D}$	$74.56^{A£}$	70.95^{AB}	63.30^{CD}	73.92^{A}	
Dalbergia	8 yrs	43.86^{klmno}	51.91^{ijkl}	54.38^{ghij}	43.41^{lmno}	48.39^{E}
sissoo	14 yrs	83.33^{e}	55.62^{ghij}	49.37^{ijklm}	29.37^{pq}	54.42^{D}
	25 yrs	67.24^{f}	35.36^{op}	101.45^{cd}	80.31^{e}	71.09^{C}
	$Mean_{T \times D}$	64.81^{BC}	47.63^{E}	68.40^{ABC}	51.03^{E}	
Gmelina	8 yrs	19.87^{rs}	7.65^{t}	8.91^{t}	5.36^{t}	10.44^{G}
arborea	14 yrs	35.73^{op}	20.65^{qr}	10.94^{st}	11.31^{st}	19.66^{F}
	25 yrs	118.43^{b}	79.31^{e}	62.06^{fgh}	83.53^{e}	85.83^{B}
	$Mean_{T \times D}$	58.01^{D}	35.87^{F}	27.30^{G}	33.40^{FG}	
Recent dump		7.99^{h}	5.67^{h}	8.61^{h}	3.85^{h}	6.53^{D}

	Azadirachta indica	***Dalbergia sissoo***	***Gmelina arborea***	
Mean tree	70.69^{A}	57.97^{B}	38.64^{C}	
	8 yrs	14 yrs	25 yrs	
Mean year	35.28^{C}	41.10^{B}	69.82^{A}	
	0–15 cm	15–30 cm	30–45 cm	45–60 cm
Mean depth	60.01^{A}	46.90^{B}	48.56^{B} 47.89^{B}	47.89^{B}

Means followed by the same letter(s) are not significantly different ($p < 0.05$).
£Capital letters have been used to differentiate the means. T = Tree species (represents the significant difference among the tree species). D = Depth (represents the significant difference at different T = Tree species (represents the significant difference among the tree species). D = Depth (represents the significant difference at different depths).

for soil restored over 25 years compared to recent restorations (Table 12.4). Among the three tree species, *A. indica* registered the maximum carbon stock. Mukhopadhyay and Masto (2016) assessed the evolution of the C pool in rehabilitated coal mine soils of a tropical climate, revealing a SOC of <0.50% after five years of afforestation, with SOC stock linearly rising with afforestation age. Notably, SOC stock increase in secondary forests is contingent upon soil characteristics and vegetation types (Paz et al. 2016). Factors influencing mine soil attribute modifications include afforestation age, plant species, ambient climate, and the underlying bedrock (Shrestha et al. 2013). In the study of reclaimed mine soil (0–30 cm) within the dry tropical climate of India, Tripathi et al. (2016) documented 13.70 Mg C ha^{-1} after four years of afforestation. Ahirwal and Maiti (2017) noted a C accrual rate of 1.9 Mg C ha^{-1} $year^{-1}$ after 14 years of reclamation, while Tripathi et al. (2014) observed a rate of 1.35 Mg C ha^{-1} $year^{-1}$ over 19 years of reclamation in Singrauli, India.

Table 12.5 C sequestration (Mg C ha^{-1}) of soil with respect to recent dumps with the reclamation age under three tree species.

Depth (cm)		**0–15**	**15–30**	**30–45**	**45–60**	**Mean$_{T\times Y}$**
Azadirachtaindica	8 yrs	51.01^{efgh¶}	57.74^{ef}	15.29^{p}	37.85^{jklm}	40.47^{E}
	14 yrs	49.71^{efghij}	47.61^{fghij}	39.16^{jklm}	34.26^{lmno}	42.68^{E}
	25 yrs	98.90^{c}	90.49^{bcde}	109.63^{abc}	138.10^{ijkl}	109.28^{A}
	Mean $_{T\times D}$	66.54$^{A£}$	65.28^{A}	54.69^{BC}	70.07^{A}	
Dalbergia	8 yrs	35.86^{klmn}	46.24^{ghijk}	45.77^{fghijk}	39.55^{ijklm}	41.86^{E}
sissoo	14 yrs	75.33^{d}	49.95^{efgh}	40.76^{hijkl}	25.52^{o}	47.89^{D}
	25 yrs	59.25^{e}	29.69^{mno}	92.84^{c}	76.46^{d}	64.56^{C}
	Mean $_{T*D}$	56.81^{B}	41.96^{E}	59.79^{B}	47.18^{DE}	
Gmelina	8 yrs	11.87^{p}	1.98^{q}	0.30^{q}	1.50^{q}	3.91^{G}
arborea	14 yrs	27.73^{no}	14.98^{p}	2.33^{q}	7.46^{pq}	13.12^{F}
	25 yrs	110.43^{b}	73.64^{d}	53.45^{efg}	79.68^{q}	79.30^{B}
	Mean$_{T\times D}$	50.01^{CD}	30.20^{F}	18.69^{G}	29.55^{F}	

	Azadirachta indica	***Dalbergia sissoo***	***Gmelina arborea***	
Mean tree	64.15^{A}	51.44^{B}	32.11^{C}	
	8 yrs	14 yrs	25 yrs	
Mean year	28.75^{C}	34.57^{B}	84.38^{A}	
	0–15 cm	15–30 cm	30–45 cm	45–60 cm
Mean depth	57.79^{A}	45.81^{C}	44.39^{C}	48.93^{B}

Means followed by the same letter(s) are not significantly different ($p < 0.05$).
£Capital letters have been used to differentiate the means. T = Tree species (represents the significant difference among the tree species). D = Depth (represents the significant difference at different depths).

The highest carbon sequestration was noted for *A. indica*, followed sequentially by *D. sissoo* and *G. arborea* (Table 12.5). The rate of SOC sequestration in this study aligns closely with data from various global regions. For instance, Frouz et al. (2009) documented SOC sequestration rates ranging between 0.1 and 1.2 Mg C ha^{-1} year^{-1} in 22- to 32-year-old mine soils in the Czech Republic. Similarly, Yuan et al. (2017) indicated a range of 0.2–2.8 Mg C ha^{-1} year^{-1} in 22- to 25-year-old reclaimed mine soils in China. In Polish mine soils, Pietrzykowski and Krzaklewski (2010) noted an accumulation rate of 1.5 Mg C ha^{-1} year^{-1}. While direct comparisons across varied sites might be challenging, the rate of SOC accumulation is influenced by factors such as site age, vegetation type, and climatic conditions (Vinduskova and Frouz 2013). In our study, the highest mean carbon sequestration was found in a 25-year-old restored dump. In the Indian tropical climate's 16-year-old *D. sissoo* plantation on a post-mining site, a SOC stock of 9.82 Mg C ha^{-1} was reported (Mukhopadhyay and Masto 2016).

Variability in SOC stocks of rehabilitated mine soil can be attributed to the diversity of plant species and the inherent quality of mine soil. For instance, a noticeable increase in carbon stock was observed with *A. indica* 25 years post-restoration. Over time, the decomposition of litter and residual plant matter releases nutrients via mineralization, thereby enhancing soil nutrient levels and C concentrations. Pietrzykowski and Daniels (2014) found a carbon stock of 69.4 Mg C ha^{-1} in 25-year-old regenerated post-mining locales in Poland.

Tripathi et al. (2014) noted that revegetated mine soil could effectively absorb atmospheric CO_2. This study revealed a consistent rise in TOC content with the age of reclamation, suggesting the positive influence of time on TOC development in rehabilitated soils. Numerous chronosequence studies have demonstrated the ongoing accumulation of SOC in restored soils, highlighting its potential for carbon storage (Singh et al. 2015). The diminished TOC content in newly reclaimed lands can be attributed to escalated SOC mineralization, soil erosion resulting from inadequate soil aggregation, increased soil compaction, SOC leaching, and reduced litter deposition during the early stages of restoration (Shrestha and Lal 2006). Among the observed species, *A. indica* recorded the highest SOC accumulation, followed by *D. sissoo* and *G. arborea*. This indicates a substantial deposit of organic residues in restored mine soils and a greater prevalence of humic-like substances in mature mine soils. A study involving a 16-year-old regenerated mine spoil with *D. sissoo* revealed an ecosystem C pool growth rate of 3.78 Mg C ha^{-1} $year^{-1}$. In contrast, a mixed forest stand on a 14-year-old afforested coal mine dump showcased a potential carbon storage of 5.20 Mg C ha^{-1} $year^{-1}$ (Ahirwal and Maiti 2017).

Deng et al. (2016) reported that SOC was positively related to both litter and root biomass following afforestation. The establishment of forestland can decrease artificial disturbance and subsequently enhance the SOC (Singh and Seema 2017). This is related to a different mechanism of SOC sequestration for the topsoil and subsoil. This increase in SOC in topsoil may be associated with a higher carbon input and lower rates of SOC decomposition following afforestation. The mechanisms associated with the SOC accumulation in the subsoil are different from those in the topsoil. Our results showed that soil C represented an important part of the total C accumulated at post-mining sites. The rate of soil C accumulation was similar to that reported from other restored forests (Akala and Lal 2000; Garten 2002). Several studies have shown that reforestation could improve soil conditions in mine areas by increasing the quantity of organic matter and available nutrients (Reversat et al. 1988; Parrotta 1992).

Plant productivity has been regarded as the most important process influencing organic matter adding to the soil (Zak et al. 1994; Liao and Boutton 2008). Therefore, annual inputs of organic matter through plant residues would increase with stand age due to the increasing tree productivity (Zak et al. 1994). Gonzalez-Sangregorio et al. (1991) found that soil TOC could increase from 2 to 14 mg g^{-1} mostly during the first year after seeding of lignite mine spoil in Spain, and reaching 23 mg g^{-1} at the end of the third year in the surface soil. Soil TOC had shown a slight increase after plantation establishment for eight years in this study, suggesting that soil quality was improved. According to the study by Yang et al. (2010) TOC increase was 2.3% in *Robiniapseudoacacia* plantation, 2.1% in *Ziziphus jujube var. spinosa* plantation, and 1.9% in *Ulmus. pumila* plantation, which also implied species difference in accumulation soil TOC. The increased C may be because of the above litter fall decomposition which contributed to the organic matter increasing in soil. Some species can accumulate nutrients more than others, or they can affect the accumulation of a certain nutrient, and others cannot (Singh et al. 2004). Our analysis of variance also indicated TOC difference in coal mine spoil due to species and restoration age.

12.8 Insights from Spectroscopic Analysis on Soil Organic Carbon Characteristics in Restored Mine Soils

Higher E4/E6 ratios signify a predominance of aliphatic carbon compounds, while lower E4/E6 ratios suggest a greater presence of aromatic carbon functional groups (McDonald et al. 2004). Factors such as molecular dimensions, ambient pH, oxygen levels, carbon composition, carboxylic group concentrations, and the genesis and maturity of the humic substance affect the E4/E6 ratio (Stevenson 1994). Across the chronosequence, mean E4/E6 ratios for HAs decreased for all three tree species, with the minimum observed for *D. sissoo*. Absorbance intensities at 465 nm relate to moderately complex organic carbon molecules, encompassing carbohydrates and proteins. In contrast, the 665 nm band is indicative of intricate, humified biomolecules (D'Orazio and Modelli 2010). The E4/E6 value has an inverse relationship with aromatic C=C absorption intensity; a lower value signifies increased aromaticity, molecular compaction, and a higher molecular mass (Chen et al. 2002). The reduced E4/E6 ratios in 25-year-old restored mine soils and for *D. sissoo* imply an increase in aromaticity with chronosequence progression (Naidja et al. 2002). In our analysis, a higher E4/E6 ratio was observed in 8-year restored soil compared to 25-year reclaimed soil (Table 12.6). This suggests that in the 25-year-old restored soil, SOC molecules evolved, showing more abundant, condensed aromatic C variants with increased humification.

The FT-IR spectra are efficient in pinpointing major functional groups and vast classes of biomolecules in SOM, such as carbohydrates, lignins, cellulose, and amino acids, leveraging the vibrational attributes of their structural bonds. In this study, the FT-IR spectral techniques primarily aimed to deepen insights into the molecular features of SOC in mine terrains. Thus, by dissecting and chemically analyzing the organic matterof restored soil, we can deduce optimal management or plant types for soil quality enhancement at initial phases, expediting revegetation. Schiavo et al. (2007) employed FTIR for assessing the HA of a terrain deteriorated by clay extraction, vegetated with diverse species, noting species-induced variances in the fraction's composition. The FT-IR technique in our context sought to assess the "relative presences" of vast functional group classes in mine soils. Specific absorption bands identified in the mine soils (Figure 12.1) included broadband from 2983 to 3000 cm^{-1} (indicative of CH_3, CH_2, and NH_3stretch); a peak within 1862–1820 cm^{-1}, signifying C=O stretch in cyclic and mixed anhydrides; bands above 3400 cm^{-1} marking O—H and N—H stretches; a 1695 cm^{-1} peak denoting COOH vibrations; a 1500–1490 cm^{-1} peak for aromatic carbonyl groups; 873–728 cm^{-1} linked to aromatic C—H groups; and a proximate 1500 cm^{-1} peak echoing N—H, C=N (amide II band), and C=C.

After 25 years of reclamation, similar absorption bands were observed for all tree species. However, the intensities at 1695 cm^{-1} in HA of *D. sissoo* for surface soil of 25-year-old restored soil were higher, indicating the presence of – COOH vibrations, ascribed to the higher complexity of the HA fraction (Schiavo et al. 2007). In 25-year restored soil, the

Table 12.6 Influence of tree species and years of reclamation on E_{465}/E_{665} at two depths.

	Azadirachta indica				*Dalbergia sissoo*				*Gmelina arborea*			
Depth (cm)	**8 yrs**	**14 yrs**	**25 yrs**	**Mean T × D**	**8 yrs**	**14 yrs**	**25 yrs**	**Mean T × D**	**8 yrs**	**14 yrs**	**25 yrs**	**Mean T × D**
0–15	3.73^{bc}	3.87^{b}	3.52^{bcd}	$3.71^{A£}$	2.89^{ef}	3.46^{bcd}	3.39^{cd}	3.25^{B}	3.40^{cd}	3.44^{bcd}	2.74^{f}	3.19^{B}
15–30	3.55^{bcd}	3.83^{bc}	3.51^{bcd}	3.63^{A}	2.84^{ef}	3.13^{def}	3.51^{bcd}	3.16^{B}	3.21^{de}	3.21^{de}	4.27^{a}	3.56^{A}
Mean T × Y	3.64^{AB}	3.85^{A}	3.52^{BC}		2.87^{D}	3.29^{C}	3.45^{BC}		3.30^{C}	3.32^{C}	3.50^{BC}	

	Azadirachta indica	*Dalbergia sissoo*	*Gmelina arborea*
Mean tree	3.67^{A}	3.20^{C}	3.38^{B}
	8 yrs	14 yrs	25 yrs
Mean year	3.27^{B}	3.49^{A}	3.49^{A}
	0–15 cm	15–30 cm	
Mean depth	3.38^{A}	3.45^{A}	

Means followed by the same letter(s) are not significantly different ($p < 0.05$).
£Capital letters have been used to differentiate the means. T = Tree species (represents the significant difference among the tree species). D = Depth (represents the significant difference at different depths).

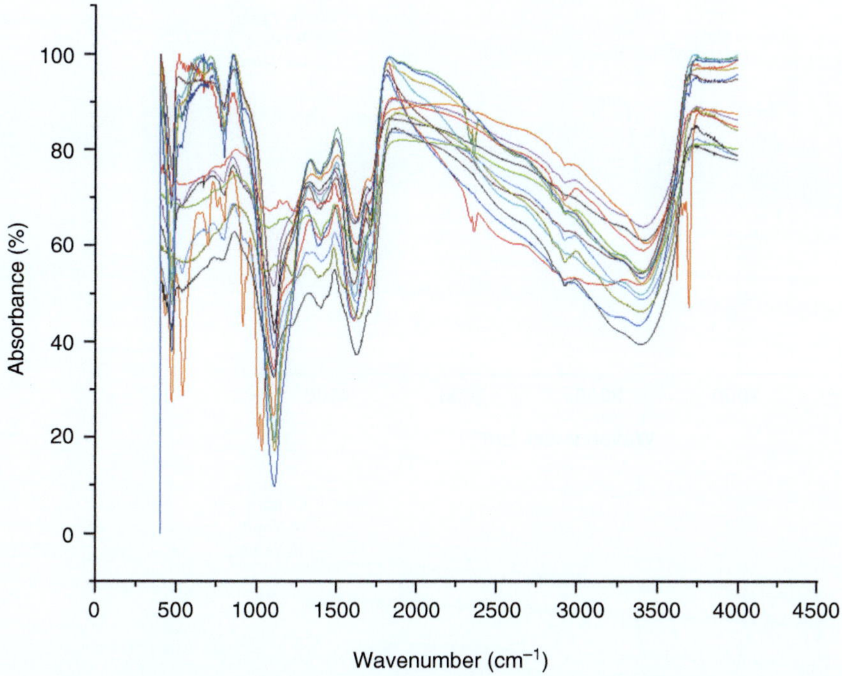

Figure 12.1 Spectral pattern of humic acid extracted from restored coalmine soil.

intensities were lowest in the 2920, –2850 cm^{-1} and highest in the 1685, 1425, and 1285 cm^{-1} regions, suggesting a greater contribution of aromatic compounds than of aliphatic structures to OM composition. Low plant residue input in the soil in 8-year restored soil may have caused a lower proportional contribution of aromatic groups. The functional group aromatic C=C hydrogen bonded C=O double bond conjugated with carbonyl and COO vibrations varied from 1618 to 1624 $cm^{-1}$and increased in the sequence *D. sissoo* > *G. arborea* > *A. indica*, indicating the lowest concentration of recalcitrant compounds in *A. indica*. Probably, due to the revegetation, the aromatic functional group in the 25-year-old restored soil was high, suggesting the presence of recalcitrant compounds, rich in unsaturated bonds and also the input of labile compounds due to biomass addition to the soil.

Although the absorption spectra of HA samples extracted from various chronosequence mine soils and tree species manifested consistent peaks, the intensity of absorption varied notably (Figures 12.2–12.5). Notably, in species such as *A. indica*, *D. sissoo*, and *G. arborea*, there was a marked increase in the relative proportions of aromatic groups through the chronosequence. Specifically, in *D. sissoo* and *G. arborea*, functional groups such as COO—, —C—NO_2, C=C, and aromatic C—H exhibited higher concentrations in 25-year-old reclaimed soils compared to 8-year-old soils. This observation aligns with the findings of Chaudhuri et al. (2015), suggesting that SOC molecules in mature mine soils predominantly comprise aromatic carbonyl and C—H groups.

Further, the FT-IR spectra in restored mine soils for the aforementioned tree species highlight an amplifying trend in the relative proportions of aromatic groups through the chronosequence, pointing to an increased presence of stable SOC components in aged

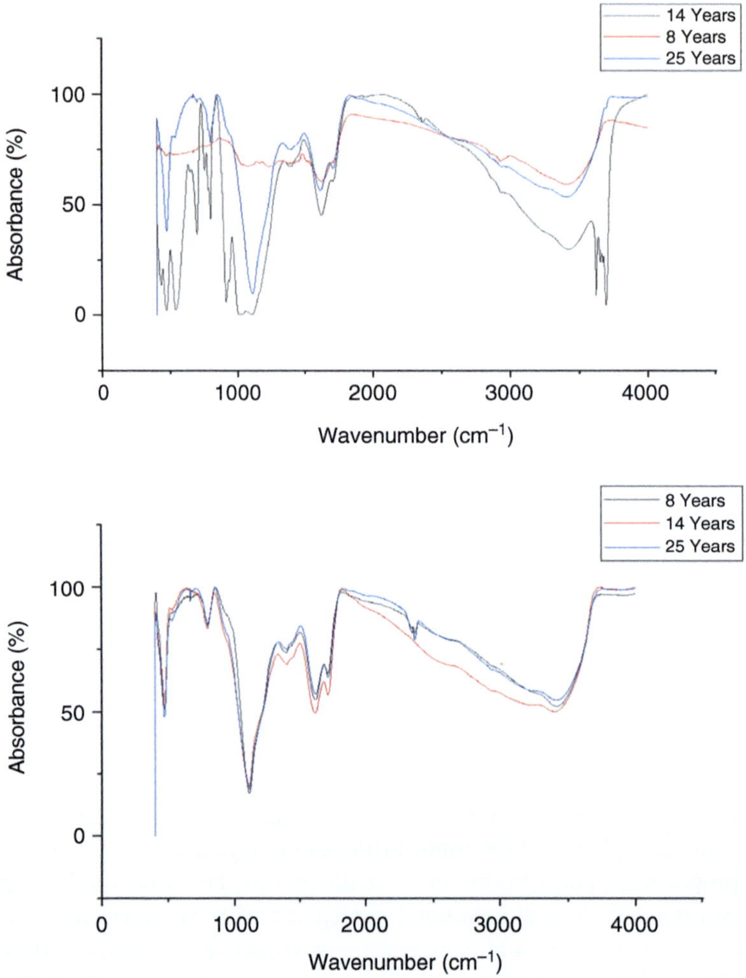

Figure 12.2 Spectral pattern of humic acid extracted from soil under *Dalbergia sissoo* at 0–15 and 15–45 cm depth.

mine soils. Such a pattern could be attributed to the relatively lower litter deposition during the initial restoration years, which culminated in lesser organic matter formation. However, as the restoration period extended, an accumulation of plant residues in the soil became evident (Liu et al. 2013). In species such as *D. sissoo* and *G. arborea*, higher concentrations of functional groups were observed in 25-year-old soils versus 8-year-old soils, corroborating insights derived from the E4/E6 ratio. Over time, the SOC molecules in older mine soils enriched in aromatic C species, showed increased substitutions by quinones and ketonic entities, resonating with the observed spectral ratios.

The spectroscopic indices used in this study were used to obtain complementary information about the structure and composition of the broad classes of functional groups associated with the organic molecules in the mine soils and gain a holistic view of the overall SOC molecular characteristics in the chronosequence. Results indicated that the SOC

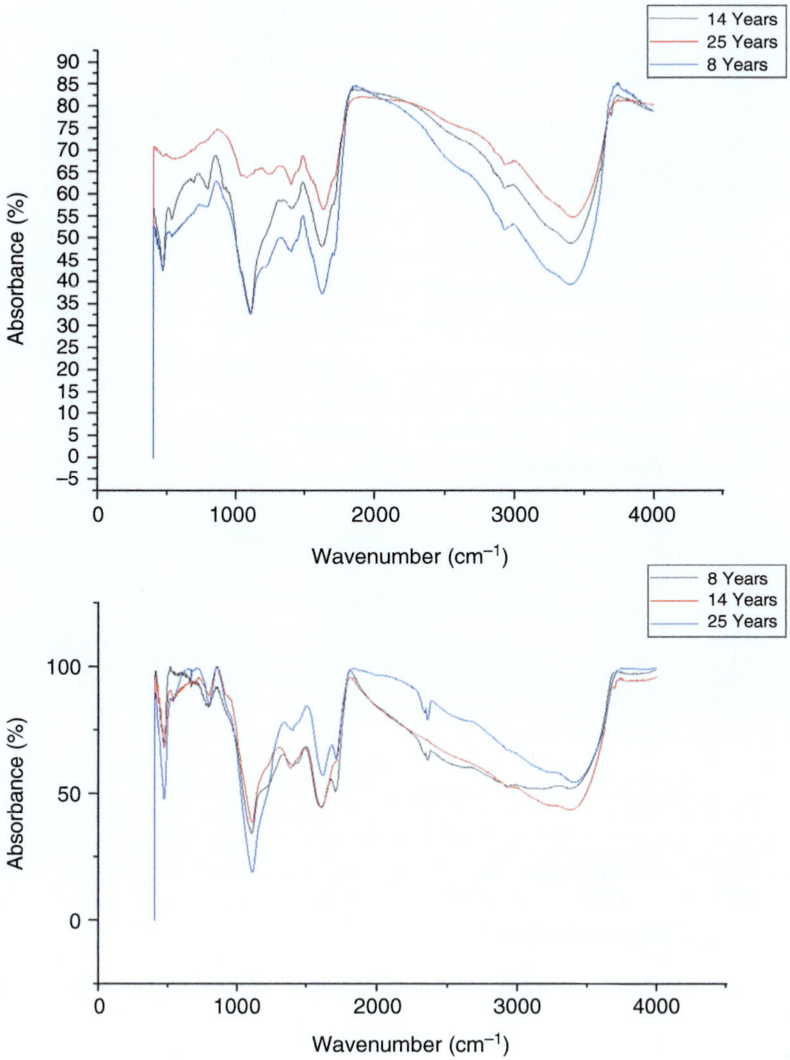

Figure 12.3 Spectral pattern of humic acid extracted from soil under *Azadirachta indica* at 0–15 and 15–45 cm depth.

molecules in the older mine soils were comprised of highly humified polyaromatic and polycondensed species with higher proportions of O-containing and N-containing functional groups. Significant ($r = -0.36$, $p < 0.01$) negative relationships were found between the spectroscopic index E_{465}/E_{665}, and resistant carbon indicating that humification of SOC molecules probably has progressed with increased aromatization of the SOC molecules. According to Harrison (2008) lignin, a complex polymer breaks down during decomposition to give more aromatic compounds. Also, Lignin decomposition products of microbial transformation, including phenolic aldehydes and acids, undergo enzymatic conversion to quinones. The quinones then polymerize in the presence of amino compounds to form

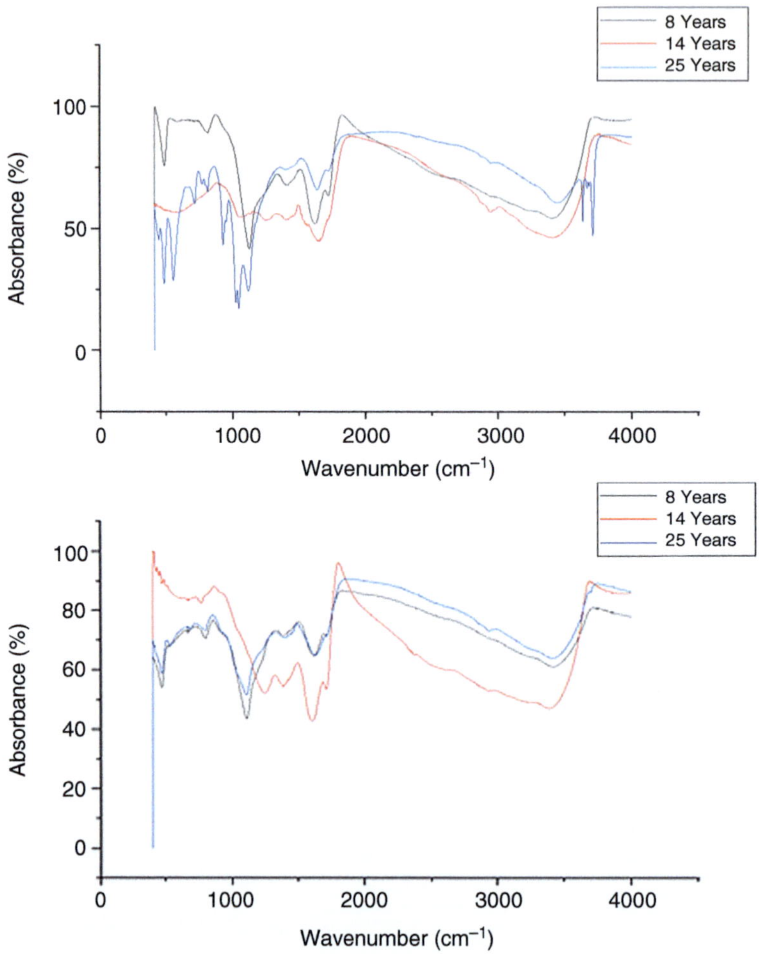

Figure 12.4 Spectral pattern of humic acid extracted from soil under *Gmelina arborea* at 0–15 and 15–45 cm depth.

large humus-like molecules. The result obtained complimented with above finding, *D. sissoo* has medium N content in its leaf residue and lower lignin content, polyphenol/N ratio decomposes easily and forms complex with amino acid to form humus and hence higher aromaticity in the case of *Dalbergia sisso*. A similar result was found by Zhou (2019).

12.9 Conclusion

Success in the reclamation of coal mine sites, especially when the topsoil has been discarded, depends on the rapid formation of a surface soil horizon containing high SOM content. The residual SOM content in the topsoil coupled with input litter quality and quantity determines the amount of SOC sequestered in the soil. The factors and variables

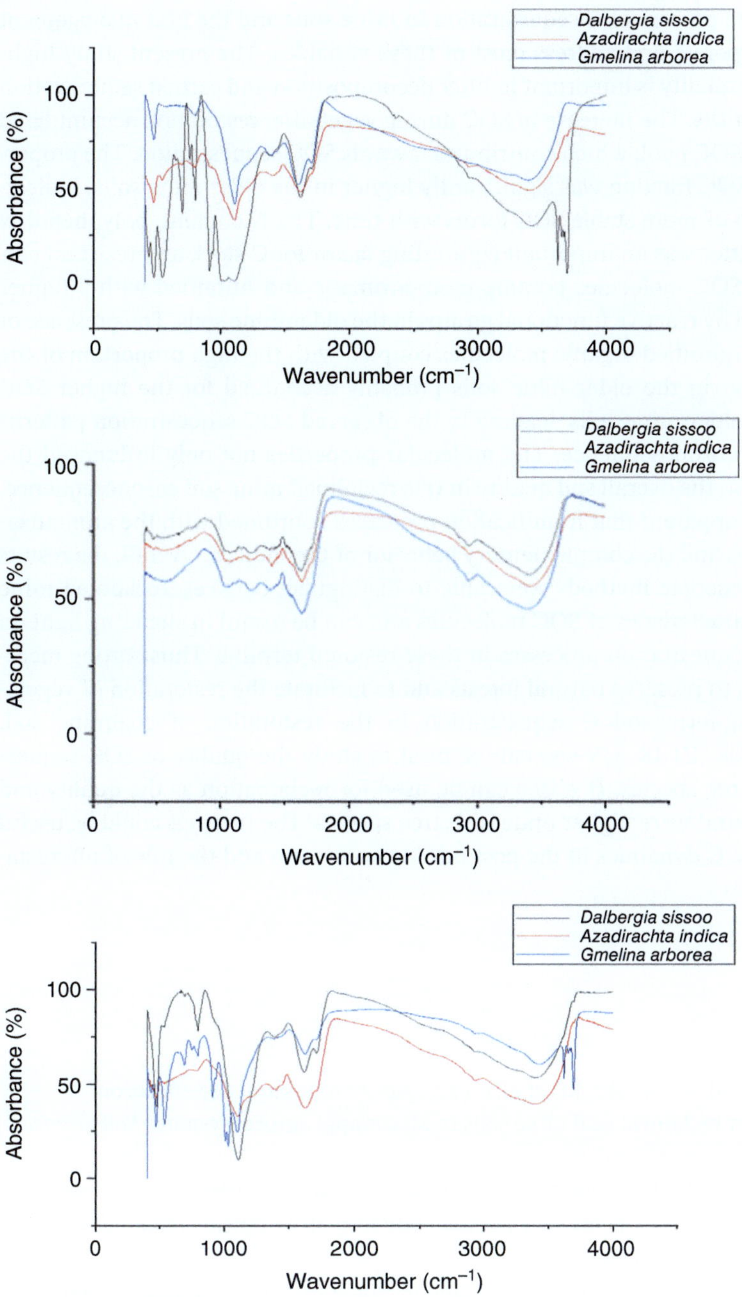

Figure 12.5 Spectral pattern of humic acid extracted from soil under three tree species for surface soil after 8, 14, and 25 years of restoration.

stated above affect the process of C sequestration in mine soils and the best management practice would incorporate and address most of these variables. The present study highlights that plant litter quality is important in litter decomposition and carbon sequestration rather than litter quantity. The increase in SOC during vegetation restoration accumulated mainly as a resistant SOC pool, which contributes towards SOC sequestration. The proportion of the resistant SOC fraction was significantly higher in the older mine soils, indicating the accumulation of more stable SOC forms with time. The N content, polyphenol/N leaf lignin/N in the litter was an important controlling factor for C stock and resistant carbon. Over time, the SOC molecules became more aromatic and humified with a higher degree of substitution by reactive functional groups in the older mine soils. The presence of highly aromatic and humified organic molecules coupled with the high proportion of the resistant SOC fraction in the older mine soils probably accounted for the higher SOC sequestration in the older mine soils, leading to the observed SOC sequestration patterns along this mine soil chronosequence. The molecular properties not only influenced the SOC dynamics but also the overall soil quality in this reclaimed mine soil chronosequence. Therefore, it became apparent that humification processes continued with the aromatization of SOC molecules and the complementary behavior of the indices. Overall, the results indicated that spectroscopic methods were able to distinguish between reclaimed mine soils based on the characteristics of SOC molecules and can be useful in shedding light on SOC dynamics and sequestration processes in these restored terrains. Thus, strong measures need to be taken to preserve natural forests and to facilitate the restoration of vegetation to increase long-term soil C sequestration in the restoration of coalmine soil. Spectroscopic methods (FT-IR, UV-vis) can be used to study the quality of SOC sequestered. Among three tree species, *D. sissoo* can be used for reclamation as the quality and quantity of carbon found were higher under this tree species. The findings could be useful for understanding the C dynamics in the post-mining ecosystem and the role of afforestation in C sequestration.

References

Adeli, A., McLaughlin, M.R., Brooks, J.P. et al. (2013). Age chronosequence effects on restoration quality of reclaimed coal mine soils in Mississippi agroecosystems. *Soil Science* 178: 335–343.

Ahirwal, J. and Maiti, S.K. (2017). Assessment of carbon sequestration potential of revegetated coal mine overburden dumps: a chronosequence study from dry tropical climate. *Journal of Environment Management* 201: 369–377.

Akala, V.A. and Lal, R. (2000). Potential of mine land reclamation for soil organic carbon sequestration in Ohio. *Land Degradation Development* 11: 289–297.

Alexandrova, L.N. (1980). *Soil Organic Matter and Its Transformation Processes*. Moscow, Russia: Leningrad Nauka, 287p.

An, S.S., Mentler, A., Mayer, H., and Blum, W.E.H. (2010). Soil aggregation, aggregate stability, organic carbon and nitrogen in different soil aggregate fractions under forest and shrub vegetation on the Loess Plateau, China. *Catena* 81: 226–233.

Bernhard-Reversat, F. (1988). Soil nitrogen mineralization under a Eucalyptus plantation and a natural Acacia forest in Senegal. *Forest Ecology and Management* 23: 233–244.

Carrera, A.L., Mazzarino, M.J., Bertiller, M.B. et al. (2009). Plant impacts on nitrogen and carbon cycling in the Monte Phytogeographical Province. *Argentina Journal of Arid Environment* 73: 192–201.

Chaudhuri, S., Pena-Yewtukhiw, E.M., McDonald, L.M., and Skousen, J. (2015). Soil organic carbon molecular properties: effects of time since reclamation in a mine soil chronosequence. *Land Degradation and Development* 26: 237–248. https://doi.org/10.1002/ldr.2202.

Chen, J., Gu, B., Leboeuf, E.J. et al. (2002). Spectroscopic characterization of the structural and functional properties of natural organic matter fractions. *Chemosphere* 48: 59–68.

Cotrufo, M.F., Wallenstein, M.D., Boot, C.M. et al. (2013). The Microbial Efficiency-Matrix Stabilization (MEMS) framework integrates plant litter decomposition with soil organic matter stabilization: do labile plant inputs form stable soil organic matter? *Global Change Biology* 19 (4): 988–995.

D'Orazio, V., Mondelli, D., and Senesi, N. (2010). Dissolved organic matter (DOM) evolution during the composting process of coffee by-products. In: *Proceedings of the 15th IHSS Meeting "Advances in Natural Organic Matter and Humic Substances Research"*, vol. 2 (ed. J.A. Gonzales-Perez et al.), 389–392.

Deng, Q., Cheng, X., Hui, D. et al. (2016). Soil microbial community and its interaction with soil carbon and nitrogen dynamics following afforestation in central China. *Science of the Total Environment* 541: 230–237.

Devi, N.B. and Yadava, P.S. (2010). Influence of climate and litter quality on litter decomposition and nutrient release in sub-tropical forest of Northeast India. *Journal of Forest Research* 21 (2): 143–150.

Dutta, R.K. and Agrawal, M. (2001). Litterfall, litter decomposition and nutrient release in five exotic plant species planted on coal mine spoils. *Pedobiologia* 45: 298–312.

Frouz, J., Cienciala, E., Pizl, V., and Kalcic, J. (2009). Carbon storage in post-mining forest soil, the role of tree biomass and soil bioturbation. *Biogeochemistry* 94: 111–121.

Garten, C.T. (2002). Soil carbon storage beneath recently established tree plantations in Tennessee and South Carolina, USA. *Biomass Bioenergy* 23: 93–102.

Gonzalez-Sangregorio, M.V., Trasar-Ceepda, M.C., Leiros, M.C. et al. (1991). Early stages of lignitemine soil genesis: changes in biochemical properties. *Soil Biology and Biochemistry* 23: 589–595.

Harrison RB (2008) Composting and formation of humic substances. Brian D. Fath, S.E. Jorgensen Encyclopedia of Ecology 713–719. Elsevier https://doi.org/10.1016/B978-00804 5405-4.00262-7

Horodecki, P. and Jagodziński, A.M. (2017). Tree species effects on litter decomposition in pure stands on afforested post-mining sites. *Forest Ecology and Management* 406: 1–11. https://doi.org/10.1016/j.foreco.2017.09.059.

Hyvönen, R., Agren, G.I., Linder, S. et al. (2007). The likely impact of elevated CO_2, nitrogen deposition, increased temperature and management on carbon sequestration in temperate and boreal forest ecosystems: a literature review. *New Phytology* 173: 463–480.

Ingram, L.J., Schuman, G.E., Stahl, P.D., and Spackman, L.K. (2005). Microbial respiration and organic carbon indicate nutrient cycling recovery in reclaimed soils. *Soil Science Society of American Journal* 69: 1737–1745.

Jha, P., De, A., Lakaria, B.L. et al. (2012). Soil carbon pools, mineralization and fluxes associated with land use change in vertisols of Central India. *National Academy Science Letters* 35: 475–483.

Krishna, M.P. and Mohan, M. (2017). Litter decomposition in forest ecosystems: a review. *Energy Ecology and Environment* 2: 236–249.

Lajtha, K., Townsend, K.L., Kramer, M.G. et al. (2014). Changes to particulate versus mineral-associated soil carbon after 50 years of litter manipulation in forest and prairie experimental ecosystems. *Biogeochemistry* 119: 341–360.

Liao, J.D. and Boutton, T.W. (2008). Soil microbial biomass response to woody plant invasion of grassland. *Soil Biology and Biochemistry* 40: 1207–1216.

Liu, X., Li, L., Qi, Z. et al. (2017). Land-use impacts on profile distribution of labile and recalcitrant carbon in the ili river valley, northwest China. *Science Total Environment* 586: 1038–1045.

Liu, Y., Su, Y.Q., Zhang, L.L. et al. (2013). Study on dynamic change of organic carbon in young *Robinia pseudoacacia* plantation in Loess Plateau. *Journal of Nanjing Forestry University* 37: https://doi.org/10.3969/j.issn.1000-2006.2013.

von Lützow, M.V., Kögel-Knabner, I., Ekschmitt, K. et al. (2007). SOM fractionation methods: relevance to functional pools and to stabilization mechanisms. *Soil Biology and Biochemistry* 39: 2183–2207.

Macdonald, S.E., Landhäusser, S.M., Skousen, J. et al. (2015). Forest restoration following surface mining disturbance: challenges and solutions. *New forests* 46: 703–732. https://doi.org/10.1007/s11056-015-9506-4.

Maiti, S.K. (2012). *Ecorestoration of the Coalmine Degraded Lands*. Springer Science & Business Media.

Mcdonald, S., Bishop, A.G., Prenzler, P.D., and Robards, K. (2004). Analytical chemistry of freshwater humic substances. *Analytica Chimica Acta* 527: 105–124.

Mukhopadhyay, S. and Masto, R.E. (2016). Carbon storage in coal mine spoil by *Dalbergia sissoo*Roxb. *Geoderma* 284: 204–213. https://doi.org/10.1016/j.geoderma.2016.09.004.

Naidja, A., Huang, P.M., Anderson, D.W., and Van Kessel, C. (2002). Fourier transform infrared, UV-visible, and X-ray diffraction analysis of organic matter in humin, humic acid, and fulvic acid fractions in soil exposed to elevated CO_2 and N-fertilization. *Applied Spectroscopy* 56: 318–324.

Palm, C.A., Gachengo, C.N., Delve, R.J. et al. (2001). Organic inputs for soil fertility management in tropical agroecosystems: application of an organic resource database. *Agriculture, Ecosystems &Environment* 83 (1–2): 27–42.

Pandey, D., Agrawal, M., Bohra, J.S. et al. (2014). Recalcitrant and labile carbon pools in a sub-humid tropical soil under different tillage combinations: a case study of rice-wheat system. *Soil Tillage Research* 143: 116–122.

Parrotta, J.A. (1992). The role of plantation forests in rehabilitating degraded tropical ecosystems. *Agriculture Ecosystem Environment* 41: 115–133.

Parton, W., Silver, W.L., and Burke, I.C. (2007). Global-scale similarities in nitrogen release patterns during long-term decomposition. *Science* 315: 361–364.

Paz, C.P., Goosem, M., Bird, M. et al. (2016). Soil types influence predictions of soil carbon stock recovery in tropical secondary forests. *Forest Ecology Management* 376: 74–83.

Pietrzykowski, M. and Daniels, W.L. (2014). Estimation of carbon sequestration by pine (*Pinus sylvestris* L.) ecosystems developed on reforested post-mining sites in Poland on differing mine soil substrates. *Ecological Engineering* 73: 209–218.

Pietrzykowski, M. and Krzaklewski, W. (2010). Potential for carbon sequestration in reclaimed mine soil on reforested surface mining areas in Poland. *Nature Science* 2: 1015–1021.

Prach, K. and Hobbs, R.J. (2008). Spontaneous succession versus technical reclamation in the restoration of disturbed sites. *Restoration Ecology* 16: 363–366. g/10.1007/s11056-015-.

Ruess, R.W. and McNaughton, S.J. (1987). Grazing and the dynamics of nutrient and energy regulate microbial processes in the Serengeti grasslands. *Oikos* 49: 101–110.

Schiavo, J.A., Canellas, L.P., and Martins, M.A. (2007). Revegetação de cava de extração de argila com Acacia mangium. I – Atributosquímicos do solo, ácidosfúlvicos e húmicos. *Revista Brasileira de Ciência do Solo* 31: 1153–1162.

Shrestha, R.K. and Lal, R. (2006). Ecosystem carbon budgeting and soil carbon sequestration in reclaimed mine soil. *Environment International* 32: 781–796.

Shrestha, R.K., Lal, R., and Rimal, B. (2013). Soil carbon fluxes and balances and soil properties of organically amended no-till corn production systems. *Geoderma* 197: 177–185.

Singh, A.P., Singh, S.K., Rai, S., and Kumar, M. (2020). Soil carbon dynamics in relation to soil surface management and cropping system. *Carbon Management in Tropical and Sub-Tropical Terrestrial Systems* 159–172.

Singh, P., Ghosh, A.K., Kumar, S. et al. (2022a). A molecular and spectroscopic approach to reclamation of coal mine soil using tree species: a case study of Gevra mining area, Korba, India. *J Soil Sci Plant Nutr* 22: 2205–2220. https://doi.org/10.1007/s42729-022-00803-1.

Singh, P., Ghosh, A.K., Kumar, S. et al. (2022b). Influence of input litter quality and quantity on carbon storage in post-mining forest soil after 14 years of reclamation. *Ecological Engineering* 178: 106575.

Singh, P., Ghosh, A.K., and Kumar, S. (2022c). Soil organic matter formation and sequestration in rehabilitated mine soil: the role of input litter quality and quantity. *Indian Forester* 148 (3): 280–284.

Singh, N., Raghubanshi, A.S., and Singh, J.S. (2004). Impact of native tree plantation on mine spoil in a dry tropical environment. *For Ecology and Management* 187: 49–60.

Singh, P., Ram, S., Jayant, H., and Ghosh, A.K. (2015). Redevelopment of soil carbon pools and biological properties on restored mine spoils under plantation. *Journal of Pure Applied Microbiology* 9: 3031–3037.

Singh, P. and Seema (2017). Ecorestoration of coalmine overburden dump. *Environment Conservation Journal* 18 (1 &2): 117–119.

Singh, K.P., Singh, P.K., and Tripathi, S.K. (1999). Litterfall, litter decomposition and nutrient release patterns in four native tree species raised on coal mine spoil at Singrauli, India. *Biology and Fertility of Soils* 29: 371–378.

Smolander, A., Lvula, T., and Kitunen, V. (2008). Response of litter decomposition and soil C and N transformations in a Norway spruce thinning stand to removal of logging residue. *Forest Ecology Management* 1080–1086.

Sroka, K., Chodak, M., Klimek, B., and Pietrzykowski, M. (2018). Efect of black alder (*Alnus glutinosa*) admixture to Scots pine (*Pinus sylvestris*) plantations on chemical and microbial properties of sandy mine soils. *Applied Soil Ecology* 124: 62–68. https://doi.org/10.1016/j.apsoil.2017.10.031.

Stevenson, F.J. (1994). *Humus Chemistry: Genesis, Composition and Reactions*, vol. 496. New York: John Wiley and Sons Publications.

Tripathi, N., Singh, R.S., and Hills, C.D. (2016). Soil carbon development in rejuvenated Indian coal mine spoil. *Ecology England* 90: 482–490.

Tripathi, N., Singh, R.S., and Nathanail, C.P. (2014). Mine spoil acts as a sink of carbon dioxide in Indian dry tropical environment. *Science of the Total Environment* 468: 1162–1171.

Vinduskova, O. and Frouz, J. (2013). Soil carbon accumulation after open-cast coal and oil shale mining in Northern Hemisphere: a quantitative review. *Environment Earth Science* 69: 1685–1698.

Waksman, S.A. and Tenney, F. (1928). Composition of natural organic materials and their decomposition in soil. *Soil Science* 26: 155–171.

Walela, C., Daniel, H., Wilson, B. et al. (2014). The initial lignin: nitrogen ratio of litter from above and below ground sources strongly and negatively influenced decay rates of slowly decomposing litter carbon pools. *Soil Biology and Biochemistry* 77: 268–275.

Xiang, H., Luo, X., Zhang, L. et al. (2022). Forest succession accelerates soil carbon accumulation by increasing recalcitrant carbon stock in subtropical forest topsoils. *Catena* 212: 106030.

Yang, D., Zeng, D., Mao, R., Li, L. (2010). Restoration Effect of Young Plantations on Magnesite Mine Spoil in Northeast China. *4th International Conference on Bioinformatics and Biomedical Engineering*, Chengdu, China, pp. 1–5.

Yuan, Y., Zhao, Z., Zhang, P. et al. (2017). Soil organic carbon and nitrogen pools in reclaimed mine soils under forest and cropland ecosystems in the Loess Plateau. *China, Ecology England* 102: 137–144.

Zak, D.R., Tilman, D., Parmenter, R.R. et al. (1994). Plant production and soil microorganisms in late-successional ecosystems: a continental-scale study. *Ecology* 75: 2333–2347.

Zhang, H., Wang, B., and Xu, M. (2008). Effects of inorganic fertilizer inputs on grain yields and soil properties in a long-term wheat–corn cropping system in South China. *Communications in Soil Science and Plant Analysis* 39 (11–12): 1583–1599.

Zhang, H. and Zhou, Z. (2018). Recalcitrant carbon controls the magnitude of soil organic matter mineralization in temperate forests of northern China. *Forest Ecosystem* 3: 211–220.

Zhou, G., Xu, S., Ciais, P. et al. (2019). Climate and litter C/N ratio constrain soil organic carbon accumulation. *National Science Review* 6 (4): 746–757.

13

Ecosystem Services from Rehabilitated Waste Dumpsites

ML Dotaniya[1], CK Dotaniya[2], Kuldeep Kumar[3], RK Yadav[4], RK Doutaniya[5], HM Meena[6], DK Yadav[7], Saurabh Shukla[8], AO Shirale[7], and Manju Lata[9]

[1] *ICAR-Directorate of Rapeseed-Mustard Research, Bharatpur, India*
[2] *Department of Soil Science and Agricultural Chemistry, SKRAU, Bikaner, India*
[3] *ICAR Indian Institute of Soil and Water Conservation, Dehradun, RS Kota, India*
[4] *Agriculture University Kota, India*
[5] *Department of Agronomy, SKN College of Agriculture, Jobner, India*
[6] *ICAR-Central Arid Zone Research Institute, Jodhpur, India*
[7] *ICAR-Indian Institute of Soil Science, Bhopal, India*
[8] *Faculty of Civil Engineering, Institute of Technology, Shri Ramswaroop Memorial University, Lucknow, India*
[9] *University of Rajasthan, Jaipur, India*

13.1 Introduction

Ecosystem services are key to human life's survival on the earth. These are defined by the sense of utilization and source of services. Any advantageous resource that ecosystems or wildlife offer to humans is considered an ecosystem service (Berthet et al. 2022). The advantages may be minor or major, direct or indirect based on the utility. In broader terms, ecosystem services are provider (facilitator of food and water), regulatory (climatic events and disease), supportive (oxygen and nutrient transformations), and cultural services like recreational and spiritual. Any advantage derived for humans from nature is categorized as a provisioning service (EPA 2000). Beyond food, these services encompass resources like wood, fuel obtained from wood, potable water, oils, and plants suitable for textile and material production, as well as therapeutic advantages. Plants contribute to air purification and water filtration, bacteria facilitate waste decomposition, bees play a crucial role in flower pollination, and tree roots prevent soil erosion by anchoring it in place. These interdependent processes collectively contribute to the cleanliness, sustainability, functionality, and resilience of ecosystems (Pérez-Ruzafa et al. 2020). Regulating services, as another facet, encompasses the advantages derived from ecosystem processes that moderate natural phenomena (Feitosa et al. 2019). These services involve pollination, decomposition, water purification, erosion, and flood control, as well as carbon storage and climate regulation. However, the ecosystem services of some of the researchers are classified based on energy, living, based on sources, and functionality (Sánchez-Moyano et al. 2017).

Biodiversity and Ecosystem Services on Post-Industrial Land, First Edition.
Edited by Vimal Chandra Pandey.

Over a period, the increasing population at the global level demanded huge volumes of foodstuff, medicine for a cure, water for agriculture and industry, space for colonization, etc., putting immanence pressure on natural resources. Shrinking the services providing period imposed different challenges for food for humans and animals (Van Raamsdonk et al. 2020). The emergence of consumer culture has sparked an increase in the manufacture and disposal of goods. Disposable and single-use items, in particular, play a substantial role in the overall generation of waste (Kumar et al. 2021). The rapid pace of industrialization has prompted a growth in production and manufacturing processes. Technological advancements leading to shorter product lifecycles have heightened concerns regarding electronic waste, commonly known as "e-waste." The totality of waste generation is increasing with the pace of time and needs scientific management. Contamination in soil and water natural resources is losing the viability and sustainability of the ecosystem (Hsieh et al. 2021). However, different types of pollution are affecting the ecosystems, but waste generation and its poor management by society are the main reasons. India's urban population produces 188 500 tons of MSW per day (TPD), or 68.8 million tons of MSW yearly. The amount of waste produced in Indian cities ranges from 200 to 870 g a day, depending on the population and lifestyle. The amount of garbage generated per person in India is rising at a pace of 1.3% annually. Cities in western India produce the least amount of rubbish per person, at 440 g per day; whereas, East India, North India, and South India follow with 500, 520, and 560 g per day, respectively. Goa generates the most garbage per day – 620 g – among smaller states, while Manipur generates the least – 220 g per day. Gujarat tops the list of larger states with 395 g per day, followed by Madhya Pradesh and Odisha with 400 g. Indian states like Andhra Pradesh (570 g per day), Jammu & Kashmir (600 g per day), and Tamil Nadu (630 g per day) are among the states with the highest per capita trash generation. Of all the Union Territories, the trash produced per person is highest in the Andaman and Nicobar Islands (870 g per day), and lowest in Lakshadweep Islands (340 g per day). The largest Union Territory, Delhi, generates 650 g of garbage per person each day. The population of the area has a direct bearing on the amount of MSW produced; larger towns and metropolitan areas produce more MSW than do smaller cities (Figure 13.1). Of all Indian cities, Kolkata produces the most MSW (1520 TPD, or 4.2 million TPY) (Saha et al. 2017).

The increasing population needs 333 million tons of food grains by 2050 to feed the 1.67 billion Indians. However, the soil partial factor productivity is going down, and adverse effects on soil health are also observed (Dotaniya et al. 2023a). Long-term application of poor-quality resources through fertilizers, organic manure, irrigation water, pesticides, and insecticides leads to poor health and crop yield potential (Geissen et al. 2021). Inadequate waste management adversely affects the quality of soil, water, and air, resulting in the degradation of ecosystems (Meena et al. 2023a). Reports highlighted instances of wildlife encountering plastic garbage, leading to animals either ingesting it or becoming entangled, rendering them particularly susceptible to harm. Disturbance of saline ecosystems by mediating the soil pH, organic C, biodiversity, and population leads to loss of revenue in terms of crop yield and species (Röthig et al. 2023).

Management of these ecosystems by the addition of organic and chemical soil amendments enhances the crop production potential and also the biodiversity of the regions (Meena et al. 2023b). The application of gypsum for the reclamation of saline soils is the key

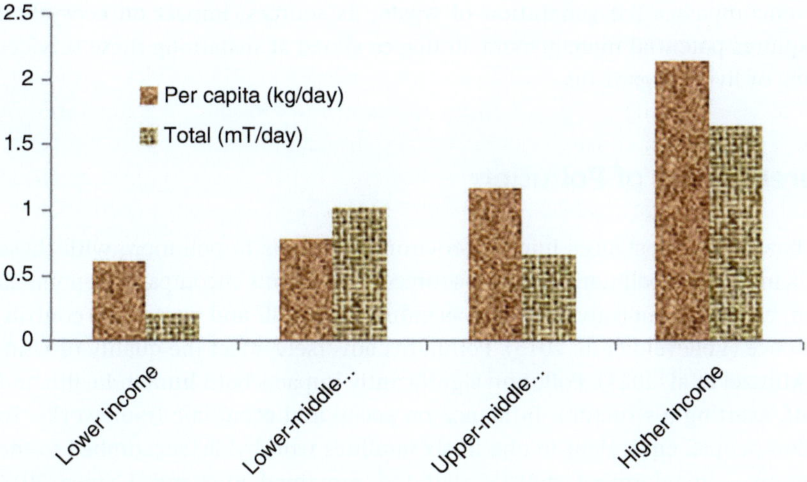

Figure 13.1 Generation of MSW by different income groups.

component to improve soil properties and also the production potential of most crops (Dotaniya et al. 2023b). Fertilizer efficiency was also improved by the addition of soil amendments (Rashmi et al. 2023). The excessive salt concentration from soil solution adsorbed by soil particles and the plant uptake kinetics of essential nutrients are enhanced (Sheoran et al. 2021). These ecosystems are very important for the quality of life of the animals. However, discharging the industrial effluent, sewage, and sludge contamination reduces the auto-cleaning capacity of water bodies. In recent times, society has grown to appreciate the services offered by wetlands, such as water quality enhancement, flood control, wildlife habitat, and recreational opportunities. Despite this heightened recognition, owners of wetlands often find it challenging to derive profits from these services, as the benefits generated are freely enjoyed by the public at large. Wetlands stand as one of the most endangered ecosystems in the United States (Mitsch et al. 2015). With over half of our wetlands lost in the contiguous United States, it becomes imperative to recognize their vital role. A concise overview of the benefits provided by wetlands underscores their importance to humans, emphasizing the urgency of conservation and restoration initiatives (Heimlich et al. 1998).

Effectively managing waste is a complex challenge that demands collaborative efforts from individuals, communities, industries, and governments. Tackling the root causes, adopting sustainable practices, and instilling a shared sense of responsibility can pave the way toward a future with reduced waste and lessened environmental impact (Dotaniya et al. 2018c). Investing in cutting-edge technologies has the potential to enhance waste management and recycling procedures. Additionally, the principle of a circular economy advocates for the re-use and recycling of materials, thereby minimizing overall waste (Kirchherr et al. 2023). Taking decisive action now is crucial to guarantee a cleaner, healthier planet for the well-being of future generations. Communities and nations bear financial burdens associated with the expenses of waste management and cleaning efforts. Embracing resource conservation through recycling and waste reduction not only contributes to environmental well-being but also yields positive economic impacts (Tseng et al. 2018).

This chapter encompasses the generation of waste, its sources, impact on ecosystem services, and explores potential management strategies aimed at sustaining these services for the well-being of living organisms.

13.2 Different Types of Pollution

The release of hazardous substances into the environment leads to pollution, with these harmful agents being termed pollutants. Natural sources of pollutants encompass phenomena like volcanic ash, while human activities, such as industrial runoff and waste, also contribute to their presence (Lelieveld et al. 2015). Pollutants adversely affect the quality of land, water, and air (Münzel et al. 2023). Pollution significantly impacts both human health and the environment, exerting a structural influence on social and economic frameworks. In 2019, nine million people, equivalent to one in six fatalities worldwide, succumbed to the effects of pollution – an alarming statistic that has remained unchanged since 2015 (Carrington 2017; Beil 2017; Fuller et al. 2022). The majority of these deaths were attributed to air pollution (Dickie 2022). A literature assessment reveals that levels of human-induced chemical pollution have exceeded planetary thresholds, posing a global threat to entire ecosystems (Persson et al. 2022). Based on the ecosystem structure, pollution is classified into:

13.2.1 Air Pollution

Air pollution is the discharge of harmful substances into the atmosphere, encompassing gases, particulate matter, and biological load. Primary contributors to this issue include the combustion of fossil fuels, car exhaust, industrial pollutants, and agricultural practices (Plutino et al. 2022). Prolonged exposure to these sources induces significant distress in the surrounding environments (Schraufnagel 2020). Inhaling polluted air leads to human respiratory illnesses, accelerates climate change, causes substantial harm to plants, and results in a decline in air quality. Many essential ecological processes, such as primary production (plant growth) and bio-geochemical cycling, are impacted by air pollution (Figure 13.2). These changes impact ecosystem services, which in turn impacts the advantages that humans receive from the environment (Jones et al. 2014). These advantages include having access to wood, having clean water to drink, and having a greater appreciation for the natural world (Mills et al. 2013).

The process of acidification, which disrupts ecosystems and diminishes biodiversity, is triggered by the deposition of sulfur dioxide (SO_2), nitrogen oxides (NO_X), and nitrous oxide (N_2O). This deposition alters the chemical composition of soils, lakes, rivers, and marine waters (Mitra and Zaman 2021). As SO_2 emissions have significantly decreased in recent decades, NH_3, and NO_X have contributed proportionally more to the acidity of soil and surface water (EEA 2019). United Nations Economic Commission for Europe directed norms for the protection of biodiversity (Table 13.1). To protect forests, $10\,000\,\mu g\,m^{-3}$ per hour from April to September is the crucial ozone exposure limit (UNECE 2022). Access to data on emissions, measurements, and modeling, as well as information on how air pollution affects crops, materials, ecosystems, and human health, are made possible by the different conventions of UNECE.

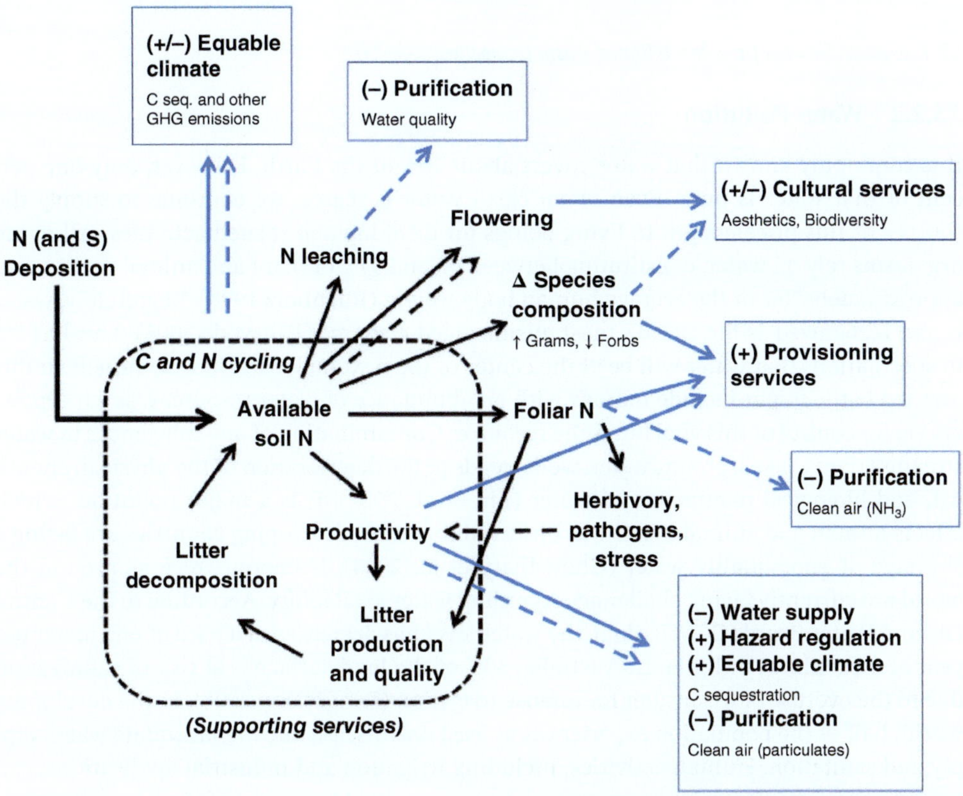

Figure 13.2 Eutrophical model of ecological services mediated by pollution. Here, black arrow-process-based impacts, solid lines (synergistic relationships), dotted lines (antagonistic negative link), and blue solid arrows indicated ecosystem services either increasing (+) or decreasing (–). *Source:* Adapted from Jones et al. (2014).

Table 13.1 Guidelines for ensuring air quality and protecting vegetation.

Pollutant gases	Averaging period	Standard type and concentration	Comments
Ozone (O_3)	AOT40[a] accumulated from May to July	EU target value: 18 000 $\mu g\,m^{-3}$ per hour	Averaged over five years[b]
		EU long-term objective: 6000 $\mu g\,m^{-3}$ per hour	
	AOT40[a] accumulated from April to September	Critical level for the protection of forests: 10 000 $\mu g\,m^{-3}$ per hour	Defined under the UNECE Air Convention
Oxides of nitrogen (NO_X)	Calendar year	EU critical level: 30 $\mu g\,m^{-3}$	
Sulfur dioxide (SO_2)	Winter	EU critical level: 20 $\mu g\,m^{-3}$	1 October to 31 March
	Calendar year	EU critical level: 20 $\mu g\,m^{-3}$	

[a] AOT40 is an indication of accumulated O_3 exposure, expressed in $\mu g\,m^{-3}$ per hour (at 40 ppb).
[b] Only yearly AOT40 values, and not five-year mean.
Source: Adapted from UNECE (2022).

13.2.2 Water Pollution

It is commonly known that water covers about 70% of the Earth. However, only one percent of that water is pure. Even when clean water is scarce, we continue to supply the sources of this priceless gift to living beings for drinking and related activities. All living organisms rely on water, constituting between 50 and 97% of plant and animal matter, and approximately 70% of the average human body weight (Buchholz 1998). Regrettably, water is also recognized as the world's most mismanaged resource (Fakayode 2005). This has led to speculations that water will be at the center of the next major world war. Targets in this awaited battle might include nations with an abundance of water resources, as superpowers vie for control of this vital but finite resource. Contamination of any substances in water streams (rivers, lakes, groundwater, wells) leads to the deterioration of the physical, chemical, and biological parameters of water (Lin et al. 2022). It is a major pollution, which affects human and animal life. It was observed that most developing countries are facing a shortage of good-quality water (Chowdhary et al. 2020). Numerous regions around the world are currently facing challenges regarding water availability. According to the Central Ground Water Board (CGWB) data, the water level has decreased at a pace of one meter per year at most places in India. In Australia, 33% of the land surface is at risk of salinization due to the overuse of freshwater for intense irrigation (Rengasamy 2006). In the developing world, half of the population experiences at least one disease due to inadequate water supply and sanitation. Human activities, including irrigation and industrial applications, can have detrimental effects on downstream ecosystems due to water use (Minhas et al. 2021). Additionally, chemical contamination of freshwater poses a serious threat to ecosystems, causing significant harm. By contaminating drinking water, water pollution can have a direct negative effect on human health by causing diseases and both acute and chronic poisoning (Xu et al. 2022). The financial effects of hospital stays and diseases must also account for the time that family members miss from work. Future water shortages would put the world's population at serious risk, affecting everything from food grain production to general health and sanitation. Water pollution may be a point source or a non-point source. Among the sources non-point sources from agricultural fields are contributing a major chunk (Figure 13.3). Balance application of agricultural inputs based on scientific recommendation may be the right way to minimize water pollution (Dotaniya et al. 2018a, b).

13.2.3 Soil Pollution

Soil pollution minimizes the function of soil due to the addition of poor-quality substances, faulty management systems, over-exploitation of soil fertility, etc. As stated by Singh and Singh (2020), soil contamination is a worldwide issue that poses a significant threat to both ecosystems and human health. Soils play a crucial role in providing various ecosystem services necessary for sustaining life on the Earth (Stavi et al. 2016). However, the alarming reality is that the quality of our soils and the invaluable benefits they offer are diminishing rapidly (Jie et al. 2002). Soil pollutants are predominantly a result of human activities (Peña 2022). Various practices, including the use of fertilizers and pesticides, improper disposal of wastewater, burning of fossil fuels, and the application of plastics, medications, and fertilizers, can introduce both organic and inorganic pollutants into the soil matrix (Gautam et al. 2023).

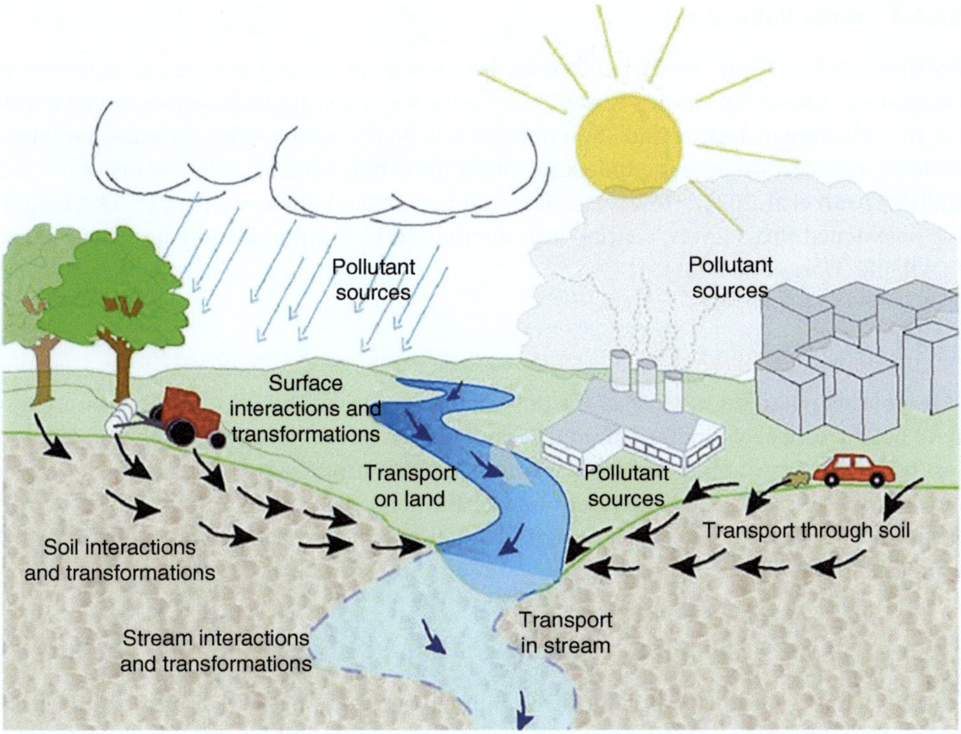

Figure 13.3 A major source of water pollution. *Source:* Aqua Terra Consultants (2004)/US Environmental Protection Agency/Public Domain.

Once these contaminants become integrated into the soil matrix, the cleanup process becomes challenging and costly. Moreover, they may pose significant health risks throughout the food chain. Moreover, there is evidence linking these inorganic and organic contaminants to severe human disorders. Chemical pollutants such as polychlorinated biphenyls, polycyclic aromatic hydrocarbons, insecticides, and herbicides have been associated with various human diseases (Jianrong et al. 2022). Additionally, trace toxic metals like arsenic, lead, and cadmium have been linked to several types of cancer (Rahman et al. 2015). Long-term application of industrial effluent accumulated a significant amount of metal in soil. Dotaniya et al. (2014b) reported that tannery-irrigated fields of Kanpur, India, accumulated 25–30 times more total Cr than good-quality irrigated water. Another study was conducted to see if the continuous application of sewage enhanced the C content in soil, whereas it also increased the biological load in soil (Dotaniya et al. 2018a). However, sewage mixed with industrial effluent mediated the soil pH, soil hardness, C content, biological diversity, and population resulting in poor crop yield (Singh et al. 2012; Koutroubas et al. 2023). These contaminated soils reduced the soil's ecological services. Poor soil aggregates due to lower amount of biological activities and minimum secretion of low molecular organic acids are the key effects of industrial effluent irrigation during crop cultivation (Soni et al. 2023). These acids are the rhizospheric food material for the soil biota. Increasing soil pollution reduced the soil fertility and crop yield of major crops yield (Tindwa and Singh 2023).

13.2.4 Noise Pollution

As time advances and modern lifestyles become more widespread, noise pollution is increasingly impacting various ecosystems. Excessive or disruptive loudness can be harmful to both human health and the environment. Major contributors to noise pollution include construction activity, industrial machinery, noisy recreational events, and road traffic (Tiwari et al. 2023). However, noise pollution comes with several drawbacks, including heightened stress levels, hearing loss, disruptions in communication, and disturbance to wildlife (Garg et al. 2021).

13.2.5 Light Pollution

It is ignoring pollution in most of the developing countries. The lower level of awareness does not attract people's attention to this pollution issue. However, this pollution leads to loss of biodiversity of birds and small creatures. Higher artificial light interrupts the breeding activities in birds and adversely affects the food web of an ecosystem (Bará and Falchi 2023). Directly and indirectly, light pollution reduces the system's productivity and ecological services.

13.2.6 Thermal Pollution

The discharge of heated compounds from human activities, particularly in the industrial belt of a region with inadequate controls, leads to temperature fluctuations in aquatic bodies (Zhang et al. 2023a). The primary contributor to this thermal pollution is the release of warm water by power plants, industrial operations, and other human activities, resulting in lowered oxygen levels, adverse effects on aquatic species, and disruptions to aquatic ecosystems. Over time, alterations in water temperature significantly impact the breeding and survival activities of aquatic organisms (Adla et al. 2022).

13.3 Risk Assessment of Metal Pollution

Risk assessment is a structural process comprising of identification, analysis, and evaluation of potential associated with a particular activity, project, or situation. It is time and location-based and needs integrated tools and techniques. It helps decide for minimizing the associated risk in an activity related to metals (Cr, Cd, Ni, Pb, Hg, As, Se). Sometimes people neglect the risk factor of a project/activity during the installation may lead the vast damage to the environment and human resources all over the globe. The selection of the appropriate risk assessment method depends on factors such as the complexity of the activity, available data, resources, and the level of detail required for decision-making (Saha et al. 2017). It is essential to conduct risk assessments regularly and integrate risk management into the overall planning and decision-making processes to ensure a proactive and well-informed approach to risk management (Saha et al. 2023). Across the scientific fraternity, many methods are involved in the identification of metal from a natural system. Modern tools and techniques are advocating for precision to understand the gravity of metal toxicity. Modifications in the method are needed for the correct identification of metals and their biotoxic activities in soil (Rout and Sikdar 2017). Metagenomic analysis methods also strengthen the metal risk assessment under organic carbon-stabilized or amended soil (Hu et al. 2023). Different types of metal risk assessment methods are described below:

1) **Qualitative risk assessment:** It is used to describe the metal risk into low, medium, and high based on the metal toxicity and assess the metal for carcinogenic effects on the living system. It is easier and quicker and needs a lower amount of investment and skill (Latosińska et al. 2021). It has having lower precision rate and is mostly used for the primary classification of contaminated soils in broader groups.
2) **Semi-quantitative risk assessment:** It is an advanced method of metal assessment. It has numerical values to describe metal-contaminated areas. It needs skilled labor with semi-advanced scientific instruments. For example, the classification of Cr-contaminated soils into 0–1 (class I), 1–2 (class II), etc. (Dotaniya et al. 2017b).
3) **Quantitative risk assessment:** This is based on a thorough survey related to land patterns, historical records, primary and secondary data collection, and mathematical theories. In this method, a wide range of data, statistics procedures, and models are used for the calculation of metal risk assessment. It needs specialized manpower with advanced tools and techniques (Li 2018). Performing this method needs a huge amount of infrastructural investment running costs, and time.
4) **Scenario-based risk assessment:** It is a mixed type of method. Application of different methods are used for the project based on the complexity of the project. It has multiple interconnected risks associated with a particular part of activities (Zhang et al. 2023b).

13.4 Source of Contamination and Its Identification

Metal toxicity depends on the sources and spreading media in delineating boundaries. It was also observed that the identification of different sources is key to the remediation process and prevention of a metal entry in virgin areas (Dotaniya et al. 2022a; Dotaniya et al. 2024a). In a major class of the metal pollution classification, it is classified into (i) Point sources: these refer to specific, identifiable, and localized locations or facilities from which pollutants are directly discharged into the environment (Dotaniya et al. 2022b). These sources typically have a well-defined point of emission, making it relatively easier to monitor and control the release of pollutants (Table 13.2). It includes industrial effluent, municipal discharge, sewage, spill, and household chimneys. However, (ii) diffuse source: it refers to a pollutant source that is extensive and dispersed and cannot be easily traced back to a single source. Because of its dispersed nature, this source frequently covers bigger areas and might be more difficult to govern and monitor (Briffa et al. 2020); for example, agricultural activities, urban runoff, atmospheric deposition, erosion, etc.

Environmental assessments and pollution monitoring are critical in detecting metal contamination sources and levels, allowing for the creation of appropriate remediation techniques and regulatory measures to protect human health and the environment (Solanki et al. 2020).

Table 13.2 Pollution sources, identification method, and involved environmental process.

Type of sources	Source of metal	Identification	Environmental process
Point source	Natural sources	Geological surveys, metals analysis in the soil or bedrock	Leading metal contamination to natural
Point source	Industrial discharges	Monitoring industrial processes, analyzing industrial wastewater, and conducting site assessments can help identify the presence of metals and their sources	Release metal contaminants into the environment
Point source	Mining process	Monitoring of water bodies nearby to mining areas, effluent sampling, and analysis	Release significant amounts of heavy metals into the environment
Point source	Landfills and waste sites	Site assessments, soil and water sampling, and analysis of leachate waste materials	Improper disposal of electronic waste can lead to contamination of surrounding areas
Diffuse source	Agricultural activities	Soil and water sampling and analysis, input analysis	Chemicals used during crop production, sewage, and sludge application release metal in soil
Diffuse source	Urban runoff	Monitoring runoff, analyzing sediment and water samples, identifying potential point sources of metals	Metal reaches into nearby water bodies
Diffuse source	Atmospheric deposition	Air quality monitoring, analysis of dust and particulate matter samples, and mapping and identifying sources of atmospheric metal contamination	Deposited onto land and water surfaces

Laboratory examinations of soil, water, sediment, air, and biological materials are routinely used in identification methods. Metal concentrations are measured using techniques such as inductively coupled plasma mass spectrometry (ICP-MS), X-ray fluorescence (XRF), atomic absorption spectroscopy (AAS), and chemical extraction assays (Dotaniya et al. 2018c).

13.5 Effect on Ecological Services

13.5.1 Effect on Soil Health

Soil bacteria, fungi, actinomycetes, algae, and other microorganisms play a crucial role in the natural cycling of elements such as C, N, P, S, and more. These microorganisms actively facilitate the decomposition of organic materials and the transformation of nutrients in the environment (Dotaniya et al. 2014a). Soil pollution leads to infertility and contamination of poor-quality substances impairs the mineralization kinetics of essential plant nutrients

(Aqeel et al. 2014). It was also observed that the addition of tannery (containing Cr) contaminated effluent in soil and reduced the mineralization of soil organic matter and soil enzymatic activities (Nikolova et al. 2023; Dotaniya et al. 2024b). It was also observed that the addition of 300 mg kg^{-1} lead in black soil decreased the uptake kinetic of micronutrients (Mn, Zn, Cu, Ni) in spinach (Dotaniya et al. 2019a). In environments contaminated with heavy metals, the presence of non-essential metal stress has the potential to induce root exudation, thereby enhancing external detoxification (Kochian et al. 2004). Both low- and high-molecular-weight chemicals, such as amino acids, organic acids, sugars, and phenolics, present in root exudates, play a significant role in rhizospheric processes (Bais et al. 2006; Nagar et al. 2022). The high-molecular-weight substances encompass polysaccharides and proteins, with low-molecular-weight organic acids (LMWOA), identified as the most prevalent and metal-reactive among them (Koo et al. 2010). LMWOA, in particular, stands out as a crucial component contributing to enhanced element acquisition (Zhang et al. 2021). Heavy metals exert a significant impact on the activity of soil enzymes (Dotaniya et al. 2016). Changes in soil fertility and composition can be caused by industrial discharges into the environment, overuse of fertilizers, and introduction of hazardous materials. The reduced fertility of the soil interferes with the cycling of nutrients, which negatively impacts ecosystems' ability to support plant growth (Alengebawy et al. 2021). This in turn affects the availability of food and habitat for a wide variety of species. One mechanism through which they influence soil enzyme function involves directly disrupting the spatial structure of the enzyme's active groups. Additionally, the synthesis and metabolism of microbial enzymes are hindered due to the inhibition of microorganisms' growth and reproduction (Chu 2018). Conditions marked by heavy metal pollution, significant suppression of soil respiration, and severe instances of heavy metal contamination can collectively impede soil microbial activity. This poses a considerable threat to the overall functioning of the soil ecosystem. Over time, microbial populations that have adapted to flourish in polluted environments tend to be more prevalent in soil contaminated with heavy metals (Kozdroj and Jdvan 2001). Salati and Moore (2010) observed that the utilization of untreated sewage for irrigation leads to environmental issues, including soil pollution, accumulation of heavy metals, and microbiological hazards.

13.5.2 Environmental Qualities

Different pollutions reducing the quality of the atmospheric parameters are mediated by the ecosystem services. The different advantages that humans derive from ecosystems are collectively referred to as ecological services, or ecosystem services (Zhu et al. 2021). These services are intimately tied to the environment and essential to human well-being. Pollination, water purification, climate regulation, flood control, fertility of the land, biodiversity preservation, recreational and aesthetic values, medicinal resources, etc., are all directly impacted by environmental pollution. Pollination is essential to agricultural output because it produces fruits, vegetables, and nuts. Water passing through wetlands and forests is naturally filtered to remove toxins and contaminants. The global climate is regulated by carbon dioxide, which is absorbed and stored by forests and seas. Additionally, through photosynthesis, trees release oxygen, reducing the effects of climate change and maintaining a stable atmosphere with sufficient oxygen. But woods and wetlands serve as natural buffers, soaking up extra water during intense downpours and lowering the chance of floods. For ecosystems to remain stable, adaptable to environmental changes, and a source of genetic resources for both agriculture and medicine, biodiversity is essential (Kusi et al. 2020). Many plant and animal species in ecosystems that are a source of resources and therapeutic compounds performed worse when environmental pollution increased. Taking action to reduce greenhouse gas (GHG) emissions, advocating for the use of renewable energy sources, and formulating strategies for adapting to climate change are crucial steps in mitigating the impact of climate change on ecosystems (Dotaniya et al. 2017a). Taking action to reduce GHG emissions, advocating for the use of renewable energy sources, and formulating strategies for adapting to climate change are crucial steps in mitigating the impact of climate change on ecosystems (Meena and Dotaniya 2017).

13.5.3 Water Parameters

While the use of treated wastewater for irrigation carries the potential for both positive and negative environmental impacts, proper planning and management can yield favorable outcomes. Opting for the re-use of treated wastewater when appropriate is a sound approach, and disposal methods should be designed to minimize potential harm to the environment and public health (Andeobu et al. 2023). Addressing chemical concerns related to wastewater re-use in agriculture involves managing water-quality issues such as elevated salt levels, heavy metals, minerals, organic debris, and hazardous organic compounds (Qadir et al. 2007). The United Arab Emirates confronts unprecedented challenges concerning its water resources, including prohibitively high drinking water costs, elevated salt levels in developed groundwater, a shortage of groundwater supplies, and restrictions on water re-use. The nation's water system is experiencing escalating pressure due to growing water demand. Demonstrating the urgent need for water efficiency is a particularly challenging issue. In 2019, the government allocated $1.6 billion for energy and water initiatives. However, without proactive measures, the UAE's Environment 2030 assessment predicts the depletion of both brackish and groundwater sources from the aquifer system within the next 50 years (Al-Mulla 2011; Hamedi et al. 2023).

13.6 Important Rehabilitated Waste Dumpsites

One of the biggest environmental problems facing the world today is heavy metal contamination of soil. Experts estimate that more than 20 million hectares of China's farmland – or 20% of the country's total landmass – are poisoned (Wei and Chen 2001). Reclamation initiatives involving local governments, state governments, and environmental organizations are frequently joint ventures. Usually, the rehabilitation process entails recycling programs, appropriate waste disposal, and putting policies in place to stop additional environmental deterioration. In most countries, rehabilitation programmers are very much affected by the social structure, scientific and development process, and insolvency of government agencies. Some of the famous rehabilitation programs are well known at international levels as model case studies. These case studies demonstrate the complexities of metal-polluted sites as well as the numerous approaches and strategies employed to remediate them (Dotaniya et al. 2022). A thorough understanding of the site's unique conditions, pollutant characteristics, and long-term environmental goals is required for successful remediation (Brunner et al. 2023). It also emphasizes the significance of continuing monitoring and adaptive management in ensuring the efficacy and sustainability of remediation initiatives. Here are some case studies of metal-polluted sites and their remediation efforts:

13.6.1 Case Study: Bunker Hill Superfund Site, Idaho, United States

Overview: The Bunker Hill Superfund Site is one of the largest and most complex metal-contaminated sites in the United States. It was historically a major lead and zinc mining and smelting area.

Contaminants: The site was contaminated with lead, zinc, cadmium, and other heavy metals, along with arsenic.

Remediation efforts: The Bunker Hill Superfund Site has been undergoing repair for decades. Remediation actions have included removing and replacing contaminated soil, capping and encapsulating certain locations, treating groundwater, and implementing engineering controls to prevent additional contamination migration. To stabilize soils and decrease erosion, phytoremediation and revegetation have also been used.

Results: Significant progress has been made in decreasing the dangers associated with metal contamination at the site through remediation activities. However, given the site's complexity and substantial contamination, full restoration and cleanup are still ongoing.

13.6.2 Case Study: Yangjiazhangzi Abandoned Lead-Zinc Mine, China

Overview: The Yangjiazhangzi mine is an abandoned lead-zinc mine in China. It was a significant source of metal pollution due to historical mining and smelting activities.

Contaminants: The site was heavily contaminated with lead, zinc, and other heavy metals.

Remediation efforts: Excavation and removal of contaminated soil, installation of containment structures for storing hazardous waste, and application of stabilizing measures to prevent future metal dispersion were all part of the Yangjiazhangzi mine's remediation activities.

Results: Metal concentrations in soil and water have decreased as a result of cleanup operations. Furthermore, the use of containment measures has aided in the management of contamination spread to nearby areas.

13.6.3 Case Study: Silvermines Mining District, Ireland

Overview: The Silvermines Mining District is a historic mining area in Ireland, known for its lead and zinc deposits.

Contaminants: The site was contaminated with lead, zinc, and other metals due to mining and processing activities.

Remediation efforts: The Silvermines Mining District's remediation initiatives have included the stabilization of tailings and waste dumps, the treatment of mine water to remove metals, and revegetation to prevent erosion and metal uptake by plants.

Results: Metal concentrations in soil and water have decreased as a result of cleanup activities, and the measures employed have helped avoid further contamination and lessen threats to human health and the environment.

13.7 Ecosystem Services from Rehabilitated Waste Dumpsites

Any advantageous resource that ecosystems or wildlife offer to humans is considered an ecosystem service. The advantages may be minor or major, directly or indirectly classified based on the need and opportunities. It was observed that rehabilitated waste dumping sites are used for greenery purposes by improving the upper layer of soil or by adding soil amendments (Hobbs et al. 2008). Some of the sites are planted with fruit trees providing food for humans, animals, and birds. This may be in different forms. However, this is the main ecosystem service for the living creatures. Cultivation of mustard crops on metal-contaminated soil is a food for human beings and it also improves the soil health parameters by removing metal concentration from soils (Dotaniya et al. 2022c). Many medicinal and aromatic crop plants can be cultivated at these rehabilitated sites. In addition to producing lumber, forestry operations may also yield minimally processed goods including firewood, charcoal, wood chips, and roundwood that is utilized raw (FAO 2020). Ecosystem services improvement in waste dumping sites may be used for the storage of water, infrastructure development, and recycling plants after primary treatments of the waste. Many municipal dumping sites segregate the waste into different parts and based on the utility categorize waste into recycling or non-recycling waste. Preparation of compost from the MSW compost is a popular ecosystem service in most of the developing countries (Meena et al. 2021). Pollutant contamination in compost is the main deciding factor for its further application in crop or forest cultivation. Reclaimed soils are also used for the cultivation of energy crops as an energy source. It was also observed that polluted soils or water are used for the cultivation of non-food crops like flower crops. Shrinking natural resources for producing different types of life-supporting food, fodder, medicine, and infrastructure materials attracted the mindset of the researchers to work on the waste dumping sites for their potential uses (Adla et al. 2022). Across the globe, many pilot research projects are initiated for the recreation of wasteland as a wealth land land (Lal 2004). Targets for improving the use of ecosystem services for sustainable tourism are also included in UN Sustainable Development Goal 14, particularly for Small Island Developing States (UNDP 2020). Need of scientific manpower, the initial huge money in infrastructure and installation of machinery, and awareness among the people are the key to adopting the concept of rehabilitation of wastelands (Khan et al. 2022).

13.8 Improvement in Ecological Services by Improving the Waste Discharge Process

13.8.1 Treatment of Waste

Treatment of waste generated from industries, households, municipal corporations, and food and fiber cottages is needed before the disposal of natural resources. It was observed that a significant volume of the generated waste is dumped into natural resources as raw or partially treated (Jadeja et al. 2022). Many river segments across the country suffer from contamination due to untreated or partially treated household sewage from cities and towns, industrial effluents from nearby industries, issues with the maintenance and operation of sewage and industrial effluent treatment facilities, a lack of dilution, and other sources of non-point pollution.

The situation has worsened with rapid industrialization and urbanization. A March 2021 report from the Central Pollution Control Board (CPCB) revealed that urban areas in the nation produce approximately 72368 million liters of sewage per day (MLD), with only 31841 MLD being treatable (CPCB 2023). Additionally, the Ministry of Housing & Urban Affairs's Smart Cities Mission and the Atal Mission for Rejuvenation & Urban Transformation (AMRUT) are two efforts that construct sewerage infrastructure (GOI 2024). ETPs (effluent treatment plants) or common effluent treatment plants (CETPs) and sewage treatment plants (STPs) must be installed by industrial units and local bodies in compliance with the regulations outlined in the Environment (Protection) Act, 1986, and the Water (Prevention & Control of Pollution) Act, 1974. Before discharging their sewage or effluent into rivers or other bodies of water, they must treat it to the required environmental requirements. Under the terms of these Acts, non-compliance is punishable by the CPCB, State Pollution Control Boards (SPCBs), and Pollution Control Committees (PCCs). Industries are also expected to foster wastewater re-use and recycling, decrease wastewater creation through technical developments, and aim for Zero Liquid Discharge (ZLD) whenever possible (Ricky et al. 2022).

13.8.2 Addition of Organic Matter in Soil

Organic matter is the main soil amendment to reduce or mediate the soil health parameters for improving ecosystem services of an ecosystem. The structure of the soil, its ability to hold onto moisture, the variety and activity of soil organisms – both helpful and harmful to crop production – and the availability of nutrients are just a few of the features that are greatly impacted by organic matter (Gao et al. 2023). Low-molecular-weight organic acids (LMWOAs), such as those highlighted by Hinsinger et al. (2006) and Haoliang et al. (2007), play a significant role in influencing the bioavailability and mobility of metals in soils. Processes that contribute to altering metal and nutrient availability include acidification, changes in redox conditions, and chelating ability, as outlined by Seshadri et al. (2015). Montiel-Rozas et al. (2016) observed a correlation between the exudation of malic acid and increased Cu availability in soils with reduced organic matter concentration. The notable stability of malate-metal complexes, as indicated by Qin et al. (2004), contributes to the efficacy of malic acid in desorbing Cu from soil particles. While organic or in combination with inorganic soil amendments are frequently employed in the remediation of contaminated soils (Ciadamidaro et al. 2015; Hattab et al. 2015; Montiel-Rozas et al. 2015; Dotaniya et al. 2019b); their influence on the exudation of low-molecular-weight organic acids (LMWOA) by roots has not been extensively explored (Dotaniya et al. 2023d). Organic matter addition through pressmud, FYM, crop residues are acted as a source of food for soil biota and improved the soil physico-chemical properties of soil (Dotaniya and Nagar 2022). It was also observed that the addition of organic matter reduces pollutant toxicity and improves the plant nutrient supply capacity of the soil (Haroun et al. 2023).

13.8.3 Metal Immobilization and Remediation Process

In soil and water pollution, heavy metals reduce the ecosystem services potential over a period. However, organic pollutants are more persistent in the ecosystem; and adversely affect the system activities (Saha et al. 2017; Solanki et al. 2020). In the metal remediation process, different soil microbes and plants are involved in minimizing the metal toxicity by

Table 13.3 Naturally occurring different phytodecontamination processes in plants.

Plant name	Metal accumulated/remediated	References
Brassica juncea	Lead, Cadmium	Meers et al. (2004), Kumar et al. (2015)
Helianthus annuus	Lead, Chromium	Hseu et al. (2012), Maestri et al. (2010)
Salix spp.	Copper, Zinc	Wang et al. (2007), Leštan et al. (2008)
Zea mays	Cadmium, Lead	Sheng et al. (2008), Chen et al. (2012)
Populus spp.	Arsenic, Mercury, Selenium	Singh et al. (2010), Pilon-Smits (2005)
Vetiveria zizanioides	Cadmium, Lead	Wang et al. (2013), Ma et al. (2001)
Eichhornia crassipes	Copper, Chromium	Tripathi et al. (2010), RamaKrishna et al. (2004)
Alyssum spp.	Nickel, Zinc	Brooks et al. (1998), Whiting et al. (2000)
Eucalyptus spp.	Cadmium, Zinc	Han et al. (2014), Nkrumah et al. (2006)

either immobilizing or removing it from the system (Dotaniya et al. 2023c). However, the phytoremediation mechanism is cheaper, eco-friendly, and requires less skill to perform the process. Table 13.3 lists some common plants used in phytoremediation and the specific metals they are known to accumulate or remediate. Please note that the efficacy of phytoremediation can vary depending on site-specific conditions and the concentration of contaminants (Saha et al. 2023).

13.9 Strategies for Improving Ecological Services

Enhancing the natural processes that ecosystems provide to support life on Earth is a necessary step toward improving ecological services. Following are a few tactics to enhance ecological services:

i) **Conservation and restoration of ecosystems:** Maintain and safeguard current natural environments and their ecosystems. Create the systems that are increasing an ecosystem's biodiversity. Execute restoration initiatives, such as replanting or wetland restoration, to revitalize damaged ecosystems.
ii) **Sustainable land use practices:** Adopt sustainable farming methods that prevent soil erosion, preserve biodiversity, and utilize less hazardous chemicals. using cover crops to lessen soil erosion and enhance soil health. Encourage the use of sustainable forestry methods that preserve the health of forests and take into account long-term ecological effects. Agricultural farming enhances ecological services depending on soil properties.
iii) **Biodiversity conservation:** Concern for biodiversity is necessary to safeguard ecological services in an environmentally sound way. Nonetheless, occasionally the kind of resources and their maintenance are crucial. Put policies in place to preserve and protect threatened species. Create and preserve natural passageways that enable

species migration between fragmented environments. Utilize the biodiversity protection procedure to improve an ecosystem's interconnectedness and system activity.

iv) **Water conservation and management:** It is the origin of all living things. Water regulates most ecological services either directly or indirectly. Nevertheless, it is mismanaged everywhere in the world. Use effective water management techniques to guarantee that ecosystems have access to clean water. Humans are discarding garbage into bodies of water without considering the welfare of aquatic life. It has also been noted that the supply of drinking water is impacted by diminishing water resources. Preserve and replenish wetlands, as they are essential for both water filtration and managing flooding.

v) **Climate change mitigation:** The primary cause of global climate change is an increase in air temperature. Additionally, it was noted that rising GHG levels also intensified climate occurrences. It decreased the potential for primary product output and increased crop plant respiration rate. Encourage programs aimed at lowering GHG emissions to slow down global warming. To reduce the influence on the environment, encourage the use of renewable energy sources and energy efficiency. Different crop management techniques are now advised to farmers to reduce the impact of climate change on various ecosystem components.

vi) **Reduce pollution by awareness programs:** Put policies in place to cut down on the majority of pollutants. But the main ones are soil, water, and air pollution. Promote the use of eco-friendly technologies and practices in both households and industries. Involve local communities in sustainable resource management and conservation initiatives. Encourage cooperation among local communities, non-governmental organizations, and governments to accomplish shared ecological objectives. Increase public knowledge of the value of ecological services and how human activity affects ecosystems. Inform localities about eco-friendly lifestyle choices and how they contribute to maintaining natural equilibrium.

vii) **Research and policy guidelines:** Make scientific research investments to get deeper insights into ecosystems and their functions. The majority of developing countries are investing a larger portion of their GDP in ecosystem research and development. It was also noted that an ecosystem's ability to sustain itself depends on effective waste management. Create monitoring programs to keep an eye on changes in ecosystem health, biodiversity, and the effects of human activity. Support and encourage the creation of laws and policies that advance conservation and sustainable development. Implement legislation that safeguards natural ecosystems and sanctions actions that endanger the environment. To promote overall environmental health, offer habitat, and improve air quality, and integrate green spaces and ecological characteristics into urban development.

viii) **International collaboration and exposure:** Rehabilitating an ecosystem can help restore its lost variety. Programs for wildlife maintain food webs and chains to improve ecological services. When introducing foreign biota to new areas, a comprehensive strategy is required to sustain and increase variety over time. To solve transboundary environmental concerns and advance sustainable development globally, foster international cooperation. Through these partnerships, ecological services can be improved and restored, promoting a more wholesome and long-lasting coexistence of human activity and the environment.

13.10 Challenges Faced During Rehabilitation

- Rehabilitation of waste dumping sites for potential use needs a huge amount of money. Restricted financial resources can hinder the execution of thorough rehabilitation plans, resulting in incomplete or delayed initiatives.
- Effective rehabilitation demands proficiency in diverse disciplines such as ecology, engineering, and environmental science. Insufficient expertise may lead to suboptimal rehabilitation strategies, potentially resulting in unintended negative consequences.
- Achieving tangible results in ecological rehabilitation often necessitates considerable timeframes. Impatience or the anticipation of immediate outcomes can prompt the premature abandonment of rehabilitation efforts, causing discouragement among stakeholders.
- Areas with historical pollution or contamination may demand specialized remediation techniques. Legacy pollution can endure, presenting ongoing challenges for rehabilitation and potentially constraining the success of restoration endeavors. Additionally, local communities might resist or oppose rehabilitation efforts due to concerns about alterations in land use, economic impacts, or perceived disruptions.
- Shifting climate conditions can introduce additional challenges to rehabilitation efforts. Nevertheless, monitoring and evaluating the progress of rehabilitation is essential for adaptive management. Inadequate monitoring and evaluation can impede the identification of success factors or the need for adjustments, thereby limiting the effectiveness of rehabilitation strategies.

13.11 Conclusions

Rehabilitation of degraded areas, whether impacted by industrial activities, pollution, or other environmental factors, is a complex process fraught with numerous challenges. Following are common hurdles encountered in rehabilitation efforts. One of the most important markers of the resilience and health of an ecosystem is biodiversity. The improvement of ecological services is greatly aided by tactics like preserving natural ecosystems, creating and upholding protected areas, and putting various species preservation at the forefront of conservation initiatives. By incorporating these strategies into policies, practices, and individual behaviors, we can strive to enhance ecological services and secure the long-term sustainability of ecosystems. Embracing behaviors that enhance ecosystem health and function, support biodiversity, and ensure the sustainable use of natural resources is imperative for improving ecological services. Collaborative efforts among governments, communities, businesses, and conservation organizations are essential for achieving significant improvements in ecological services. Tackling these challenges necessitates a holistic and adaptive approach, encompassing the involvement of stakeholders from various sectors, clear communication, effective planning, and a sustained commitment to long-term environmental stewardship. Promoting public awareness and education is essential for nurturing a sense of rehabilitation process.

References

Adla, K., Dejan, K., Neira, D., and Dragana, Š. (2022). Degradation of ecosystems and loss of ecosystem services. In: *One Health* (ed. C. Joana, A.I. Prata, and T.R.-S. Ribeiro), 281–327. Academic Press.

Alengebawy, A., Abdelkhalek, S.T., Qureshi, S.R., and Wang, M.Q. (2021). Heavy metals and pesticides toxicity in agricultural soil and plants: ecological risks and human health implications. *Toxics* 9 (3): 42.

Al-Mulla, M. UAE State of the Water Report. In Proceedings of the 2nd Arab Water Forum, Cairo, Egypt, 20–23 November 2011.

Andeobu, L., Wibowo, S., and Grandhi, S. (2023). Environmental and health consequences of e-waste dumping and recycling carried out by selected countries in Asia and Latin America. *Sustainability* 15 (13): 10405.

Aqeel, M., Jamil, Mohd., & Yusoff, I. (2014). Soil Contamination, Risk Assessment and Remediation. InTech. doi: https://doi.org/10.5772/57287

Aqua Terra Consultants 2004 Lecture 2: Introduction to HSPF and the Model Application Processes. WWW document, http://www.epa.gov/waterscience/basins/training.htm

Bais, H., Weir, T., Perry, L. et al. (2006). The role of root exudates in rhizosphere interactions with plants and other organisms. *Annual Review of Plant Biology* 57: 233–266.

Bará, S. and Falchi, F. (2023). Artificial light at night: a global disruptor of the night-time environment. *Philosophical Transactions of the Royal Society B* 378: 20220352.

Beil L (2017). "Pollution killed 9 million people in 2015". *Science News*. Retrieved 16 December 2023.

Berthet, E.T., Bretagnolle, V., and Gaba, S. (2022). Place-based social-ecological research is crucial for designing collective management of ecosystem services. *Ecosystem Services* 55: 101426.

Briffa, J., Sinagra, E., and Blundell, R. (2020). Heavy metal pollution in the environment and their toxicological effects on humans. *Heliyon* 6 (9): e04691.

Brooks, R.R., Lee, J., Reeves, R.D., and Jaffre, T. (1998). Detection of nickeliferous rocks by analysis of herbarium specimens of indicator plants. *Journal of Geochemical Exploration* 64: 69–81.

Brunner, P.H., Unterberger, A., and Fellner, J. (2023). Improving waste management by focussing on goals: A mini review of the publication sector. *Waste Management and Research: The Journal for a Sustainable Circular Economy* 41 (12): 1717–1727.

Buchholz, R.A. (1998). *Principles of Environmental Management. The Greening of Business*, 2e. London: Prentice-Hall.

Carrington, D. (2017). "Global pollution kills 9m a year and threatens 'survival of human societies'". *The Guardian*. Retrieved December 22, 2023.

Chen, H., Zhang, P., Duan, T., and Feng, Y. (2012). Heavy metal concentrations in soils and plant accumulation in a restored manganese mineland in Guangxi, South China. *Environmental Earth Sciences* 65: 853–860.

Chowdhary, P., Bharagava, R. N., Mishra, S., and Khan, N. (2020). Role of industries in water scarcity and its adverse effects on environment and human health. Vertika Shukla, Narendra Kumar Environmental Concerns and Sustainable Development, 235–256. Springer. doi:https://doi.org/10.1007/978-981-13-5889-0_12

Chu, D. (2018). Effects of heavy metals on soil microbial community. *IOP Conference Series: Earth and Environmental Science* 113: 012009. https://doi.org/10.1088/1755-1315/113/1/012009.

Ciadamidaro, L., Puschenreiter, M., Santner, J. et al. (2015). Assessment of trace element phytoavailability in compost amended soils using different methodologies. In: *Journal of Soils and Sediments* (ed. P.N. Owens and Z. Xu). Springer https://doi.org/10.1007/s11368-015-1283-3.

CPCB (2023). Hazardous and other Wastes (Management & Transboundary Movement) Rules, 2016. https://cpcb.nic.in/rules/acssed on 29 December, 2023.

Dickie, G. (2022). "Pollution killing 9 million people a year, Africa hardest hit - study". Reuters. Retrieved 20 June 2023.

Dotaniya, M.L., Amita, S., Nagar, M.C. et al. (2023d). Can application of pressmud mediated plant nutrient dynamics under lead contaminated soils of Indian Vertisol? *Bulletin of Environmental Contamination and Toxicology* 110: 44. https://doi.org/10.1007/s00128-023-03690-z.

Dotaniya, M.L., Aparna, K., Dotaniya, C.K. et al. (2018b). Role of soil enzymes in sustainable crop production. In: *Enzymes in Food Biotechnology* (ed. Khudus et al.), 569–589. Springer International.

Dotaniya, M.L., Dotaniya, C.K., Yadav, D.K. et al. (2023a). Chromium dynamics in the soil-plant continuum. In: *Chromium in Plants and Environment* (ed. Kumar et al.). Switzerland: Springer Nature https://doi.org/10.1007/978-3-031-44029-8.

Dotaniya, M.L., Meena, M.D., Choudhary, R.L. et al. (2023b). Management of plant nutrient dynamics under alkaline soils through graded application of pressmud and gypsum. *PLOSONE* 18 (8): e0288784. https://doi.org/10.1371/journal.pone.0288784.

Dotaniya, M.L., Meena, V.D., Dotaniya, C.K. et al. (2022a). Engineered biochar as adsorbent for removal of heavy metals from soil medium. In: *Engineered Biochar* (ed. S. Ramola, D. Mohan, O. Masek, et al.). Singapore: Springer https://doi.org/10.1007/978-981-19-2488-0_9.

Dotaniya, M.L., Meena, V.D., Lata, M., and Meena, B.L. (2017a). Climate change impact on agriculture: adaptation strategies. In: *Climate Change & Sustainable Agriculture* (ed. P.S. Kumar, M. Kanwat, P.D. Meena, et al.), 27–38. New Delhi: New India Publishing Agency.

Dotaniya, M.L., Meena, V.D., Rajendiran, S. et al. (2018a). Impact of long-term application of Patranala sewage on carbon sequestration and heavy metal accumulation in soils. *Journal of the Indian Society of Soil Science* 66 (3): 310–317. https://doi.org/10.5958/0974-0228.2018.00038.5.

Dotaniya, M.L., Meena, V.D., Rajendiran, S. et al. (2017b). Geo-accumulation indices of heavy metals in soil and groundwater of Kanpur, India under long term irrigation of tannery effluent. *Bulletin of Environmental Contamination and Toxicology* 98 (5): 706–711. https://doi.org/10.1007/s00128-016-1983-4.

Dotaniya, M.L., Meena, V.D., Saha, J.K. et al. (2022b). Use of poor quality water for sustainable crop production in changing scenario of climate change. *Environment, Development and Sustainability* https://doi.org/10.1007/s10668-022-02365-9.

Dotaniya, M.L., Meena, V.D., Saha, J.K. et al. (2018c). Environmental impact measurements: tool and techniques. In: *Handbook of Ecomaterials* (ed. L. Martínez, O. Kharissova, and B. Kharisov). Cham: Springer https://doi.org/10.1007/978-3-319-48281-1_60-1.

Dotaniya, M.L. and Nagar, M.C. (2022). Can pressmud improve nickel availability in lead contaminated soils? *National Academy Science Letters* 46 (12): https://doi.org/10.1007/s40009-022-01160-4.

Dotaniya, M.L., Nagar, M.C., Amita, S. et al. (2024a). Pressmud subdue phytoremediation indices in lead-contaminated soils: a human health perspective. *Bulletin of Environmental Contamination and Toxicology* 112: 16. https://doi.org/10.1007/s00128-023-03848-9.

Dotaniya, M.L., Nagar, M.C., Amita, S. et al. (2023c). Management of acid and alkaline phosphatase, dehydrogenase activity by sugarcane industry waste under lead contamination – a case study of Indian Vertisol. *PLOSONE* 18 (5): e0286223. https://doi.org/10.1371/journal.pone.0286223.

Dotaniya, M.L., Pipalde, J.S., Jain, R.C. et al. (2019a). Can lead and nickel interaction affect plant nutrient uptake pattern in spinach (*Spinacia oleracea*)? *Agricultural Research* 9: 358–364.

Dotaniya, M.L., Pipalde, J.S., Jain, R.C. et al. (2022). Nickel-mediated lead dynamics and their interactive effect on lead partitioning and phytoremediation indices in spinach. *Environmental Monitoring and Assessment* 194: 334.

Dotaniya, M.L., Rajendiran, S., Elanchezhian, R. et al. (2024b). Effect of chromium and cations concentration on stress enzymes in spinach crop grown under vertisol of central India. *Journal of Agriculture and Food Research* https://doi.org/10.1016/j.jafr.2024.100969.

Dotaniya, M.L., Rajendiran, S., Kirti, S. et al. (2022c). Immobilization of chromium bioavailability through application of organic waste to Indian mustard (*Brassica juncea*) under chromium contaminated Indian soils. *Environmental Monitoring and Assessment* 195 (1): https://doi.org/10.1007/s10661-022-10625-4.

Dotaniya, M.L., Rajendiran, S., Meena, V.D. et al. (2016). Influence of chromium contamination on carbon mineralization and enzymatic activities in Vertisol. *Agricultural Research* 6 (1): 91–96. https://doi.org/10.1007/s40003-016-0242-6.

Dotaniya, M.L., Saha, J.K., Meena, V.D. et al. (2014b). Impact of tannery effluent irrigation on heavy metal build up in soil and ground water in Kanpur. *Agrotechnology* 2 (4): 77.

Dotaniya, M.L., Saha, J.K., Rajendiran, S. et al. (2019b). Reducing chromium uptake through application of calcium and sodium in spinach. *Environmental Monitoring and Assessment* 191: 754. https://doi.org/10.1007/s10661-019-7948-4.

Dotaniya, M.L., Thakur, J.K., Meena, V.D. et al. (2014a). Chromium pollution: a threat to environment. *Agricultural Reviews* 35 (2): 153–157.

EEA (2019). European Environment Agency's: EEA's 2019 EU air quality report. https://www.ersnet.org/news-and-features/interstitial-lung-diseases/eeas-2019-eu-air-quality-report-finds-air-pollution-caused-400000-premature-deaths-in-2016/

EPA (2000). Environmental Protection Agency Precious Heritage. Adams, J.S., L.S. Kutner and B.A. Stein, ed. New York: Oxford University Press, 2000. Sustaining Life. Chivian, E. and A. Bernstein, ed. New York, Oxford University Press, 2008. UN Millennium Ecosystem Assessment - A Framework for Assessment Document

Fakayode, S.O. (2005). Impact assessment of industrial effluent on water quality of the receiving Alaro River in Ibadan, Nigeria. *AJEAM-RAGEE* 10: 1–13.

FAO (2020). Global Forest Resources Assessment 2020 – Key findings. FAO. 2020. https://doi.org/10.4060/ca8753en. ISBN 978-92-5-132581-0. S2CID 130116768.

Feitosa, I.B., Huszar, V.L.M., Domingues, C.D. et al. (2019). Sarmento. Plankton Community interactions in an Amazonian floodplain lake, from bacteria to zooplankton. *Hydrobiologia* 831: 55–70.

Fuller, R., Landrigan, P.J., Balakrishnan, K. et al. (2022). Pollution and health: a progress update. *The Lancet Planetary Health* 6 (6): e535–e547.

Gao, J., Han, H., Gao, C. et al. (2023). Organic amendments for in situ immobilization of heavy metals in soil: a review. *Chemosphere* 335: 139088.

Garg, N., Chauhan, B.S., and Singh, M. (2021). Normative framework of noise mapping in India: Strategies, Implications and Challenges ahead. *Acoustics Australia* 49: 23–41.

Gautam, K., Sharma, P., Dwivedi, S. et al. (2023). A review on control and abatement of soil pollution by heavy metals: emphasis on artificial intelligence in recovery of contaminated soil. *Environmental Research* 115592.

Geissen, V., Silva, V., Lwanga, E.H. et al. (2021). Cocktails of pesticide residues in conventional and organic farming systems in Europe – legacy of the past and turning point for the future. *Environmental Pollution* 1 (278): 116827.

GOI 2024 Swachh Bharat 2.0 and AMRUT 2.0: Towards Cleaner and Water Secure Cities acessed on Jan 3, 2024 https://www.investindia.gov.in/team-india-blogs/swachh-bharat-20-and-amrut-20-towards-cleaner-and-water-secure-cities-part-1-2

Hamedi, A.F.H., Kandhan, K., Liu, Y. et al. (2023). Wastewater irrigation: a promising way for future sustainable agriculture and food security in the United Arab Emirates. *Water* 15 (12): 2284. https://doi.org/10.3390/w15122284.

Han, Y., Wang, P., Zhao, D. et al. (2014). Cadmium tolerance and accumulation in eight potential energy crops. *International Journal of Phytoremediation* 16: 53–62.

Haoliang, L., Chongling, Y., and Jingchun, L. (2007). Low-molecular-weight organic acids exuded by Mangrove (*Kandelia candel* (L.) Druce) roots and their effect on cadmium species change in the rhizosphere. *Environmental and Experimental Botany* 61: 159–166.

Haroun, M., Xie, S., Awadelkareem, W. et al. (2023). Influence of biofertilizer on heavy metal bioremediation and enzyme activities in the soil to revealing the potential for sustainable soil restoration. *Scientific Reports* 13: 20684.

Hattab, N., Motelica-Heino, M., Faure, O., and Bouchardon, J.L. (2015). Effect of fresh and mature organic amendments on the phytoremediation of technosols contaminated with high concentrations of trace elements. *Journal of Environmental Management* 159: 37–47.

Heimlich R, Wiebe K, Claassen R, Gadsby D, House R (1998). Wetlands and Agriculture: Private Interests and Public Benefits, USDA Economic Research Service Report Agricultural Economic Report No. (AER-765) 99 pp.

Hinsinger, P., Plassard, C., and Jaillard, B. (2006). Rhizosphere: a new frontier for soil biogeochemistry. *Journal of Geochemical Exploration* 88: 210–213. https://doi.org/10.1016/j.gexplo.2005.08.041.

Hobbs, P.R., Sayre, K., and Gupta, R. (2008). The role of conservation agriculture in sustainable agriculture. *Philosophical Transactions of the Royal Society, B: Biological Sciences* 363: 543–555.

Hseu, Z.Y., Bau, D.T., Tsai, J.H. et al. (2012). Phytoremediation potential of six plant species from a Pb/Zn mine tailing. *Journal of Soils and Sediments* 12: 265–277.

Hsieh, H.H., Chuang, M.H., Shih, Y.Y. et al. (2021). Eutrophication and hypoxia in tropical Negombo Lagoon. Sri Lanka. *Frontiers in Marine Science* 8: 678832.

Hu, M., Zhou, S., Xiong, X. et al. (2023). Dynamics of soil microbial communities involved in carbon cycling along three successional forests in southern China. *Frontiers in Microbiology* Sec. Terrestrial Microbiology 14: https://doi.org/10.3389/fmicb.2023.1326057.

Jadeja, N.B., Banerji, T., Kapley, A., and Kumar, R. (2022). Water pollution in India – current scenario. *Water Security* 16: 100119.

Jianrong, C., Yanjun, L., and Sujie, Y. (2022). The concentrations and sources of PAHs and PCBs in soil from an oil field and estuary in the Yellow River Delta, China. *Frontiers in Environmental Science* 10: 1028299.

Jie, C., Jing, Z.C., Man, Z.T., and Zi, T.G. (2002). Soil degradation: a global problem endangering sustainable development. *Journal of Geographical Sciences* 12: 243–252. https://doi.org/10.1007/bf02837480.

Jones, M.L.M., Provins, A., Harper-Simmonds, L. et al. (2014). Using the Ecosystems Services Approach to value air quality.

Khan, A.H., Sharholy, M., Alam, P. et al. (2022). Evaluation of cost benefit analysis of municipal solid waste management systems. *Journal of King Saud University, Science* 34 (4): 101997.

Kirchherr, J., Yang, N.-H.N., Schulze-Spüntrup, F. et al. (2023). Conceptualizing the circular economy (revisited): an analysis of 221 definitions. *Resources, Conservation and Recycling* 194: 107001.

Kochian, L.V., Hoekenga, O.A., and Pineros, M.A. (2004). How do crop plants tolerate acid soils? Mechanisms of aluminum tolerance and phosphorous efficiency. *Annual Review of Plant Biology* 55: 459–493.

Koo, B., Chen, W., Chang, A. et al. (2010). A root exudates based approach to assess the long-term phytoavailability of metals in biosolidsamended soils. *Environmental Pollution* 158: 2582–2588.

Koutroubas, S.D., Antoniadis, V., Damalas, C.A. et al. (2023). Municipal sewage sludge effects on maize yield, nitrogen use efficiency, and soil properties. *Journal of Soil Science and Plant Nutrition* 23: 1209–1221. https://doi.org/10.1007/s42729-022-01115-0.

Kozdroj, J. and Jdvan, E. (2001). Structural diversity of microbial communities in arable soils of a heavily industrialised area determined by PCR-DGGE fingerprinting and FAME profiling. *Applied Soil Ecology* 17 (1): 31–42.

Kumar, P.B.A.N., Dushenkov, V., Motto, H., and Raskin, I. (2015). Phytoextraction: the use of plants to remove heavy metals from soils. *Environmental Science and Technology* 29: 1232–1238.

Kumar, R., Verma, A., Shome, A. et al. (2021). Impacts of plastic pollution on ecosystem services, sustainable development goals, and need to focus on circular economy and policy interventions. *Sustainability* 13: 9963.

Kusi, K.K., Khattabi, A., Mhammdi, N., and Lahssini, S. (2020). Prospective evaluation of the impact of land use change on ecosystem services in the Ourika watershed, Morocco. *Land Use Policy* 97: 104796.

Lal, R. (2004). Soil carbon sequestration impacts on global climate change and food security. *Science* 304: 1623–1627.

Latosińska, J., Kowalik, R., and Gawdzik, J. (2021). Risk assessment of soil contamination with heavy metals from municipal sewage sludge. *Applied Sciences* 11 (2): 548.

Lelieveld, J., Evans, J.S., Fnais, M. et al. (2015). The contribution of outdoor air pollution sources to premature mortality on a global scale. *Nature* 525: 367–371.

Leštan, D., Luo, C., Li, X. et al. (2008). Comparison of wild plants for phytoremediation of lead-contaminated soils from a woody shrubbery in an urban brownfield. *Journal of Environmental Sciences* 20: 1085–1092.

Li, F. (2018). *Heavy Metal in Urban Soil: Health Risk Assessment and Management*. InTech https://doi.org/10.5772/intechopen.73256.

Lin, L., Yang, H., and Xu, X. (2022). Effects of water pollution on human health and disease heterogeneity: a review. *Frontiers in Environmental Science* 10: 880246.

Ma, L.Q., Komar, K.M., Tu, C. et al. (2001). A fern that hyperaccumulates arsenic. *Nature* 409: 579.

Maestri, E., Marmiroli, M., Visioli, G., and Marmiroli, N. (2010). Metal tolerance and hyperaccumulation: costs and trade-offs between traits and environment. *Environmental and Experimental Botany* 68: 1–13.

Meena, V.D. and Dotaniya, M.L. (2017). Climate change, water scarcity and sustainable agriculture for food security. In: *Climate Change & Sustainable Agriculture* (ed. P.S. Kumar, M. Kanwat, P.D. Meena, et al.), 123–142. New Delhi: New India Publishing Agency.

Meena, M.D., Dotaniya, M.L., Meena, M.K. et al. (2021). Maturity indices as an index to evaluate the quality of sulphur enriched municipal solid waste compost using variable byproduct of sulphur. *Waste Management* 126: 180–190.

Meena, M.D., Dotaniya, M.L., Meena, B.L. et al. (2023a). Municipal solid waste: opportunities, challenges and management policies in India: a review. *Waste Management Bulletin* 1 (1): 4–18. https://doi.org/10.1016/j.wmb.2023.04.001.

Meena, M.K., Yadav, B.L., Dotaniya, M.L. et al. (2023b). Can plant nutrient dynamics and Indian mustard yield be influenced by addition of organic amendments under saline sodic soils? *Communications in Soil Science and Plant Analysis* 55 (1): https://doi.org/10.1080/00103624.2023.2283000.

Meers, E., Hopgood, M., Lesage, E. et al. (2004). Enhanced phytoextraction: in search for EDTA alternatives. *International Journal of Phytoremediation* 6 (2): 95–109.

Mills, G., Wagg, S., and Harmens, H. (2013). *Ozone Pollution: Impacts on Ecosystem Services and Biodiversity*. ICP Vegetation Programme Coordination Centre.

Minhas, P.S., Saha, J.K., Dotaniya, M.L. et al. (2021). Wastewater irrigation in India: current status, impacts and response options. *Science of the Total Environment* 808: 1–17.

Mitra, A. and Zaman, S. (2021). Impact of acidification on ecosystem services of Sundarban estuaries. In: *Estuarine Acidification*. Cham: Springer https://doi.org/10.1007/978-3-030-84792-0_6.

Mitsch, W.J., Bernal, B., and Hernandez, M.E. (2015). Ecosystem services of wetlands. *International Journal of Biodiversity Science, Ecosystem Services and Management* 11 (1): 1–4.

Montiel-Rozas, M.M., Madejon, E., and Madejon, P. (2015). Evaluation of phytostabilizer ability of three ruderal plants in mining soils restored by application of organic amendments. *Ecological Engineering* 83: 431–436.

Montiel-Rozas, M.M., Madejon, E., and Madejon, P. (2016). Effect of heavy metals and organic matter on root exudates (low molecular weight organic acids) of herbaceous species: an assessment in sand and soil conditions under different levels of contamination. *Environmental Pollution* 216: 273–281.

Münzel, T., Hahad, O., Daiber, A., and Landrigan, P.J. (2023). Soil and water pollution and human health: what should cardiologists worry about? *Cardiovascular Research* 119 (2): 440–449. https://doi.org/10.1093/cvr/cvac082.

Nagar, M.C., Dotaniya, M.L., Amita, S. et al. (2022). Pressmud overcome lead toxicity by improving spinach biomass in lead contaminated soils. *Environmental Monitoring and Assessment* 195 (1): –107. https://doi.org/10.1007/s10661-022-10718-0.

Nikolova, R., Boteva, S., Kenarova, A. et al. (2023). Enzyme activities in soils under heavy metal pollution: a case study from the surroundings of a non-ferrous metal plant in Bulgaria. *Biotechnology and Biotechnological Equipment* 37 (1): 49–57.

Nkrumah, P.N., Baker, A.J.M., Chaney, R.L., and Erskine, P.D. (2006). Zinc and cadmium uptake by hyperaccumulator *Thlaspi caerulescens* grown in nutrient solution. *Soil Science Society of America Journal* 70: 1918–1924.

Peña, A. (2022). A comprehensive review of recent research concerning the role of low molecular weight organic acids on the fate of organic pollutants in soil. *Journal of Hazardous Materials* 434: 128875. https://doi.org/10.1016/j.jhazmat.2022.128875.

Pérez-Ruzafa, A., Morkune, R., Marcos, C. et al. (2020). Can an oligotrophic coastal lagoon support high biological productivity? Sources and pathways of primary production. *Marine Environmental Research* 153: 104824.

Persson, L., Almroth, B.M.C., Collins, C.D. et al. (2022). *Environmental Science and Technology* 56 (3): 1510–1521.

Pilon-Smits, E.A.H. (2005). Phytoremediation. *Annual Review of Plant Biology* 56: 15–39.

Plutino, M., Bianchetto, E., Durazzo, A. et al. (2022). Rethinking the connections between ecosystem services, pollinators, pollution, and health: focus on air pollution and its impacts. *International Journal of Environmental Research and Public Health* 19 (5): 2997.

Qadir, M., Sharma, B.R., Bruggeman, A. et al. (2007). Non-conventional water resources and opportunities for water augmentation to achieve food security in water scarce countries. *Agricultural Water Management* 87: 2–22.

Qin, F., Shan, X.-Q., and Wei, B. (2004). Effects of low-molecular-weight organic acids and residence time on desorption of Cu, Cd, and Pb from soils. *Chemosphere* 57: 253–263.

Rahman, M.M., Dong, Z., and Naidu, R. (2015). Concentrations of arsenic and other elements in groundwater of Bangladesh and West Bengal, India: potential cancer risk. *Chemosphere* 139: 54–64.

RamaKrishna, K., Venkateswarlu, K., and Rambabu, C. (2004). Screening of aquatic plants for phytoremediation of chromium. *Journal of Phytological Research* 17: 163–167.

Rashmi, I., Kumawat, A., Munawery, A. et al. (2023). *Soil Amendments: An Ecofriendly Approach for Soil Health Improvement and Sustainable Oilseed Production*. IntechOpen https://doi.org/10.5772/intechopen.106606.

Rengasamy, P. (2006). World salinization with emphasis on Australia. *Journal of Experimental Botany* 57: 1017–1023.

Ricky, R., Shanthakumar, S., Ganapathy, G.P., and Chiampo, F. (2022). Zero liquid discharge system for the tannery industry—an overview of sustainable approaches. *Recycling* 7 (3): 31.

Röthig, T., Trevathan-Tackett, S.M., Voolstra, C.R. et al. (2023). Human-induced salinity changes impact marine organisms and ecosystems. *Global Change Biology* 29 (17): 4731–4749. https://doi.org/10.1111/gcb.16859.

Rout, B.K. and Sikdar, B.K. (2017). Hazard identification, risk assessment, and control measures as an effective tool of occupational health assessment of hazardous process in an iron ore pelletizing industry. *Indian Journal of Occupational and Environmental Medicine* 21 (2): 56–76.

Saha, J.K., Dotaniya, M.L., and Coumar, M.V. (2023). Risk assessment and remedition of contamination soil. In: *Recent Advances of Soil Research: IISS Contribution* (ed. P. Jha et al.), 64–79. Indian Society of Soil Science.

Saha, J.K., Rajendiran, S., Coumar, M.V. et al. (2017). *Soil Pollution – An Emerging Threat to Agriculture*, 380. Singapore: Springer https://doi.org/10.1007/978-981-10-4274-4.

Salati, S. and Moore, F. (2010). Assessment of heavy metal concentration in the Khoshk River water and sediment, Shiraz, Southwest Iran. *Environmental Monitoring and Assessment* 164: 677–689.

Sánchez-Moyano, J.E., García-Asencio, I., Donázar-Aramendía, I. et al. (2017). BENFES, a new biotic index for assessing ecological status of soft-bottom communities. Towards a lower taxonomic complexity, greater reliability, and less effort. *Marine Environmental Research* 132: 41–50.

Schraufnagel, D.E. (2020). The health effects of ultrafine particles. *Experimental and Molecular Medicine* 52: 311–317.

Seshadri, B., Bolan, N., and Naidu, R. (2015). Rhizosphere-induced heavy metal (loid) transformation in relation to bioavailability and remediation. *Journal of Soil Science and Plant Nutrition* 15: 524–548.

Sheng, X.-F., Xia, J.-J., Jiang, C.-Y. et al. (2008). Characterization of heavy metal-resistant endophytic bacteria from rape (Brassica napus) roots and their potential in promoting the growth and lead accumulation of rape. *Environmental Pollution* 156: 1164–1170.

Sheoran, P., Kumar, A., Singh, A. et al. (2021). Pressmud alleviates soil sodicity stress in a rice–wheat rotation: Effects on soil properties, physiological adaptation and yield-related traits. *Land Degradation and Development* 32 (9): 1–14.

Singh, S., Ashraf, N., Hussain, I. et al. (2010). Lead-induced genotoxicity in the root cells of Allium sativum L. *Biologia Plantarum* 54: 111–114.

Singh, P.K., Deshbhratar, P.B., and Ramteke, D.S. (2012). Effects of sewage wastewater irrigation on soil properties, crop yield and environment. *Agricultural Water Management* 103: 100–104.

Singh, S.P. and Singh, M.K. (2020). Soil pollution and human health. In: *Plant Responses to Soil Pollution* (ed. P. Singh, S.K. Singh, and S.M. Prasad). Singapore: Springer.

Solanki, P., Dotaniya, M.L., Khanna, N. et al. (2020). Recent advances in bioremediation for clean-up of inorganic pollutant-contaminated soils. In: *Frontiers in Soil and Environmental Microbiology* (ed. S.K. Nayak and S.K. Mishra), 299–310. New York: CRC Press.

Soni, A., Kumar, A., Kumar, A. et al. (2023). Impact of industrial effluent on physico-chemical properties of soil in Unnao District of Uttar Pradesh, India. *International Journal of Plant and Soil Science* 35 (20): 1300–1305.

Stavi, I., Bel, G., and Zaady, E. (2016). Soil functions and ecosystem services in conventional, conservation, and integrated agricultural systems. A review. *Agronomy for Sustainable Development* 36: 32.

Tindwa, H.J. and Singh, B.R. (2023). Soil pollution and agriculture in sub-Saharan Africa: state of the knowledge and remediation technologies. *Frontiers in Soil Science* 2: 1101944.

Tiwari, S.K., Kumaraswamidhas, L.A., and Garg, N. (2023). Assessment of noise pollution and associated subjective health complaints in Jharia Coalfield, India: a structural equation model analysis. *Noise Mapping* 10 (1): 20220172. https://doi.org/10.1515/noise-2022-0172.

Tripathi, R.D., Srivastava, S., Mishra, S. et al. (2010). Lead detoxification by coontail (*Ceratophyllum demersum* L.) involves induction of phytochelatins and antioxidant system in response to its accumulation. *Chemosphere* 78: 199–206.

Tseng, M.L., Wong, W.P., and Soh, K.L. (2018). An overview of the substance of resource, conservation and recycling. *Resources, Conservation and Recycling* 136: 367–375.

UNDP (2020). "Goal 14 targets". UNDP. Archived from the original on 30 September 2020. Retrieved 24 September 2020.

UNECE (2022). Annual Report-2022. United Nations Economic Commission For Europe. https://unece.org/sites/default/files/2023-04/UNECE_AR%202022_Web.pdf, accessed on December 21, 2023.

Van Raamsdonk, L.W., van der Zande, M., Koelmans, A.A. et al. (2020). Current insights into monitoring, bioaccumulation, and potential health effects of microplastics present in the food chain. *Food* 9: 72.

Wang, X., Chen, J., Tan, Q. et al. (2007). Zinc accumulation and compartmentation in root, transport into xylem, and absorption into leaf cells in the hyperaccumulator Sedum alfredii. *The New Phytologist* 174: 4–14.

Wang, X., Sato, T., Xing, B. et al. (2013). An approach to characterizing human exposure to lead contaminated soil in active smelting regions, China. *Science of the Total Environment* 442: 39–45.

Wei, C. and Chen, T. (2001). Hyperaccumulators and phytoremediation of heavy metal contaminated soil: a review of studies in China and abroad. *Acta Ecologica Sinica* 21: 1196–1203.

Whiting, S.N., Leake, J.R., McGrath, S.P., and Baker, A.J.M. (2000). Hyperaccumulation of Zn by *Thlaspi caerulescens* can ameliorate Zn toxicity in the rhizosphere and shoots of the hyperaccumulator. *The New Phytologist* 145: 429–437.

Xu, X., Wang, Q., and Li, C. (2022). The impact of dependency burden on urban household health expenditure and its regional heterogeneity in China: based on quantile regression method. *Frontiers in Public Health* 10: 876088. https://doi.org/10.3389/fpubh.2022.876088.

Zhang, J., Chen, M., Gao, S. et al. (2023a). Examining impacts of policy, technology, and climate extremes on thermoelectric power production and river thermal pollution in the Midwest and Northeast of the United States. *Frontiers in Environmental Science* 11: 1212211.

Zhang, J., Liu, Z., Tian, B. et al. (2023b). Assessment of soil heavy metal pollution in provinces of China based on different soil types: from normalization to soil quality criteria and ecological risk assessment. *Journal of Hazardous Materials* 441: 129891.

Zhang, L., Qiao, J., Cui, H. et al. (2021). Using low molecular weight organic acids to enhance microbial degradation of polycyclic aromatic hydrocarbons: current understanding and future perspectives. *Water* 13 (4): 446. https://doi.org/10.3390/w13040446.

Zhu, Q., Guo, J., Guo, X. et al. (2021). Relationship between ecological quality and ecosystem services in a red soil hilly watershed in southern China. *Ecological Indicators* 121: 107119.

14

Harnessing Aromatic Plants for Phytoremediation

A Scented Solution for Ecology and Economy

Vimal Chandra Pandey[1], Valeria Ancona[2], D.D. Tewari[3], and Sanat Kumar Dwibedi[4]

[1] *CSIR-National Botanical Research Institute, Lucknow, Uttar Pradesh, India*
[2] *Water Research Institute of the Italian National Research Council, Bari, Italy*
[3] *Gayatri Vidyapeeth P.G. College, Rishi Bhoomi, Risia, Bahraich, Uttar Pradesh, India*
[4] *Odisha University of Agriculture and Technology, Bhubaneswar, India*

14.1 Introduction

Rapid urbanization, industrialization, and intensified agricultural practices, coupled with irresponsible management of wastes across the globe, result in the release of diverse contaminants into the soil, water, and air. This has severe repercussions on environmental quality contributing to the occurrence of numerous contaminated sites. Numerous locations worldwide experience pollution due to natural and anthropogenic reasons (Pandey et al. 2019a). Land contamination and deterioration are estimated to be around 20% of the overall land cover, corresponding to over 100 million acres. In Europe, such contaminated sites are of 340 000. Nearly one-third of these polluted sites have already been recognized and around 15% have been recuperated (European Environmental Agency 2020). The cost of management of these contaminated sites in Europe is about €6.5 billion per year.

Contaminated soil poses substantial hazards to both the environment and livelihood, rendering them unsuitable for agricultural use due to the possible accumulation of heavy metals within the food chain. Consequently, the escalating prevalence of polluted soil has emerged as a potential global concern demanding urgent attention and immediate intervention. The polluting heavy metals enter into the environment through natural weathering of rocks and minerals, soil erosion, and volcanic eruptions while anthropogenic activities such as mining, steel, aluminum, textile and tanning industries, electroplating of metals, agricultural use of pesticides and fertilizers (phosphate), municipal sludge, chromite, red mud and fly ash dumping, coal-fired power plants, etc., play a significant role in their accumulation. Such contaminated sites are very often abandoned and pose serious health and ecological hazards in neighboring areas (Pandey et al. 2019a). These sites pollute air, water, and soil with elements such as As, Pb, Hg, Cd, Cr, Ni, and Se, which are highly toxic and contribute to illnesses such as cancer, skin and cardiac diseases, genetic disorders, mesothelioma, and pneumoconiosis.

Biodiversity and Ecosystem Services on Post-Industrial Land, First Edition.
Edited by Vimal Chandra Pandey.

Contaminated soils, water, and dumpsites cannot be directly used for agriculture. There is an immediate need for the formulation and implementation of sustainable rehabilitation strategies aligned with the principles of ecological transition. Various remediation options although available should critically be sorted based on the type and extent of contamination, site-specific conditions, costs, and regulatory requirements. Among the strategies for remediation, the utilization of phyto-technologies for the remediation of contaminated areas stands as an essential technical solution. Plant-assisted bioremediation (PABR) represents a form of phytoremediation designed to restore multi-contaminated areas. It relies on the synergistic actions of plant species and rhizospheric micro-organisms to promote the containment, degradation, or transformation of pollutants (Ancona et al. 2021). The exudates released through the root system play a crucial role in shaping microbial-plant interactions forming the foundation for processes involved in decontamination of soils. Typically, plant species for PABR or phytoremediation are sorted based on their resistance offered to the contaminant concentration and toxicity, their adaptability to stressful soil conditions such as extreme pH, high salinity, and nutrient deficiencies, and their ability to influence soil microbial composition and functioning. Among the diverse array of plant species suitable for PABR or traditional phytoremediation applications, the selection of plant species is made depending on their ability to grow in contaminated sites, inoculation of microbes, and appropriate soil amendments. Aromatic grasses from the *Poaceae* family have been the subject of interest to assess the phytoremediation potential in both inorganic and organic pollution (Pandey and Maiti 2020; Pandey et al. 2023). The essential oil prepared from the aromatic crops through distillation process does not contain any pollutants (Gupta et al. 2013; Pandey et al. 2023). On the contrary, cultivation of edible crops such as cereals, pulses, oilseeds and vegetables, and fodder crops accumulate toxic heavy metals and are not suitable for healthy consumption (Figure 14.1).

In this chapter, attempts have been undertaken to clarify the information on the suitability of aromatic grasses for the phytoremediation of contaminated soil encompassing the general aspects of aromatic grasses. Additionally, it illustrates the most recent findings on phytoremediation capabilities and the diverse applications of aromatic grasses.

14.2 Aromatic Grasses

Aromatic grasses frequently used for production of essential oils, perfumes, insect repellents, soaps, cosmetics, and many more, are perennial grasses. Aromatic grasses initiate ecological succession. They fall under "nurse species" categories and thus play a significant role in ecosystem goods and services (Pandey et al. 2019b, 2019c, 2023). Their phytoremediation properties have been well established due to their notable traits of high capacity for biomass development, low nutrient requirements, and their ability to establish even in degraded soils (Pandey and Maiti 2020; Pandey and Bajpai, 2019). They play a crucial role in restoring soil quality and structure, reducing soil erosion, and enhancing soil biodiversity. Additionally, their extensive coverage of soil surface renders their cropping more sustainable than agricultural practices. Aromatic grasses are used *in situ* for the remediation of dump sites and contaminated wastelands because of their perennial rapidly growing habits with high biomass and multiple harvests, easy and quick propagation, prolific root

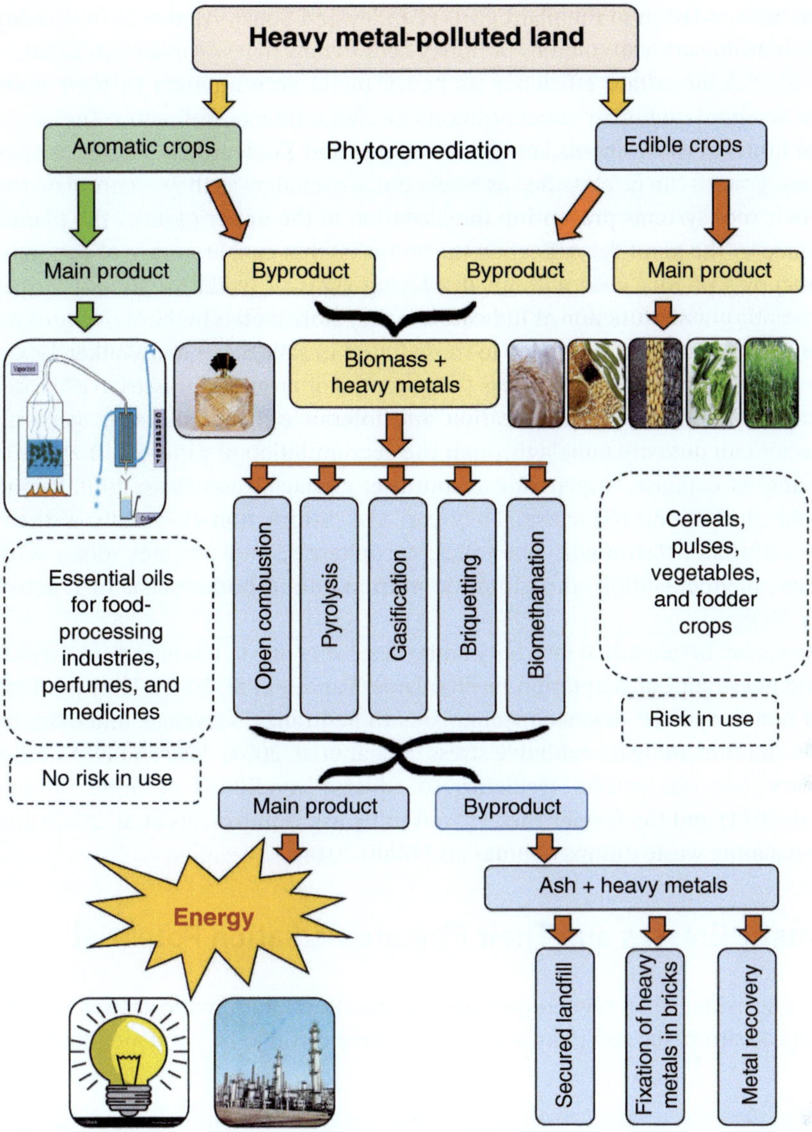

Figure 14.1 An informative diagram emphasizing the phytoremediation advantage of aromatic crops over edible crops in contaminated land. *Source:* Gupta et al. (2013)/with permission of American Chemical Society.

systems, tolerant to abiotic and biotic stress conditions such as extreme pH and temperature, salinity, deficit and excess soil moisture, heavy metal and chemical toxicity, high metal stabilization, and unpleasant and unpalatable to livestock (Pandey et al. 2015b; Pandey and Singh 2015; Verma et al. 2014; Pandey et al. 2023), and disease-pest infestations. Some additional benefits include carbon sequestration, biodiversity conservation, ecosystem goods and services for society and aesthetics (Pandey and Souza-Alonso 2019). Essential oil derived from aromatic grasses is completely free from any sort of heavy metal in the soil and subsequent accumulation in the plant parts (Pandey and Souza-Alonso 2019; Pandey et al. 2023). Their aroma attracts colonies of honey bees to new hives (Topal et al. 2020).

Aromatic grasses demonstrate efficiency as heavy metal accumulators in their body parts. They can accumulate higher concentrations of contaminants well above the maximum allowable limits in roots, shoots, and leaves (Gołda and Korzeniowska 2016; Pandey et al. 2023). These grasses can be classified as heavy metal excluders as they accumulate the pollutants in their root systems preventing translocation to the upper parts of the plants, which is facilitated by the plant detoxification mechanisms that enable storage of contaminants in the vacuoles of root cells without displaying signs of oxidative stress (Fatima et al. 2018). Aromatic grasses function as indicators if they store metals in the above-ground plant tissues in concentrations comparable to those found in soil (Baker and Walker 1990).

The numerous studies conducted to assess the potential of aromatic grasses in phytoremediation have exhibited metal detoxification and tolerance mechanisms. In general, these plant species can detoxify metals through the accumulation of different enzymatic antioxidants such as catalase, superoxide dismutase, guaiacol, ascorbate, glutathione peroxidases, and glutathione reductase. They can also utilize non-enzymatic antioxidants such as ascorbate, flavonoids, phenolics, tocopherols, and prolines along with metallothionines, phytochelatins, and glutathione to maintain homeostasis of reactive oxygen species (ROS).

Aromatic grasses are better suited to aridity and soil salinity due to alterations in various biochemical and physiological adaptation mechanisms (Pandey et al. 2023). They produce enzymatic and non-enzymatic defence mechanisms to neutralize, scavenge, and detoxify high ROS levels and thus mitigate oxidative stress (Dagar et al. 2004). *Vetiveria zizanioides* and *Cymbopogon flexuosus* exhibit resilience to adverse conditions of mine tailings (Srivastava et al. 2014) and the former thrives well in fly ash dumps (Das et al. 2013) and chromite asbestos mine waste dumps (Kumar and Maiti 2014).

14.3 Aromatic Grasses and Their Phytoremediation Potential

The latest scientific evidence on phytoremediation capabilities of aromatic grasses such as vetiver, lemongrass, citronella, and palmarosa are depicted in Table 14.1 and are illustrated as follows:

14.3.1 Vetiver Grass

Vetiver, also known as *Khus*, refers to plant species like *Chrisopogon zizanioides* or *Vetiveria zizanioides*, which are categorized under the Poaceae family. The average value of oil yield ranges from 1 to 1.5% on a dry-weight basis. Originating from India, Vetiver is a perennial herbaceous species that has been cropped in several countries due to its non-invasive nature (Truong et al. 2008). Due to its attributes such as quick growth, high biomass production, and tolerance to numerous stressors like heavy metals, pH variations, salinity, sodicity, etc. Vetiver has been proven to be highly attractive and promising for phytoremediation purposes (Danh et al. 2009; Aibibu et al. 2010; Singh et al. 2017). Numerous studies have illustrated its effectiveness in removing heavy metals from soil or water (Gravand et al. 2021; Otunola et al. 2022a). As suggested by Roongtanakiat et al. (2007), the use of

Table 14.1 Condensed overview of research on the use of aromatic grasses for phytoremediation.

Plant species	Contaminant	References
Vetiver		
Chrysopogon zizanioides	Atrazine	Zhang et al. (2023)
Chrysopogon zizanioides	Cu, Al	Linggawati et al. (2022)
Chrysopogone zizanioides	Cu, Cr	Purnama et al. (2022)
Chrysopogon zizanioides	Cu, Ni, Pb, Zn, Mn, and As	Kereeditse et al. (2023)
Vetiveria zizanioides	Crude oil	Kiamarsi et al. (2021)
Vetiver grass	Al, Mn	Otunola et al. (2023)
Vetiver grass	Cr, Cu, Co, Ni, and Zn	Otunola et al. (2022a)
Lemongrass		
Cymbopogon citratus	Cu	Das and Maiti (2009)
Cymbopogon citratus	Ni	Lee et al. (2014)
Cymbopogon citratus	Pb, Cd, Zn	Hassan (2016)
Cymbopogon flexuosus	Cr	Patra et al. (2015)
Cymbopogon flexuosus	Multi-metals	Jha and Kumar (2017)
Cymbopongon flexuosus	Cd, Cr, Ni, and Pb	Lal et al. (2013)
Cymbopogon citratus	Pb	Sobh et al. (2014)
Cymbopogon flexuosus	Multi-metals	Khilji and Sajid (2020)
Cymbopogon spp.	Multi-metals	Pandey et al. (2020)
Citronella		
Cymbopogon winterianius	Cd	Boruah et al. (2000)
Cymbopogon winterianius	Cr	Sinha et al. (2013)
Palmarosa		
Cymbopogon martinii	Multi-metals	Pandey et al. (2015a)

Vetiver can be a viable strategy for the recovery of heavy metals in ancient landfills and for promoting land rehabilitation.

This plant species, like other aromatic crops, is well recognized for its ability to accumulate heavy metals primarily in its root system than in its epigean plant tissues (Gravand et al. 2021; Otunola et al. 2022a). Nevertheless, several authors have observed that Vetiver can be considered an effective hyperaccumulator of heavy metals exhibiting elevated translocation factors for certain heavy metals (Banerjee et al. 2016; Gautam and Agrawal 2017; Ng et al. 2016). Through regular cultivation and removal from the contaminated sites, the heavy metal concentration in the contaminated land is reduced (Lone and Gaffar 2021). The employability of Vetiver in swampy land for wastewater remediation of heavy metals has been tested and validated (Zhang et al. 2014). Singh et al. (2017) have demonstrated the accumulation of As in Vetiver without symptoms of metal toxicity owing to its improved plant antioxidative capacity under a hydroponic system. Purnama et al. (2022) evidenced microbial-assisted phytoremediation of Cr and Cu from textile-polluted wastewater by the use of *Bacillus* or *Pseudomonas aeruginosa* combined with Vetiver (*C. zizanioides*) in hydroponic experiments. Both the contaminants were primarily concentrated within the root system. Due to the configuration of the root structure, which prevents the translocation of heavy metals in the above-ground plant parts, Vetiver demonstrated a more significant metal uptake in the roots than in the leaves (Hasan et al. 2017; Chintani et al. 2021). Various authors, including Melato et al. (2012) and Banerjee et al. (2016), have argued in favor of the Vetiver grass for heavy metal-induced stress tolerance in soils. These elucidations encompass the metal chelation processes mediated by phenolic compounds, and enhanced enzyme activity (glutathione S-transferase, catalase, and guaiacol peroxidase), sequestration, and storage of metals in the cell wall. Moreover, Banerjee et al. (2016) outlined the detoxification mechanisms against ROS in the shoots of *Khus*.

Linggawati et al. (2022) studied the potential of *Chrysopogon zizanioides* in phytoremediation of liquid batik waste contaminated by Cr and Al. This plant species accumulated more Cu than Al, mostly in the roots. The anatomical analysis of Vetiver induced under liquid batik waste treatment evidenced more notable modifications in the roots than in the leaves. Studies carried out under wastewater treatments evidenced less uptake of Fe than other metals (Zn and Cu) due to the development of iron plaques into the roots of Vetiver which prevents the upward movement of Fe in the plant (Yang et al. 2019; Kiiskila et al. 2019).

Different amelioration strategies have been evaluated to improve the Vetiver capabilities for the recovery of heavy metal-contaminated soils and water. The application of organic amendments (compost, biochar, and manure), organic acids, or direct and alternating current has been already reported (Andra et al. 2009; Chiu et al. 2005; Ng et al. 2016; Siyar et al. 2020). In the presence of EDTA, the Vetiver plant exhibited Pb tolerance through the creation of complexes among the phytochelatins (Andra et al. 2009). It has also been demonstrated that the application of EDTA notably enhanced the uptake of Zn and Cu in Vetiver (Lai and Chen 2004) and improved phytoextraction of Pb (Wilde et al. 2005). Vetiver grass in a pot experiment with vermicompost amendment was capable of tolerating and growing under high concentrations of heavy metal pollution such as Mn, Zn, Pb, Cu, and Cd. Vermicompost promoted high biomass development which led to a reduction in pH and an increase in heavy metal bioavailability (Jayashree et al. 2011). Vetiver demonstrated remediation of high concentrations of heavy metals in polluted mine tailings by applying manure compost and sewage (Chiu et al. 2006). These amendments led to a substantial reduction of Pb uptake in Vetiver and simultaneously enhanced the uptake of Zn and Cu. The incorporation of clay minerals such as attapulgite (palygorskite) and bentonite, proved highly useful in enhancing the capability of Vetiver for phytoremediation of Al and Mn-contaminated water (Otunola et al. 2022b). Utilizing red mud obtained from alumina production from bauxite has been effective in promoting robust growth of Vetiver and improving its tolerance to several heavy metals including Cd, Cr, Cu, Fe, Mn, Mg, Ni, Pb, and Zn. Gautam and Agrawal (2017) have observed the metal exclusion property of Vetiver in red mud with the bioaccumulation factors (BAFs) less than 1.0 for all heavy metals investigated. This phenomenon could be attributed to the reduced bioavailability of heavy metals due to the adsorption on red mud.

Some researchers have described the efficiency of Vetiver in the uptake and removal of organic contaminants, including polychlorinated biphenyls (PCBs), polycyclic aromatic hydrocarbons (PAHs), crude oil, pesticides, and antibiotics (Zhang et al. 2023; Chen et al. 2020; Noshadi et al. 2019). About Atrazine, a well-known herbicide, Vetiver can absorb it through the synthesis of Atrazine and glutathione conjugates in the leaves (Marcacci et al. 2005). In a more recent study, Zhang et al. (2023) successfully identified the metabolites of Atrazine produced by the pollutant degradation facilitated by Vetiver. Additionally, the study identified the main associated Atrazine degrading bacterial genera including *Arthrobacter*, *Bradyrhizobium*, *Nocardioides*, and *Rhodococcus*.

The fruitful cultivation of Vetiver and selected bacterial consortium for enhancing phytoremediation of crude oil-contaminated soil has been demonstrated by Kiamarsi et al. (2021). In this study, *Vetiveria zizanioides* was combined with five native bacterial isolates (*Pseudomonas resinovorans*, BH10; *Plantibacter auratus*, MU15; *Bacillus subtilis*, BA13; *Staphylococcus pasteuri*, BK37, and *Bacillus atrophaeus*, BI47) to rehabilitate the soil polluted by crude oil at different concentrations (2, 4, 6, 8, 10, and 12% w_{oil}/w_{soil}). In combination with the bacterial consortium, the phytoremediation efficiency of Vetiver increased, particularly at high crude oil levels (>C6) (Kiamarsi et al. 2021).

Vetiver can remediate 200 ppm of phenol within four days without involving microbes (Singh et al. 2008a). Phenrat et al. (2015) reported the complete removal of TNT (40 mg L^{-1}) from a hydroponics system with Vetiver plants. Ethidium bromide (EtBr), a mutagenic toxic compound used across the world in laboratories for DNA staining, should be disposed of carefully. But if released accidentally then Vetiver can accumulate 0.7 μg of EtBr per kg of dry weight (Danh et al. 2009). On a similar note, Vetiver can remediate Acrylamide, a neurotoxin and carcinogen, used to synthesize polyacrylamide gels for RNA and DNA synthesis (Paz-Alberto et al. 2007).

Tetracycline, a widely used antibiotic, if released unsafely could contaminate soil and water, induce tolerance in microbes and destabilize microbial diversity (Depledge 2011; Sengupta et al. 2016). Vetiver has been able to accumulate it as observed in hydroponic experiments (Datta et al. 2013). Vetiver grass is reported to absorb Naproxen, a micropollutant drug used to treat inflammation and pain up to 60% from the contaminated soil and water (Marsidi et al. 2016).

Radioactive wastes released from nuclear power reactors and weapon testing sites are a threat to flora and fauna and need urgent remediation. Shaw and Bell (1991) and Singh et al. (2008b) have reported effective remediation of Strontium (^{90}Sr) and Cesium (^{137}Cs) from the contaminated environment. Vetiver can grow in heavily polluted oil drilling sites. Its growth is more affected in diesel-polluted sites than in hydraulic oil-polluted sites (Rajaei and Seyedi 2018). In fly ash dump sites, Vetiver can grow profusely and can remediate heavy metals in one and a half years as observed in pot experiments (Chakraborty and Mukherjee 2011). As maximum heavy metal contaminants remain in roots and a very negligible quantity is translocated to shoots, thus on grazing, bioaccumulation of heavy metals in the next tropic level is minimized. Nutrient loads in water and soil, such as nitrogen and phosphorus, are removed faster by engaging Vetiver (Hart et al. 2003). Similarly, boron, fluoride, and aluminum could be also removed by Vetiver (Ruiz et al. 2013) and thus are regarded as nature's gift to mankind.

Due to its long leaves, profuse tillering ability, and deep roots Vetiver can survive in degraded lands under stress environments. Its sharp leaves with strong midrib fold inward under moisture deficit stress conditions retain their original position and reopen for

Table 14.2 Rooted remediation: Vetiver vs. other plants in surface and subsurface contaminant clean-up.

Plant	Maximum root depth
Indian mustard	Upto 12 Inches
Grasses	Upto 48 Inches
Poplar tree	Upto 180 Inches (15 ft)
Vetiver grass	Upto 236.16 Inches (19.68 ft)

Source: Adapted from Pandey and Praveen (2020).

transpiration on the availability of soil moisture (Liao et al. 2003). Although stomata are present on both sides, during moisture stress, the folding of leaves prevents water loss from the outer side. Its profuse root system retains moisture, conserves soil and supports plants in dry environments and undulating terrains (Pandey and Praveen 2020). Its roots can penetrate underground up to 6 m and thus can be used for the removal of surface and subsurface contaminants (Table 14.2). Its deep root system helps in nutrient mining from the deeper underground zones and helps in the inversion of the leached nutrients. Its fast-growing ability enables intercropping in polluted land with significantly lower accumulation of contaminants in food crops (Pandey and Praveen 2020). The roots penetrate vertically without interfering with the roots of the adjacent intercrops. The extremely fast-growing roots of Vetivers reach the contaminated soil zone and remediate it both at the surface and subsurface before the proliferation of the root systems of the companion crops or intercrops.

Vetiver can be propagated vegetatively only as its flowers are sterile and do not set seeds. It does not have rhizomes and stolon and hence is mostly propagated by tillers, root division and slips. Tissue culture, ratooning, budding, stem cutting, and micropropagation are also other means of propagation. Vetiver slips are produced in light sandy soils and uprooted easily for fresh plantation. It grows rapidly and becomes able to remediate heavy metals and restore the environment within 4–5 months as compared to trees and shrubs engaged for the same cause. It can withstand high acidity and alkalinity with a pH ranging from 3 to 10.5; high salinity (EC = 8 dScm), sodicity (ESP = 33%), magnesium in soil; and extremely elevated levels of heavy metals, herbicides and pesticides in soil. It is native to India but can grow in tropical and subtropical areas dispersed along riverbanks, degraded lands, wastelands, marshy lands, mines tailings, fly ash, and red mud dump sites. It is said to be an omnipresent species with wide ecological distribution and can grow in areas with rainfall from 200 to 3000 mm. There are reports of its surviving in rainfall up to 5000 mm in Sri Lanka and −10 °C in the United States.

Vetiver, like its close family lines such as maize, sugarcane, and sorghum, is a C4 plant with a specialized Kranz anatomy using carbon dioxide more effectively than typical C3 plants. It is unresponsive to day length variations, allowing it to grow and bloom throughout the year. Although it does not establish itself easily under a shady environment after its establishment, it survives in deep shade (Pandey and Praveen 2020).

14.3.2 Lemongrass

Lemongrass is the common name used to indicate a group of 50 species among which *Cymbopogon flexuosus* and *Cymbopogon citratus* are the most recognized ones under the *Poaceae family*. Its leaves are utilized for green tea and stems are added in soups, poultry and seafood dishes (Jasha 2014). Essential oil produced from its leaves and inflorescence is used for making perfumes and medicines (Verma et al. 2014; Pandey et al. 2023). The typical oil production varies from 0.36 to 1% based on the fresh weight. It has anti-carcinogenic (Kumar et al. 2008), antibiotic, anti-microbial, antioxidant (Anand et al. 2011), anti-inflammatory (Figueirinha et al. 2010), and insecticidal (Rajeshwara Rao et al. 2015) properties. Polyphenol, a natural antioxidant derived from lemongrass fights against cancer, and cardiovascular and degenerative diseases (Ogura et al. 2008). It is mostly cultivated for its medicinal properties but its phytoremediation properties are also well verified. It is generally propagated from the narrow bulb-like base with roots by separating the slips and also from its seeds.

Numerous authors evidenced the capabilities of lemongrass in the phytoremediation of heavy metal pollutants such as As, Cu, Cr, Fe, Mn, Ni, Pb, etc. *Cymbopogon citratus* exhibits capabilities in the recovery of wastewater enriched with B by promoting lead absorption (Sobh et al. 2014). It is recognized as an effective biosorbent for removing Pb^{3+}, Cd^{2+}, and Zn^{2+} from industrial effluents and Ni^{2+} attributed to hydroxyl and carbonyl groups on its surface. Lemongrass exhibited phytostabilization of Cu, Fe, and Mn while promoting the accumulation of As, Cd, Cr, Ni, Pb, and Zn based on the calculated values of BAF and transfer factor (TF). Proline synthesis was induced in lemongrass by Cd (Handique and Handique 2009) and its accumulation was detected in the young leaves. The efficiency of lemongrass for phytoremediation of As and Cr in contaminated sites was also demonstrated by Jha and Kumar (2017). Lemongrass grown in Pb-contaminated soils accumulated more Pb in roots (0.73%) than shoot (0.13%) with bioaccumulation factor < 1 and translocation factor < 1 (Mwansa 2018).

A study performed by Pandey et al. (2020) highlighted the morphological development and enhanced chlorophyll content of lemongrass when cultivated using tannery sludge by 15%. Additionally, the application of tannery sludge was found to increase antioxidant activities and metal uptake. In an investigation by Pandey et al. (2020), it was observed that different lemongrass varieties, namely Shikhar, Krishna, Suwarna, and Chirharit, exhibit phytoextraction capabilities for Pb and Ni in tannery sludge-contaminated soils. The percentage of essential oils of each lemongrass variety was influenced by the concentrations of tannery sludge. The essential oil content of lemongrass can be significantly impacted by the inoculation of michorrhizhal fungi enriched with heavy metals (Lermen et al. 2015).

Approximately, 130 varieties of distinct chemicals are utilized in the leather industries, which are the primary contributors to environmental pollution (Bareen and Tahira 2011). The remediation of heavy metals in the tannery sludge mixed soil by lemongrass was $Cr > Cd > Cu > Mg > Fe > Zn > Ni > Pb$ (Khilji and Sajid 2020). Pandey et al. (2019a) reported significant phytoremediation potential of lemongrass for Ni and Pb while Suwarna was the most promising variety with the maximum reduction potential of 70.07, 85.29, and 77.36% for Cr, Pb, and Cd, respectively. Similarly, red mud released as waste from bauxite can be remediated by mixing with soil, cow dung manure, or sewage sludge. In an experiment by

Gautam et al. (2016), the best remediation result was observed at 5% (w/w) addition of red mud in the presence of organic manure and sewage sludge. In a pot culture experiment, a mixture of garden soil, cow dung manure, and overburden from coal mines significantly increase soil health and nutrient availability and enhanced lemongrass biomass by 38.6% with the mixture of cow dung manure and overburden at an 80 : 20 ratio (Pandey et al. 2021). In a separate experiment conducted by Kumar et al. (2022) to study the effects of fly ash and vermicompost amendment on growth characteristics, and essential oil content of lemongrass revealed a relative mixture at 10 and 90% of these two resulting in significantly higher biomass yield and essential oil content of 69.7 $t\,ha^{-1}$ and 429.3 $kg\,ha^{-1}$, respectively. Combined application of NPK fertilizers @ 120–90–90 $kg\,ha^{-1}$, mycorrhiza and compost reduced concentration of Cr, Cu, Ni, and Pb in fly ash amended soils and increased the yield of lemongrass biomass and essential oil (*Cymbopogon citratus*) (Ultra Jr 2020).

Soil contaminated with the spillover petroleum hydrocarbons from the mechanic workshop contains elevated levels of heavy metals like Cd, Cu, Ni, Pb, Mn, and Zn than adjacent areas above the permissible recommended levels. Lemongrass cultivation could significantly reduce the Cd, Ni and Pb by 83, 67, and 95%, respectively while 99.3% of total petroleum hydrocarbon decreased to 31.2% after 96 days of cultivation of lemongrass (Oguche et al. 2022). Moreover, industrial greywater contaminated with heavy metals and toxic chemicals can be recycled and reused for agriculture after remediation by lemongrass (Rinitha 2022).

14.3.3 Citronella

Citronella is a common name that is often used for perennial aromatic plant species belonging to the Poaceae family having citronella oils, which are used as insect repellents and Ayurveda medicines. Among the citronella species, *Cymbopogon winterianus* (Java citronella) originating from Malaysia is used for deriving essential oils. Although citronella resembles lemongrass, the former has wider leaves and the base of the leaf (pseudostem) is reddish. Citronella contains geraniol (40.06%), citronellal (27.44%), and citronellol (10.45%) as a primary constituent of essential oil, which is used by cosmetics, food flavoring and insect repellent industries. Considering the fresh weight, the average oil percentages can range from 0.4 to 1.1%. The essential oil in citronella exhibits antigenotoxic, antioxidant, antimicrobial and antiparasitic characteristics (Sinha et al. 2011).

According to Jisha et al. (2017), citronella can be considered a valued phytostabilizer in sites polluted with heavy metals. Studies performed by several authors (Boruah et al. 2000; Jisha et al. 2017) reported the use of *Cymbopogon winterianus* for phytoremediation of heavy metals like Cd, As, Ni, Fe, and Cu. Sulastri et al. (2022) evidenced higher Cd phytoremediation capabilities of citronella than other aromatic grasses (vetiver, lemongrass, patchouli, and basil) due to its higher dry weight of shoots and roots. The high levels of Cd accumulation in the roots of citronella can cause its high elevated mortality (Pandey et al. 2019). Citronella grows profusely with adequate moisture accumulation in the root cells and a higher concentration of Pb increases essential oil concentration indicating its ability to remediate Pb (Antea et al. 2020). Boruah et al. (2000) observed a threshold limit of 200 ppm of Cd in soil that restricts citronella growth. Even 50 and 100 ppm exhibited pronounced retardation of plant growth. They reported the profile of Cd accumulation in the line of root > stem > leaf sheath > leaf. Cd remediation by citronella has also been reported by Sulastri et al. (2022).

14.3.4 Palmarosa

Palmarosa typically refers to species of *Cymbopogon martini* with an average percentage of oil ranging from 0.7 to 1.1% based on fresh weight. The oil is derived from the floral shoots and above-ground plant parts through a hydrodistillation process. This essential oil in India is available in two major varieties i.e., Rusha grass and Ginger grass. The oil from Rusha grass, known as Rusha oil or North Indian geranium oil, has a high geraniol content (75.9%) while the Ginger grass is inferior with low geraniol content. The oil is used for blending soap and flavoring tobacco.

Pandey et al. (2015a) demonstrated the capability of Palmarosa for the phytostabilization of Ni, Cr, and Cd in tannery sludge-amended soil. Jain et al. (2020) demonstrated that biochar can assist *Cymbopogon martini* in the stabilization of highly acidic mine waste. Biochar application had a positive impact on the metal tolerance index of palmarosa, increasing it by 54% (Rouhani et al. 2023). Among the heavy metals, Cr exhibited the highest accumulation of heavy metals in epigean plant tissues (roots and shoots, respectively) in Palmarosa followed by Ni, Pb, and Cd. An experiment conducted by Pandey et al. (2015b) to study the effect of tannery sludge on Palmarosa revealed enhanced plant biomass, production of essential oil, and buildup of heavy metals. The uptake of heavy metals by roots and shoots followed the same order (Cr > NI > Pb > Cd) and the translocation factor was <1 for all trace elements and the bioaccumulation factor was >1 in the case of heavy metals. Tannery sludge boosted plant development with higher metal buildup in roots and minimal in shoots.

14.3.5 Sweetgrass

Sweetgrass (*Hierochloe odorata*), also known as vanilla grass or holy grass, is a scented perennial grass native to North America and North Eurasia, distributed in cool and moist climates. Many North Americans burn it in ceremonies to invite good spirits. It is considered a symbol of peace, spirituality, and positive energy in North America. Its natural habitat is swampy moist lands, meadows, marshes, river banks, and Savannas but it can survive in any soil, sandy to clay with pH from 6 to 8 but with a minimum half day of sun. The sweet pleasant aroma in sweetgrass is from coumarin compounds in plants. Indians call it "a grass that never dies." Blackfeet drinks sweetgrass tea to stop vaginal bleeding. Sweetgrass is also fed by rodents and small ruminants thereby securing food and maintaining ecological balance even in stressful environments. As it is propagated from the rootstocks and rhizomes as well as seeds, it naturally spreads in an extensive area within a short span and reaches the climax population. Sweetgrass like other grasses can contribute to phytoremediation in the tissues of its root system but it has not yet been extensively researched to explore its effectiveness in different agroclimatic zones and soil types, the types of contaminants it can address, the optimal environmental conditions for its application, and its capabilities and limitations.

14.4 Manifold Usages of Aromatic Grasses

Aromatic grasses offer versatile positive impacts on ecological, economic, and social aspects as depicted in Figure 14.2. Concerning ecological aspects, these grasses can be employed to combat various kinds of land quality degradation processes such as soil

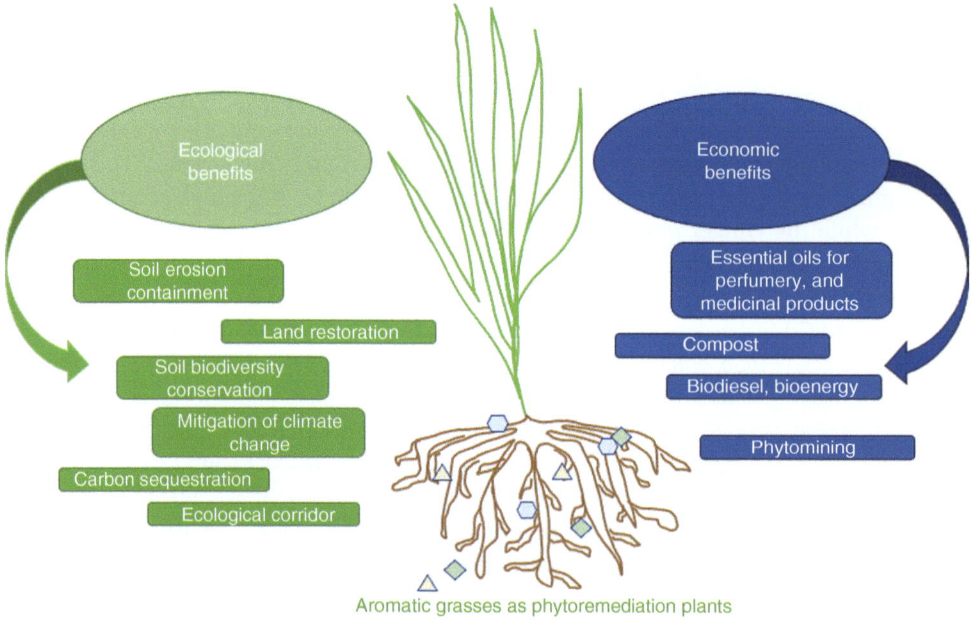

Figure 14.2 Attainable environmental and economic benefits of engaging aromatic grasses in phytoremediation endeavor.

erosion, desertification, and soil contamination to mitigate climate change, promote carbon sequestration, safeguard and conserve biodiversity, provide restoration for wild fauna, and generate ecological corridors. Aromatic grasses contribute positively to economic aspects through the synthesis of cosmetic, medicinal, and bioenergetic products. Various studies (Pandey et al. 2019a; Lone and Gaffar 2021) have demonstrated that essential oils obtained through the steam distillation process for phytoremediation do not contain residues of contaminants. This makes them safe for the creation of perfumes, personal care products, soaps, and detergents, do not represent a risk for humans and livestock but rather a profitable market opportunity that can allow the economic rehabilitation of degraded areas. Furthermore, after the extraction of essential oil, the residual biomass from aromatic grasses can be utilized for producing on-site composting, serving as valuable organic fertilizer (Pandey et al. 2019a, 2023). Aromatic grasses such as *Cympogon flexuosus* and *Cympogon citratus* have been verified as efficient in production of biofuel for diesel-powered engines (Dhinesh et al. 2016) and biogas produced through co-digestion of animal manures (Owamah et al. 2014), respectively.

14.5 Aromatic Grasses Rooted in UN-SDGs

The process of phytoremediation of contaminated land using aromatic grasses for achieving the United Nations' Sustainable Development Goals (UN-SDGs) involves several key steps, viz. (i) thorough assessment of the contaminated site to determine the nature and extent of pollutants, (ii) the selection of aromatic grass species in the targeted environment, (iii) planting and establishment of the aromatic grasses in the targeted sites for root absorption, accumulation, and in some cases, transformation of pollutants present in soil and water, (iv) monitoring and analysis of contaminants in the plant parts and surrounding environment, (v) adaptation and optimization strategy based on monitoring results, refining planting densities, and implementing supplementary techniques for enhancing phytoremediation, and (vi) long-term maintenance plan to ensure continued health and effectiveness of the aromatic plants.

By integrating aromatic grasses into the phytoremediation process, it is possible to contribute to the overall objectives of the UN-SDGs. Goal 1 of "poverty alleviation" is fulfilled with the creation of job opportunities for local communities and economic benefits, especially in the areas of contaminated sites being restored. Goal 2 of "zero hunger" is addressed with the remediation of the contaminated land by integration of aromatic plants that can restore agricultural lands and food security by enabling safe and healthy crops. Goal 3 of "good health and well-being" could be achieved through phytoremediation of harmful contaminants in soil and water, thereby improving public health and well-being. Phytoremediation by aromatic plants could be used to treat water contaminated with pollutants, contributing to the availability of "clean and safe water resources" which fulfil Goal 6 of UN-SDG. Aromatic plants engaged in phytoremediation could be used to generate affordable clean and green 2G bioethanol, which can satisfy the Goal 7 of "affordable and clean energy." Aromatic plants used in phytoremediation can sequester atmospheric carbon dioxide and contribute to climate change mitigation. Additionally, the process can help restore ecosystems affected by pollution and contribute to overall environmental health, thereby fulfilling Goal 13 of "climate action." Goals 14 and 15 emphasize "life below water" and "life on land" by significant elimination of contamination from the land and water simultaneously or by addressing the contaminants on land first and subsequently, its effects are felt in the form of clean water. Figure 14.3 illustrates a schematic diagram showcasing the remediation process of soil and water contamination through the cultivation of aromatic plants, aligned with the United Nations Sustainable Development Goals (UN-SDG) framework. With the integration of phytoremediation into environmental strategies, countries and communities can work towards accomplishing several SDGs at the same time, nurturing a future characterized by sustainability and resilience.

14.6 Conclusion and Prospects

Vetiver, Lemongrass, Citronella, and Palmarosa stand out as perennial grasses well known for their remarkable aroma, and resilience to unfavorable soil conditions. Their effective application in remediation of lands affected by nutrient deficiency, aridity, and contamination makes them invaluable. These grasses not only contribute to land restoration but also generate lucrative yields in return, especially essential oils. These oils are devoid of contaminants and, consequently, possess no harm to humans and livestock. Several studies conducted in recent years demonstrated that aromatic grasses, particularly Vetiver and Lemongrass, showcase significant proficiency in the phytoremediation of various heavy metals. These grasses not only display tolerance to metals but also utilize detoxification mechanisms that make them capable of accumulating, indicating, or stabilizing heavy metals. Figure 14.4 illustrates a schematic diagram depicting the primary steps of

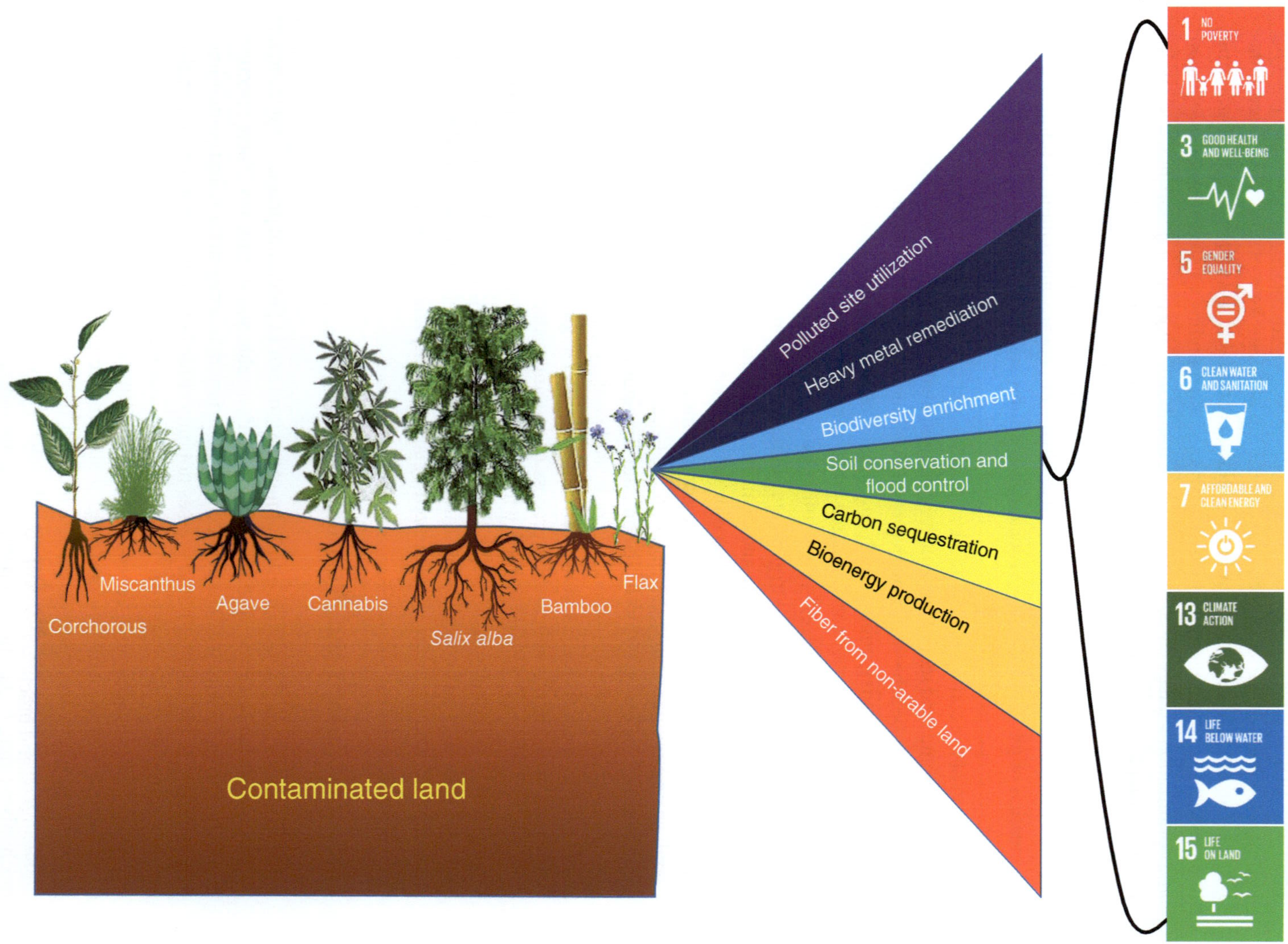

Figure 14.3 Essence of sustainability: Aromatic grasses and the UN-SDG framework.

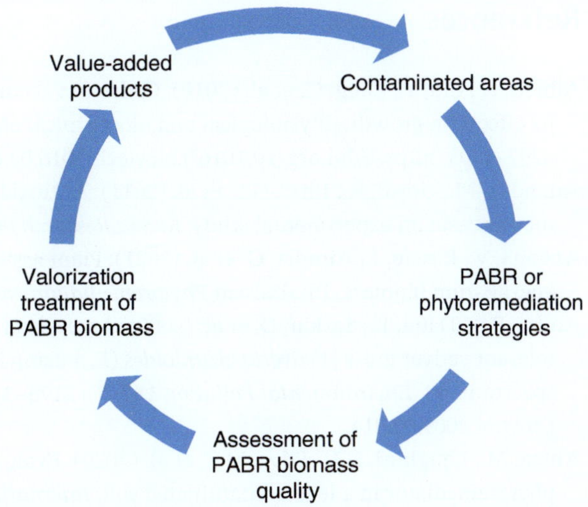

Figure 14.4 A schematic representation illustrating the key stages involved in engaging aromatic grasses for plant-assisted bioremediation (PABR)/ phytoremediation purposes.

cultivating the aromatic grasses in contaminated areas. Prior to assigning the biomass from PABR/phytoremediation to energy valorization processes, it is crucial to assess quality concerning the presence of contaminants in different plant tissues, encompassing shoots, and roots that can be treated.

Pandey et al. (2019a) have intriguingly suggested the utilization of aromatic grasses for the implementation of ecologically sound phytoremediation strategies on polluted sites and a diverse range of waste disposal areas. In the initial step of phytomanagement, the engagement of perennial scented grasses, known for stress tolerance and ability to thrive on unfavorable soil conditions, facilitates the improvement of soil physical, chemical, and microbiological characteristics in contaminated sites. Essentially, these aromatic grasses function as pioneering species. Subsequently, the remaining biomass of these fragrant grasses post-oil extraction can be utilized for fertilization in subsequent cultivation of aromatic crops that are more sensitive to stress conditions.

Extensive exploration of the capability of heavy metal phytoremediation capability of aromatic grasses has been studied, primarily conducting research in laboratories at a small scale. However, limited attention has yet been made to assess the ability and suitability of various aromatic herbs in degrading organic pollutants. Most of the experiments are confined to pots in the greenhouses in regulated environments while large-scale testing of the remediation measures by engaging these aromatic grasses in a natural environment is more acceptable and replicable. Thus, there is a compelling need for comprehensive research, adopting a multidisciplinary research process encompassing biochemical, biomolecular, chemical, microbiological, and other relevant approaches. The objective is to decipher the interactions between plants and microbes that form the basis for PABR processes. Special emphasis should be placed on the soil rhizosphere, where the secretion of root exudates triggers the reorganization of soil microbial community structure and impacts metabolic manifestation. While phytoremediation is gaining popularity among environmentalists, policymakers, and practitioners, the exploration of valuable plant-based phytoremediation emerges as a secure and encouraging approach for repurposing contaminated terrain, and production of essential oils devoid of metals with an economically viable return.

References

Aibibu, N., Liu, Y., Zeng, G. et al. (2010). Cadmium accumulation in *Vetiveria zizanioides* and its effects on growth, physiological and biochemical characters. *Bioresource Technology* 101: 6297–6303. https://doi.org/10.1016/j.biortech.2010.03.028.

Anand, K.M., Goyal, R., Bhat, G.S. et al. (2011). Antioxidant property of a novel lemongrass oil mouthwash: an experimental study. *Recent Research in Science and Technology* 3 (10): 14–18.

Ancona, V., Rascio, I., Aimola, G. et al. (2021). Plant assisted bioremediation: soil recovery and energy from biomass. In: *Assisted Phytoremediation*, vol. 2, 25–47. Elsevier.

Andra, S.S., Datta, R., Sarkar, D. et al. (2009). Analysis of phytochelatin complexes in the lead tolerant vetiver grass [*Vetiveria zizanioides* (L.)] using liquid chromatography and mass spectrometry. *Environmental Pollution* 157 (7): 2173–2183. https://doi.org/10.1016/j.envpol.2009.02.014.

Antea, M., Longhini, K.L., Silva, M.T. et al. (2020). Evaluation of citronella plant potential as a phytoremediator in a lead contaminated soil. *International Journal of Development Research* 10 (9): 39944–39947.

Baker, A. and Walker, P. (1990). Ecophysiology of metal uptake by tolerant plants. *Heavy Metal Tolerance in Plants: Evolutionary Aspect* 2: 155–165.

Banerjee, R., Goswami, P., Pathak, K., and Mukherjee, A. (2016). Vetiver grass: an environment clean-up tool for heavy metal contaminated iron ore mine-soil. *Ecological Engineering* 90: 25–34. https://doi.org/10.1016/j.ecoleng.2016.01.027.

Bareen, F. and Tahira, E.S.A. (2011). Metal accumulation potential of wild plants in tannery effluent contaminated soil of Kasur, Pakistan: field trials for toxic metal cleanup using *Suaeda fruticosa*. *Journal of Hazardous Material* 186 (1): 443–450. https://doi.org/10.1016/j.jhazmat.2010.11.022.

Boruah, H.P.D., Handique, A.K., and Borah, G.C. (2000). Response of Java citronella (*Cymbopogon winterianus* Jowitt) to toxic heavy metal cadmium. *Indian Journal of Experimental Biology* 38: 1267–1269.

Chakraborty, R. and Mukherjee, A. (2011). Vetiver can grow on coal fly ash without DNA damage. *International Journal of Phytoremediation* 13 (2): 206–214.

Chen, X., Wong, J., Wang, J., and Wong, M. (2020). Vetiver grass-microbe interactions for soil remediation. *Critical Review in Environmental Science and Technology* 51 (9): 897–938. https://doi.org/10.1080/10643389.2020.1738193.

Chintani, Y.S., Butarbutar, E.S., Nugroho, A.P., and Sembiring, T. (2021). Uptake and release of chromium and nickel by vetiver grass (*Chrysopogon zizanioides* (L.) Roberty). *SN Applied Sciences* 3: 1–13. https://doi.org/10.1007/s42452-021-04298-w.

Chiu, K.K., Ye, Z.H., and Wong, M.H. (2005). Enhanced uptake of As, Zn, and Cu by *Vetiveria zizanioides* and *Zea mays* using chelating agents. *Chemosphere* 60 (10): 1365–1375. https://doi.org/10.1016/j.chemosphere.2005.02.035.

Chiu, K.K., Ye, Z.H., and Wong, M.H. (2006). Growth of *Vetiveria zizanioides* and *Phragmities australis* on Pb/ Zn and Cu mine tailings amended with manure compost and sewage sludge: a greenhouse study. *Bioresource Technology* 97: 158–170. https://doi.org/10.1016/j.biortech.2005.01.038.

Dagar, J.C., Tomar, O.S., Kumar, Y., and Yadav, R. (2004). Growing three aromatic grasses in different alkali soils in semi-arid regions of Northern India. *Land Degradation & Development* 15: 143–151. https://doi.org/10.1002/ldr.598.

Danh, L.T., Truong, P., Mammucari, R. et al. (2009). Vetiver grass, *Vetiveria zizanioides*: a choice plant for phytoremediation of heavy metals and organic wastes. *International Journal of Phytoremediation* 11 (8): 664–691. https://doi.org/10.1080/15226510902787302.

Das, M., Agarwal, P., Singh, R., and Adholeya, A. (2013). A study of abandoned ash ponds reclaimed through green cover development. *International Journal of Phytoremediation* 15 (4): 320–329. https://doi.org/10.1080/15226514.2012.702801.

Das, M. and Maiti, S.K. (2009). Growth of *Cymbopogon citratus* and *Vetiveria zizanioides* on Cu mine tailings amended with chicken manure and manure-soil mixtures: a pot scale study. *International Journal of Phytoremediation* 11 (8): 651–663.

Datta, R., Das, P., Smith, S. et al. (2013). Phytoremediation potential of vetiver grass [*Chrysopogon zizanioides* (L.)] for tetracycline. *International Journal of Phytoremediation* 15 (4): 343–351.

Depledge, M. (2011). Reduce drug waste in the environment. *Nature* 478 (7367): 36–136.

Dhinesh, B., Lalvani, J.I.J., Parthasarathy, M., and Annamalai, K. (2016). An assessment on performance, emission and combustion characteristics of single cylinder diesel engine powered by *Cymbopogon flexuosus* biofuel. *Energy Conversion and Management* 117: 466–474.

European Environmental Agency (2020). https://www.eea.europa.eu/highlights/soil-contamination-widespread-in-europe.

Fatima, K., Imran, A., Amin, I. et al. (2018). Successful phytoremediation of crude-oil contaminated soil at an oil exploration and production company by plants bacterial synergism. *International Journal of Phytoremediation* 20 (7): 675–681.

Figueirinha, A., Cruz, M.T., Fransisco, V. et al. (2010). Anti-inflammatory activity of *Cymbopogon citratus* leaf infusion in lipopolysaccharide-stimulated dendritic cells: contribution of the polyphenols. *Journal of Medicinal Food* 13 (3): 681–690. https://doi.org/10.1089/jmf.2009.0115.

Gautam, M. and Agrawal, M. (2017). Influence of metals on essential oil content and composition of Lemongrass (*Cymbopogon citratus* (DC.) Stapf.) grown under different levels of red mud in sewage sludge amended soil. *Chemosphere* 175: 315–322. https://doi.org/10.1016/j.chemosphere.2017.02.065.

Gautam, M., Pandey, D., and Agrawal, M. (2016). Phytoremediation of metals using lemongrass (*Cymbopogon citratus* (D.C.) Stapf.) grown under different levels of red mud in soil amended with biowastes. *International Journal of Phytoremediation* 555–562. https://doi.org/10.1080/15226514.2016.1267701.

Gołda, S. and Korzeniowska, J. (2016). Comparison of phytoremediation potential of three grass species in soil contaminated with cadmium. *Ochrona Srodowiska i Zasobów Naturalnych* 27 (1): 8–14.

Gravand, F., Rahnavard, A., and Pour, G.M. (2021). Investigation of vetiver grass capability in phytoremediation of contaminated soils with heavy metals (Pb, Cd, Mn, and Ni). *Soil and Sediment Contamination: An International Journal* 30 (2): 163–186. https://doi.org/10.1080/15320383.2020.1819959.

Gupta, A.K., Verma, S.K., Khan, K., and Verma, R.K. (2013). Phytoremediation using aromatic plants: a sustainable approach for remediation of heavy metals polluted sites. *Environmental Science & Technology* 47 (18): 10115–10116. https://doi.org/10.1021/es403469c.

Handique, G.K. and Handique, A.K. (2009). Proline accumulation in lemongrass (*Cymbopogon flexuosus* Stapf.) due to heavy metal stress. *Journal of Environmental Biology* 30 (2): 299–302.

Hart, B., Cody, R., and Truong, P. (2003). Hydroponic vetiver treatment of post septic tank effluent. *International Conference of Vetiver Exhibition* 3: 121–131.

Hasan, S.N.M.S., Kusin, F.M., Lee, A.L.S. et al. (2017). Performance of vetiver grass (*Vetiveria zizanioides*) for phytoremediation of contaminated water. *MATEC Web Conference* 103: https://doi.org/10.1051/matecconf/201710306003.

Hassan, E. (2016). Comparative study on the biosorption of Pb (II), Cd(II) and Zn (II) using Lemon grass (*Cymbopogon citratus*): kinetics, isotherms and thermodynamics. *Chemistry International* 2 (2): 89–102.

Jain, S., Khare, P., Mishra, D. et al. (2020). Biochar aided aromatic grass [*Cymbopogon martini* (Roxb.) Wats.] vegetation: a sustainable method for stabilization of highly acidic mine waste. *Journal of Hazardous Materials* 390: 121799. https://doi.org/10.1016/j.jhazmat.2019.121799.

Jasha, M.A.H. (2014). Trace and essential elements analysis in Cymbopogon citratus (DC.) Stapf samples by graphite furnace-atomic absorption spectroscopy and its health concern. *Journal of Toxicology*. https://doi.org/10.1155/2014/690758.

Jayashree, S., Rathinamala, J., and Lakshmanaperumalsamy, P. (2011). Determination of heavy metal removal efficiency of *Chrysopogon zizanioides* (Vetiver) using textile wastewater contaminated soil. *Journal of Environmental Science and Technology* 4 (5): 543–551.

Jha, A.K. and Kumar, U. (2017). Studies on removal of heavy metals by *Cymbopogon flexuosus*. *International Journal of Agricultural Environment and Biotechnology* 10 (1): 89.

Jisha, C.K., Baudh, K., and Shukla, S.K. (2017). Phytoremediation and bioenergy production efficiency of medicinal and aromatic plants. *Phytoremediation Potential of Bioenergy Plants* 1: 287–304. https://doi.org/10.1007/978-981-10-3084-0_11.

Kereeditse, T.T., Pheko-Ofitlhile, T., Ultra, V.U. Jr., and Dinake, P. (2023). Effects of heavy metals on the yield of essential oil from vetiver grass cultivated in mine tailings amended with edta and arbuscular mycorrhizal fungi. *Natural Product Communications* 18 (3): 1–12. https://doi.org/10.1177/1934578X231164813.

Khilji, S.A. and Sajid, Z.A. (2020). Phytoremediation potential of lemongrass (*Cymbopogon flexuosus* stapf.) grown on tannery sludge contaminated soil. *Applied Ecology and Environmental Research* 18 (6): 7703–7715. https://doi.org/10.15666/aeer/1806_77037715.

Kiamarsi, Z., Kafi, M., Soleimani, M. et al. (2021). Evaluating the bio-removal of crude oil by vetiver grass (*Vetiveria zizanioides* L.) in interaction with bacterial consortium exposed to contaminated artificial soils. *International Journal of Phytoremediation* 5: 483–492. https://doi.org/10.1080/15226514.2021.1954876.

Kiiskila, J.D., Sarkar, D., Panja, S. et al. (2019). Remediation of acid mine drainage-impacted water by vetiver grass (*Chrysopogon zizanioides*): a multiscale long-term study. *Ecological Engineering* 129: 97–108. https://doi.org/10.1016/j.ecoleng.2019.01.018.

Kumar, A., Jnanesha, A.C., Verma, R.K. et al. (2022). Phytoremediation, eco-restoration, and adaptive response of lemongrass (*C. flexuosus* Wats) grown on fly ash and vermicompost improved quality essential oil yield. *Biochemical Systematics and Ecology* 104: 104457. https://doi.org/10.1016/j.bse.2022.104457.

Kumar, A. and Maiti, S.K. (2014). Effect of organic manures on the growth of *Cymbopogon citratus* and *Chrysopogon zizanioides* for the phytoremediation of chromite-asbestos mine waste: a pot scale experiment. *International Journal of Phytoremediation*. https://doi.org/10.1080/15226514.2014.910174.

Kumar, A., Malik, F., Bhushan, S. et al. (2008). An essential oil and its major constituent iso intermedeol induce apoptosis by increased expression of mitochondrial cytochrome c and apical death receptors in human leukemia HL-60 cells. *Chemico-Biological Interactions* 171 (3): 332–347.

Lai, H.Y. and Chen, Z.S. (2004). Effect of EDTA on solubility of cadmium, zinc and lead and their uptake by rainbow pink and vetiver grass. *Chemosphere* 55: 421–430. https://doi.org/10.1016/j.chemosphere.2003.11.009.

Lal, K., Yadav, R.K., Kaur, R. et al. (2013). Productivity, essential oil yield, and heavy metal accumulation in lemon grass (*Cymbopogon flexuosus*) under varied wastewater–groundwater irrigation regimes. *Industrial Crops and Products* 45: 270–278.

Lee, L.Y., Lee, X.J., Chia, P.C. et al. (2014). Utilisation of *Cymbopogon citratus* (lemon grass) as biosorbent for the sequestration of nickel ions from aqueous solution: equilibrium, kinetic, thermodynamics and mechanism studies. *Journal of Taiwan Institute of Chemical Engineers* 45 (4): 1764–1772. https://doi.org/10.1016/j.jtice.2014.02.002.

Lermen, C., Morelli, F., Gazim, Z.C. et al. (2015). Essential oil content and chemical composition of *Cymbopogon citratus* inoculated with arbuscular mycorrhizal fungi under different levels of lead. *Industrial Crops and Products* 76: 734–738.

Liao, X., Shiming, L., Yinbao, W. and Zhisan, W. (2003). Studies on the abilities of *Vetiveria zizanioides* and *Cyperus alternifolius* for pig farm wastewater treatment. In: Proceedings of the Third International Vetiver Conference; Guangzhou, China; October 2003

Linggawati, A., Maryani, M., Nugroho, A.P., and Rachmawati, D. (2022). Anatomical and histochemical responses of vetiver grass (*Chrysopogon zizanioides* L. Roberty) to phytoremediation ability of liquid batik waste. *Environment and Natural Resources Journal* 20 (4): 359–368. https://doi.org/10.32526/ennrj/20/202100232.

Lone, I.A. and Gaffar, M. (2021). Phytoremediation potential of medicinal and aromatic plants. In: *Medicinal and Aromatic Plants* (ed. T. Aftab and K.R. Hakeem). Switzerland: *Springer Nature* https://doi.org/10.1007/978-3-030-58975-2_29;Marcacci, S., Raveton, M., Ravanel, P., and Schwitzguebel, J.P. (2005). The possible role of hydroxylation in the detoxification of atrazine in mature vetiver (*Chrysopogon zizanioides* nash) grown in hydroponics. *Zeitschrift fur Z Naturforsch C J Biosci* 60 (5-6): 427–434. https://doi.org/10.1515/znc-2005-5-611.

Marsidi, N., Nye, C.K., Abdullah, S.R., Abu Hassan, H. and Halmi, M.I. (2016). Phytoremediation of naproxen in waste water using *Vetiver zizaniodes. Journal of Engineering Science Technology* V 11 (8): 1086–1097.

Melato, F., Regnier, T., McCrindle, R., and Mokgalaka, N. (2012). Impact of metals on secondary metabolites production and plant morphology in vetiver grass (*Chrysopogon zizanioides*). *South African Journal of Chemistry* 65 (1): 178–183.

Mishra, B. and Chandra, M. (2022). Evaluation of phytoremediation potential of aromatic plants: a systematic review. *Journal of Applied Research on Medicinal and Aromatic Plants* 31: 100405. https://doi.org/10.1016/j.jarmap.2022.100405.

Mwansa, M. (2018). Evaluation of the phytoremediation potential of lemongrass (*Cymbopogon citratus*) and vetiver grass (*Chrysopogon zizanioides*) in lead contaminated soils of Kabwe. Master's Thesis. The University of Zambia,

Ng, C.C., Boyce, A.N., Rahman, M.M., and Abas, M.R. (2016). Effects of different soil amendments on mixed heavy metals contamination in vetiver grass. *Bulletin of Environmental Contamination and Toxicology* 97 (5): 695–701. https://doi.org/10.1007/s00128-016-1921-5.

Noshadi, M., Foroutani, A., and Sepaskhah, A. (2019). Evaluation of HYDRUS-1D and modified PRZM-3 models for tribenuron methyl herbicide transport in soil profile under vetiver cultivation. *International Journal of Phytoremediation* 21 (9): 878–891. https://doi.org/10.1080/15226 514.2019.15836 32.

Oguche, J., Agbo, S., Oyegoke, T., and Emenru, D. (2022). Phytoremediation of used engine oil contaminated soil using lemongrass. *International Journal of Engineering Science and Application* 6 (4): 95–102.

Ogura, R., Ikeda, N., and Yukietal, K. (2008). Geno toxicity studies on green tea catechin. *Food and Chemical Toxicology* 46 (6): 2190–2200. https://doi.org/10.1016/j.fct.2008.02.016.

Otunola, B.O., Aghoghovwia, M.P., Thwala, M. et al. (2022b). Influence of clay mineral amendments characteristics on heavy metals uptake in vetiver grass (*Chrysopogon zizanioides* L. Roberty) and Indian Mustard (*Brassica juncea* L. Czern). *Sustainability* 14: 5856. https://doi.org/10.3390/su14105856.

Otunola, B.O., Aghoghovwia, M.P., Thwala, M. et al. (2023). Improving capacity for phytoremediation of vetiver grass and Indian mustard in heavy metal (Al and Mn) contaminated water through the application of clay minerals. *Environmental Science and Pollution Research* 30: 53577–53588. https://doi.org/10.1007/s11356-023-26083-5.

Otunola, B.O., Aghoghovwia, M.P., Thwala, M., and Ololade, O.O. (2022a). Heavy metal phytoremediation potential of vetiver grass and Indian mustard update on enhancements and research opportunities. *Water Air & Soil Pollution* 233: 154. https://doi.org/10.1007/s11270-022-05620-x.

Owamah, H.I., Alfa, M.I., and Dahunsi, S.O. (2014). Optimization of biogas from chicken droppings with *Cymbopogon citratus*. *Renewable Energy* 68: 366–371.

Pandey, V.C., Ancona, V., Roy, M., and Ranđelović, D. (2023). *Aromatic Plant-Based Phytoremediation*. Elsevier https://doi.org/10.1016/C2022-0-00341-X.

Pandey, V.C. and Bajpai, O. (2019). Phytoremediation: from theory towards practice. In: *Phytomanagement of Polluted Sites* (ed. V.C. Pandey and K. Bauddh), 1–49. Amsterdam: Elsevier https://doi.org/10.1016/B978-0-12-813912-7.00001-6.

Pandey, J., Chand, S., Pandey, S., and Rajkumari and Patra, D.D. (2015a). Palmarosa [*Cymbopogon martinii* (Roxb.) Wats.] as a putative crop for phytoremediation, in tannery sludge polluted soil. *Ecotoxicology and Environmental Safety* 122: 296–302. https://doi.org/10.1016/j.ecoenv.2015.08.005.

Pandey, V.C. and Maiti, D. (2020). Perennial grasses in phytoremediation—challenges and opportunities. In: *Phytoremediation Potential of Perennial Grasses*. Elsevier. https://doi.org/10.1016/B978-0-12-817732-7.00001-8.

Pandey, V.C., Pandey, D.N., and Singh, N. (2015b). Sustainable phytoremediation based on naturally colonizing and economically valuable plants. *Journal of Cleaner Production* 86: 37–39. https://doi.org/10.1016/j.jclepro.2014.08.030.

Pandey, V.C. and Praveen, A. (2020). *Vetiveria zizanioides* (L.) Nash–more than a promising crop in phytoremediation. In: *Phytoremediation Potential of Perennial Grasses* (ed. V.C. Pandey and D.P. Singh), 31–62. Amsterdam: Elsevier. https://doi.org/10.1016/B978-0-12-817732-7.00002-X.

Pandey, V.C., Rai, A., and Korstad, J. (2019c). Aromatic crops in phytoremediation: from contaminated to waste dump sites. In: *Phytomanagement of Polluted Sites* (ed. V.C. Pandey and K. Bauddh), 255–275. Amsterdam: Elsevier. https://doi.org/10.1016/B978-0-12-813912-7.00009-0.

Pandey, V.C. and Singh, N. (2015). Aromatic plants versus arsenic hazards in soils. *Journal of Geochemical Exploration* 157: 77–80.

Pandey, B., Singh, S., Roy, L.B. et al. (2021). Phytostabilization of coal mine overburden waste, exploiting the phytoremedial efficacy of lemongrass under varying level of cow dung manure. *Ecotoxicology and Environmental Safety* 208: 111757. https://doi.org/10.1016/j.ecoenv.2020.111757.

Pandey, V.C., Rai, A., and Korstad, J. (2019). Aromatic crops in phytoremediation: from contaminated to waste dump sites. In: *Phytomanagement of Polluted Sites* (ed. V.C. Pandey and K. Bauddh). Elsevier.

Pandey, V.C. and Souza-Alonso, P. (2019). Market opportunities in sustainable phytoremediation. In: *Phytomanagement of Polluted Sites* (ed. V.C. Pandey and K. Bauddh), 51–82. Amsterdam: Elsevier. https://doi.org/10.1016/B978-0-12-813912-7.00002-8.

Pandey, J., Verma R.K. and Singh, S. (2019a). Screening of most potential candidate among different lemongrass varieties for phytoremediation of tannery sludge contaminated sites. *International Journal of Phytoremediation* 21 (6): 600–609. https://doi.org/10.1080/15226514.2018.1540538.

Pandey, J., Verma, R.K., and Singh, S. (2019b). Suitability of aromatic plants for phytoremediation of heavy metal contaminated areas: a review. *International Journal of Phytoremediation* 21 (5): 405–418. https://doi.org/10.1080/15226514.2018.1540546.

Pandey, J., Verma, R.K., and Singh, S. (2020). Trace element accumulation potential in lemongrass varieties (*Cymbopogon* species) and prediction through regression model equations followed by path analysis: a field study. *Chemosphere* 257: 127102. https://doi.org/10.1016/j.chemosphere.2020.127102.

Patra, H.K., Marndi, D.S., and Mohanty, M. (2015). Chromium toxicity, physiological responses and tolerance potential of lemon grass (*Cymbopogon flexuosus* Nees ex steud. wats.). *Annals of Plant Sciences* 4 (05): 1080–1084.

Paz-Alberto, A.M., Sigua, G.C., Baui, B.G., and Prudente, J.A. (2007). Phytoextraction of lead contaminated soil using vetivergrass (*Vetiveria zizanioides* L.), cogongrass (*Imperata cylindrica* L.) and carabaograss (*Paspalum conjugatum* L.). *Environmental Science and Pollution Research* 14 (7): 498–504.

Phenrat, T., Teeratitayangkul, P., Imthiang, T., Sawasdee, Y., Wichai, S., Piangpia, T., Naowaopas, J. and Supanpaiboon, W. (2015). Laboratory-scaled Developments and Field-scaled Implementations of Using Vetiver Grass to Remediate Water and Soil Contaminated with Phenol and Other Hazardous Substances from Illegal Dumping at Nong-Nea Subdistrict, Phanom Sarakham District, Chachoengsao Province, Thailand. In: The Sixth International Conference on Vetiver, Vietnam, Danang, 3-5.

Purnama, C.L.R., Rossiana, N., Butar Butar, E.S., and Sembiring, T. (2022). Association of plant and microorganisms on textile wastewater remediation. *Environment and Ecology Research* 10 (6): 648–659. https://doi.org/10.13189/eer.2022.100603.

Rajaei, S. and Seyedi, S.M. (2018). Phytoremediation of petroleum contaminated soils by *Vetiveria zizanioides* (L.) Nash. *Clean–Soil, Air, Water* 1800244. https://doi.org/10.1002/clen.201800244.

Rajeshwara Rao, B.R., Adinarayana, G., Rajput, D.K. et al. (2015). Essential oil profiles of different parts of East Indian lemongrass {*Cymbopogon flexuosus* (Nees ex Steud.) Wats.}. *Journal of Essential Oil Research* 27 (3): 225–231. https://doi.org/10.1080/10412905.2015.1007218.

Rinitha, P. (2022). Grey water treatment by phytoremediation technique – a comparative study using vetiver grass and lemon grass. *International Journal of Engineering Research & Technology* 10 (6): 1–5. https://doi.org/10.17577/IJERTCONV10IS06031.

Roongtanakiat, N., Tangruangkiat, S., and Meesat, R. (2007). Utilization of vetiver grass (*Vetiveria zizanioides*) for removal of heavy metals from industrial wastewaters. *Science Asia* 33 (4): 397–403. https://doi.org/10.2306/scienceasia1513-1874.2007.33.397.

Rouhani, A., Skousen, J., and Tack, F.M.G. (2023). An overview of soil pollution and remediation strategies in coal mining regions. *Minerals* 13: 1064. https://doi.org/10.3390/min13081064.

Ruiz C., Rodríguez O., Luque O. and Alarcón M. (2013). Effecto del vetiver (Chrysopogon zizanioides L.) en la reducción del flúor y otros compuestos contaminantes en aguas de consumo humano. Caso: Caserío Guarataro, estado Yaracuy, Venezuela. The second Latin America International Conference on the Vetiver System, October 3–5, 2013, Medillin, Colombia.

Sengupta, A., Sarkar, D., Das, P. et al. (2016). Tetracycline uptake and metabolism by vetiver grass (*Chrysopogon zizanioides* L. Nash). *Environmental Science and Pollution Research* 23 (24): 24880–24889.

Shaw, G. and Bell, J.N.B. (1991). Competitive effects of potassium and ammonium on caesium uptake kinetics in wheat. *Journal of Environmental Radioactivity* 13 (4): 283–296.

Singh, S., Eapen, S., Thorat, V. et al. (2008a). Phytoremediation of 137cesium and 90strontium from solutions and low-level nuclear waste by *Vetiveria zizanoides*. *Ecotoxicology and Environtal Safety* 69 (2): 306–311.

Singh, S., Melo, J.S., Eapen, S., and D'souza, S.F. (2008b). Potential of vetiver (*Vetiveria zizanoides* L. Nash) for phytoremediation of phenol. *Ecotoxicology and Environmental Safety* 71 (3): 671–676.

Singh, S., Suvarna, S., Kiran, K., and Fulzele, P. (2017). Investigation of arsenic accumulation and biochemical response of in vitro developed *Vetiveria zizanoides* plants. *Ecotoxicology and Environmental Safety* 145: 50–56. https://doi.org/10.1016/j.ecoenv.2017.07.013.

Sinha, S., Biswas, D., and Mukherjee, A. (2011). Antigenotoxic and antioxidant activities of palmarosa and citronella essential oils. *Journal of Ethnopharmacology* 137 (3): 1521–1527. https://doi.org/10.1016/j.jep.2011.08.046.

Sinha, S., Mishra, R.K., Sinam, G. et al. (2013). Comparative evaluation of metal phytoremediation potential of trees, grasses, and flowering plants from tannery-wastewater-contaminated soil in relation with physicochemical properties. *Soil Sediment Contamination* 22 (8): 958–983. https://doi.org/10.1080/15320383.2013.770437.

Siyar, R., Ardejani, F.D., Farahbakhsh, M. et al. (2020). Potential of vetiver grass for the phytoremediation of a real multi-contaminated soil, assisted by electrokinetic. *Chemosphere* 246: 125802. https://doi.org/10.1016/j.chemosphere.2019.125802.

Sobh, M., Moussawi, M.A., Rammal, W. et al. (2014). Removal of lead (II) ions from wastewater by using Lebanese *Cymbopogon citratus* (lemon grass) stem as adsorbent. *American Journal of Phytomedicine and Clinical Therapeutics* 2 (9): 1070–1080.

Srivastava, N.K., Ram, L.C., and Masto, R.E. (2014). Reclamation of overburden and lowland in coal mining area with fly ash and selective plantation: a sustainable ecological approach. *Ecological Engineering* 71: 479–489. https://doi.org/10.1016/j.ecoleng.2014.07.062.

Sulastri, Y.S., Mukhlis, Damanik, R.I., and Sabrina, T. (2022). Study of aromatic plants on cadmium exposure through anatomical structure, shoot root ratio, essential oil and heavy metals. *Asian Journal of Plant Sciences* 21 (4): 565–573. https://doi.org/10.3923/ajps.2022.565.573.

Topal, E., Cornea-Cipcigan, M., Tunca, R.I. et al. (2020). The use of medicinal aromatic plants against bee diseases and pests. *Bee Studies* 12: 5–11. https://doi.org/10.51458/BSTD.2021.2.

Truong, P., Van, T.T., and Pinners, E. (2008). *Vetiver System Applications Technical Reference Manual*, 89. The Vetiver Network International.

Ultra, V. Jr. (2020). Growth and yield of lemongrass (*Cymbopogon citratus*) in fly ash with nutrient amendments and Mycorrhiza for three-ratoon period. *International Journal of Phytoremediation*. https://doi.org/10.1080/15226514.2020.1786005.

Verma, S.K., Singh, K., Gupta, A.K. et al. (2014). Aromatic grasses for phytomanagement of coal fly ash hazards. *Ecological Engineering* 73: 425–428. https://doi.org/10.1016/j.ecoleng.2014.09.106.

Wilde, E.W., Brigmon, R.L., and Dunn, D.L. (2005). Phytoextraction of lead from firing range soil by vetiver grass. *Chemosphere* 61: 1451–1457. https://doi.org/10.1016/j.chemosphere.2005.04.059.

Yang, Y., Ge, Y., Tu, P. et al. (2019). Phytoextraction of Cd from a contaminated soil by tobacco and safe use of its metal enriched biomass. *Journal of Hazardous Materials* 363: 385–393. https://doi.org/10.1016/j.jhazm at.2018.09.093.

Zhang, X., Gao, B., and Xia, H. (2014). Effect of cadmium on growth, photosynthesis, mineral nutrition and metal accumulation of bana grass and vetiver grass. *Ecotoxicology and Environmental Safety* 106: 102–108. https://doi.org/10.1016/j.ecoenv.2014.04.025.

Zhang, F., Sun, S., Rong, Y. et al. (2023). Enhanced phytoremediation of atrazine-contaminated soil by vetiver (*Chrysopogon zizanioides* L.) and associated bacteria. *Environmental Science and Pollution Research* 30: 44415–44429. https://doi.org/10.1007/s11356-023-25395-w.

Index

Biodiversity and Ecosystem Services on Post-Industrial Land, First Edition.
Edited by Vimal Chandra Pandey.